THE MATRIX ANALYSIS OF VIBRATION

THE MATRIX ANALYSIS OF VIBRATION

BY

R.E.D. BISHOP

Kennedy Research Professor in the University of London and Fellow of University College

G.M.L. GLADWELL

Professor of Civil Engineering in the University of Waterloo, Ontario, Canada

S. MICHAELSON

Director of the Computing Laboratory, University of Edinburgh

CAMBRIDGE UNIVERSITY PRESS

CAMBRIDGE

LONDON · NEW YORK · MELBOURNE

CAMBRIDGE UNIVERSITY PRESS
Cambridge, New York, Melbourne, Madrid, Cape Town, Singapore, São Paulo, Delhi

Cambridge University Press
The Edinburgh Building, Cambridge CB2 8RU, UK

Published in the United States of America by Cambridge University Press, New York

www.cambridge.org
Information on this title: www.cambridge.org/9780521042574

First published 1965
Reissued 1979
This digitally printed version 2008

A catalogue record for this publication is available from the British Library

Library of Congress Catalogue Card Number: 65–10106

ISBN 978-0-521-04257-4 hardback
ISBN 978-0-521-09885-4 paperback

CONTENTS

Chapter 7 The solution of linear equations and the inversion of matrices

Chapter 8 Iterative methods for characteristic value problems

Chapter 9 Direct methods for characteristic value problems

PREFACE

The introduction of high-speed electronic computers into engineering has opened the way for the solution of vibration problems of great complexity. One of the most suitable ways of expressing a problem for computational analysis is to use matrices. This book is concerned with the matrix formulation of the equations of motion of vibrating systems, and with techniques for the solution of matrix equations.

The purely computational side of practical vibration analysis naturally divides into two parts. The first is the construction of a mathematical model for the vibrating system, that is the setting up of equations governing its motion. The second is the process of solving the mathematical equations and extracting the properties of the solutions. This book is mainly concerned with the second part of the analysis. It will be assumed, on the whole, that the reader is acquainted with the various ways of setting up the equations of motion of vibrating systems.

This book can be regarded as a sequel to an earlier volume.† The first, as its title suggests, is concerned mainly with the mechanics—the physical side—of vibration: this book is concerned more with the mathematical aspects of the subject. The connection between the two books is loose and they may be read independently of each other.

The fact that the book is concerned with the mathematical side of vibration theory, rather than the physical, throws emphasis on examples. These are provided at the end of each section and are arranged roughly in order of increasing difficulty. The reader should regard the working of examples as an essential part of the process of mastering the contents of the book, and it is for this reason that the answers are given in greater detail than is usual.

The subject-matter of this book falls quite naturally into two parts. The first, Chapters 1–6, is almost entirely the work of two of us (R. E. D. B. and G. M. L. G.) and it is concerned with matrix theory and its application to vibration problems. Its scope has purposely been restricted so that it does not provide a complete treatment of vibration theory. It is intended, rather, to link together various aspects of the subject and to serve as an introduction to more specialised treatises on matrix theory, vibration or aeroelasticity.‡

The second part of the book, Chapters 7–9, of which the third one of us (S. M.) wrote the first draft, is concerned with the numerical solution of matrix equations. It is sometimes imagined that, because comprehensive computer programmes are available for their solution, matrix equations no longer present any difficulty. This is by no means the case as the reader may easily verify by some practice. When

† R. E. D. Bishop and D. C. Johnson, *The Mechanics of Vibration* (Cambridge University Press, 1960).

‡ See, for example, W. L. Ferrar, *Algebra* (Oxford University Press, 2nd ed., 1957); R. A. Frazer, W. J. Duncan and A. R. Collar, *Elementary Matrices* (Cambridge University Press, 1955); R. L. Bisplinghoff, H. Ashley and R. L. Halfman, *Aeroelasticity* (Addison–Wesley, 1955).

complications arise during the process of solving equations, it is essential that the computer user should know the procedure which the computer follows to solve the equations and how the various kinds of difficulty which arise may be overcome. It is to provide this kind of information that the second part of the book was written. Again, it does not pretend to be an exhaustive treatment of the subject, but rather a discussion of one or two typical computer methods. Throughout most of this part of the book the methods are described so that they may also be used with a desk calculating machine. This is done so that the reader may work the examples, Although it is extremely tedious to solve some of these examples using only a desk machine (they may take up to five hours), their working should be considered an essential part of the vibration analyst's apprenticeship.

We should be most grateful to be notified of any errors in the text and examples.

R. E. D. B.
G. M. L. G.
S. M.

May 1963

CHAPTER 1

NOTATION AND ELEMENTARY PROPERTIES OF MATRICES

Now, Jurymen, hear my advice...
All kinds of vulgar prejudice
I pray you set aside:
With stern judicial frame of mind
From bias free of every kind
This trial must be tried. *Trial by Jury*

The analysis of systems having finite freedom is found to require manipulation of linear algebraic equations.† These equations are clumsy in form, particularly if the system to which they relate has more than three or four degrees of freedom, so that some form of algebraic shorthand becomes desirable. Matrix notation serves this purpose. The essence of matrix methods is orderliness, and this feature makes the techniques particularly helpful in the setting out of problems for numerical solution by digital computers or by semi-skilled human computers.

The theme of this presentation of vibration theory is that sets of linear equations may be written concisely in matrix form, by the use of certain conventions. The matrices can then be manipulated (for instance to give other matrices) according to certain rules, re-interpretation at any stage being possible through the conventions. The purpose of this first chapter is merely to provide a background in matrix algebra before the treatment of the dynamical problems is presented.

1.1 Matrix notation and preliminary definitions

A matrix is an array of numbers (or 'elements'), the positions as well as the magnitudes of which have significance. It is thus like a determinant, but with one essential difference. A determinant has a single numerical value that is arrived at by a process of reduction. A matrix may not be reduced in this way, but remains an array of numbers. Rules of addition and multiplication can be formulated for matrices and the nature of these will be explained—the nature of the rules is not to be regarded as obvious and is a matter of definition.

The numbers which form a matrix‡ are usually enclosed in square brackets, thus,

$$\begin{bmatrix} a_{11} & a_{12} & a_{13} \\ a_{21} & a_{22} & a_{23} \end{bmatrix}.$$

† R. E. D. Bishop and D. C. Johnson, *The Mechanics of Vibration* (Cambridge University Press, 1960), ch. 3.

‡ The elements may be real or complex numbers, functions, differential operators or even other matrices. The theory which will be presented will be general, but at first the reader may, if he wishes, consider matrices whose elements are real numbers. All the matrices occurring in the examples will be of this type.

While this array has two rows and three columns, it is more usual in vibration theory to deal with 'square', 'row', or 'column' matrices. A square matrix has equal numbers of rows and columns, thus

$$\begin{bmatrix} a_{11} & a_{12} \\ a_{21} & a_{22} \end{bmatrix}.$$

A matrix formed by a single row is represented in the same way; for instance, a row matrix having three elements would be of the form

$$[a_{11} \quad a_{12} \quad a_{13}].$$

Sometimes commas are used to separate the elements for the sake of compactness. A column matrix, on the other hand, is of the form

$$\begin{bmatrix} a_{11} \\ a_{21} \\ a_{31} \end{bmatrix}.$$

Plainly, a large column matrix takes up a wasteful amount of space on a printed page, so it is common to write down the array in a horizontal line, using curly brackets; thus the column matrix with these elements can be written

$$\{a_{11} \quad a_{21} \quad a_{31}\},$$

where, again, commas are sometimes placed between the elements. Column and row matrices are sometimes called column and row 'vectors'.

The matrix

$$\begin{bmatrix} a_{11} & a_{12} & \cdots & a_{1n} \\ a_{21} & a_{22} & \cdots & a_{2n} \\ \cdots & \cdots & \cdots & \cdots \\ a_{m1} & a_{m2} & \cdots & a_{mn} \end{bmatrix} \tag{1.1.1}$$

has m rows and n columns. Its typical element may be denoted by a_{ij} and the entire matrix will be referred to by the single symbol $\mathbf{A}$. This matrix $\mathbf{A}$ is said to be 'of order m by n'; such a statement may be made symbolically by the equation

$$\mathbf{A} = \mathbf{A}(m, n). \tag{1.1.2}$$

However, as was pointed out above, such matrices seldom occur in vibration analysis. The more common matrices are square, row, and column matrices. In these the distinction between rows and columns does not occur and they are spoken of as being 'of order n'.

By thinking of a square matrix as a sort of double-entry table, we see that it might be used to display various quantities that are met with in the vibration analysis of systems having n degrees of freedom. Thus the inertia coefficients a_{rs} of a system could be set out in a matrix. So, too, could stability coefficients c_{rs}, receptances α_{rs} or the constants which relate generalised coordinates to principal coordinates.†

† Bishop and Johnson, *The Mechanics of Vibration*, ch. 3.

There are, in fact, several possibilities. *Why* it is desirable to display these quantities in matrix form is something we shall explain.

Before coming to a discussion of how matrices are manipulated, we shall find it convenient to define a few particular types of matrix.

1.1.1 Null matrix

The name 'null matrix' is given to a matrix all of whose elements are zero. This array of noughts is denoted by the symbol **0**.

1.1.2 Unit matrix

A square matrix is said to be a 'unit matrix' if all the elements in the principal diagonal ('top left to bottom right') are unity, and the remainder are all zero. Thus the unit matrix of order three is

$$\begin{bmatrix} 1 & 0 & 0 \\ 0 & 1 & 0 \\ 0 & 0 & 1 \end{bmatrix}.$$

In general, if **A** is a square matrix of order n such that

$$a_{ij} = \begin{Bmatrix} 0 & \text{for} & i \neq j \\ 1 & \text{for} & i = j \end{Bmatrix}, \tag{1.1.3}$$

then **A** is the unit matrix $\mathbf{I}_n$ (or, sometimes, simply **I**). This equation is sometimes stated in another way by saying that a_{ij} is equal to the so-called 'Kronecker Delta' δ_{ij}—a quantity which has the property indicated in equation (1.1.3).

1.1.3 Diagonal matrix

If all the elements of a square matrix other than those in the principal diagonal are zero, the matrix is a 'diagonal matrix'. Thus a typical diagonal matrix of order three is

$$\begin{bmatrix} -1 & 0 & 0 \\ 0 & 2 & 0 \\ 0 & 0 & 4 \end{bmatrix},$$

while the unit matrix $\mathbf{I}_3$ is another, but of a particular kind.

1.1.4 Transposed matrix

Suppose that some given matrix is rearranged in such a way that its rows are made into columns; the matrix is said to be 'transposed'. Consider, for example, the two matrices

$$\begin{bmatrix} 1 & 4 & -7 \\ 3 & -2 & 1 \end{bmatrix} \quad \text{and} \quad \begin{bmatrix} 1 & 3 \\ 4 & -2 \\ -7 & 1 \end{bmatrix},$$

the second of these is found from the first by transposition, and the first is the second transposed. To generalise this definition somewhat, we may say that the transpose $\mathbf{A}'$ of a matrix $\mathbf{A}$ is the matrix which has rows identical with the columns of $\mathbf{A}$. We shall retain this convention of denoting transposition by a dash (or 'prime'). Thus if the element in row i and column j of $\mathbf{A}$ is a_{ij} then the element in row i and column j $\mathbf{A}'$ is a_{ji}, and if

$$\mathbf{A} = \mathbf{A}(m, n) \quad \text{then} \quad \mathbf{A}' = \mathbf{A}'(n, m). \tag{1.1.4}$$

One particular case of transposition will occur frequently in this book; it is the transposition from a column matrix to a row matrix and vice versa. Thus if

$$\mathbf{A} = \begin{bmatrix} a_1 \\ a_2 \\ a_3 \end{bmatrix} \quad \text{then} \quad \mathbf{A}' = [a_1 \quad a_2 \quad a_3]. \tag{1.1.5}$$

1.1.5 Symmetric matrix

A square matrix $\mathbf{A}$ is said to be 'symmetric' if it is such that $a_{ij} = a_{ji}$ for all values of i and j. Such an array, of which the unit matrix is an example, is its own transpose. The process of transposing a square matrix may be visualised as the act of turning it over, using its principal diagonal as an axis; with a *symmetric* matrix, this produces no change. Thus a typical symmetric matrix is

$$\begin{bmatrix} 1 & 2 & -3 \\ 2 & 0 & 0 \\ -3 & 0 & 4 \end{bmatrix}.$$

1.1.6 Skew matrices

A square matrix $\mathbf{A}$ is 'skew' if $a_{ij} = -a_{ji}$ and if the elements a_{ii} are not all zero, as in

$$\begin{bmatrix} 1 & 2 & -4 \\ -2 & 0 & 1 \\ 4 & -1 & 3 \end{bmatrix}.$$

But if the $a_{ij} = -a_{ji}$ and the elements a_{ii} are all zero, then the matrix is said to be 'skew symmetric' as with

$$\begin{bmatrix} 0 & 2 & -4 \\ -2 & 0 & 1 \\ 4 & -1 & 0 \end{bmatrix}.$$

EXAMPLES 1.1

1. Describe the following matrices:

(*a*) $[0 \quad -1 \quad 4 \quad 3]$, (*b*) $\begin{bmatrix} 3 & 2 & 5 \\ 2 & 4 & 9 \\ 5 & 9 & -16 \end{bmatrix}$, (*c*) $\{1 \quad 9 \quad -4 \quad 2\}$,

(*d*) $\begin{bmatrix} 0 & 4 & -1 & 2 \\ -4 & 0 & 2 & 0 \\ 1 & -2 & 0 & 0 \end{bmatrix}$, (*e*) $\begin{bmatrix} -1 & -2 & -9 \\ 2 & 3 & 14 \\ 9 & -14 & 4 \end{bmatrix}$.

2. Write down and describe the transposes of each of the matrices in Ex. 1.1.1.

3. What is the greatest number of distinct elements which may be possessed by the following matrices of order n: (*a*) a symmetric matrix, (*b*) a skew matrix, (*c*) a skew-symmetric matrix?

1.2 The addition, subtraction and equality of matrices; scalar multipliers

It has been explained that matrices are merely arrays of numbers or other objects. If it is proposed to add matrices, then clearly some rule for addition must be framed. The rule that is used stipulates that the operation of addition can be performed only on matrices having the same order.

The sum of two matrices **A** and **B** is defined as a third matrix **C** which is such that

$$c_{ij} = a_{ij} + b_{ij} \tag{1.2.1}$$

for all the values of i and j. Thus

$$\begin{bmatrix} 3 & -1 \\ 6 & -8 \\ -1 & 4 \end{bmatrix} + \begin{bmatrix} 7 & 5 \\ 3 & -12 \\ 0 & 2 \end{bmatrix} = \begin{bmatrix} 10 & 4 \\ 9 & -20 \\ -1 & 6 \end{bmatrix}, \tag{1.2.2}$$

and this equation implies that six relations between the elements are satisfied individually. In general, the addition of two matrices of order m by n implies the satisfaction of $m \times n$ relations between the elements.

Since the addition of matrices involves only the addition of corresponding elements, it is clear that the associative and commutative laws of addition hold for matrices. That is to say (*a*) the additions

$$(\mathbf{A}+\mathbf{B})+\mathbf{C} \quad \text{and} \quad \mathbf{A}+(\mathbf{B}+\mathbf{C})$$

produce the same matrix and (*b*)

$$\mathbf{A}+\mathbf{B} \quad \text{and} \quad \mathbf{B}+\mathbf{A}$$

also produce the same matrix.

As with addition, the operation of subtraction can be performed only with matrices of the same order. The difference of two matrices **A** and **B** is defined to be the matrix **D** such that

$$d_{ij} = a_{ij} - b_{ij} \tag{1.2.3}$$

for all the values of i and j. This may be illustrated by the matrix equation

$$\begin{bmatrix} -2 & 3 & 1 \\ 0 & -1 & 7 \end{bmatrix} - \begin{bmatrix} -2 & -3 & 5 \\ 4 & 3 & 1 \end{bmatrix} = \begin{bmatrix} 0 & 6 & -4 \\ -4 & -4 & 6 \end{bmatrix}, \qquad (1.2.4)$$

which implies the satisfaction of six separate subtraction sums. The subtraction of one matrix of order m by n from another requires the satisfaction of $m \times n$ relations between the elements.

The statement that two matrices are 'equal' can now be given a meaning which follows logically from the definition of subtraction. Two matrices **A** and **B** are said to be equal if their difference is the null matrix. That is

$$\mathbf{A} = \mathbf{B} \quad \text{if} \quad a_{ij} - b_{ij} = 0 \qquad (1.2.5)$$

for all values of i and j.

The matrix equation

$$\begin{bmatrix} a_{11} & a_{12} \\ a_{21} & a_{22} \end{bmatrix} = \begin{bmatrix} b_{11} & b_{12} \\ b_{21} & b_{22} \end{bmatrix} \qquad (1.2.6)$$

yields four 'scalar' equations between the elements:

$$a_{11} = b_{11}; \quad a_{12} = b_{12}; \quad a_{22} = b_{22}; \quad a_{21} = b_{21}. \qquad (1.2.7)$$

Conversely, the four scalar equations

$$a_1 = b_1; \quad a_2 = b_2; \quad a_3 = b_3; \quad a_4 = b_4 \qquad (1.2.8)$$

may be represented by

$$\left.\begin{aligned} &\begin{bmatrix} a_1 & a_2 \\ a_3 & a_4 \end{bmatrix} = \begin{bmatrix} b_1 & b_2 \\ b_3 & b_4 \end{bmatrix}, \\ \text{or} \quad &[a_1 \quad a_2 \quad a_3 \quad a_4] = [b_1 \quad b_2 \quad b_3 \quad b_4], \\ \text{or} \quad &\begin{bmatrix} a_1 \\ a_2 \\ a_3 \\ a_4 \end{bmatrix} = \begin{bmatrix} b_1 \\ b_2 \\ b_3 \\ b_4 \end{bmatrix}. \end{aligned}\right\} \qquad (1.2.9)$$

A set of ordinary equations like (1.2.8) can always be represented by a single equation between matrices.

The question of what is meant by the multiplication of one matrix by another will be dealt with in the next section; the process is a little more complicated than those of addition and subtraction. But the multiplication of a matrix by a number (or 'scalar coefficient') is a simple matter which can be dealt with at once.

The multiplication of a matrix by a number k, say, means that *every* element is multiplied by k.† That is, if

$$\mathbf{C} = k\mathbf{A} \quad \text{then} \quad c_{ij} = ka_{ij} \qquad (1.2.10)$$

† Here is another point at which the superficial resemblance that exists between matrices and determinants may be misleading, for if each element of an $n \times n$ determinant is multiplied by k the determinant is multiplied by k^n.

for all values of i and j so that there are as many separate relations satisfied as there are elements in **A** or **C**. This is illustrated by the equations

$$(-3)\times\begin{bmatrix}1 & 5\\ 0 & -7\end{bmatrix} = (-1)\times\begin{bmatrix}3 & 15\\ 0 & -21\end{bmatrix} = \begin{bmatrix}-3 & -15\\ 0 & 21\end{bmatrix}. \tag{1.2.11}$$

If a matrix **A** has a scalar multiplier k of this sort, no distinction is made between the products $k\mathbf{A}$ and $\mathbf{A}k$.

EXAMPLES 1.2

1. If $\mathbf{A} = \begin{bmatrix}1 & 5 & 9\\ 2 & 6 & 3\\ -1 & 4 & -3\end{bmatrix}$, $\mathbf{B} = \begin{bmatrix}6 & 1 & 8\\ -2 & -6 & -3\\ 0 & 5 & 2\end{bmatrix}$

find $\mathbf{A}+\mathbf{B}$ and $\mathbf{A}-2\mathbf{B}$.

2. Is it possible to define $\mathbf{A}+\mathbf{B}$ when:

(*a*) **A** has 2 rows and **B** has 4 rows;
(*b*) **A** has 2 columns and **B** has 4 columns;
(*c*) **A** has 2 rows and **B** has 4 columns?

3. Show that any square matrix can be represented as the sum of a symmetric matrix and a skew symmetric matrix in one and only one way. Express the matrix **A** of Ex. 1.2.1 in this way.

1.3 Multiplication of matrices; linear algebraic equations

It will be shown in this section how the adoption of matrix notation permits sets of linear algebraic equations of an important class to be written down in a condensed form. To do this, however, it is necessary to frame a suitable rule for multiplying one matrix by another. At first sight this rule may seem rather arbitrary and it is only later that its usefulness will be apparent.

When dealing with the addition of matrices we found that two matrices could be added only when they were of the same order. There is a similar condition for the multiplication of matrices. When two matrices **A** and **B** satisfy this condition they are said to be 'conformable' for multiplication. The condition is as follows: The matrices **A** and **B** are conformable for multiplication in the order $\mathbf{A}\times\mathbf{B}$ if the number of columns of the matrix **A** is equal to the number of rows of the matrix **B**. The two matrices

$$\mathbf{A}(3,3) = \begin{bmatrix}a_{11} & a_{12} & a_{13}\\ a_{21} & a_{22} & a_{23}\\ a_{31} & a_{32} & a_{33}\end{bmatrix}; \qquad \mathbf{B}(3,2) = \begin{bmatrix}b_{11} & b_{12}\\ b_{21} & b_{22}\\ b_{31} & b_{32}\end{bmatrix} \tag{1.3.1}$$

satisfy this condition and can therefore be multiplied together in the order $\mathbf{A}\times\mathbf{B}$. Such a product will be referred to as **A** 'postmultiplied' by **B** or **B** 'premultiplied' by **A**. In this example the product $\mathbf{B} \times \mathbf{A}$ may not be formed. For an example of two matrices which may be multiplied together both ways we may take

$$\mathbf{A} = \mathbf{A}(2,3) = \begin{bmatrix} 1 & 7 & 6 \\ 2 & -4 & 3 \end{bmatrix}; \qquad \mathbf{B} = \mathbf{B}(3,2) = \begin{bmatrix} 1 & -7 \\ 4 & 9 \\ 11 & 2 \end{bmatrix}. \tag{1.3.2}$$

In this case not only is the number of columns of **A** equal to the number of rows of **B**, but also the number of columns of **B** is equal to the number of rows of **A**.

Suppose now that

$$\mathbf{A} = \mathbf{A}(p,n), \quad \mathbf{B} = \mathbf{B}(n,q), \tag{1.3.3}$$

then the product $\mathbf{A} \times \mathbf{B}$ is defined to be a matrix **C** whose *order* is given by the equation

$$\mathbf{A}(p,n) \times \mathbf{B}(n,q) = \mathbf{C}(p,q). \tag{1.3.4}$$

Thus the product $\mathbf{A} \times \mathbf{B}$ of the matrices in (1.3.1) is a matrix $\mathbf{C}(3,2)$, while the products $\mathbf{A} \times \mathbf{B}$ and $\mathbf{B} \times \mathbf{A}$ of the matrices in (1.3.2) are respectively $\mathbf{C}(2,2)$ and $\mathbf{C}(3,3)$. The matrices **A** and **B** of (1.3.2) illustrate an important property of matrices, viz. that even where $\mathbf{A} \times \mathbf{B}$ and $\mathbf{B} \times \mathbf{A}$ may both be formed the products are not always identical. In this case the product matrices are not even of the same order.

Before giving the general rule for finding the *value* of the product of two conformable matrices we shall first consider the product of a row matrix and a column matrix. If

$$\mathbf{A} = [a_1 \quad a_2 \quad \dots \quad a_n], \qquad \mathbf{B} = \begin{bmatrix} b_1 \\ b_2 \\ \vdots \\ b_n \end{bmatrix}, \tag{1.3.5}$$

then the product $\mathbf{A} \times \mathbf{B}$ is defined to be the 1×1 matrix, or in other words the scalar quantity,

$$\mathbf{C} = a_1 b_1 + a_2 b_2 + \dots + a_n b_n. \tag{1.3.6}$$

Equation (1.3.4) shows that this is the only case in which the product of two matrices is a scalar.

If now

$$\mathbf{A} = \mathbf{A}(p,n) = \begin{bmatrix} a_{11} & a_{12} & \dots & a_{1n} \\ a_{21} & a_{22} & \dots & a_{2n} \\ \dots & \dots & \dots & \dots \\ a_{p1} & a_{p2} & \dots & a_{pn} \end{bmatrix}, \qquad \mathbf{B} = \mathbf{B}(n,q) = \begin{bmatrix} b_{11} & b_{12} & \dots & b_{1q} \\ b_{21} & b_{22} & \dots & b_{2q} \\ \dots & \dots & \dots & \dots \\ b_{n1} & b_{n2} & \dots & b_{nq} \end{bmatrix}, \tag{1.3.7}$$

then the term c_{ij} in the ith row and jth column of the product **AB** is defined to be the product of the matrices

$$[a_{i1} \quad a_{i2} \quad \dots \quad a_{in}], \qquad \begin{bmatrix} b_{1j} \\ b_{2j} \\ \vdots \\ b_{nj} \end{bmatrix} \tag{1.3.8}$$

that is

$$c_{ij} = a_{i1}b_{1j} + a_{i2}b_{2j} + \dots + a_{in}b_{nj} = \sum_{k=1}^{n} a_{ik}b_{kj}. \tag{1.3.9}$$

The matrices (1.3.8) are respectively the ith row of **A** and the jth column of **B**. Thus the product **AB** of the matrices in equation (1.3.1) is

$$\mathbf{C} = \begin{bmatrix} a_{11}b_{11}+a_{12}b_{21}+a_{13}b_{31}, & a_{11}b_{12}+a_{12}b_{22}+a_{13}b_{32} \\ a_{21}b_{11}+a_{22}b_{21}+a_{23}b_{31}, & a_{21}b_{12}+a_{22}b_{22}+a_{23}b_{32} \\ a_{31}b_{11}+a_{32}b_{21}+a_{33}b_{31}, & a_{31}b_{12}+a_{32}b_{22}+a_{33}b_{32} \end{bmatrix}. \tag{1.3.10}$$

The following examples illustrate some of the important properties of matrix multiplication.

Consider the two matrices

$$\mathbf{A} = \begin{bmatrix} 1 & 1 & 1 \\ 2 & 2 & 2 \\ 5 & 5 & 5 \end{bmatrix}; \qquad \mathbf{B} = \begin{bmatrix} 3 & 4 & 2 \\ -2 & -1 & -1 \\ -1 & -3 & -1 \end{bmatrix}. \tag{1.3.11}$$

Since **A** and **B** are square matrices of the same order, they are conformable for multiplication in both ways. The product $\mathbf{A} \times \mathbf{B}$ is found to be

$$\mathbf{A} \times \mathbf{B} = \begin{bmatrix} 3-2-1, & 4-1-3, & 2-1-1 \\ 6-4-2, & 8-2-6, & 4-2-2 \\ 15-10-5, & 20-5-15, & 10-5-5 \end{bmatrix} = \mathbf{0}, \tag{1.3.12}$$

the symbol **0** representing the null matrix. This one example *proves* that a matrix equation of the type $\mathbf{AB} = \mathbf{0}$ does not necessarily imply that either $\mathbf{A} = \mathbf{0}$ or $\mathbf{B} = \mathbf{0}$.

We have already proved that the products **AB** and **BA** of two matrices are not always identical—in technical language this is described by saying that matrix multiplication is in general non-commutative. The matrices of (1.3.11) may be used to prove that, for the two products **AB** and **BA** to be equal, it is not sufficient that **A** and **B** be square matrices of the same order. For although $\mathbf{AB} = \mathbf{0}$ the product **BA** is

$$\mathbf{BA} = \begin{bmatrix} 3+8+10, & 3+8+10, & 3+8+10 \\ -2-2-5, & -2-2-5, & -2-2-5 \\ -1-6-5, & -1-6-5, & -1-6-5 \end{bmatrix} = \begin{bmatrix} 21 & 21 & 21 \\ -9 & -9 & -9 \\ -12 & -12 & -12 \end{bmatrix}. \tag{1.3.13}$$

If two matrices, **A** and **B**, do possess the special property **AB** = **BA**, they are said to 'commute' or to be 'permutable'. The two matrices

$$\mathbf{A} = \begin{bmatrix} 3 & -3 & 0 \\ 0 & 0 & -2 \\ 0 & 0 & -3 \end{bmatrix}, \qquad \mathbf{B} = \begin{bmatrix} 3 & 3 & 0 \\ 0 & 6 & 2 \\ 0 & 0 & 9 \end{bmatrix}, \tag{1.3.14}$$

are permutable, as the reader may verify for himself. The unit matrix $\mathbf{I}_n$ commutes with any square matrix of order n, and any pair of diagonal matrices commute.

Consider a row matrix postmultiplied by a square matrix. The result is a row matrix, as in the equation

$$[1 \quad 2 \quad 1] \begin{bmatrix} 0 & -1 & 4 \\ 1 & 2 & -6 \\ 3 & 0 & 1 \end{bmatrix} = [5 \quad 3 \quad -7]. \tag{1.3.15}$$

A row matrix postmultiplied by a column matrix is a number, or scalar (see (1.3.6)). A special case of this is a row matrix postmultiplied by its own transpose. The result is the sum of the squares of the elements, as in the equation

$$[a_1 \quad a_2 \quad a_3] \begin{bmatrix} a_1 \\ a_2 \\ a_3 \end{bmatrix} = a_1^2 + a_2^2 + a_3^2. \tag{1.3.16}†$$

A row matrix may also be premultiplied by a column matrix and the rule (1.3.4) implies that

$$\mathbf{A}(n, 1) \times \mathbf{B}(1, n) = \mathbf{C}(n, n), \tag{1.3.17}$$

so that the result is a square matrix. Thus

$$\begin{bmatrix} 1 \\ 2 \\ 3 \end{bmatrix} [-4 \quad 3 \quad -2] = \begin{bmatrix} -4 & 3 & -2 \\ -8 & 6 & -4 \\ -12 & 9 & -6 \end{bmatrix}. \tag{1.3.18}$$

It will be noted that the product here has the special property that the elements in any row are proportional to the elements in any other, and the same may be said of the columns. It can, in fact, be shown that any matrix which has only one independent row (or column) can always be expressed in the form of a row matrix premultiplied by a column matrix; this matter will not be pursued here, however.

The value of matrix notation as a means of diminishing the labour of writing out linear algebraic equations can readily be seen by considering a typical set of such equations. Let a set of variables $x_1, x_2, \ldots, x_n$ be related to a set $y_1, y_2, \ldots, y_n$ through a set of coefficients a_{ij} as in the relations

$$\left.\begin{aligned} a_{11}x_1 + a_{12}x_2 + \ldots + a_{1n}x_n &= y_1, \\ a_{21}x_1 + a_{22}x_2 + \ldots + a_{2n}x_n &= y_2, \\ &\cdots\cdots\cdots \\ a_{n1}x_1 + a_{n2}x_2 + \ldots + a_{nn}x_n &= y_n. \end{aligned}\right\} \tag{1.3.19}$$

† When the elements of the matrices are complex numbers this expression is complex and is of little importance compared with the real quantity $[a_1 \quad a_2 \quad a_3]\{\bar{a}_1 \quad \bar{a}_2 \quad \bar{a}_3\} = |a_1|^2 + |a_2|^2 + |a_3|^2$.

By the rule for equality of matrices, this set of equations may be written in the matrix form

$$\begin{bmatrix} a_{11}x_1+a_{12}x_2+\dots+a_{1n}x_n \\ a_{21}x_1+a_{22}x_2+\dots+a_{2n}x_n \\ \dots\dots\dots\dots\dots\dots \\ a_{n1}x_1+a_{n2}x_2+\dots+a_{nn}x_n \end{bmatrix} = \begin{bmatrix} y_1 \\ y_2 \\ \dots \\ y_n \end{bmatrix}. \tag{1.3.20}$$

But the left-hand matrix is the product

$$\begin{bmatrix} a_{11} & a_{12} & \dots & a_{1n} \\ a_{21} & a_{22} & \dots & a_{2n} \\ \dots & \dots & \dots & \dots \\ a_{n1} & a_{n2} & \dots & a_{nn} \end{bmatrix} \begin{bmatrix} x_1 \\ x_2 \\ \dots \\ x_n \end{bmatrix}. \tag{1.3.21}$$

That is to say equations (1.3.19) may be written in the matrix form

$$\mathbf{Ax} = \mathbf{y}, \tag{1.3.22}$$

where $\mathbf{A}$ is the square matrix and $\mathbf{x}$, $\mathbf{y}$ are column matrices.

This last use of matrix notation is so useful and commonly employed that it is worth while to illustrate it with a couple of simple examples. The reader can verify that

$$\begin{bmatrix} -1 & 0 \\ 1 & 1 \end{bmatrix} \begin{bmatrix} 2 \\ 1 \end{bmatrix} = \begin{bmatrix} -2 \\ 3 \end{bmatrix} \tag{1.3.23}$$

and

$$\begin{bmatrix} 2 & -3 & 4 \\ 1 & 0 & 1 \\ 1 & 0 & -1 \end{bmatrix} \begin{bmatrix} -1 \\ 2 \\ 3 \end{bmatrix} = \begin{bmatrix} 4 \\ 2 \\ -4 \end{bmatrix}. \tag{1.3.24}$$

EXAMPLES 1.3

1. If $\mathbf{A} = \begin{bmatrix} 3 & 2 & 5 \\ 2 & 4 & 9 \\ 5 & 9 & -16 \end{bmatrix}$ and $\mathbf{B} = \begin{bmatrix} -1 & -2 & -9 \\ 2 & 3 & 14 \\ 9 & -14 & 4 \end{bmatrix}$

find **AB** and **BA**.

2. Consider whether any or all of $\mathbf{A}+\mathbf{B}$, $\mathbf{AB}$ and $\mathbf{BA}$ may be formed when:

(*a*) $\mathbf{A} = \mathbf{A}(3, 2)$, $\mathbf{B} = \mathbf{B}(3, 2)$;
(*b*) $\mathbf{A} = \mathbf{A}(3, 5)$, $\mathbf{B} = \mathbf{B}(5, 4)$;
(*c*) $\mathbf{A} = \mathbf{A}(3, 5)$, $\mathbf{B} = \mathbf{B}(5, 3)$;
(*d*) $\mathbf{A} = \mathbf{A}(3, 5)$, $\mathbf{B} = \mathbf{B}(4, 5)$.

Under what conditions can $\mathbf{A}+\mathbf{B}$, $\mathbf{AB}$ and $\mathbf{BA}$ all be formed?

3. Write the equations

$$\begin{aligned} x+7y-3z &= 4,\\ y+z &= 2,\\ 3x-4y+11z+t &= 9,\\ 5x+4t &= 12, \end{aligned}$$

as a simple matrix equation of the type (1.3.22).

4. If $\mathbf{A}(\theta) = \begin{bmatrix} \cos\theta & -\sin\theta \\ \sin\theta & \cos\theta \end{bmatrix}$

show that $$\mathbf{A}(\theta)\,\mathbf{A}(\phi) = \mathbf{A}(\theta+\phi).$$

1.4 Further considerations of matrix multiplication

Consider the product of three matrices such as **ABC**. By this is meant the continued product of **A** postmultiplied by **B** to form **AB**, which matrix is then postmultiplied by the matrix **C**. Clearly this is possible only if **B** is conformable with **A** and if **C** is conformable with the product **AB**.

It may be shown by means of the general expression (1.3.9) for the element in a product matrix, that the associative law holds good for matrices. That is to say, subject to the obvious requirements of conformability,

$$\mathbf{ABC} = (\mathbf{AB})\times\mathbf{C} = \mathbf{A}\times(\mathbf{BC}). \tag{1.4.1}$$

In the same way

$$\mathbf{ABCD} = (\mathbf{AB})\times\mathbf{C}\times\mathbf{D} = \mathbf{A}\times\mathbf{B}\times(\mathbf{CD}) = (\mathbf{AB})\times(\mathbf{CD}),\ \text{etc.} \tag{1.4.2}$$

But it must be noted that in all cases: (*a*) this is subject to the requirements of conformability, and (*b*) the order of multiplication must be preserved. It is *not* necessarily true that $\mathbf{AB}\times\mathbf{D}\times\mathbf{C}$ would be equal to, say $\mathbf{AB}\times\mathbf{C}\times\mathbf{D}$, as we saw in equations (1.3.12) and (1.3.13).

Apart from the non-commutative nature of matrix multiplication all the laws of algebra apply to matrices in their usual form. Thus, not only does the associative law (1.4.1) hold good, but also the distributive law. That is to say

$$\mathbf{A}(\mathbf{B}+\mathbf{C})\,\mathbf{D} = \mathbf{ABD}+\mathbf{ACD}. \tag{1.4.3}$$

A square matrix may be raised to a positive power. If a matrix **A** is multiplied by itself r times, the resultant matrix is defined as $\mathbf{A}^r$. If, then

$$\mathbf{A} = \begin{bmatrix} 3 & 1 & 0 \\ 0 & 2 & 0 \\ -1 & 0 & 1 \end{bmatrix}, \tag{1.4.4}$$

the reader will find that

$$\mathbf{A}^2 = \begin{bmatrix} 9 & 5 & 0 \\ 0 & 4 & 0 \\ -4 & -1 & 1 \end{bmatrix}, \quad \mathbf{A}^3 = \begin{bmatrix} 27 & 19 & 0 \\ 0 & 8 & 0 \\ -13 & -6 & 1 \end{bmatrix}, \tag{1.4.5}$$

and so on.

Let two matrices **A** and **B** be conformable when multiplied in the order **AB** so that

$$\mathbf{AB} = \mathbf{P}. \tag{1.4.6}$$

When transposed, **A** and **B** must be conformable when multiplied in the order **B′A′** and their product is the transpose of **P**, so that

$$\mathbf{B'A'} = \mathbf{P'}. \tag{1.4.7}$$

The validity of this result can be confirmed by forming the expression for the typical element by means of equation (1.3.9).

To illustrate this property of matrices, consider the following set of algebraic equations

$$\left.\begin{aligned} 3x+4y+5z &= 2,\\ x-2y+7z &= -7,\\ -2x+y &= -5.\end{aligned}\right\} \tag{1.4.8}$$

It has been explained that this can be written

$$\begin{bmatrix} 3 & 4 & 5\\ 1 & -2 & 7\\ -2 & 1 & 0\end{bmatrix}\begin{bmatrix} x\\ y\\ z\end{bmatrix} = \begin{bmatrix} 2\\ -7\\ -5\end{bmatrix}. \tag{1.4.9}$$

But it can equally well be expressed in the form

$$[x \quad y \quad z]\begin{bmatrix} 3 & 1 & -2\\ 4 & -2 & 1\\ 5 & 7 & 0\end{bmatrix} = [2 \quad -7 \quad -5]. \tag{1.4.10}$$

This 'transposition rule' which states that the transpose of the product of two matrices is the reversed product of the transposed matrices, can clearly be extended. Thus if

$$\mathbf{ABC} = \mathbf{PC} = \mathbf{Q}, \quad \text{then} \quad \mathbf{Q'} = \mathbf{C'P'} = \mathbf{C'B'A'}, \tag{1.4.11}$$

and so on for any number of factors.

The product of a matrix with its own transpose is a symmetric matrix. If

$$\mathbf{A'A} = \mathbf{P}, \tag{1.4.12}$$

then, by the transposition rule,

$$\mathbf{P'} = (\mathbf{A'A})' = \mathbf{A'}(\mathbf{A'})' = \mathbf{A'A} = \mathbf{P}. \tag{1.4.13}$$

For instance

$$\begin{bmatrix} 3 & 2 & 1\\ -1 & 2 & -5\end{bmatrix}\begin{bmatrix} 3 & -1\\ 2 & 2\\ 1 & -5\end{bmatrix} = \begin{bmatrix} 14 & -4\\ -4 & 30\end{bmatrix}. \tag{1.4.14}$$

This result may be extended as follows. If $\mathbf{A}$ is a symmetric matrix and

$$\mathbf{B}'\mathbf{A}\mathbf{B} = \mathbf{B}'\mathbf{P} = \mathbf{Q}, \tag{1.4.15}$$

then

$$\mathbf{Q}' = \mathbf{P}'(\mathbf{B}')' = \mathbf{P}'\mathbf{B} = \mathbf{B}'\mathbf{A}'\mathbf{B} = \mathbf{B}'\mathbf{A}\mathbf{B} = \mathbf{Q}. \tag{1.4.16}$$

That is to say the product $\mathbf{B}'\mathbf{A}\mathbf{B}$ is a symmetric matrix, since it is equal to its own transpose.

EXAMPLES 1.4

1. $\mathbf{A}$ and $\mathbf{B}$ are the matrices given in Ex. 1.3.1 and

$$\mathbf{C} = \begin{bmatrix} 1 & 0 & -2 \\ 4 & 1 & 6 \\ 2 & -1 & 3 \end{bmatrix}.$$

Find $(\mathbf{AB})\mathbf{C}$ and verify equation (1.4.1) for these matrices.

2. Express

$$ax^2 + by^2 + cz^2 + 2fyz + 2gzx + 2hxy$$

as the product of three matrices.

3. Show that the matrix $\mathbf{A}(\theta)$ of Ex. 1.3.4 satisfies

$$(\mathbf{A}(\theta))^r = \mathbf{A}(r\theta)$$

for all positive integers r.

4. Find *sufficient* conditions on $\mathbf{A}$ and $\mathbf{B}$ for the equation

$$(\mathbf{AB})^r = \mathbf{A}^r\mathbf{B}^r$$

to hold for all positive integers r.

1.5 Elementary properties of determinants

It is clear that any square matrix has an associated determinant. The determinant has the same array of numbers as the matrix, but they are written between vertical bars instead of brackets. As we shall see, the determinant of a matrix is of great importance, so that a brief review of determinant theory will be given in this section.

A determinant is an array of numbers having the familiar form

$$\Delta \text{ or } |\mathbf{A}| = \begin{vmatrix} a_{11} & a_{12} & \cdots & a_{1n} \\ a_{21} & a_{22} & \cdots & a_{2n} \\ \cdots & \cdots & \cdots & \cdots \\ a_{n1} & a_{n2} & \cdots & a_{nn} \end{vmatrix}. \tag{1.5.1}$$

The array is necessarily square with equal numbers of rows and columns and, unlike a matrix, the determinant has a value. The value of a determinant must be defined and there are several ways of doing this, all leading to the same result. The method we shall adopt will be to define determinants by working upwards from determinants of lower order. We shall define the values of determinants of order 2 and 3 and shall then show that a determinant of order 4 may be expressed as the sum of 4 determinants of order 3; a determinant of order 5 may be expressed as the sum of 5 determinants of order 4 (i.e. 20 determinants of order 3), and so on. It will be found that this method of defining determinants is unsuitable for evaluating large determinants. However, it will lead to the formulation of certain general rules which will in turn provide simpler methods of evaluating large determinants.

The value of the 2×2 determinant

$$|\mathbf{A}| = \begin{vmatrix} a_{11} & a_{12} \\ a_{21} & a_{22} \end{vmatrix} \tag{1.5.2}$$

is given by

$$|\mathbf{A}| = a_{11}a_{22} - a_{12}a_{21}. \tag{1.5.3}$$

If $|\mathbf{A}|$ is the 3×3 determinant

$$|\mathbf{A}| = \begin{vmatrix} a_{11} & a_{12} & a_{13} \\ a_{21} & a_{22} & a_{23} \\ a_{31} & a_{32} & a_{33} \end{vmatrix}, \tag{1.5.4}$$

then

$$|\mathbf{A}| = a_{11}a_{22}a_{33} + a_{21}a_{32}a_{13} + a_{31}a_{12}a_{23} - a_{31}a_{22}a_{13} - a_{11}a_{32}a_{23} - a_{21}a_{12}a_{33}. \tag{1.5.5}$$

A simple way of remembering the terms and their signs is to write the first row again below the last and beneath this to repeat the second row. The three diagonals sloping down to the right give the positive terms while the three diagonals sloping down to the left give the negative terms. Thus

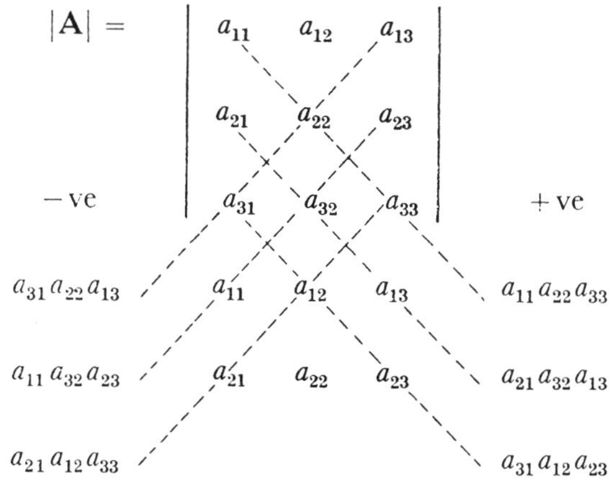

(1.5.6)†

† This rule may not be used for the evaluation of determinants of order greater than three, and although there are similar rules for such determinants they are very complicated. See J. D. Bankier, 'The diagrammatic expansion of determinants', *American math. Mon.* **68** (Oct. 1961), pp. 788–90. We prefer to use an entirely different approach.

The following example illustrates the working of the method

$$
\begin{array}{rccccr}
|\mathbf{A}| = & & 1 & 9 & 7 & \\
 & & -3 & 4 & 2 & \\
-\text{ve} & & -5 & 6 & 8 & +\text{ve} \\
-140 & & 1 & 9 & 7 & 32 \\
12 & & -3 & 4 & 2 & -126 \\
\underline{-216} & & & & & \underline{-90} \\
-344 & & & & & -184
\end{array}
$$

Thus $$|\mathbf{A}| = -184-(-344) = 160. \tag{1.5.7}$$

The six terms in the expression for a 3×3 determinant may be arranged in various ways. In particular they may be grouped together as follows:

$$|\mathbf{A}| = a_{11}(a_{22}a_{33}-a_{23}a_{32})-a_{12}(a_{21}a_{33}-a_{23}a_{31})+a_{13}(a_{21}a_{32}-a_{22}a_{31}), \tag{1.5.8}$$

which may be written

$$|\mathbf{A}| = a_{11}(+1)\begin{vmatrix} a_{22} & a_{23} \\ a_{32} & a_{33} \end{vmatrix} + a_{12}(-1)\begin{vmatrix} a_{21} & a_{23} \\ a_{31} & a_{33} \end{vmatrix} + a_{13}(+1)\begin{vmatrix} a_{21} & a_{22} \\ a_{31} & a_{32} \end{vmatrix}. \tag{1.5.9}$$

This way of writing the determinant is called its expansion 'along the first row'. It is seen that the expansion consists of the sum of the products of the elements of the first row and certain determinants with signs attached. These signed determinants are called the 'cofactors' of the elements they multiply and are denoted by A_{11}, A_{12}, A_{13}.

Thus

$$A_{11} = +\begin{vmatrix} a_{22} & a_{23} \\ a_{23} & a_{33} \end{vmatrix}; \qquad A_{12} = -\begin{vmatrix} a_{21} & a_{23} \\ a_{31} & a_{33} \end{vmatrix}; \qquad A_{13} = +\begin{vmatrix} a_{21} & a_{22} \\ a_{31} & a_{32} \end{vmatrix}, \tag{1.5.10}$$

so that $$|\mathbf{A}| = a_{11}A_{11}+a_{12}A_{12}+a_{13}A_{13}. \tag{1.5.11}$$

If the determinant of (1.5.7) is evaluated in this way it is found that

$$\begin{aligned} |\mathbf{A}| &= \begin{vmatrix} 4 & 2 \\ 6 & 8 \end{vmatrix} - 9\begin{vmatrix} -3 & 2 \\ -5 & 8 \end{vmatrix} + 7\begin{vmatrix} -3 & 4 \\ -5 & 6 \end{vmatrix} \\ &= (32-12)-9(-24+10)+7(-18+20) \\ &= 20+126+14 = 160, \end{aligned} \tag{1.5.12}$$

which agrees with the previous result.

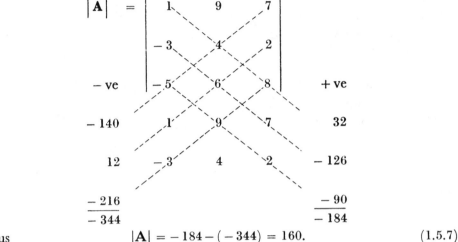

A 3×3 determinant may be expanded in a similar way along any row or column, so that there are in all six ways of grouping the terms. The reader may verify that two of the remaining five ways are

$$|\mathbf{A}| = a_{21}(-1)\begin{vmatrix} a_{12} & a_{13} \\ a_{32} & a_{33} \end{vmatrix} + a_{22}(+1)\begin{vmatrix} a_{11} & a_{13} \\ a_{31} & a_{33} \end{vmatrix} + a_{23}(-1)\begin{vmatrix} a_{11} & a_{12} \\ a_{31} & a_{32} \end{vmatrix}, \tag{1.5.13}$$

$$|\mathbf{A}| = a_{13}(+1)\begin{vmatrix} a_{21} & a_{22} \\ a_{31} & a_{32} \end{vmatrix} + a_{23}(-1)\begin{vmatrix} a_{11} & a_{12} \\ a_{31} & a_{32} \end{vmatrix} + a_{33}(+1)\begin{vmatrix} a_{11} & a_{12} \\ a_{21} & a_{22} \end{vmatrix}. \tag{1.5.14}$$

The first expression is the expansion 'along the second row' and is written

$$|\mathbf{A}| = a_{21}A_{21} + a_{22}A_{22} + a_{23}A_{23}. \tag{1.5.15}$$

The second is the expansion 'down the third column' and is written

$$|\mathbf{A}| = a_{13}A_{13} + a_{23}A_{23} + a_{33}A_{33}. \tag{1.5.16}$$

The reader will see that in every case the cofactor A_{ij} of a_{ij} is given by the equation

$$A_{ij} = (-1)^{i+j}\Delta_{ij}, \tag{1.5.17}$$

where Δ_{ij} is the determinant that is formed from $\mathbf{A}$ when the elements in the ith row and jth column are deleted.

Up to this point we have considered only 2×2 and 3×3 determinants; we shall now make use of the concept of a cofactor to define the value of an $n \times n$ determinant. If $|\mathbf{A}|$ is the $n \times n$ determinant given by (1.5.1) we define the cofactor A_{ij} of a_{ij} by equation (1.5.17). The quantity Δ_{ij}, which is called the 'minor' of a_{ij}, is obtained in exactly the same way as for the 3×3 determinant and is now a determinant of order $(n-1)$. The value of $|\mathbf{A}|$ is now given by the following procedure, which is called the 'Laplace Expansion'.† Select any row or column and sum the products of the successive elements and their cofactors. The same value is obtained whatever row or column is chosen.‡ Thus if the second row is chosen the rule gives

$$|\mathbf{A}| = a_{21}A_{21} + a_{22}A_{22} + \dots + a_{2n}A_{2n}, \tag{1.5.18}$$

while the same numerical value would be obtained from, say, the third-column expansion

$$|\mathbf{A}| = a_{13}A_{13} + a_{23}A_{23} + \dots + a_{n3}A_{n3}. \tag{1.5.19}$$

These last two equations may be compared with (1.5.15) and (1.5.16). In general for the expansion along row i the rule gives

$$|\mathbf{A}| = \sum_{k=1}^{n} a_{ik}A_{ik}, \tag{1.5.20}$$

and for the expansion down column j

$$|\mathbf{A}| = \sum_{k=1}^{n} a_{kj}A_{kj}. \tag{1.5.21}$$

† This term is sometimes used in a wider sense. See W. L. Ferrar, *Algebra* (Oxford University Press, 2nd ed., 1957), p. 36, for the generalisation of the procedure described below.

‡ This has been proved for 3×3 determinants—see, for example (1.5.11), (1.5.15) and (1.5.16). Using the result for determinants of order 3 the corresponding result may be proved for 4×4 determinants, and so on.

Since the cofactors A_{ij} are themselves determinants we have done what we set out to do and have expressed the $n \times n$ determinant $|\mathbf{A}|$ in terms of determinants of order $(n-1)$.

The Laplace expansion will give the value of a determinant of any order, but in practice the expansion is often too complicated. Thus the evaluation of the relatively small determinant

$$\begin{vmatrix} 1 & 3 & -5 & 4 & 1 \\ 2 & 4 & 7 & 9 & -4 \\ -10 & 2 & 9 & -1 & 2 \\ 17 & 13 & 6 & 5 & 3 \\ 9 & 1 & 7 & 6 & 4 \end{vmatrix}, \tag{1.5.22}$$

by the Laplace expansion, would require the calculation of five 4×4 determinants, and each of these calculations would entail the calculation of four 3×3 determinants—twenty 3×3 determinants in all. The evaluation of a 6×6 determinant would entail the calculation of 120 3×3 determinants. This should be sufficient to show that the evaluation of large determinants (a 16×16 determinant would not be unusual in practical vibration analysis) by this method is a tremendous task. The purpose of the remainder of this section is to state some rules which lead to simpler ways of evaluating large determinants. Using these rules we may evaluate a 5×5 determinant like (1.5.22) by performing about forty operations of multiplication between two numbers and forty operations of addition.

The familiar rules of determinants, which will now be enunciated, can be shown to follow the Laplace expansion.

(*a*) The value of a determinant remains the same when the rows are changed to columns and the columns to rows. Thus the determinants

$$\begin{vmatrix} 1 & 1 & 2 \\ 0 & -1 & 2 \\ 3 & 0 & 1 \end{vmatrix}, \quad \begin{vmatrix} 1 & 0 & 3 \\ 1 & -1 & 0 \\ 2 & 2 & 1 \end{vmatrix}$$

are equal. This rule conforms to our previous assertion that the same numerical value is obtained whether a determinant is expanded along a row or down a column.

The following eleven rules will be formulated in terms of the rows of a determinant. They apply equally to the columns of a determinant and this will be indicated by placing the word 'column' in parentheses after each occurrence of the word 'row'.

(*b*) If every element in a row (column) is zero, the determinant is zero. This may be seen from the expansion along the row (column) concerned.

(*c*) If all the elements but one in a row (column) are zero then the determinant is equal to that element multiplied by its cofactor.

(*d*) If all the elements of a determinant below (above) the principal diagonal are zero the determinant equals the product of the diagonal terms. Thus by the systematic use of rule (*c*) the value of

$$\begin{vmatrix} 4 & 6 & -3 & 5 \\ 0 & 2 & 17 & 49 \\ 0 & 0 & 3 & 114 \\ 0 & 0 & 0 & 7 \end{vmatrix}$$

is seen to be

$$4\times\begin{vmatrix} 2 & 17 & 49 \\ 0 & 3 & 114 \\ 0 & 0 & 7 \end{vmatrix} = 4\times 2\times\begin{vmatrix} 3 & 114 \\ 0 & 7 \end{vmatrix} = 4\times 2\times 3\times 7. \qquad (1.5.23)$$

(*e*) If all the elements of a row (column) are multiplied by some constant factor, the value of the determinant is multiplied by that factor. This may be demonstrated by expanding the determinant along the row (column) concerned.

(*f*) The value of the determinant changes sign if two rows (two columns) are interchanged. Thus it will be found that

$$\begin{vmatrix} 1 & 1 & 2 \\ 0 & -1 & 2 \\ 3 & 0 & 1 \end{vmatrix} = -\begin{vmatrix} 3 & 0 & 1 \\ 0 & -1 & 2 \\ 1 & 1 & 2 \end{vmatrix} = 11. \qquad (1.5.24)$$

This rule is obvious for 2×2 determinants; using the rule for these it may be proved for 3×3 determinants, and so on.

(*g*) The value of a determinant is zero if all the elements of a row (column) are identical with the corresponding elements of another row (column), or are multiples of them. Consider, for instance, the determinant

$$\Delta = \begin{vmatrix} 1 & 1 & 5 \\ 0 & -1 & 0 \\ 3 & 0 & 15 \end{vmatrix}, \qquad (1.5.25)$$

in which the third-column elements are five times the first-column elements. By rule (*e*)

$$\Delta = 5\begin{vmatrix} 1 & 1 & 1 \\ 0 & -1 & 0 \\ 3 & 0 & 3 \end{vmatrix}, \qquad (1.5.26)$$

but if columns one and three are now interchanged, the determinant must change sign (by rule (*f*)), although its structure is unchanged; the only possibility is that $\Delta = 0$.

(*h*) If each element in any one row (column) appears as the sum or the difference of two or more quantities, the determinant may be written as the sum (or difference) of two or more determinants of the same order. This rule may be proved by expanding the determinant along the row or column concerned. Thus the elements of the first row of the determinant

$$\begin{vmatrix} 4 & 6 \\ 1 & 3 \end{vmatrix}$$

may be split up to give

$$\begin{vmatrix} 2 & 1 \\ 1 & 3 \end{vmatrix} + \begin{vmatrix} 2 & 5 \\ 1 & 3 \end{vmatrix},$$

or, alternatively, to give

$$\begin{vmatrix} 10 & 7 \\ 1 & 3 \end{vmatrix} - \begin{vmatrix} 6 & 1 \\ 1 & 3 \end{vmatrix}.$$

(i) The value of a determinant remains the same if all the elements of any row (column) are multiplied by the same constant and are then added to the corresponding elements of any other row (column). Let $m \times$ (column 3) be added to column 1 of the determinant

$$\begin{vmatrix} a_{11} & a_{12} & a_{13} \\ a_{21} & a_{22} & a_{23} \\ a_{31} & a_{32} & a_{33} \end{vmatrix}.$$

The resulting determinant is

$$\begin{vmatrix} a_{11}+ma_{13} & a_{12} & a_{13} \\ a_{21}+ma_{23} & a_{22} & a_{23} \\ a_{31}+ma_{33} & a_{32} & a_{33} \end{vmatrix},$$

which, by rule (h), is equal to

$$\begin{vmatrix} a_{11} & a_{12} & a_{13} \\ a_{21} & a_{22} & a_{23} \\ a_{31} & a_{32} & a_{33} \end{vmatrix} + m \begin{vmatrix} a_{13} & a_{12} & a_{13} \\ a_{23} & a_{22} & a_{23} \\ a_{33} & a_{32} & a_{33} \end{vmatrix}. \tag{1.5.27}$$

Since the second determinant has two identical columns it is zero, and the required result is proved.

This last rule is the basis of many of the techniques for evaluating large determinants, and as such is extremely important. It has a number of simple corollaries which will now be stated.

(j) The value of a determinant remains the same if a linear combination of some of the rows (columns) is added to one row (column). Using this rule we may reduce any determinant to the type considered in rule (d). This is the so-called process of 'triangulation'. Suppose it is required to evaluate the determinant

$$\begin{vmatrix} 1 & 4 & 5 & 9 \\ 2 & 8 & 9 & 8 \\ -10 & 4 & 5 & 16 \\ 9 & 13 & 2 & 5 \end{vmatrix}.$$

If $2 \times$ (row 1) is subtracted from row 2 the first element in row 2 becomes zero. If in addition $10 \times$ (row 1) is added to row 3 and $9 \times$ (row 1) is subtracted from row 4 the determinant (still unchanged in value) becomes

$$\begin{vmatrix} 1 & 4 & 5 & 9 \\ 0 & 0 & -1 & -10 \\ 0 & 44 & 55 & 106 \\ 0 & -23 & -43 & -76 \end{vmatrix}.$$

By rule (*c*) the value of this determinant is the value of the 3×3 determinant in the lower right-hand corner. Alternatively, its value may be found by further use of rule (*j*). First interchange columns 2 and 3 so that there is a non-zero in the second diagonal position. This changes the sign of the determinant (rule (*f*)) so that the original determinant is

$$-\begin{vmatrix} 1 & 5 & 4 & 9 \\ 0 & -1 & 0 & -10 \\ 0 & 55 & 44 & 106 \\ 0 & -43 & -23 & -76 \end{vmatrix}.$$

If $55 \times$ (row 2) and $-43 \times$ (row 2) are added to rows 3 and 4 respectively the determinant becomes

$$-\begin{vmatrix} 1 & 5 & 4 & 9 \\ 0 & -1 & 0 & -10 \\ 0 & 0 & 44 & -444 \\ 0 & 0 & -23 & 354 \end{vmatrix},$$

whose value is

$$-(-1)\begin{vmatrix} 44 & -444 \\ -23 & 354 \end{vmatrix}.$$

If the determinant is reduced still further by adding $\frac{23}{44} \times$ (row 3) to row 4 it becomes

$$-\begin{vmatrix} 1 & 5 & 4 & 9 \\ 0 & -1 & 0 & -10 \\ 0 & 0 & 44 & -444 \\ 0 & 0 & 0 & \frac{5364}{44} \end{vmatrix},$$

and by rule (*d*) the value of the determinant is

$$-1 \times -1 \times 44 \times \tfrac{5364}{44} = 5364. \tag{1.5.28}$$

This triangulation process may be used to evaluate determinants of *any* order.

(*k*) If there is a linear relation between the rows (columns) of a determinant then its value is zero. This rule follows from the last. If the rows (columns) are linearly related then it is possible to find a linear combination of some of them which when added to one row (column) produces a row (column) of noughts. Thus in the determinant

$$\begin{vmatrix} 1 & 1 & 2 \\ -1 & 0 & 3 \\ 5 & 2 & -5 \end{vmatrix}$$

$$\text{row } 3 = 2 \times (\text{row } 1) - 3 \times (\text{row } 2), \tag{1.5.29}$$

so that if $-2 \times (\text{row } 1) + 3 \times (\text{row } 2)$ is added to row 3 the result is

$$\begin{vmatrix} 1 & 1 & 2 \\ -1 & 0 & 3 \\ 0 & 0 & 0 \end{vmatrix},$$

and the value of this determinant is clearly zero.

To *test* whether a determinant is zero or not we can use the triangulation process. The determinant will be zero if (and only if) at some stage of the triangulation a row of noughts is reached. Thus the converse of the rule (k) is also true, namely: if the value of a determinant is zero then its rows (columns) are linearly related.

The following rule is concerned with the cofactors of a determinant and is a corollary of rule (g).

(l) If the elements of one row (column) are multiplied by the cofactors of the corresponding elements of another row (column) and the results are added, the sum is zero.

For instance, for the determinant

$$\begin{vmatrix} 3 & -1 & 1 \\ 2 & 2 & 3 \\ 1 & -3 & 1 \end{vmatrix},$$

the sum of the products of the elements of the second row and the cofactors of the third is

$$2\begin{vmatrix} -1 & 1 \\ 2 & 3 \end{vmatrix} + 2(-1)\begin{vmatrix} 3 & 1 \\ 2 & 3 \end{vmatrix} + 3\begin{vmatrix} 3 & -1 \\ 2 & 2 \end{vmatrix}$$

$$= 2(-5) + 2(-7) + 3(8) = 0. \tag{1.5.30}$$

The reason for this is that the expansion is identical with that of

$$\begin{vmatrix} 3 & -1 & 1 \\ 2 & 2 & 3 \\ 2 & 2 & 3 \end{vmatrix},$$

and this determinant is zero because it has two identical rows. The general result for the rows of a determinant is

$$\sum_{k=1}^{n} a_{ik} A_{jk} = 0 \quad (i \neq j). \tag{1.5.31}$$

This expresses the fact that the sum of the products of row i and the cofactors of row j is zero. This equation may be combined with (1.5.20) to give

$$\sum_{k=1}^{n} a_{ik} A_{jk} = |\mathbf{A}|\,\delta_{ij} = \begin{Bmatrix} |\mathbf{A}| & \text{if} & i=j \\ 0 & \text{if} & i \neq j \end{Bmatrix}. \tag{1.5.32}$$

The corresponding result for columns is

$$\sum_{k=1}^{n} a_{ki} A_{kj} = |\mathbf{A}|\,\delta_{ij} = \begin{Bmatrix} |\mathbf{A}| & \text{if} & i=j \\ 0 & \text{if} & i \neq j \end{Bmatrix}. \tag{1.5.33}$$

We conclude this section by discussing another way, which is often used, of defining the value of a determinant. This definition is a generalisation of equation (1.5.5), and although it is not often of much value as a practical means of evaluation it is important for theoretical work.

Equation (1.5.5) expresses the 3×3 determinant (1.5.4) as the sum of certain products, with signs attached. Each of the six products contains three terms of which just one comes from each row and just one from each column. It may easily be proved, by using the definition we have adopted—see equations (1.5.20) and (1.5.21)—that this result may be generalised quite naturally to give the value of a determinant of any order. In fact, leaving aside for a moment the way in which the signs are determined, we may write

$$|\mathbf{A}| = \sum_{i,j,\ldots,t} \pm a_{i1} a_{j2} \ldots a_{tn}, \tag{1.5.34}$$

where $i, j, \ldots, t$ denote the n numbers $1, 2, \ldots, n$ in some order and the summation extends to all possible sets of such numbers.

The way in which the sign accompanying a given term is obtained may be explained as follows:

All rearrangements of the numbers $1, 2, \ldots, n$ may be obtained by successive interchanges of two numbers. For an example, take $n = 4$. The arrangement 2, 3, 4, 1 may be obtained from 1, 2, 3, 4 as follows

$$\underline{1}, 2, 3, \underline{4} \rightarrow \underline{4}, \underline{2}, 3, 1 \rightarrow 2, \underline{4}, \underline{3}, 1 \rightarrow 2, 3, 4, 1,$$

where the underlined numbers are the ones which are to be interchanged next. Similarly, 2, 1, 4, 3 may be obtained by making the interchanges

$$\underline{1}, \underline{2}, 3, 4 \rightarrow 2, 1, \underline{3}, \underline{4} \rightarrow 2, 1, 4, 3.$$

The reader will notice that there are other ways of obtaining 2, 3, 4, 1 beside that which has been given, but that they all involve an *odd* number of interchanges.

It is found that all the rearrangements (or permutations) of $1, 2, \ldots, n$ may be divided into two classes, those which may be obtained from $1, 2, \ldots, n$ by making an even number of interchanges, and those which may be obtained by making an odd number. The two classes of permutations are called even and odd respectively. For example, 2, 3, 4, 1 is an odd permutation of 1, 2, 3, 4 and 2, 1, 4, 3 is an even permutation.

The rule for the sign accompanying the term $a_{i1}a_{j2}\dots a_{tn}$ is simply 'the sign is positive if $i, j, \dots, t$ is an even permutation and negative if it is an odd permutation'.

Finally, the reader may verify that because the value of a determinant remains the same when the rows are changed to columns and the columns to rows (rule (*a*)), the value of a determinant may equally well be expressed in the form

$$|\mathbf{A}| = \sum_{i,j,\dots,t} \pm a_{i1}a_{j2}\dots a_{tn}, \tag{1.5.35}$$

where $i, j, \dots, t$ have the same meaning as before and the signs are determined in the same way as before.

EXAMPLES 1.5

1. Evaluate the original 4×4 determinant considered under rule (*j*) above by expanding it along its first column and calculating the relevant 3×3 determinants. Verify that the result (1.5.28) is obtained.

2. Prove that

$$\begin{vmatrix} a & b & c \\ a^2 & b^2 & c^2 \\ b+c & c+a & a+b \end{vmatrix} = (a+b+c)\,(b-c)\,(c-a)\,(a-b).$$

3. Evaluate the determinant (1.5.22) by triangulation and show that its value is 122796.

4. Prove that the determinant of a skew-symmetric matrix of odd order is zero.

5. Prove that if $\mathbf{A}$ and $\mathbf{B}$ are square matrices of order n then

$$|\mathbf{AB}| = |\mathbf{A}|\,.\,|\mathbf{B}|.$$

[*Hint.* Consider the matrix

$$\mathbf{C} = \begin{bmatrix} a_{11} & a_{12} & \dots & a_{1n} & 0 & 0 & \dots & 0 \\ a_{21} & a_{22} & \dots & a_{2n} & 0 & 0 & \dots & 0 \\ \dots & \dots & \dots & \dots & \dots & \dots & \dots & \dots \\ a_{n1} & a_{n2} & \dots & a_{nn} & 0 & 0 & \dots & 0 \\ -1 & 0 & \dots & 0 & b_{11} & b_{12} & \dots & b_{1n} \\ 0 & -1 & \dots & 0 & b_{21} & b_{22} & \dots & b_{2n} \\ \dots & \dots & \dots & \dots & \dots & \dots & \dots & \dots \\ 0 & 0 & \dots & -1 & b_{n1} & b_{n2} & \dots & b_{nn} \end{bmatrix}.$$

By adding suitable multiples of rows $n+1$ to $2n$ to rows 1 to n show that

$$|\mathbf{C}| = |\mathbf{AB}|.$$

To prove that $|\mathbf{C}| = |\mathbf{A}|\,.\,|\mathbf{B}|$ use induction on n together with the Laplace expansion. Clearly

$$\begin{vmatrix} a_{11} & 0 \\ 0 & b_{11} \end{vmatrix} = a_{11}\,.\,b_{11},$$

so that the required result holds for $n = 1$. Now suppose that $|\mathbf{C}| = |\mathbf{A}|\,.\,|\mathbf{B}|$ holds for any two matrices of order $n-1$. Expand $|\mathbf{C}|$ along its first row and then expand the rth of the minors of order $2n-1$ so obtained along its $n-r+1$th row and use the result assumed for $n-1$.]

6. State whether the following are even or odd permutations:

(*a*) 4321; (*b*) 645213; (*c*) 25431; (*d*) 234561.

7. Derive the expression (1.5.34) from rule (*h*).

1.6 The inverse matrix

Let **A** be a square matrix of order n. The problem of finding the 'inverse' of **A** is the problem of finding a matrix **R** which is such that

$$\mathbf{AR} = \mathbf{RA} = \mathbf{I}_n. \tag{1.6.1}$$

When there is such a matrix **R** it is commonly denoted by $\mathbf{A}^{-1}$.

This problem of finding the inverse of a matrix is one of considerable importance. Consider, for instance, the equation

$$\mathbf{Ax} = \mathbf{y}, \tag{1.6.2}$$

which is the matrix representation of equations (1.3.19). If the matrix **A** has an inverse **R** and we can find it then we can premultiply both sides of the equation by **R** to give

$$\mathbf{RAx} = \mathbf{Ry}, \tag{1.6.3}$$

which reduces to

$$\mathbf{x} = \mathbf{Ry} \quad (\text{or } \mathbf{A}^{-1}\mathbf{y}), \tag{1.6.4}$$

because $\mathbf{RAx} = \mathbf{Ix} = \mathbf{x}$. But if we denote the elements of **R** by r_{ij} and expand this equation to give n relations of the form (1.3.19), then we have

$$\left.\begin{aligned} x_1 &= r_{11}y_1 + r_{12}y_2 + \ldots + r_{1n}y_n, \\ x_2 &= r_{21}y_1 + r_{22}y_2 + \ldots + r_{2n}y_n, \\ &\cdots\cdots\cdots\cdots\cdots\cdots\cdots\cdots \\ x_n &= r_{n1}y_1 + r_{n2}y_2 + \ldots + r_{nn}y_n, \end{aligned}\right\} \tag{1.6.5}$$

which is thus the solution of equations (1.3.19).

Practical methods of finding the inverse of a large matrix form the subject of Chapter 7; in this section we are concerned with the *theory* of inverse matrices.

Two questions are relevant to the problem of finding an inverse: 'Under what conditions does a square matrix have an inverse?' and 'If a matrix has an inverse, how can it be found?'

Consider the problem in relation to the 2×2 matrix

$$\mathbf{A} = \begin{bmatrix} a_{11} & a_{12} \\ a_{21} & a_{22} \end{bmatrix}. \tag{1.6.6}$$

The matrix **R** which we are seeking will be of the form

$$\mathbf{R} = \begin{bmatrix} r_{11} & r_{12} \\ r_{21} & r_{22} \end{bmatrix}, \tag{1.6.7}$$

and the equation (1.6.1) will be

$$\begin{bmatrix} a_{11}r_{11} + a_{12}r_{21} & a_{11}r_{12} + a_{12}r_{22} \\ a_{21}r_{11} + a_{22}r_{21} & a_{21}r_{12} + a_{22}r_{22} \end{bmatrix} = \begin{bmatrix} r_{11}a_{11} + r_{12}a_{21} & r_{11}a_{12} + r_{12}a_{22} \\ r_{21}a_{11} + r_{22}a_{21} & r_{21}a_{12} + r_{22}a_{22} \end{bmatrix} = \begin{bmatrix} 1 & 0 \\ 0 & 1 \end{bmatrix}. \tag{1.6.8}$$

If we equate the first and third of these matrices we obtain the two sets of equations

$$\left.\begin{aligned} a_{11}r_{11}+a_{12}r_{21} &= 1, \quad a_{11}r_{12}+a_{12}r_{22} = 0, \\ a_{21}r_{11}+a_{22}r_{21} &= 0, \quad a_{21}r_{12}+a_{22}r_{22} = 1. \end{aligned}\right\} \tag{1.6.9}$$

The left-hand set involves only r_{11} and r_{21}, the right-hand set involves only r_{12} and r_{22}. These equations have a solution only if

$$(a_{11}a_{22}-a_{12}a_{21}) \neq 0, \tag{1.6.10}$$

i.e.

$$|\mathbf{A}| \neq 0. \tag{1.6.11}$$

When this condition is satisfied the solution is

$$r_{11} = \frac{a_{22}}{|\mathbf{A}|}, \quad r_{21} = \frac{-a_{21}}{|\mathbf{A}|}, \quad r_{12} = \frac{-a_{12}}{|\mathbf{A}|}, \quad r_{22} = \frac{a_{11}}{|\mathbf{A}|}; \tag{1.6.12}$$

so that

$$\mathbf{R} = \mathbf{A}^{-1} = \frac{1}{|\mathbf{A}|}\begin{bmatrix} a_{22} & -a_{12} \\ -a_{21} & a_{11} \end{bmatrix}. \tag{1.6.13}$$

The reader may now verify that the equations obtained by equating the second and third matrices in (1.6.8) have a solution only if $|\mathbf{A}| \neq 0$ and that when this condition is satisfied the solution is that given in (1.6.12). We have therefore answered both our questions as far as 2×2 matrices are concerned.

For an example, take $\mathbf{A}$ to be the matrix

$$\mathbf{A} = \begin{bmatrix} 2 & 4 \\ -1 & 3 \end{bmatrix}, \tag{1.6.14}$$

so that $|\mathbf{A}| = 10$. The inverse as given by (1.6.13) is

$$\mathbf{R} = \mathbf{A}^{-1} = \tfrac{1}{10}\begin{bmatrix} 3 & -4 \\ 1 & 2 \end{bmatrix}. \tag{1.6.15}$$

To solve the 'inverse' problem for a square matrix of arbitrary order we shall employ the concept of a 'cofactor' introduced in the previous section. Let $\mathbf{A}$ be the matrix

$$\mathbf{A} = \begin{bmatrix} a_{11} & a_{12} & \dots & a_{1n} \\ a_{21} & a_{22} & \dots & a_{2n} \\ \dots & \dots & \dots & \dots \\ a_{n1} & a_{n2} & \dots & a_{nn} \end{bmatrix} \tag{1.6.16}$$

and let A_{ij} be the cofactor of the element a_{ij}. The matrix

$$\hat{\mathbf{A}} = \begin{bmatrix} A_{11} & A_{21} & \dots & A_{n1} \\ A_{12} & A_{22} & \dots & A_{n2} \\ \dots & \dots & \dots & \dots \\ A_{1n} & A_{2n} & \dots & A_{nn} \end{bmatrix}, \tag{1.6.17}$$

which is the transpose of the matrix whose elements are the cofactors of **A** is called the 'adjoint' of **A**. Thus the matrix

$$\mathbf{A} = \begin{bmatrix} 1 & 1 & 2 \\ 0 & -1 & 2 \\ 3 & 0 & 1 \end{bmatrix} \tag{1.6.18}$$

has for its adjoint the matrix

$$\hat{\mathbf{A}} = \begin{bmatrix} -1 & -1 & 4 \\ 6 & -5 & -2 \\ 3 & 3 & -1 \end{bmatrix}. \tag{1.6.19}$$

Every square matrix **A** has an adjoint matrix $\hat{\mathbf{A}}$; the existence of $\hat{\mathbf{A}}$ is quite independent of whether $|\mathbf{A}|$ is, or is not, zero. The matrix in the last example had a non-zero determinant; the matrix

$$\mathbf{A} = \begin{bmatrix} 1 & 1 & 2 \\ -1 & 0 & 3 \\ 5 & 2 & -5 \end{bmatrix} \tag{1.6.20}$$

has a zero determinant: the adjoint of this matrix is

$$\hat{\mathbf{A}} = \begin{bmatrix} -6 & 9 & 3 \\ 10 & -15 & -5 \\ -2 & 3 & 1 \end{bmatrix}. \tag{1.6.21}$$

This example incidentally provides an example of the general result—which we shall not prove at this stage—that if $|\mathbf{A}|$ is zero then so is $|\hat{\mathbf{A}}|$.

The matrix

$$\begin{bmatrix} a_{22} & -a_{12} \\ -a_{21} & a_{11} \end{bmatrix},$$

which occurred in (1.6.13), can now be identified as the adjoint of the matrix **A** given in (1.6.6). Thus for the particular case of the 2×2 matrix, the inverse, when it exists, is given by the equation

$$\mathbf{R} = \mathbf{A}^{-1} = \frac{\hat{\mathbf{A}}}{|\mathbf{A}|}. \tag{1.6.22}$$

We shall now show that this equation gives the inverse of a matrix of arbitrary order.

Returning to the general case we consider the product $\mathbf{A}\hat{\mathbf{A}}$ of the matrices **A** and $\hat{\mathbf{A}}$. The rule (1.3.9) states that the element c_{ij} of the product of **A** and **B** is

$$c_{ij} = \sum_{k=1}^{n} a_{ik} b_{kj}. \tag{1.6.23}$$

If **B** is identified with $\hat{\mathbf{A}}$ then $b_{kj} = A_{jk}$ so that the element c_{ij} in the product $\mathbf{A}\hat{\mathbf{A}}$ is

$$c_{ij} = \sum_{k=1}^{n} a_{ik} A_{jk}. \tag{1.6.24}$$

But the sum on the right is just the one which occurs in (1.5.32). Therefore

$$c_{ij} = \begin{cases} |\mathbf{A}| & \text{if } i = j, \\ 0 & \text{if } i \neq j, \end{cases} \tag{1.6.25}$$

that is
$$\mathbf{A}\hat{\mathbf{A}} = |\mathbf{A}|\,\mathbf{I}. \tag{1.6.26}$$

We may deal in a similar way with the product $\hat{\mathbf{A}}\mathbf{A}$. The element d_{ij} of the product $\mathbf{BA}$ is

$$d_{ij} = \sum_{k=1}^{n} b_{ik}a_{kj} = \sum_{k=1}^{n} A_{ki}a_{kj} = |\mathbf{A}|\,\delta_{ij}, \tag{1.6.27}$$

because of (1.5.33). Therefore
$$\hat{\mathbf{A}}\mathbf{A} = |\mathbf{A}|\,\mathbf{I}. \tag{1.6.28}$$

Equations (1.6.26) and (1.6.28), which hold for every square matrix and its adjoint, show that a square matrix commutes with its adjoint. When $|\mathbf{A}|$ is zero these equations give
$$\mathbf{A}\hat{\mathbf{A}} = \mathbf{0} = \hat{\mathbf{A}}\mathbf{A}. \tag{1.6.29}$$

When $|\mathbf{A}|$ is not zero they may be divided by it to give

$$\mathbf{A}.\frac{\hat{\mathbf{A}}}{|\mathbf{A}|} = \frac{\hat{\mathbf{A}}}{|\mathbf{A}|}.\mathbf{A} = \mathbf{I}, \tag{1.6.30}$$

which shows that the inverse of $\mathbf{A}$ is given by (1.6.22). If $|\mathbf{A}|$ is not zero we say that $\mathbf{A}$ is 'non-singular'. It can be proved—though we shall not do it here—that, if $\mathbf{A}$ is non-singular, then the matrix $\mathbf{A}^{-1}$ given by (1.6.22) is the only one which satisfies (1.6.1). This means that the inverse is unique.

We have shown that when $\mathbf{A}$ is non-singular it possesses an inverse. It can also be shown that $\mathbf{A}$ does not have an inverse when $|\mathbf{A}| = 0$, i.e. when $\mathbf{A}$ is 'singular'. To prove this we make use of the result (Ex. 1.5.5) that if $\mathbf{A}$ and $\mathbf{B}$ are square matrices of the same order then
$$|\mathbf{AB}| = |\mathbf{A}|.|\mathbf{B}|. \tag{1.6.31}$$

Suppose that $|\mathbf{A}| = 0$ and $\mathbf{A}$ possesses an inverse. There is then a matrix $\mathbf{R}$ such that
$$\mathbf{AR} = \mathbf{I}. \tag{1.6.32}$$

But then
$$|\mathbf{A}|.|\mathbf{R}| = 0.|\mathbf{R}| = |\mathbf{I}| = 1, \tag{1.6.33}$$

which is impossible. Therefore $\mathbf{A}$ cannot have an inverse.

We have now obtained a complete answer to our two questions. *The necessary and sufficient condition for the square matrix* $\mathbf{A}$ *to have an inverse is that it shall be non-singular. When this condition is satisfied the inverse is given by equation* (1.6.22).

To illustrate the computation of an inverse matrix take

$$\mathbf{A} = \begin{bmatrix} 1 & 4 & 0 \\ -2 & 2 & 1 \\ 3 & 1 & 0 \end{bmatrix}, \tag{1.6.34}$$

for which $|\mathbf{A}| = 11$. The adjoint of $\mathbf{A}$ is

$$\hat{\mathbf{A}} = \begin{bmatrix} -1 & 0 & 4 \\ 3 & 0 & -1 \\ -8 & 11 & 10 \end{bmatrix}, \tag{1.6.35}$$

so that

$$\mathbf{A}^{-1} = \tfrac{1}{11}\begin{bmatrix} -1 & 0 & 4 \\ 3 & 0 & -1 \\ -8 & 11 & 10 \end{bmatrix}. \tag{1.6.36}$$

It is clear that the inverse of the matrix $\mathbf{A}^{-1}$ is the matrix $\mathbf{A}$. For the inverse of $\mathbf{A}^{-1}$ is *the* matrix $\mathbf{S}$ which satisfies

$$\mathbf{A}^{-1}\mathbf{S} = \mathbf{S}\mathbf{A}^{-1} = \mathbf{I}, \tag{1.6.37}$$

and $\mathbf{S} = \mathbf{A}$ satisfies this equation. Thus the reader may verify, by finding the adjoint of $\mathbf{A}^{-1}$, that the inverse of the matrix in equation (1.6.36) is given by (1.6.34).

At the beginning of this section it was shown how the inverse of the matrix $\mathbf{A}$ may be used to solve equations (1.3.19) for the unknown quantities $x_1, x_2, \ldots, x_n$. We shall now show that the matrix solution (1.6.4) of these equations is equivalent to Cramer's Rule for the solution. To do this we substitute the values

$$r_{11} = \frac{A_{11}}{|\mathbf{A}|}, \quad r_{12} = \frac{A_{21}}{|\mathbf{A}|}, \quad \ldots, \quad r_{1n} = \frac{A_{n1}}{|\mathbf{A}|}, \tag{1.6.38}$$

obtained from (1.6.22) into equation (1.6.5). The expression for x_1 becomes

$$x_1 = \frac{1}{|\mathbf{A}|}\{y_1 A_{11} + y_2 A_{21} + \ldots + y_n A_{n1}\}, \tag{1.6.39}$$

which is the expansion 'down the first column' of the determinant

$$x_1 = \frac{1}{|\mathbf{A}|}\begin{vmatrix} y_1 & a_{12} & a_{13} & \ldots & a_{1n} \\ y_2 & a_{22} & a_{23} & \ldots & a_{2n} \\ \ldots & \ldots & \ldots & \ldots & \ldots \\ y_n & a_{n2} & a_{n3} & \ldots & a_{nn} \end{vmatrix}. \tag{1.6.40}$$

Similarly, the expression for x_2, viz.

$$x_2 = \frac{1}{|\mathbf{A}|}\{y_1 A_{12} + y_2 A_{22} + \ldots + y_n A_{n2}\} \tag{1.6.41}$$

is the expansion 'down the second column' of

$$x_2 = \frac{1}{|\mathbf{A}|}\begin{vmatrix} a_{11} & y_1 & a_{13} & \ldots & a_{1n} \\ a_{21} & y_2 & a_{23} & \ldots & a_{2n} \\ \ldots & \ldots & \ldots & \ldots & \ldots \\ a_{n1} & y_n & a_{n3} & \ldots & a_{nn} \end{vmatrix}. \tag{1.6.42}$$

Equations (1.6.40) and (1.6.42) are just the expressions ior x_1 and x_2 given by Cramer's Rule.

EXAMPLES 1.6

1. Find the inverse of the matrix

$$\mathbf{A} = \begin{bmatrix} 1 & 4 & 0 & 2 \\ 1 & 6 & 2 & 1 \\ 0 & 3 & 7 & 2 \\ 1 & -1 & 0 & 4 \end{bmatrix}$$

and verify that equation (1.6.1) is satisfied.

2. Use the inverse of the matrix $\mathbf{A}$ in the previous example to solve the equations

$$\begin{aligned} x+4y+2t &= 3, \\ x+6y+2z+t &= -4, \\ 3y+7z+2t &= 1, \\ x-y+4t &= 2. \end{aligned}$$

3. Prove that the inverse of a diagonal matrix is also a diagonal matrix.

4. Prove that if there is a matrix $\mathbf{R}$ such that

$$\mathbf{AR} = \mathbf{I},$$

then $\mathbf{RA} = \mathbf{I}$ also and $\mathbf{R}$ is the inverse of $\mathbf{A}$. [The importance of this result is that if a matrix is found, by any means, which satisfies $\mathbf{AR} = \mathbf{I}$ then it will be *the* inverse of $\mathbf{A}$.]

5. Prove that if $\mathbf{A}$ is singular then so is $\hat{\mathbf{A}}$.

6. If $\mathbf{A}$ is non-singular prove that

$$|\hat{\mathbf{A}}| = |\mathbf{A}|^{n-1}.$$

Use this result to prove that when $\mathbf{A}$ is non-singular

$$\hat{\hat{\mathbf{A}}} = |\mathbf{A}|^{n-2}\mathbf{A},$$

where $\hat{\hat{\mathbf{A}}}$ denotes the adjoint of $\hat{\mathbf{A}}$.

1.7 Further considerations of matrix division and the properties of inverse matrices

The division of one real number a by another, b, may be interpreted as ab^{-1} or $b^{-1}a$. Whichever process is implied, a/b has a unique meaning. The same is not true of matrices, since, in general,

$$\mathbf{AB}^{-1} \neq \mathbf{B}^{-1}\mathbf{A}. \tag{1.7.1}$$

Thus the 'division of $\mathbf{A}$ by $\mathbf{B}$' has no unique meaning unless $\mathbf{A}$ and $\mathbf{B}^{-1}$ happen to commute. In referring to 'matrix division', therefore, we shall mean either premultiplication or postmultiplication by an inverse matrix; which of these two processes is implied will be clear from the context.

The cancellation of common non-zero factors is one of the familiar processes of algebra. The question of whether or not cancellation is permissible in matrix algebra is not quite so simple, partly because the product of two matrices is not always commutative and partly because the inverse of a singular matrix does not exist.

Consider the equation

$$\mathbf{AB} = \mathbf{AC}, \tag{1.7.2}$$

where the common factor is the matrix **A**. Let **A** be non-singular so that $\mathbf{A}^{-1}$ exists. If the two sides of the equation are premultiplied by $\mathbf{A}^{-1}$, then it is found that

$$\mathbf{A}^{-1}\mathbf{AB} = \mathbf{A}^{-1}\mathbf{AC}, \tag{1.7.3}$$

so that

$$\mathbf{B} = \mathbf{C}. \tag{1.7.4}$$

Similarly, if $\mathbf{BA} = \mathbf{CA}$ where **A** is non-singular, then $\mathbf{B} = \mathbf{C}$. The conclusion to be drawn from this is that a common factor may be cancelled in matrix algebra provided that (*a*) it is non-singular, and (*b*) it is similarly located on the two sides of the equation concerned. It will be seen, furthermore, that the cancellation is not permitted if either (or both) of these two conditions is not fulfilled.

The matrix equation

$$\mathbf{AB} = \mathbf{CD} \tag{1.7.5}$$

may be treated in a similar way. Suppose that **A** is not singular so that $\mathbf{A}^{-1}$ exists. If the equation is premultiplied by $\mathbf{A}^{-1}$ it gives

$$\mathbf{A}^{-1}\mathbf{AB} = \mathbf{A}^{-1}\mathbf{CD}, \tag{1.7.6}$$

or

$$\mathbf{B} = \mathbf{A}^{-1}\mathbf{CD}. \tag{1.7.7}$$

This is a striking result when it is remembered that the symbols do not represent numbers, but arrays which are manipulated by the rules which have been explained. The result is far from obvious and cannot be inferred from the analogy between matrix algebra and ordinary algebra.

We turn now to some of the elementary properties of matrices in relation to their inverses. Suppose that the product **P** of two non-singular square matrices is given by

$$\mathbf{P} = \mathbf{AB}. \tag{1.7.8}$$

Premultiplying both sides of this equation by $\mathbf{B}^{-1}\mathbf{A}^{-1}$, we get

$$\mathbf{B}^{-1}\mathbf{A}^{-1}\mathbf{P} = \mathbf{B}^{-1}\mathbf{A}^{-1}\mathbf{AB} = \mathbf{I}. \tag{1.7.9}$$

This equation shows that **P** has an inverse and that it is given by the equation

$$\mathbf{P}^{-1} = \mathbf{B}^{-1}\mathbf{A}^{-1}. \tag{1.7.10}$$

The process of reversing the order of multiplication while taking the inverse matrices is called that of 'reciprocation' and the result can easily be generalised to cover any number of factors. Thus if

$$\mathbf{P} = \mathbf{ABC}, \tag{1.7.11}$$

then

$$\mathbf{P}^{-1} = \mathbf{C}^{-1}\mathbf{B}^{-1}\mathbf{A}^{-1}, \tag{1.7.12}$$

provided that the members concerned are square and non-singular.

Let

$$\mathbf{A} = \begin{bmatrix} 2 & 4 \\ -1 & 3 \end{bmatrix}, \quad \mathbf{B} = \begin{bmatrix} 3 & 2 \\ -4 & 1 \end{bmatrix}; \tag{1.7.13}$$

then
$$\mathbf{A}^{-1} = \tfrac{1}{10}\begin{bmatrix} 3 & -4 \\ 1 & 2 \end{bmatrix}; \qquad \mathbf{B}^{-1} = \tfrac{1}{11}\begin{bmatrix} 1 & -2 \\ 4 & 3 \end{bmatrix}. \tag{1.7.14}$$

The product $\mathbf{P} = \mathbf{AB}$ is
$$\mathbf{P} = \begin{bmatrix} 2 & 4 \\ -1 & 3 \end{bmatrix}\begin{bmatrix} 3 & 2 \\ -4 & 1 \end{bmatrix} = \begin{bmatrix} -10 & 8 \\ -15 & 1 \end{bmatrix}. \tag{1.7.15}$$

If $\mathbf{B}^{-1}$ is postmultiplied by $\mathbf{A}^{-1}$ it is found that
$$\mathbf{B}^{-1}\mathbf{A}^{-1} = \tfrac{1}{110}\begin{bmatrix} 1 & -2 \\ 4 & 3 \end{bmatrix}\begin{bmatrix} 3 & -4 \\ 1 & 2 \end{bmatrix} = \tfrac{1}{110}\begin{bmatrix} 1 & -8 \\ 15 & -10 \end{bmatrix}, \tag{1.7.16}$$

and this last matrix is, in fact, the inverse of $\mathbf{P}$.

In vibration theory, one is frequently concerned with symmetric matrices. It will now be shown that the inverse of a symmetric matrix is itself symmetric. Let $\mathbf{A}$ be a non-singular symmetric matrix whose inverse is $\boldsymbol{\alpha}$; $\mathbf{A}\boldsymbol{\alpha} = \mathbf{I}$ so that, by the transposition rule (§ 1.4),
$$\boldsymbol{\alpha}'\mathbf{A}' = \boldsymbol{\alpha}'\mathbf{A} = \mathbf{I}' = \mathbf{I}. \tag{1.7.17}$$

If, now, the equation $\boldsymbol{\alpha}'\mathbf{A} = \mathbf{I}$ is postmultiplied by the matrix $\boldsymbol{\alpha}$, it is found that
$$\boldsymbol{\alpha}' = \boldsymbol{\alpha}, \tag{1.7.18}$$
so that $\boldsymbol{\alpha}$ is symmetric.

A non-singular square matrix may be raised to a negative power. Negative powers of $\mathbf{A}$ are defined by raising the inverse matrix $\mathbf{A}^{-1}$ to positive powers. That is to say
$$\mathbf{A}^{-r} = (\mathbf{A}^{-1})^{r}. \tag{1.7.19}$$

Thus, the matrix $\mathbf{B}$ of equations (1.7.13) gives
$$\mathbf{B}^{-1} = \tfrac{1}{11}\begin{bmatrix} 1 & -2 \\ 4 & 3 \end{bmatrix}; \qquad \mathbf{B}^{-2} = \tfrac{1}{121}\begin{bmatrix} -7 & -8 \\ 16 & 1 \end{bmatrix}. \tag{1.7.20}$$

EXAMPLES 1.7

1. Find three 3×3 matrices $\mathbf{A}$, $\mathbf{B}$ and $\mathbf{C}$ such that
$$\mathbf{AB} = \mathbf{AC} \quad \text{but} \quad \mathbf{B} \neq \mathbf{C}.$$

2. Show that if $\mathbf{AB} = \mathbf{BA}$ and $\mathbf{B}$ is non-singular then
$$\mathbf{AB}^{-1} = \mathbf{B}^{-1}\mathbf{A}.$$

3. Prove that
$$(\mathbf{A}')^{-1} = (\mathbf{A}^{-1})'$$
for any non-singular matrix $\mathbf{A}$.

1.8 Preliminary discussion of linear equations

In this section we shall be concerned with the solution of the matrix equation
$$\mathbf{Ax} = \mathbf{0}. \tag{1.8.1}$$

This is the matrix representation of equations (1.3.19) for the particular case in which the y's on the right are all zero. The matrix $\mathbf{A}$ is square, $\mathbf{x}$ is a column matrix and $\mathbf{0}$ represents a zero column matrix. This equation is of great importance in vibration theory; we shall now examine it in a preliminary way, reserving a more detailed discussion until later.

If $\mathbf{A}$ is non-singular, then $\mathbf{A}^{-1}$ exists and both sides of equation (1.8.1) may be multiplied by it. By this means it is found that

$$\mathbf{x} = \mathbf{0}. \tag{1.8.2}$$

In this case, then, $\mathbf{x}$ is a null matrix and the solution is often described by the unattractive adjective 'trivial'.

Of much more interest is the case where $\mathbf{A}$ is singular. The adjoint $\hat{\mathbf{A}}$ exists and the equation can be premultiplied by it. This leads to the result

$$|\mathbf{A}|\,\mathbf{x} = \mathbf{0} \tag{1.8.3}$$

in conformity with equation (1.6.28). But, since $|\mathbf{A}|$ is zero, $\mathbf{x}$ need not also be zero (i.e. composed of zero elements). Neither can it be taken arbitrarily because it may happen that if a definite value is assigned to one of its elements, the remainder are uniquely determined.

Consider, for example, the equations

$$\left.\begin{aligned} 3x_1+4x_2+x_3 &= 0,\\ x_1-2x_2+3x_3 &= 0,\\ x_1+3x_2-x_3 &= 0. \end{aligned}\right\} \tag{1.8.4}$$

Since the determinant of the coefficients is zero there must be a linear relation between its rows and therefore between the equations (§ 1.5, rule (k)). In fact

$$\textbf{equation } (1) = \textbf{equation } (2) + 2\times\textbf{equation } (3). \tag{1.8.5}$$

Suppose the value 1 is assigned to x_2. By choosing any two of the three equations for x_1 and x_3 we find that

$$x_1 = \frac{-7}{4}; \quad x_3 = \frac{5}{4}. \tag{1.8.6}$$

The first point to note about the solutions of (1.8.1) is that if $\mathbf{x}$ is a solution and θ is any constant then $\theta\mathbf{x}$ is also a solution. This means that if there is one solution $\mathbf{x}$ there is a whole infinity of solutions corresponding to all the possible values of θ. We shall, however, lump all these solutions together and speak of a 'solution' as a set of ratios between the elements of $\mathbf{x}$.

Since $|\mathbf{A}| = 0$, equation (1.6.26) gives

$$\mathbf{A}\hat{\mathbf{A}} = |\mathbf{A}|.\mathbf{I} = \mathbf{0}, \tag{1.8.7}$$

where $\mathbf{0}$ now represents a null square matrix. The columns of this null matrix are the products of $\mathbf{A}$ with the columns of $\hat{\mathbf{A}}$. Therefore each column of $\hat{\mathbf{A}}$ is a solution of equation (1.8.1). There are therefore n 'solutions' (in the sense defined above) given by

$$x_1 : x_2 : \ldots : x_n :: A_{j1} : A_{j2} : \ldots : A_{jn} \tag{1.8.8}$$

for $j = 1, 2, \ldots, n$. This solution can be described by saying that the x's are proportional to the cofactors of (the elements of) the jth row of $\mathbf{A}$. In fact, however, there is really one solution since the ratios between the cofactors are the same for each value of j. This is a general property of singular matrices which will be proved in Chapter 3 (Theorem 3.5.1).

Consider the homogeneous equation

$$\begin{bmatrix} 1 & 1 & 2 \\ -1 & 0 & 3 \\ 5 & 2 & -5 \end{bmatrix} \begin{bmatrix} x_1 \\ x_2 \\ x_3 \end{bmatrix} = \mathbf{0}. \tag{1.8.9}$$

The square matrix, here, is the singular matrix $\mathbf{A}$ of equation (1.6.20). By inspection of the adjoint matrix $\hat{\mathbf{A}}$ of equation (1.6.21), it is seen that the solutions corresponding to the cofactors of the first, second and third rows are, respectively

$$\mathbf{x} = \begin{bmatrix} x_1 \\ x_2 \\ x_3 \end{bmatrix} = \theta_1 \begin{bmatrix} -6 \\ 10 \\ -2 \end{bmatrix} \text{ or } \theta_2 \begin{bmatrix} 9 \\ -15 \\ 3 \end{bmatrix} \text{ or } \theta_3 \begin{bmatrix} 3 \\ -5 \\ 1 \end{bmatrix}, \tag{1.8.10}$$

where θ_1, θ_2 and θ_3 are multiplying factors of arbitrary value. It will be seen, moreover, that these are not three distinct solutions, but just one; this is

$$x_1 : x_2 : x_3 :: 3 : -5 : 1. \tag{1.8.11}$$

There are certain degenerate forms of this result and their existence depends upon the possible structures of the matrix A. First it may happen that for some, but not all, j, all the cofactors $A_{j1}, A_{j2}, \ldots, A_{jn}$ are zero. This, too, can be illustrated. The homogeneous equations

$$\begin{bmatrix} 1 & 1 & 2 \\ -1 & 0 & 3 \\ -2 & -2 & -4 \end{bmatrix} \begin{bmatrix} x_1 \\ x_2 \\ x_3 \end{bmatrix} = \mathbf{0} \tag{1.8.12}$$

embody a singular square matrix whose adjoint is

$$\begin{bmatrix} 6 & 0 & 3 \\ -10 & 0 & -5 \\ 2 & 0 & 1 \end{bmatrix}. \tag{1.8.13}$$

The rule given in equations (1.8.8) now yields the results

$$\mathbf{x} = \begin{bmatrix} x_1 \\ x_2 \\ x_3 \end{bmatrix} = \theta_1 \begin{bmatrix} 6 \\ -10 \\ 2 \end{bmatrix} \text{ or } \theta_2 \begin{bmatrix} 0 \\ 0 \\ 0 \end{bmatrix} \text{ or } \theta_3 \begin{bmatrix} 3 \\ -5 \\ 1 \end{bmatrix}, \tag{1.8.14}$$

so that only two of the three columns give

$$x_1 : x_2 : x_3 :: 3 : -5 : 1, \tag{1.8.15}$$

the third solution being the 'trivial' one $x_1 = 0 = x_2 = x_3$. It should be noted that there is still essentially only one solution of the equations.

Secondly, it may happen that *all* the cofactors of the determinant $|\mathbf{A}|$ are zero. In this event, the solution (1.8.8) is the 'trivial' one for all values of j. Since, however, there are non-trivial solutions another approach is indicated; this problem will be taken up later, in Chapter 3.

EXAMPLES 1.8

1. Solve the equations

$$\begin{aligned} 3s+4t+u+2v &= 0, \\ 9s-6t+2u-3v &= 0, \\ s+4t+u &= 0, \\ 4s-10t+u-7v &= 0. \end{aligned}$$

2. Show that if $\mathbf{X}^{(1)} = \{x_1^{(1)}, x_2^{(1)}, \ldots, x_n^{(1)}\}$ and $\mathbf{X}^{(2)} = \{x_1^{(2)}, x_2^{(2)}, \ldots, x_n^{(2)}\}$ are two solutions of equation (1.8.1), and if θ, ϕ are arbitrary multipliers, then $\theta\mathbf{X}^{(1)}+\phi\mathbf{X}^{(2)}$ is also a solution.

3. Show that the equations

$$\begin{aligned} 3s+4t+u+2v &= 0, \\ 9s-6t+2u-3v &= 0, \\ 12s-2t+3u-v &= 0, \\ 6s-10t+u-5v &= 0, \end{aligned}$$

have non-trivial solutions, but that they cannot be obtained by the method outlined above.

CHAPTER 2

THE VIBRATION OF CONSERVATIVE SYSTEMS HAVING A FINITE NUMBER OF DEGREES OF FREEDOM

If you ask the special function
  Of our never-ceasing motion,
We reply, without compunction,
  That we haven't any notion!   *Iolanthe*

In this chapter we shall commence our discussion of the *small* oscillations of dynamical systems having a *finite* number of degrees of freedom. It will be assumed that there are no forces proportional to velocity acting on the systems; this will exclude damping and gyroscopic effects. It will mean also that the systems are 'conservative', since forces depending on displacements may always be obtained from a potential function. Our purpose in this chapter is to cast the theory of these systems into matrix form.

The reader who is already familiar with the basic theory,† and who wishes immediately to pass on to the application of matrices in it should turn directly to § 2.1. For the benefit of readers who require either an introduction to the theory or a brief review of it, however, an introductory '§ 2.0' is given which has no direct connection with matrix theory.

2.0 Introduction

The analytical determination of the vibration behaviour of a physical system may be divided into a number of stages:

(1) The choice of a mathematical model which is to be used to represent the physical system. If the system were a wing of an aircraft, for instance, it would have to be decided whether to treat it as a continuous beam in flexure and/or torsion, as a plate or a box-like structure, as some lumped-mass system, or as some other type of system.

(2) The derivation of the equations of motion of the mathematical model, and if necessary their transformation into a form suitable for calculation.

(3) The determination of the physical constants occurring in the mathematical equations. These constants—lengths, masses, stiffnesses, etc.—will have to be determined from given data, which itself will often have been determined experimentally.

(4) The solution of the mathematical equations.

† For example, see Bishop and Johnson, *The Mechanics of Vibration*, ch. 3.

This book is concerned almost wholly with the last aspect of the problem. The first two stages are discussed briefly in § 2.1 and again in Chapter 6, but the third is not discussed at all. For this important technical matter the reader is referred elsewhere.

Vibrating systems may conveniently be divided into two types, discrete and continuous. Fundamentally the distinction is mathematical, and not physical. A discrete system is one whose exact equations of motion may be expressed as a set of ordinary differential equations in a finite number of unknowns $q_1, q_2, \ldots, q_n$, functions of time t alone. For the small oscillations with which we are concerned in this book the equations will be linear and have constant coefficients. Discrete systems will then be characterised by having a finite number of natural frequencies and corresponding modes of free vibration. Although discrete systems never occur exactly in practice, many physical systems are discrete for most practical purposes. An example of such a system is a set of massive flywheels on a light shaft undergoing torsional vibration. It is possible to write down a set of linear, ordinary, differential, equations governing the angular deflections $\theta_1, \theta_2, \ldots, \theta_n$ of the flywheels and obtain from them results which conform with experiment over a very large frequency range. In other words, the system may be regarded as discrete within this range.

In general a continuous system is one whose exact equations of motion may be expressed as one or more partial differential equations (and boundary conditions) governing certain displacements which are functions of the space variables x, y, z and the time t. For small oscillations, the equations will be linear, but the coefficients will not necessarily be constant. The equations governing a continuous system may usually be expressed in an alternative form involving integral equations. Continuous systems are characterised by having an infinity of natural frequencies and corresponding modes of free vibration. A massive shaft undergoing torsional oscillation is an example of a continuous system.

It is clear that although it is possible to make an absolute distinction between discrete and continuous systems in mathematics, no such clear-cut distinction is possible in the real world. In practice the decision whether to treat a system as discrete or continuous depends on arguments involving convenience and accuracy. The system of massive flywheels on a shaft is not really discrete; for one thing, to obtain a set of equations involving $\theta_1, \theta_2, \ldots, \theta_n$, it has to be assumed that the flywheels vibrate as rigid bodies, and this is strictly untrue. An exact treatment of the system would include allowances for the deformation of the flywheels, and would involve partial differential equations. Although some such allowances might be needed for the discussion of the vibration at very high frequencies, the increased complication of the equations would not be justified by a comparable increase in accuracy for the ordinary working range of frequencies. This is the justification for considering the system to be discrete.

Sometimes it is easier to think of a system as being continuous rather than discrete. This is the case when it is possible to write down the equations of motion as one or more partial differential equations which can be solved exactly in terms of known functions. The class of systems for which this is possible is comparatively small, but includes important systems such as uniform strings, uniform shafts in torsion, uniform beams in longitudinal and transverse vibration, uniform membranes and plates with rectangular or circular boundaries. When it is impossible, or difficult, to obtain an exact solution of the partial differential equations governing a continuous system, the system is almost always reduced to discrete form. This may be done in many ways. The system may be approximated by a discrete 'lumped-mass' system (see § 6.4). Alternatively, it may be assumed that the system may vibrate only in combinations of a certain set of assumed modes; this is the Rayleigh–Ritz method which is discussed briefly in § 6.5. Yet another way, which will not be discussed at all in this book, is to obtain an approximate solution of the partial differential equations, or preferably the integral equations, using finite-difference methods. In this chapter we are concerned with the properties of the equations governing a vibrating system which has been reduced to discrete form.

Broadly speaking, there are two ways of reducing a physical system to discrete form, by using 'lumped-masses' or 'assumed modes'. In the former method the distributed mass and stiffnesses of the structure are concentrated into point masses or 'dumb-bells' (to give moments of inertia) and connecting 'springs' (elements which have elasticity but no mass). The equations of motion of a lumped-mass system may be derived from Newton's laws of motion, which state that the position of a particle $P(x, y, z)$ of mass m is related to the forces F_x, F_y, F_z acting on it by the equations

$$m\ddot{x} = F_x, \quad m\ddot{y} = F_y, \quad m\ddot{z} = F_z, \tag{2.0.1}$$

where dots denote differentiation with respect to time. The equations of motion for the system may be obtained by writing down the appropriate equations for each particle of the system.

Often it is convenient to have a formulation of the equations of motion which makes use of quantities relating to the system as a whole, not to the elements from which it is made up. For a lumped-mass system such a formulation is provided by Lagrange's equations

$$\frac{d}{dt}\left(\frac{\partial T}{\partial \dot{q}_r}\right) - \frac{\partial T}{\partial q_r} + \frac{\partial V}{\partial q_r} = Q_r \quad (r = 1, 2, \ldots, n). \tag{2.0.2}$$

Here, T and V are the kinetic and potential energies of the system and the quantities $q_1, q_2, \ldots, q_n$ are a set of generalised coordinates—a set of quantities, that is, in terms of which all the linear and angular displacements involved in the problem may be expressed. The derivation of Lagrange's equations for a lumped mass system is discussed in text-books on dynamics.†

A convenient starting-point for a discussion of 'assumed modes' methods is the formulation of the equations of motion of a system in terms of the Principle of Virtual Work.‡ For an elastic body *in equilibrium* this Principle runs as follows: 'If a body is in equilibrium under the action of prescribed external forces the virtual

† E. T. Whittaker, *Analytical Dynamics* (Cambridge University Press, 4th ed., 1937), ch. 2. A. S. Ramsey, *Dynamics*, Part II (Cambridge University Press, 2nd ed., 1944), ch. 9.

‡ For example, see I. S. Sokolnikoff, *Mathematical Theory of Elasticity* (McGraw-Hill, 2nd ed., 1956), ch. 7.

work done by these forces in any virtual displacement is equal to the change in the elastic potential energy.'† The terms 'virtual work' and 'virtual displacement' have the following meanings. The virtual work done by the forces is the work done assuming that they retain their equilibrium values unchanged throughout the displacement. A virtual displacement is any small displacement compatible with the geometric constraints on the body. For example, for a cantilever beam in flexure a virtual displacement would be one satisfying the conditions

$$v = 0 = \frac{dv}{dx}. \tag{2.0.3}$$

In the language of the Calculus of Variations, the geometric constraints are the *essential* boundary conditions for the Principle.

The Principle of Virtual Work may be applied to vibrating systems provided that it is used in conjunction with D'Alembert's Principle. This states that a vibrating system may be treated as if it were in equilibrium under the action of the external forces and the reversed mass-accelerations (the inertia forces). The Principle now becomes

$$\delta W_e + \delta W_{in} = \delta V. \tag{2.0.4}$$

Here δW_e and δW_{in} denote respectively the work done by the external forces and the inertia forces in an arbitrary virtual displacement, and δV is the change in potential energy in such a displacement. For a vibrating system the essential boundary conditions for the virtual displacements are the kinematic conditions, as opposed to the dynamic ones. For a cantilever beam in flexure the kinematic conditions are

$$v(o,t) = 0 = \frac{\partial v}{\partial x}(o,t). \tag{2.0.5}$$

Equation (2.0.4) may be used to derive the exact differential or integral equations governing the motion of a continuous system. An example of this procedure is given in § 6.5.

We shall now show how the Principle may be used in conjunction with an 'assumed mode' method to obtain an approximate discrete set of equations for the system. The equations will be found to have exactly the same form as the Lagrangian equations (2.0.2), which will thus be seen to have a wider significance than is often accorded to them.

We start by assuming that all the displacements u, v, w involved in the problem may be expressed in terms of a finite set of generalised coordinates $q_1, q_2, \ldots, q_n$ through a set of equations whose most general form will be

$$\left.\begin{aligned} u(x,y,z,t) &= u(q_1, q_2, \ldots, q_n, x, y, z), \\ v(x,y,z,t) &= v(q_1, q_2, \ldots, q_n, x, y, z), \\ w(x,y,z,t) &= w(q_1, q_2, \ldots, q_n, x, y, z), \end{aligned}\right\} \tag{2.0.6}$$

† Throughout this book the term 'potential energy' is used to signify the elastic strain energy plus the work done by external forces possessing potential. This usage conforms with that of Rayleigh and the other early investigators, but differs from that of Sokolnikoff. He uses the term 'strain energy' instead, reserving 'potential energy' for the strain energy *less* the work done by the external forces.

where $q_1, q_2, \ldots, q_n$ are functions of t alone. We shall now assume that the system is *holonomic*, that is, that the kinematic constraints on the body are such that it is possible for each coordinate q_i to be varied independently without violating the constraints: almost all systems of interest in vibration are of this type. For such systems any displacement obtained by changing one or more of the generalised coordinates from q_i to $q_i + \delta q_i$ is a virtual displacement. We shall now consider the form which the equation (2.0.4) takes when the virtual displacements are of this type.

The virtual work done by the inertia forces in an arbitrary virtual displacement $\delta u, \delta v, \delta w$ of each particle of the body will be given by an integral throughout the volume of the body. In general this will have the form

$$\delta W_{in} = -\iiint \rho(\ddot{u}\,\delta u + \ddot{v}\,\delta v + \ddot{w}\,\delta w)\,dx\,dy\,dz, \tag{2.0.7}$$

where $\rho(x, y, z)$ is the volume density of the material. For many systems, however, the mass of the body may be lumped in such a way that the triple integral may be reduced to a double or single integral. Thus for a slender beam performing transverse vibration the integral is usually approximated by

$$\delta W_{in} = -\int_0^l A\rho\ddot{v}\,dx, \tag{2.0.8}$$

where A is the cross-sectional area.

Substituting the expressions

$$\delta u = \sum_{i=1}^{n} \frac{\partial u}{\partial q_i}\delta q_i, \quad \delta v = \sum_{i=1}^{n} \frac{\partial v}{\partial q_i}\delta q_i, \quad \delta w = \sum_{i=1}^{n} \frac{\partial w}{\partial q_i}\delta q_i \tag{2.0.9}$$

in equation (2.0.7) we obtain

$$\delta W_{in} = \sum_{i=1}^{n}\left\{-\iiint \rho\left(\ddot{u}\frac{\partial u}{\partial q_i} + \ddot{v}\frac{\partial v}{\partial q_i} + \ddot{w}\frac{\partial w}{\partial q_i}\right)dx\,dy\,dz\right\}\delta q_i. \tag{2.0.10}$$

The expressions in the curly brackets may be identified as

$$-\frac{d}{dt}\left(\frac{\partial T}{\partial \dot{q}_i}\right) + \frac{\partial T}{\partial q_i},$$

where T is the kinetic energy of the body, namely

$$T = \frac{1}{2}\iiint \rho(\dot{u}^2 + \dot{v}^2 + \dot{w}^2)\,dx\,dy\,dz. \tag{2.0.11}$$

The identification follows, just as in the usual derivation of Lagrange's equations,† from the equations

$$\frac{\partial u}{\partial q_i} = \frac{\partial \dot{u}}{\partial \dot{q}_i} \tag{2.0.12}$$

and

$$\ddot{u}\frac{\partial u}{\partial q_i} = \ddot{u}\frac{\partial u}{\partial q_i} + \dot{u}\frac{\partial \dot{u}}{\partial q_i} - \dot{u}\frac{\partial \dot{u}}{\partial q_i} = \frac{d}{dt}\left(\dot{u}\frac{\partial u}{\partial q_i}\right) - \dot{u}\frac{\partial \dot{u}}{\partial q_i}$$

$$= \frac{d}{dt}\left(\dot{u}\frac{\partial \dot{u}}{\partial \dot{q}_i}\right) - \dot{u}\frac{\partial \dot{u}}{\partial q_i} = \frac{d}{dt}\left\{\frac{\partial}{\partial \dot{q}_i}(\tfrac{1}{2}\dot{u}^2)\right\} - \frac{\partial}{\partial q_i}(\tfrac{1}{2}\dot{u}^2). \tag{2.0.13}$$

† For example, see Bishop and Johnson, *The Mechanics of Vibration*, ch. 2.

Thus
$$\delta W_{in} = \sum_{i=1}^{n} \left\{ -\frac{d}{dt}\left(\frac{\partial T}{\partial \dot{q}_i}\right) + \frac{\partial T}{\partial q_i} \right\} \delta q_i. \tag{2.0.14}$$

The virtual work done by the external forces will be given by an integral over the surface of the body. In general this will have the form
$$\delta W_e = \iint (F_x \delta u + F_y \delta v + F_z \delta w)\, dS, \tag{2.0.15}$$
where dS denotes an element of surface area. Again this may be approximated by a single integral or a simple sum in some cases, just as for δW_{in} (see equation (2.0.8)). Substituting for u, v, w we obtain
$$\delta W_e = \sum_{i=1}^{n} \left\{ \iint \left(F_x \frac{\partial u}{\partial q_i} + F_y \frac{\partial v}{\partial q_i} + F_z \frac{\partial w}{\partial q_i} \right) dS \right\} \delta q_i, \tag{2.0.16}$$
which we write
$$\delta W_e = \sum_{i=1}^{n} Q_i \delta q_i. \tag{2.0.17}$$
The quantity
$$Q_i = \iint \left(F_x \frac{\partial u}{\partial q_i} + F_y \frac{\partial v}{\partial q_i} + F_z \frac{\partial w}{\partial q_i} \right) dS \tag{2.0.18}$$
is defined to be the 'generalised force' corresponding to the generalised coordinate q_i.

Substituting the expressions given by equations (2.0.14) and (2.0.17) for the quantities δW_{in} and δW_e, and writing
$$\delta V = \sum_{i=1}^{n} \frac{\partial V}{\partial q_i} \delta q_i, \tag{2.0.19}$$
we obtain
$$\delta W_e + \delta W_{in} - \delta V = 0 = \sum_{i=1}^{n} \left\{ Q_i - \frac{d}{dt}\left(\frac{\partial T}{\partial \dot{q}_i}\right) + \frac{\partial T}{\partial q_i} - \frac{\partial V}{\partial q_i} \right\} \delta q_i, \tag{2.0.20}$$
and since the quantities δq_i can, by hypothesis, be varied independently, we must have
$$\frac{d}{dt}\left(\frac{\partial T}{\partial \dot{q}_i}\right) - \frac{\partial T}{\partial q_i} + \frac{\partial V}{\partial q_i} = Q_i \quad (i = 1, 2, \ldots, n). \tag{2.0.21}$$

We now investigate the form of the energies T and V as functions of the generalised coordinates. To obtain the form which T takes we substitute the expressions
$$\dot{u} = \sum_{i=1}^{n} \frac{\partial u}{\partial q_i} \dot{q}_i, \quad \dot{v} = \sum_{i=1}^{n} \frac{\partial v}{\partial q_i} \dot{q}_i, \quad \dot{w} = \sum_{i=1}^{n} \frac{\partial w}{\partial q_i} \dot{q}_i \tag{2.0.22}$$
into the integral in equation (2.0.11). This yields
$$T = \frac{1}{2} \iiint \rho \left\{ \left(\sum_{i=1}^{n} \frac{\partial u}{\partial q_i} \dot{q}_i \right)^2 + \left(\sum_{i=1}^{n} \frac{\partial v}{\partial q_i} \dot{q}_i \right)^2 + \left(\sum_{i=1}^{n} \frac{\partial w}{\partial q_i} \dot{q}_i \right)^2 \right\} dx\, dy\, dz, \tag{2.0.23}$$
which may be written
$$T = \tfrac{1}{2}[a_{11}\dot{q}_1^2 + a_{22}\dot{q}_2^2 + \ldots + a_{nn}\dot{q}_n^2 + 2a_{12}\dot{q}_1\dot{q}_2 + \ldots + 2a_{n-1,n}\dot{q}_{n-1}\dot{q}_n], \tag{2.0.24}$$
where the quantities $a_{11}, a_{22}, a_{12}, \ldots, a_{nn}$ are given by the general relation
$$a_{ij} = \iiint \rho \left\{ \frac{\partial u}{\partial q_i}\frac{\partial u}{\partial q_j} + \frac{\partial v}{\partial q_i}\frac{\partial v}{\partial q_j} + \frac{\partial w}{\partial q_i}\frac{\partial w}{\partial q_j} \right\} dx\, dy\, dz. \tag{2.0.25}$$

The quantities a_{ij} are thus functions of the coordinates $q_1, q_2, \ldots, q_n$ but do not involve the time derivatives $\dot{q}_1, \dot{q}_2, \ldots, \dot{q}_n$. Considered as a function of the generalised coordinates and their time derivatives, it is found, in fact, that T has exactly the same form whether lumped mass or assumed modes are used.

In this book we are concerned only with *small* vibrations about some datum configuration, which will usually be one of stable or neutral equilibrium. For such vibration it is convenient to choose a set of generalised coordinates for which

$$q_1 = 0 = q_2 = \ldots = q_n \tag{2.0.26}$$

in the datum configuration. Small vibration will then correspond to vibration in which all the coordinates q are small. Equations (2.0.6) then become

$$\left.\begin{aligned} u &= \phi_0(x, y, z) + \sum_{i=1}^{n} \phi_i(x, y, z)\, q_i, \\ v &= \psi_0(x, y, z) + \sum_{i=1}^{n} \psi_i(x, y, z)\, q_i, \\ w &= \chi_0(x, y, z) + \sum_{i=1}^{n} \chi_i(x, y, z)\, q_i, \end{aligned}\right\} \tag{2.0.27}$$

where
$$\phi_i(x, y, z) = \left.\frac{\partial u}{\partial q_i}\right|_{q_1=0=q_2=\ldots=q_n}, \quad \text{etc.}, \tag{2.0.28}$$

and higher powers of the q's are neglected. The quantities a_{ij} given in equation (2.0.25) now become

$$a_{ij} = \iiint \rho(\phi_i\phi_j + \psi_i\psi_j + \chi_i\chi_j)\,dx\,dy\,dz \tag{2.0.29}$$

and are thus constants; they are called 'coefficients of inertia'.

The potential energy of the system will similarly be expressible as some integral involving only the displacements u, v, w and their space derivatives throughout the volume of the body. The integral may take various forms, for a slender beam with second moment of area I, for example, it may be expressed as

$$V = \frac{1}{2}\int_0^l EI\left(\frac{\partial^2 v}{\partial x^2}\right)^2 dx, \tag{2.0.30}$$

or
$$V = \frac{1}{2}\int_0^l v(x)\int_0^l k(x, \xi)\, v(\xi)\, dx\, d\xi, \tag{2.0.31}$$

where $k(x, \xi)$ is the so-called 'stiffness influence function' of the beam. There are similar expressions for the potential energy of more general structures.† Whatever the form of the potential energy, the substitution of the expressions (2.0.27) for the displacements will lead to the equation

$$\begin{aligned} V = c_0 + c_1 q_1 + c_2 q_2 + \ldots + c_n q_n + \tfrac{1}{2}[c_{11} q_1^2 + c_{22} q_2^2 + \ldots + c_{nn} q_n^2 \\ + 2c_{12} q_1 q_2 + \ldots + 2c_{n-1,n}\, q_{n-1} q_n], \end{aligned} \tag{2.0.32}$$

† R. L. Bisplinghoff, H. Ashley and R. L. Halfman, *Aeroelasticity* (Addison-Wesley, 1955), ch. 2.

where the coefficients c are constants. Thus if the expressions (2.0.30) and (2.0.31) are used the quantities c_{ij} are given by

$$c_{ij} = \int_0^l EI\chi_i''(x)\,\chi_j''(x)\,dx, \tag{2.0.33}$$

or

$$c_{ij} = \int_0^l \chi_i(x) \int_0^l k(x,\xi)\,\chi_j(\xi)\,dx\,d\xi. \tag{2.0.34}$$

If the datum configuration is one of stable or neutral equilibrium then

$$c_1 = 0 = c_2 = \ldots = c_n, \tag{2.0.35}$$

and if the datum level of the potential energy is taken to be zero, so that $c_0 = 0$, then

$$V = \tfrac{1}{2}[c_{11}q_1^2 + c_{22}q_2^2 + \ldots + c_{nn}q_n^2 + 2c_{12}q_1q_2 + \ldots + 2c_{n-1,n}q_{n-1}q_n]. \tag{2.0.36}$$

The constants $c_{11}, c_{22}, c_{12}, \ldots, c_{nn}$ are called the 'stability coefficients' of the system.

In the foregoing discussion it has been assumed that the system has been reduced to discrete form by using some 'lumped-mass' or 'assumed-modes' method, but nothing has been said about how the masses and springs, on the one hand, or the modes, on the other, should be chosen. A brief discussion of some aspects of these matters is given in Chapter 6. In the preceding chapters it will be assumed in general that the reduction to discrete form has been carried out *in some way or other*. The examples that are given in these chapters will almost all refer to lumped-mass systems, but the methods may be applied just as easily to discrete systems obtained by the assumed-modes approach.

2.1 The equations of motion

Consider a system having n degrees of freedom, a convenient set of generalised coordinates being the quantities $q_1, q_2, \ldots, q_n$. The datum configuration, in which all these coordinates vanish, will be assumed to be one of stable or neutral equilibrium.

It was shown in the last section that if the system executes a small motion about its datum configuration, its kinetic energy T will be a 'quadratic form'—that is, an expression of the type

$$T = \tfrac{1}{2}[a_{11}\dot{q}_1^2 + a_{22}\dot{q}_2^2 + \ldots + a_{nn}\dot{q}_n^2 + 2a_{12}\dot{q}_1\dot{q}_2 + \ldots + 2a_{n-1,n}\dot{q}_{n-1}\dot{q}_n], \tag{2.1.1}$$

in the quantities $\dot{q}_1, \dot{q}_2, \ldots, \dot{q}_n$. The quantities $a_{11}, a_{22}, a_{12}, \ldots, a_{nn}$ are constants; they are known as the 'coefficients of inertia' of the system and are such that T is positive or zero for all values of $\dot{q}_1, \dot{q}_2, \ldots, \dot{q}_n$.

The kinetic energy T can be written in the alternative form

$$T = \frac{1}{2}\sum_{r=1}^{n} \dot{q}_r(a_{r1}\dot{q}_1 + a_{r2}\dot{q}_2 + \ldots + a_{rn}\dot{q}_n), \tag{2.1.2}$$

and since

$$a_{rs} \equiv a_{sr}, \tag{2.1.3}$$

this last expression is the same as

$$T = \frac{1}{2}\sum_{r=1}^{n}\sum_{s=1}^{n} a_{rs}\dot{q}_r\dot{q}_s. \tag{2.1.4}$$

It follows from equation (2.1.2) that T may be expressed as

$$T = \tfrac{1}{2}[\dot{q}_1\ \dot{q}_2\ \cdots\ \dot{q}_n]\begin{bmatrix} a_{11} & a_{12} & \cdots & a_{1n} \\ a_{21} & a_{22} & \cdots & a_{2n} \\ \cdots & \cdots & \cdots & \cdots \\ a_{n1} & a_{n2} & \cdots & a_{nn} \end{bmatrix}\begin{bmatrix} \dot{q}_1 \\ \dot{q}_2 \\ \cdot \\ \dot{q}_n \end{bmatrix}, \qquad (2.1.5)$$

or

$$T = \tfrac{1}{2}\dot{\mathbf{q}}'\mathbf{A}\dot{\mathbf{q}}. \qquad (2.1.6)$$

In this equation $\dot{\mathbf{q}}$ represents the column matrix (or 'vector') of equation (2.1.5), $\dot{\mathbf{q}}'$ is its transpose, and $\mathbf{A}$ (which is symmetric) is the 'inertia matrix'.

The matrix $\mathbf{A}$ is said to be 'positive definite' if $\dot{\mathbf{q}}'\mathbf{A}\dot{\mathbf{q}} > 0$ for all possible non-null column vectors $\dot{\mathbf{q}}$. The equality $\dot{\mathbf{q}}'\mathbf{A}\dot{\mathbf{q}} = 0$ will then hold only if $\dot{\mathbf{q}}$ is a null vector. A necessary and sufficient condition for the symmetric matrix $\mathbf{A}$ to be positive-definite is the following:† 'Every determinant obtained from $|\mathbf{A}|$ by considering only elements in the first r rows $(r = 1, 2, \ldots, n)$ and the first r columns shall be positive.' By giving r the values $1, 2, \ldots, n$ successively we obtain:

$$\left.\begin{aligned} a_{11} > 0;\quad \begin{vmatrix} a_{11} & a_{12} \\ a_{21} & a_{22} \end{vmatrix} > 0;\quad \begin{vmatrix} a_{11} & a_{12} & a_{13} \\ a_{21} & a_{22} & a_{23} \\ a_{31} & a_{32} & a_{33} \end{vmatrix} > 0; \\ \cdots\cdots\cdots\cdots\cdots\cdots\cdots\cdots\cdots\cdots \\ \begin{vmatrix} a_{11} & a_{12} & \cdots & a_{1n} \\ a_{21} & a_{22} & \cdots & a_{2n} \\ \cdots & \cdots & \cdots & \cdots \\ a_{n1} & a_{n2} & \cdots & a_{nn} \end{vmatrix} > 0. \end{aligned}\right\} \qquad (2.1.7)‡$$

The last of these inequalities implies that a positive definite matrix is non-singular.

It may be that one coordinate, q_i say, has been so chosen that motion at q_i does not contribute to T. In this case a_{ii} will be zero and T will be said to be 'positive semi-definite'. The general definition of such a quadratic form is as follows: 'If the quadratic form for T is always positive, except for certain sets of values of the $\dot{q}_1, \dot{q}_2, \ldots, \dot{q}_n$ for which it is zero, it is said to be "positive semi-definite" or "non-negative definite".' The same term is used to describe the associated matrix $\mathbf{A}$. For such a matrix, at least one of the inequalities (2.1.7) becomes an equality. It may be shown also that when $\mathbf{A}$ is positive semi-definite it is necessarily singular, so that the last determinant in (2.1.7) and perhaps others, are zero.§

For a given small distortion of the system from its equilibrium configuration, the potential energy V will be a quadratic form

$$V = \tfrac{1}{2}[c_{11}q_1^2 + c_{22}q_2^2 + \ldots + c_{nn}q_n^2 + 2c_{12}q_1q_2 + \ldots + 2c_{n-1,n}q_{n-1}q_n] \qquad (2.1.8)$$

in the quantities $q_1, q_2, \ldots, q_n$. The quantities $c_{11}, c_{22}, c_{12}, \ldots, c_{nn}$ are called the 'coefficients of stability' or 'stiffness coefficients' of the system and satisfy

$$c_{rs} \equiv c_{sr}. \qquad (2.1.9)$$

† Ferrar, *Algebra*, p. 138. ‡ See also Ex. 2.1.3.

§ See the note below on the positive semi-definiteness of $\mathbf{C}$.

As may be seen by comparing equations (2.1.1) and (2.1.8), the energy V may be expressed in the matrix form

$$V = \tfrac{1}{2}[q_1\ q_2\ \dots\ q_n]\begin{bmatrix} c_{11} & c_{12} & \dots & c_{1n} \\ c_{21} & c_{22} & \dots & c_{2n} \\ \dots & \dots & \dots & \dots \\ c_{n1} & c_{n2} & \dots & c_{nn} \end{bmatrix}\begin{bmatrix} q_1 \\ q_2 \\ \cdot \\ q_n \end{bmatrix}, \tag{2.1.10}$$

or

$$V = \tfrac{1}{2}\mathbf{q}'\mathbf{C}\mathbf{q}. \tag{2.1.11}$$

In this equation $\mathbf{q}$ is the displacement vector, $\mathbf{q}'$ is its transpose, and the symmetric matrix $\mathbf{C}$ is the 'stiffness matrix' or 'stability matrix'.

The equilibrium of the datum configuration may be stable, neutral or unstable—and we shall not consider the last possibility. If the equilibrium is stable then $\mathbf{C}$ will be positive definite; any non-zero matrix $\mathbf{q}$ will produce a positive value of V. If the equilibrium is neutral then a zero value of V may be produced by certain non-zero matrices $\mathbf{q}$. This means that $\mathbf{C}$ is positive semi-definite. It will be proved in § 4.4 that if $\mathbf{C}$ is positive semi-definite it is necessarily singular; the proof depends on the fact that $\mathbf{C}$ is symmetric.

It was shown in the last section that the equations of motion of any system reduced to discrete form may be expressed by means of Lagrange's equations

$$\frac{d}{dt}\left(\frac{\partial T}{\partial \dot{q}_r}\right) - \frac{\partial T}{\partial q_r} + \frac{\partial V}{\partial q_r} = Q_r \quad (r = 1, 2, \dots, n). \tag{2.1.12}$$

The quantities $Q_1, Q_2, \dots, Q_n$ are the generalised forces corresponding to the selected generalised coordinates $q_1, q_2, \dots, q_n$ respectively. If the expressions (2.1.1) and (2.1.8) are substituted for T and V the equations (2.1.12) become

$$\left.\begin{aligned} a_{11}\ddot{q}_1 + a_{12}\ddot{q}_2 + \dots + a_{1n}\ddot{q}_n + c_{11}q_1 + c_{12}q_2 + \dots + c_{1n}q_n &= Q_1, \\ a_{21}\ddot{q}_1 + a_{22}\ddot{q}_2 + \dots + a_{2n}\ddot{q}_n + c_{21}q_1 + c_{22}q_2 + \dots + c_{2n}q_n &= Q_2, \\ \dots\dots\dots\dots\dots\dots\dots\dots\dots\dots\dots\dots\dots\dots \\ a_{n1}\ddot{q}_1 + a_{n2}\ddot{q}_2 + \dots + a_{nn}\ddot{q}_n + c_{n1}q_1 + c_{n2}q_2 + \dots + c_{nn}q_n &= Q_n. \end{aligned}\right\} \tag{2.1.13}$$

These equations can be written in the matrix form

$$\begin{bmatrix} a_{11} & a_{12} & \dots & a_{1n} \\ a_{21} & a_{22} & \dots & a_{2n} \\ \dots & \dots & \dots & \dots \\ a_{n1} & a_{n2} & \dots & a_{nn} \end{bmatrix}\begin{bmatrix} \ddot{q}_1 \\ \ddot{q}_2 \\ \cdot \\ \ddot{q}_n \end{bmatrix} + \begin{bmatrix} c_{11} & c_{12} & \dots & c_{1n} \\ c_{21} & c_{22} & \dots & c_{2n} \\ \dots & \dots & \dots & \dots \\ c_{n1} & c_{n2} & \dots & c_{nn} \end{bmatrix}\begin{bmatrix} q_1 \\ q_2 \\ \cdot \\ q_n \end{bmatrix} = \begin{bmatrix} Q_1 \\ Q_2 \\ \cdot \\ Q_n \end{bmatrix}, \tag{2.1.14}$$

which may be abbreviated to

$$\mathbf{A}\ddot{\mathbf{q}} + \mathbf{C}\mathbf{q} = \mathbf{Q}, \tag{2.1.15}$$

where $\ddot{\mathbf{q}}$, $\mathbf{q}$ and $\mathbf{Q}$ represent the column matrices. As we have already mentioned, $\mathbf{A}$ and $\mathbf{C}$ are both symmetric, so that

$$\mathbf{A} = \mathbf{A}'; \quad \mathbf{C} = \mathbf{C}'. \tag{2.1.16}$$

We shall conclude this section by giving some examples of systems for which **A** or **C** or both are semi-definite. We start from the positive definite system of fig. 2.1.1. This simple torsional system has three degrees of freedom since the discs are rigid and the connecting shafts are light. Its kinetic and potential energies are given by

$$\left.\begin{aligned} 2T &= I_1\dot{q}_1^2 + I_2\dot{q}_2^2 + I_3\dot{q}_3^2, \\ 2V &= kq_1^2 + k(q_2-q_1)^2 + k(q_3-q_2)^2, \end{aligned}\right\} \tag{2.1.17}$$

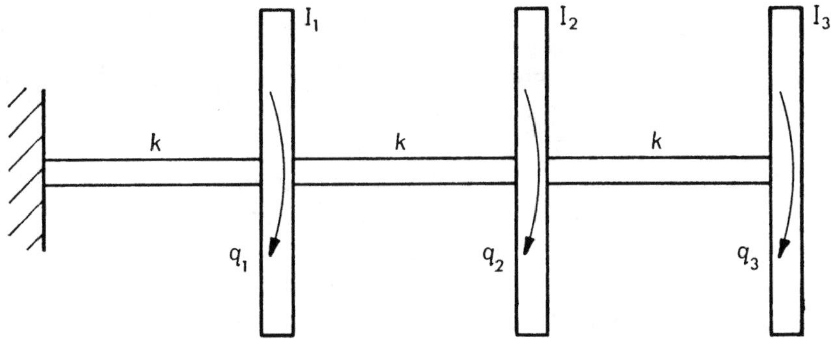

Fig. 2.1.1

so that

$$\mathbf{A} = \begin{bmatrix} I_1 & 0 & 0 \\ 0 & I_2 & 0 \\ 0 & 0 & I_3 \end{bmatrix} = \mathbf{A}'; \quad \mathbf{C} = \begin{bmatrix} 2k & -k & 0 \\ -k & 2k & -k \\ 0 & -k & k \end{bmatrix} = \mathbf{C}'. \tag{2.1.18}$$

Both **A** and **C** are non-singular and positive definite. The positive definiteness may be proved either on physical grounds—thus $V = 0$ only if $q_1 = 0$, $q_2 = q_1$ and $q_3 = q_2$—or by applying the inequalities (2.1.7).

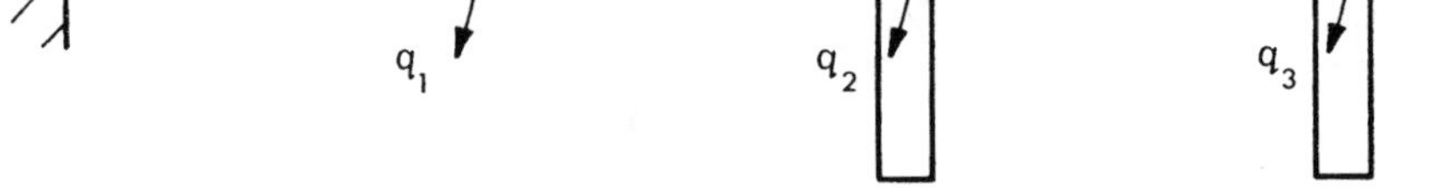

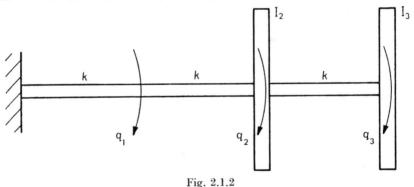

Fig. 2.1.2

If I_1 is made zero in the system of fig. 2.1.1 the system becomes that of fig. 2.1.2. This system has the same matrix **C**, but its matrix **A** is

$$\mathbf{A} = \begin{bmatrix} 0 & 0 & 0 \\ 0 & I_2 & 0 \\ 0 & 0 & I_3 \end{bmatrix}, \tag{2.1.19}$$

which is singular. The equations (2.1.13) for this system are

$$\left.\begin{aligned} 2kq_1-kq_2 &= Q_1,\\ I_2\ddot{q}_2-kq_1+2kq_2-kq_3 &= Q_2,\\ I_3\ddot{q}_3-kq_2+kq_3 &= Q_3.\end{aligned}\right\} \tag{2.1.20}$$

It is a general feature of such semi-definite systems—where **A** is singular—that one or more of the Lagrangian equations is an algebraic (rather than a differential) equation, or more generally that one can find a linear combination of the Lagrangian equations which is an algebraic equation. This matter is discussed further in § 2.11.

The torsional system of fig. 2.1.3 is not anchored, so that it can rotate as if it were a rigid body. This system has the same matrix **A** as that of fig. 2.1.1, but its potential energy V is given by

$$2V = k(q_2-q_1)^2+k(q_3-q_2)^2, \tag{2.1.21}$$

so that

$$\mathbf{C} = \begin{bmatrix} k & -k & 0 \\ -k & 2k & -k \\ 0 & -k & k \end{bmatrix}. \tag{2.1.22}$$

Clearly $V = 0$ when $q_1 = q_2 = q_3$ and **C** is semi-definite. In fact its determinant vanishes, so that the inequalities (2.1.7) are not completely satisfied. For this system the stiffness matrix is singular.

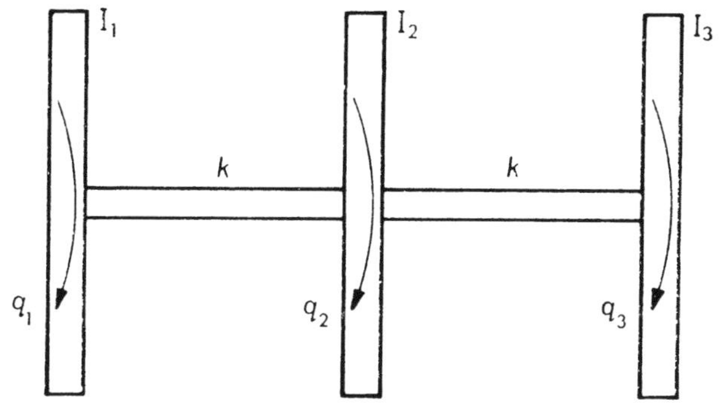

Fig. 2.1.3

If the system of fig. 2.1.3 is at rest, so that the matrix $\ddot{\mathbf{q}}$ is null, and if there is no applied torque, so that **Q** is also null, then equations (2.1.15) become simply

$$\mathbf{Cq} = \mathbf{0}. \tag{2.1.23}$$

It was shown in § 1.8 that if **C** is singular then this equation has a solution in which not all the elements of **q** are zero. The ratios between the q's are determined by the cofactors of $|\mathbf{C}|$ and, if we take those of the first row, we have

$$q_1:q_2:q_3 = k^2:k^2:k^2. \tag{2.1.24}$$

This verifies that the permissible displacements are of the 'rigid body' variety. This argument may be applied to any system to show that the matrix **C** is singular if and only if the system is unanchored. For the statement that the system is unanchored means precisely that there is a non-null solution of equation (2.1.23); and it will be shown in Chapter 3 (Theorem 3.2.1 (*b*)) that the necessary and sufficient condition for this is that **C** should be singular.

If, finally, I_1 is made zero in the system of fig. 2.1.3 so that the system is that of fig. 2.1.4 then **A** is the matrix given in equation (2.1.19) and **C** is that given in

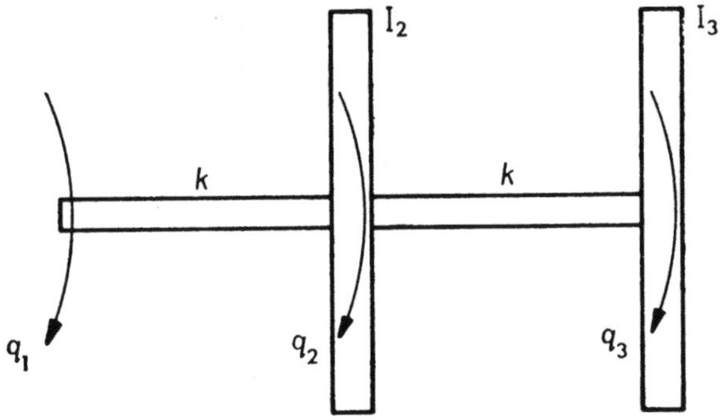

Fig. 2.1.4

equation (2.1.22). In this case both **A** and **C** are singular. The Lagrangian equations for this system are

$$\left.\begin{aligned} kq_1 - kq_2 &= Q_1, \\ I_2\ddot{q}_2 - kq_1 + 2kq_2 - kq_3 &= Q_2, \\ I_3\ddot{q}_3 - kq_2 + kq_3 &= Q_3. \end{aligned}\right\} \qquad (2.1.25)$$

The discussion of semi-definite systems will be taken up in §§ 2.11 and 2.12, and again in Chapter 4.

EXAMPLES 2.1

1. Write down the inertia and stability matrices for the systems for which:

$$(a)\quad T = m\left\{\dot{x}^2 + 13\dot{y}^2 + \frac{111}{4}\dot{z}^2 - 4\dot{x}\dot{y} + 5\dot{x}\dot{z} - 34\dot{y}\dot{z}\right\},$$

$$V = \frac{mg}{l}\left\{10x^2 + 2y^2 + \frac{29}{4}z^2 + 2xy + 6xz - 5yz\right\}.$$

$$(b)\quad T = \frac{ml^2}{2}\{9\dot{\theta}^2 + 7\dot{\phi}^2 + 2\dot{\psi}^2\},$$

$$V = \frac{mgl}{2}\{\theta^2 + \phi^2 + 19\psi^2 + \theta\phi - 7\theta\psi + \phi\psi\}.$$

Discuss the positive-definiteness of the matrices and find any rigid-body modes the systems may have.

2. A light string $OABC$ is tied to a fixed point at O and carries masses m, $2m$ and m at A, B and C respectively. The lengths OA, AB and BC are l, $2l$ and $3l$ respectively. The string moves in a vertical plane and performs small oscillations about the position of equilibrium while being acted on by a variable horizontal force P applied at C. The inclinations of OA, AB and BC are denoted by θ, ϕ and ψ respectively. Find:

(a) the kinetic and potential energies of the system;
(b) the inertia and stability matrices;
(c) the Lagrangian equations governing the forced motion.

[N.B. This system forms the basis of a number of problems in the following sections.]

3. Show that a necessary and sufficient condition for

$$U(x, y, z) \equiv ax^2+by^2+cz^2+2fyz+2gzx+2hxy$$

to be a positive-definite is that

$$a > 0; \quad \begin{vmatrix} a & h \\ h & b \end{vmatrix} > 0; \quad \begin{vmatrix} a & h & g \\ h & b & f \\ g & f & c \end{vmatrix} > 0.$$

Show also that when this condition holds then

$$b > 0; \quad c > 0; \quad \begin{vmatrix} b & f \\ f & c \end{vmatrix} > 0; \quad \begin{vmatrix} a & g \\ g & c \end{vmatrix} > 0.$$

2.2 The receptances of a system; forced harmonic vibration

Let the column matrix $\mathbf{Q}$ on the right-hand side of equation (2.1.14) take the particular form

$$\mathbf{Q} = \mathbf{\Phi} \sin \omega t = \begin{bmatrix} \Phi_1 \\ \Phi_2 \\ \vdots \\ \Phi_n \end{bmatrix} \sin \omega t. \tag{2.2.1}$$

This represents excitation of the system by driving forces

$$\Phi_1 \sin \omega t, \quad \Phi_2 \sin \omega t, \quad \dots, \quad \Phi_n \sin \omega t$$

at the coordinates $q_1, q_2, \dots, q_n$ respectively. The amplitudes $\Phi_1, \Phi_2, \dots, \Phi_n$ are real and the forces are all in phase. The matrix equation (2.1.15) can now be written

$$\mathbf{A}\ddot{\mathbf{q}}+\mathbf{C}\mathbf{q} = \mathbf{\Phi} \sin \omega t. \qquad (2.2.2)†$$

This differential equation in the matrix $\mathbf{q}$ is in many ways very similar to the ordinary differential equation

$$a\frac{d^2y}{dt^2}+cy = f \sin \omega t \tag{2.2.3}$$

in the function $y(t)$. Thus it may be proved that the general solution of (2.2.2) is

† The analysis of this section applies to positive semi-definite as well as to positive definite systems, i.e. it applies even when $\mathbf{A}$ and/or $\mathbf{C}$ are singular.

the sum of two parts—the complementary function and a particular integral. The complementary function is the general solution of the equation

$$\mathbf{A\ddot{q}}+\mathbf{Cq}=\mathbf{0}, \tag{2.2.4}$$

and thus represents the free vibration of the system. As the name suggests, a 'particular integral' is just any one solution of the equation (2.2.2). It is by a particular integral that the forced motion of the system is represented and it is this motion which is the subject of this section.

This being so, we now seek a solution of equation (2.2.2) having the form

$$\mathbf{q}=\begin{bmatrix} q_1 \\ q_2 \\ \vdots \\ q_n \end{bmatrix}=\begin{bmatrix} \Psi_1 \\ \Psi_2 \\ \vdots \\ \Psi_n \end{bmatrix}\sin\omega t=\mathbf{\Psi}\sin\omega t. \tag{2.2.5}$$

The column matrix $\mathbf{\Psi}$ represents the amplitudes of the q's in a harmonic motion whose frequency is that of the driving forces. It will thus be assumed that a solution can be found such that the displacement is always in phase with the excitation. This assumption is in accordance with the properties of equation (2.2.3) and will be shown to be justified.

Differentiation of the trial solution (2.2.5) gives

$$\mathbf{\ddot{q}}=\begin{bmatrix} \ddot{q}_1 \\ \ddot{q}_2 \\ \vdots \\ \ddot{q}_n \end{bmatrix}=-\omega^2\begin{bmatrix} \Psi_1 \\ \Psi_2 \\ \vdots \\ \Psi_n \end{bmatrix}\sin\omega t=-\omega^2\mathbf{\Psi}\sin\omega t, \tag{2.2.6}$$

which when substituted into (2.2.2) yields

$$\mathbf{A\ddot{q}}+\mathbf{Cq}=(-\omega^2\mathbf{A}+\mathbf{C})\mathbf{\Psi}\sin\omega t=\mathbf{\Phi}\sin\omega t, \tag{2.2.7}$$

or

$$(\mathbf{C}-\omega^2\mathbf{A})\mathbf{\Psi}=\mathbf{\Phi}. \tag{2.2.8}$$

The bracketed quantity $(\mathbf{C}-\omega^2\mathbf{A})$ can be written as a single matrix $\mathbf{Z}$. Thus

$$\mathbf{Z}=\begin{bmatrix} c_{11}-\omega^2a_{11} & c_{12}-\omega^2a_{12} & \cdots & c_{1n}-\omega^2a_{1n}, \\ c_{21}-\omega^2a_{21} & c_{22}-\omega^2a_{22} & \cdots & c_{2n}-\omega^2a_{2n}, \\ \cdots & \cdots & \cdots & \cdots \\ c_{n1}-\omega^2a_{n1} & c_{n2}-\omega^2a_{n2} & \cdots & c_{nn}-\omega^2a_{nn}, \end{bmatrix} \tag{2.2.9}$$

and

$$\mathbf{Z\Psi}=\mathbf{\Phi}. \tag{2.2.10}$$

For the present we shall assume that $\mathbf{Z}$ is non-singular; then $|\mathbf{Z}|$ is non-zero and $\mathbf{Z}$ possesses an inverse. This being so, both sides of equation (2.2.10) can be premultiplied by $\mathbf{Z}^{-1}$ giving

$$\mathbf{\Psi}=\mathbf{Z}^{-1}\mathbf{\Phi}. \tag{2.2.11}$$

The actual form of the inverse matrix has been discussed in § 1.6.

The symmetric matrix $\mathbf{Z}$ is sometimes called the 'dynamic stiffness matrix'. Its inverse $\mathbf{Z}^{-1}$ is the 'receptance matrix'† and we shall denote it by the alternative symbol $\boldsymbol{\alpha}$. The response matrix $\mathbf{q}$ is thus given by

$$\mathbf{q} = \boldsymbol{\alpha}\boldsymbol{\Phi}\sin\omega t. \tag{2.2.12}$$

Since the dynamic stiffness matrix $\mathbf{Z}$ is symmetric it follows that its inverse, the receptance matrix, is also (see equation (1.7.18)); its value clearly depends on the matrices $\mathbf{A}$ and $\mathbf{C}$ and upon ω.

It is not usual to require the computation of the whole receptance matrix $\boldsymbol{\alpha}$ from $\mathbf{Z}$ and, for most purposes, individual elements α_{ij} only are required. These can be obtained from equation (1.6.22), which gives

$$\boldsymbol{\alpha} = \mathbf{Z}^{-1} = \frac{\hat{\mathbf{Z}}}{|\mathbf{Z}|}, \tag{2.2.13}$$

where $\hat{\mathbf{Z}}$ is the adjoint of $\mathbf{Z}$. Since $\mathbf{Z}$ is symmetric, $\hat{\mathbf{Z}}$ is also. If the cofactor of z_{ij} in $|\mathbf{Z}|$ is represented by Z_{ij} then

$$Z_{ij} = Z_{ji} \tag{2.2.14}$$

and

$$\alpha_{ij} = \alpha_{ji} = \frac{Z_{ij}}{|\mathbf{Z}|}. \tag{2.2.15}$$

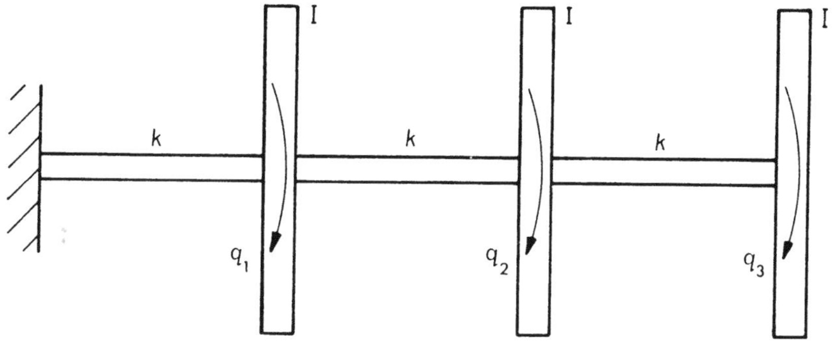

Fig. 2.2.1

For an illustration of these results, consider the torsional system of fig. 2.2.1. This consists of three rigid massive flywheels, each having polar moment of inertia I, connected by light shafts of torsional stiffness k. The left-hand end of the system is rigidly clamped and the rotational displacements of the discs are denoted by q_1, q_2 and q_3. For this system

$$2T = I\dot{q}_1^2 + I\dot{q}_2^2 + I\dot{q}_3^2, \tag{2.2.16}$$

$$\begin{aligned} 2V &= kq_1^2 + k(q_2 - q_1)^2 + k(q_3 - q_2)^2 \\ &= 2kq_1^2 + 2kq_2^2 + kq_3^2 - 2kq_1q_2 - 2kq_2q_3, \end{aligned} \tag{2.2.17}$$

† At one time the receptance matrix was called the 'mechanical admittance matrix'; this name is no longer used because the matrix is not completely analogous to the 'admittance matrix' of electric circuit analysis.

so that
$$\mathbf{A} = \begin{bmatrix} I & 0 & 0 \\ 0 & I & 0 \\ 0 & 0 & I \end{bmatrix}; \quad \mathbf{C} = \begin{bmatrix} 2k & -k & 0 \\ -k & 2k & -k \\ 0 & -k & k \end{bmatrix}. \tag{2.2.18}$$

Both $\mathbf{A}$ and $\mathbf{C}$ are non-singular and positive definite.

If torques
$$Q_1 = \Phi_1 \sin \omega t, \quad Q_2 = \Phi_2 \sin \omega t, \quad Q_3 = \Phi_3 \sin \omega t \tag{2.2.19}$$

are applied at the discs then the equations for the amplitudes of q_1, q_2 and q_3 are
$$\left.\begin{aligned} (2k - I\omega^2)\,\Psi_1 - k\Psi_2 &= \Phi_1, \\ -k\Psi_1 + (2k - I\omega^2)\,\Psi_2 - k\Psi_3 &= \Phi_2, \\ -k\Psi_2 + (k - I\omega^2)\,\Psi_3 &= \Phi_3. \end{aligned}\right\} \tag{2.2.20}$$

That is to say
$$\mathbf{Z\Psi} = \mathbf{\Phi}, \tag{2.2.21}$$
where
$$\mathbf{Z} = \begin{bmatrix} 2k - I\omega^2 & -k & 0 \\ -k & 2k - I\omega^2 & -k \\ 0 & -k & k - I\omega^2 \end{bmatrix}. \tag{2.2.22}$$

The determinant of this matrix reduces to
$$|\mathbf{Z}| = k^3 - 6k^2 I\omega^2 + 5kI^2\omega^4 - I^3\omega^6, \tag{2.2.23}$$
and the inverse of $\mathbf{Z}$ is
$$\mathbf{Z}^{-1} \equiv \boldsymbol{\alpha}$$
$$= \frac{1}{|\mathbf{Z}|} \begin{bmatrix} (2k - I\omega^2)(k - I\omega^2) - k^2 & k(k - I\omega^2) & k^2 \\ k(k - I\omega^2) & (2k - I\omega^2)(k - I\omega^2) & k(2k - I\omega^2) \\ k^2 & k(2k - I\omega^2) & (2k - I\omega^2)^2 - k^2 \end{bmatrix}. \tag{2.2.24}$$

That is to say
$$\left.\begin{aligned} q_1 &= \frac{1}{|\mathbf{Z}|}\{[(2k - I\omega^2)(k - I\omega^2) - k^2]\,\Phi_1 + k(k - I\omega^2)\,\Phi_2 + k^2\Phi_3\} \sin \omega t, \\ q_2 &= \frac{1}{|\mathbf{Z}|}\{k(k - I\omega^2)\,\Phi_1 + (2k - I\omega^2)(k - I\omega^2)\,\Phi_2 + k(2k - I\omega^2)\,\Phi_3\} \sin \omega t, \\ q_3 &= \frac{1}{|\mathbf{Z}|}\{k^2\Phi_1 + k(2k - I\omega^2)\,\Phi_2 + [(2k - I\omega^2)^2 - k^2]\,\Phi_3\} \sin \omega t. \end{aligned}\right\} \tag{2.2.25}$$

In this example, as in the general case (see (2.2.15)), each of the receptances has $|\mathbf{Z}|$ as denominator. If therefore a certain value of ω makes the determinant small then all the receptances will be large. The only case in which this will not hold is when, for a certain receptance, the chosen value of ω makes the numerator small also. Provided that this does not occur, values of ω so adjusted as to make $|\mathbf{Z}|$ tend to zero will make all the receptances tend to infinity. This *suggests* that if ω is such that $|\mathbf{Z}|$ is zero there may be a finite motion even when there is no external excitation, i.e. when $\mathbf{\Phi}$ is zero. This is the condition of 'free vibration'. When $|\mathbf{Z}|$ is zero the analysis of this section breaks down, and for this reason the subject of free vibration will be postponed to the following section.

The analysis of this section has been concerned with the motion of a system excited by harmonic forces at the coordinates $q_1, q_2, \ldots, q_n$. The final result of the analysis, equation (2.2.12), gives the responses at the coordinates $q_1, q_2, \ldots, q_n$ due to such a system of forces. There is, however, a more general problem and this will now be considered. It is to find the response at a generalised coordinate x due to a harmonic force $F \sin \omega t$ applied at a generalised coordinate y. In this wider problem both x and y are allowed to be quite general functions of the original coordinates $q_1, q_2, \ldots, q_n$. The only restriction on x and y is that they vanish at the equilibrium configuration; they are not restricted to be displacements of physical points of the system.† For example x may be the mean of the values of $q_1, q_2, \ldots, q_n$, i.e.

$$x = (1/n)(q_1 + q_2 + \ldots + q_n), \tag{2.2.26}$$

or it may be more complicated, for example

$$x = q_1^2 + 8(q_1 + q_2) - 7q_3. \tag{2.2.27}$$

We are concerned only with small vibrations, in which the second and higher powers of $q_1, q_2, \ldots, q_n$ in x can be neglected. For small vibrations therefore Taylor's expansion gives

$$x = \frac{\partial x}{\partial q_1} q_1 + \frac{\partial x}{\partial q_2} q_2 + \ldots + \frac{\partial x}{\partial q_n} q_n, \tag{2.2.28}$$

where the partial derivatives are evaluated at the datum configuration. This equation shows that, as far as small vibrations are concerned x (and y similarly) may, without loss of generality, be taken to be a *linear* combination of the q's. For example, if x is given by (2.2.27) then the Taylor expansion is

$$x = 8q_1 + 8q_2 - 7q_3. \tag{2.2.29}$$

Equation (2.2.28) is expressed by the matrix equation

$$x = \begin{bmatrix} \dfrac{\partial x}{\partial q_1} & \dfrac{\partial x}{\partial q_2} & \cdots & \dfrac{\partial x}{\partial q_n} \end{bmatrix} \cdot \begin{bmatrix} q_1 \\ q_2 \\ \vdots \\ q_n \end{bmatrix}. \tag{2.2.30}$$

The force $F \sin \omega t$ at y is equivalent to a set of forces $\Phi_1 \sin \omega t, \Phi_2 \sin \omega t, \ldots, \Phi_n \sin \omega t$ at $q_1, q_2, \ldots, q_n$ respectively. These forces may be found in the following way. The work done by $F \sin \omega t$ in a small 'displacement' is

$$\delta W = F \sin \omega t . \delta y = F \sin \omega t . \left(\frac{\partial y}{\partial q_1} \delta q_1 + \frac{\partial y}{\partial q_2} \delta q_2 + \ldots + \frac{\partial y}{\partial q_n} \delta q_n \right), \tag{2.2.31}$$

† It may be noted here that the 'normalised principal coordinates' which are introduced in §2.6 are not displacements. In fact if the coordinates q have the dimensions of length then these principal coordinates have the dimensions of $\sqrt{(\text{mass})} \times$ length. In spite of this one speaks of receptances between these coordinates.

where $\delta q_1, \delta q_2, \ldots, \delta q_n$ are the resulting small displacements at $q_1, q_2, \ldots, q_n$ respectively. But if $F\sin\omega t$ at y is equivalent to $\Phi_1\sin\omega t, \Phi_2\sin\omega t, \ldots, \Phi_n\sin\omega t$ then

$$\delta W = (\Phi_1\delta q_1 + \Phi_2\delta q_2 + \ldots + \Phi_n\delta q_n)\sin\omega t. \tag{2.2.32}$$

By equating the two expressions for δW we obtain

$$\Phi_1 = F\frac{\partial y}{\partial q_1}; \quad \Phi_2 = F\frac{\partial y}{\partial q_2}; \quad \ldots; \quad \Phi_n = F\frac{\partial y}{\partial q_n}, \tag{2.2.33}$$

so that the matrix $\mathbf{\Phi}$ is now

$$\mathbf{\Phi} = \begin{bmatrix} \Phi_1 \\ \Phi_2 \\ \vdots \\ \Phi_n \end{bmatrix} = \begin{bmatrix} \dfrac{\partial y}{\partial q_1} \\ \dfrac{\partial y}{\partial q_2} \\ \vdots \\ \dfrac{\partial y}{\partial q_n} \end{bmatrix} F. \tag{2.2.34}$$

By using this matrix $\mathbf{\Phi}$ in equation (2.2.12) and premultiplying the equation by the row matrix

$$\begin{bmatrix} \dfrac{\partial x}{\partial q_1} & \dfrac{\partial x}{\partial q_2} & \cdots & \dfrac{\partial x}{\partial q_n} \end{bmatrix}$$

in accordance with equation (2.2.30), we find that the response at x is

$$x = \begin{bmatrix} \dfrac{\partial x}{\partial q_1} & \dfrac{\partial x}{\partial q_2} & \cdots & \dfrac{\partial x}{\partial q_n} \end{bmatrix} \boldsymbol{\alpha} \begin{bmatrix} \dfrac{\partial y}{\partial q_1} \\ \dfrac{\partial y}{\partial q_2} \\ \vdots \\ \dfrac{\partial y}{\partial q_n} \end{bmatrix} F\sin\omega t. \tag{2.2.35}$$

The product of the three matrices in this equation is a scalar quantity. We shall denote it by α_{xy} and its value is

$$\alpha_{xy} = \left[\alpha_{11}\frac{\partial x}{\partial q_1}\frac{\partial y}{\partial q_1} + \alpha_{22}\frac{\partial x}{\partial q_2}\frac{\partial y}{\partial q_2} + \ldots + \alpha_{nn}\frac{\partial x}{\partial q_n}\frac{\partial y}{\partial q_n} + \alpha_{12}\frac{\partial x}{\partial q_1}\frac{\partial y}{\partial q_2}\right.$$
$$\left. + \alpha_{21}\frac{\partial x}{\partial q_2}\frac{\partial y}{\partial q_1} + \ldots + \alpha_{n-1,\,n}\frac{\partial x}{\partial q_{n-1}}\frac{\partial y}{\partial q_n} + \alpha_{n,\,n-1}\frac{\partial x}{\partial q_n}\frac{\partial y}{\partial q_{n-1}}\right]. \tag{2.2.36}$$

The right-hand side of (2.2.36) is symmetrical in x and y. Therefore

$$\alpha_{xy} = \alpha_{yx}. \tag{2.2.37}$$

If x and y are identical, equation (2.2.36) gives the direct receptance α_{xx} to be

$$\alpha_{xx} = \left[\alpha_{11}\left(\frac{\partial x}{\partial q_1}\right)^2 + \alpha_{22}\left(\frac{\partial x}{\partial q_2}\right)^2 + \ldots + \alpha_{nn}\left(\frac{\partial x}{\partial q_n}\right)^2 + 2\alpha_{12}\frac{\partial x}{\partial q_1}\frac{\partial x}{\partial q_2} + \ldots\right.$$
$$\left. + 2\alpha_{n-1,\,n}\frac{\partial x}{\partial q_{n-1}}\frac{\partial x}{\partial q_n}\right]. \tag{2.2.38}$$

To illustrate these results we return to the system of fig. 2.2.1. Suppose that torques $F \sin \omega t$ and $-2F \sin \omega t$ act on the first and second discs respectively. These are equivalent to a torque $F \sin \omega t$ at

$$y = q_1 - 2q_2. \tag{2.2.39}$$

Suppose also that it is required to find the mean displacement of the discs; this will be the displacement at

$$x = \tfrac{1}{3}(q_1 + q_2 + q_3). \tag{2.2.40}$$

For these values of x and y equation (2.2.36) gives

$$\alpha_{xy} = \tfrac{1}{3}(\alpha_{11} - 2\alpha_{22} - \alpha_{12} + \alpha_{31} - 2\alpha_{32}), \tag{2.2.41}$$

so that

$$x = \tfrac{1}{3}(\alpha_{11} - 2\alpha_{22} - \alpha_{12} + \alpha_{31} - 2\alpha_{32}) F \sin \omega t. \tag{2.2.42}$$

The reader will probably be acquainted with Maxwell's reciprocal theorem of elementary elasticity theory.† The remainder of this section will be concerned with its counterpart in the theory of forced harmonic vibration.

Let the column matrix $\mathbf{\Phi} \sin \omega t$ represent a set of generalised harmonic forces applied to the system; the forces are therefore all in phase. Let the corresponding displacements be given by the matrix $\mathbf{\Psi} \sin \omega t$. Now consider a second system of forces $\bar{\mathbf{\Phi}} \sin \omega t$, of the same type as the first, producing the displacements $\bar{\mathbf{\Psi}} \sin \omega t$. The displacements are related to the forces by the equations

$$\mathbf{\Psi} = \boldsymbol{\alpha}\mathbf{\Phi}; \quad \bar{\mathbf{\Psi}} = \boldsymbol{\alpha}\bar{\mathbf{\Phi}}. \tag{2.2.43}$$

The transpose of the first equation is

$$\mathbf{\Psi}' = \mathbf{\Phi}'\boldsymbol{\alpha} \tag{2.2.44}$$

since $\boldsymbol{\alpha}$ is symmetric; this result follows from equation (1.4.7). Premultiply the second of equations (2.2.43) by $\mathbf{\Phi}'$; we obtain

$$\mathbf{\Phi}'\bar{\mathbf{\Psi}} = \mathbf{\Phi}'\boldsymbol{\alpha}\bar{\mathbf{\Phi}} = \mathbf{\Psi}'\bar{\mathbf{\Phi}}. \tag{2.2.45}$$

This equation may be written in a symmetric form if it is noted that

$$\mathbf{\Psi}'\bar{\mathbf{\Phi}} = \bar{\mathbf{\Phi}}'\mathbf{\Psi}. \tag{2.2.46}$$

This result follows from the fact that each of the expressions is a scalar. Equations (2.2.45) and (2.2.46) may be combined to give

$$\mathbf{\Phi}'\bar{\mathbf{\Psi}} = \bar{\mathbf{\Phi}}'\mathbf{\Psi}. \tag{2.2.47}$$

This is the 'reciprocal theorem' for forced oscillations; when multiplied out it becomes

$$\Phi_1 \bar{\Psi}_1 + \Phi_2 \bar{\Psi}_2 + \ldots + \Phi_n \bar{\Psi}_n = \bar{\Phi}_1 \Psi_1 + \bar{\Phi}_2 \Psi_2 + \ldots + \bar{\Phi}_n \Psi_n. \tag{2.2.48}$$

The reciprocal theorem may usefully be applied in the same circumstances as the corresponding theorem of statics. Thus it may be employed in finding one unknown force—or displacement amplitude—by arranging that all the product terms in equation (2.2.48) shall vanish except one on each side.

† R. V. Southwell, *An Introduction to the Theory of Elasticity* (Oxford University Press, 1936), ch. 1.

To illustrate the use of the theorem, consider a system comprising n particles attached at intervals along a taut string which is devoid of mass and is stretched between two fixed points. Consider motions along the length of the string under the action of harmonic forces acting on the particles in that direction and let the displacements be $q_1, q_2, \ldots, q_n$. Suppose that a force $F \sin \omega t$ acts on particle 1 so that

$$\Phi_1 = F; \quad \Phi_2 = 0 = \Phi_3 = \ldots = \Phi_n, \tag{2.2.49}$$

and let the resulting amplitudes of displacement be $\Psi_1, \Psi_2, \ldots, \Psi_n$. Now consider a second system of excitation in which a force $(F/n) \sin \omega t$ acts at each particle and suppose that the resulting amplitudes are $\bar{\Psi}_1, \bar{\Psi}_2, \ldots, \bar{\Psi}_n$; for this motion

$$\bar{\Phi}_1 = F/n = \bar{\Phi}_2 = \ldots = \bar{\Phi}_n. \tag{2.2.50}$$

Equation (2.2.48) now shows that

$$F\bar{\Psi}_1 = \frac{F}{n} \sum_{r=1}^{n} \Psi_r. \tag{2.2.51}$$

It follows that the amplitude $\bar{\Psi}_1$ at q_1 under the second system of excitation is equal to the mean of the amplitudes under the first.

The importance of the reciprocal theorem is not that it gives results which cannot be found by straightforward methods, but rather that it may be more convenient to apply. In the above problem, for instance, we arrived at a general result which might be very tedious to establish in a given case on account of the number, spacing and masses of the particles and the possible variation of the elastic properties of the string along its length.

EXAMPLES 2.2

1. Find the receptance matrix for the system for which

$$\mathbf{A} = m\begin{bmatrix} 1 & 0 & 0 \\ 0 & 9 & 0 \\ 0 & 0 & 6 \end{bmatrix}, \quad \mathbf{C} = \frac{mg}{l}\begin{bmatrix} 2 & -1 & -2 \\ -1 & 6 & 3 \\ -2 & 3 & 11 \end{bmatrix}.$$

2. Find the angle θ when the system of Ex. 2.1.2 is acted upon by a sinusoidal force $P = F \sin \omega t$ at C.

3. Find the receptance α_{12} for the system for which

$$T = \tfrac{1}{2}I\{3\dot{q}_1^2 + 4\dot{q}_2^2 + 18\dot{q}_3^2 - 2\dot{q}_1\dot{q}_2 - 6\dot{q}_1\dot{q}_3 + 16\dot{q}_2\dot{q}_3\},$$
$$V = \tfrac{1}{2}k\{34q_1^2 + 35q_2^2 + 97q_3^2 - 4q_1q_2 - 40q_1q_3 + 110q_2q_3\}.$$

2.3 Free vibration; principal modes

In this section we shall be concerned with the solution of the matrix equation (2.2.4), viz.

$$\mathbf{A}\ddot{\mathbf{q}} + \mathbf{C}\mathbf{q} = \mathbf{0}. \tag{2.3.1}$$

Throughout this section—and indeed until § 2.11—it will be assumed that $\mathbf{A}$ and $\mathbf{C}$ are both positive definite; they will therefore both be non-singular. The changes which occur when one or both of $\mathbf{A}$ and $\mathbf{C}$ are singular will be discussed in § 2.12.

Proceeding as in § 2.2 we now seek a solution of equation (2.3.1) having the form

$$\mathbf{q} = \begin{bmatrix} q_1 \\ q_2 \\ \vdots \\ q_n \end{bmatrix} = \begin{bmatrix} \Psi_1 \\ \Psi_2 \\ \vdots \\ \Psi_n \end{bmatrix} \sin(\omega t+\theta) = \mathbf{\Psi} \sin(\omega t+\theta). \tag{2.3.2}$$

This represents a harmonic vibration of the system in which the motions at the various coordinates are all in phase. The column matrix $\mathbf{\Psi}\sin(\omega t+\theta)$ is a solution of equation (2.3.1) if

$$(\mathbf{C}-\omega^2\mathbf{A})\,\mathbf{\Psi} = \mathbf{0}, \tag{2.3.3}$$

or

$$\mathbf{Z\Psi} = \mathbf{0}, \tag{2.3.4}$$

where $\mathbf{0}$ is the null column matrix of order n. The amplitudes of the coordinates q_i are thus governed by an equation of the type discussed in § 1.8.

In § 1.8 it was shown that non-null solutions of (2.3.3) may be found provided that

$$|\mathbf{C}-\omega^2\mathbf{A}| = 0. \tag{2.3.5}$$ †

That is to say, free vibration can take place if

$$\begin{vmatrix} c_{11}-\omega^2 a_{11} & c_{12}-\omega^2 a_{12} & \cdots & c_{1n}-\omega^2 a_{1n} \\ c_{21}-\omega^2 a_{21} & c_{22}-\omega^2 a_{22} & \cdots & c_{2n}-\omega^2 a_{2n} \\ \cdots & \cdots & \cdots & \cdots \\ c_{n1}-\omega^2 a_{n1} & c_{n2}-\omega^2 a_{n2} & \cdots & c_{nn}-\omega^2 a_{nn} \end{vmatrix} = 0. \tag{2.3.6}$$

This determinant is a polynomial in ω^2. The coefficient of the highest power of ω^2 (that is, of ω^{2n}) may be found by repeated application of rule (*h*) of § 1.5. It is $|\mathbf{A}|$ which, by hypothesis, is not zero. Equation (2.3.6) is therefore an equation of the nth degree in ω^2. It can be shown to be a consequence of the assumed positive definiteness of $\mathbf{A}$ and $\mathbf{C}$ that the roots $\omega_1^2, \omega_2^2, \ldots, \omega_n^2$ are all real and positive, though not necessarily all different; this is proved in § 5.1. The quantities $\omega_1, \omega_2, \ldots, \omega_n$ are the 'natural frequencies' of the system. Practical methods for finding these frequencies are discussed in Chapters 8 and 9.

It is sometimes convenient (though we shall not do so here) to introduce the unknown λ given by

$$\lambda = 1/\omega.$$

Equations (2.3.3) and (2.3.5) may then be written in the form

$$(\mathbf{A}-\lambda^2\mathbf{C})\,\mathbf{\Psi} = \mathbf{0}, \tag{2.3.7}$$

$$|\mathbf{A}-\lambda^2\mathbf{C}| = 0. \tag{2.3.8}$$

† In this and the next few pages there are some mathematical niceties which will be glossed over for the present. For a rigorous treatment of these matters, the reader is referred to Chapter 4, and especially to § 4.4.

For each root ω_s^2 of equation (2.3.5) the matrix $\mathbf{C}-\omega_s^2\mathbf{A}$ is singular, and equation (2.3.3) has a solution which can be written

$$\boldsymbol{\Psi}^{(s)} = \begin{bmatrix} \Psi_1^{(s)} \\ \Psi_2^{(s)} \\ \vdots \\ \Psi_n^{(s)} \end{bmatrix}, \tag{2.3.9}$$

and which therefore satisfies

$$(\mathbf{C}-\omega_s^2\mathbf{A})\,\boldsymbol{\Psi}^{(s)} = \mathbf{0}. \tag{2.3.10}$$

It can be shown that when the roots $\omega_1^2, \omega_2^2, \ldots, \omega_n^2$ are all different there is at least one non-zero cofactor in each of the determinants $|\mathbf{C}-\omega_s^2\mathbf{A}|$.† This means that equation (2.3.10) falls into the category of equations for which the solution was given in §1.8; in particular the set of ratios

$$\Psi_1^{(s)} : \Psi_2^{(s)} : \ldots : \Psi_n^{(s)} \tag{2.3.11}$$

is uniquely determined. This set of ratios which, apart from an arbitrary constant multiplier, gives the column vector $\boldsymbol{\Psi}^{(s)}$, is said to define the sth 'principal mode' or 'normal mode' of the system.

As an example, consider the system of fig. 2.2.1 in which

$$\mathbf{A} = \begin{bmatrix} I & 0 & 0 \\ 0 & I & 0 \\ 0 & 0 & I \end{bmatrix}; \quad \mathbf{C} = \begin{bmatrix} 2k & -k & 0 \\ -k & 2k & -k \\ 0 & -k & k \end{bmatrix}. \tag{2.3.12}$$

The equation corresponding to (2.3.6) is

$$\begin{vmatrix} 2k-I\omega^2 & -k & 0 \\ -k & 2k-I\omega^2 & -k \\ 0 & -k & k-I\omega^2 \end{vmatrix} = 0, \tag{2.3.13}$$

or

$$I^3\omega^6 - 5kI^2\omega^4 + 6k^2I\omega^2 - k^3 = 0. \tag{2.3.14}$$

The roots of this equation in ω^2 are

$$\omega_1^2 = 0{\cdot}1980k/I, \quad \omega_2^2 = 1{\cdot}555k/I, \quad \omega_3^2 = 3{\cdot}247k/I, \tag{2.3.15}$$

so that the natural frequencies are

$$\omega_1 = 0{\cdot}4450\surd(k/I), \quad \omega_2 = 1{\cdot}247\surd(k/I), \quad \omega_3 = 1{\cdot}802\surd(k/I). \tag{2.3.16}$$

The amplitudes $\Psi_1^{(1)}$, $\Psi_2^{(1)}$ and $\Psi_3^{(1)}$ in the first mode satisfy the matrix equation obtained by substituting ω_1^2 for ω^2 in equation (2.3.3). This equation is

$$\begin{bmatrix} 2-0{\cdot}1980 & -1 & 0 \\ -1 & 2-0{\cdot}1980 & -1 \\ 0 & -1 & 1-0{\cdot}1980 \end{bmatrix} \begin{bmatrix} \Psi_1^{(1)} \\ \Psi_2^{(1)} \\ \Psi_3^{(1)} \end{bmatrix} = \mathbf{0}, \tag{2.3.17}$$

† This is proved in Chapter 4, in which the case of multiple roots ω^2 is also considered.

that is

$$\begin{bmatrix} 1\cdot8020 & -1 & 0 \\ -1 & 1\cdot8020 & -1 \\ 0 & -1 & 0\cdot8020 \end{bmatrix} \begin{bmatrix} \Psi_1'^{(1)} \\ \Psi_2'^{(1)} \\ \Psi_3'^{(1)} \end{bmatrix} = \mathbf{0}. \tag{2.3.18}$$

The general solution of such an equation was given in equation (1.8.8). Taking the Ψ''s proportional to the cofactors of the elements in the third row of the matrix we find the solution of (2.3.18) to be

$$\Psi_1'^{(1)}:\Psi_2'^{(1)}:\Psi_3'^{(1)}::\begin{vmatrix} -1 & 0 \\ 1\cdot802 & -1 \end{vmatrix}:\begin{vmatrix} -1\cdot802 & 0 \\ -1 & -1 \end{vmatrix}:\begin{vmatrix} 1\cdot802 & -1 \\ -1 & 1\cdot802 \end{vmatrix}, \tag{2.3.19}$$

that is

$$\Psi_1'^{(1)}:\Psi_2'^{(1)}:\Psi_3'^{(1)}::1:1\cdot802:2\cdot247. \tag{2.3.20}$$

The reader may verify that the same set of ratios is obtained by taking the Ψ''s proportional to the cofactors of the first or second row of the matrix. Thus the first mode of the system is characterised by the relation (2.3.20) or, in other words,

$$q_1:q_2:q_3::1:1\cdot802:2\cdot247. \tag{2.3.21}$$

The sets of ratios for the coordinates q_1, q_2 and q_3 corresponding to the second and third modes may be found in a similar way by substituting ω_2^2 and ω_3^2 respectively for ω^2 in (2.3.13). It is found that in the second mode

$$q_1:q_2:q_3::1:0\cdot445:-0\cdot802, \tag{2.3.22}$$

while

$$q_1:q_2:q_3::1:-1\cdot247:0\cdot555 \tag{2.3.23}$$

in the third mode.

EXAMPLES 2.3

1. Find the natural frequencies of the system for which

$$\mathbf{A} = ml^2\begin{bmatrix} 1 & 2 & 1 \\ 2 & 5 & 3 \\ 1 & 3 & 4 \end{bmatrix}, \quad \mathbf{C} = mgl\begin{bmatrix} 1 & 3 & 1 \\ 3 & 12 & 1 \\ 1 & 1 & 4 \end{bmatrix}.$$

2. Find the natural frequencies and principal modes of the system of Ex. 2.1.2 correct to three significant figures.

3. A certain system has the stability matrix

$$\mathbf{C} = k\begin{bmatrix} 2 & -1 & 1 \\ -1 & 9 & 2 \\ 1 & 2 & 3 \end{bmatrix}.$$

Its natural frequencies are

$$\omega_1 = \sqrt{(k/I)}, \quad \omega_2 = 2\sqrt{(k/I)}, \quad \omega_3 = 3\sqrt{(k/I)},$$

and the corresponding modes are

$$\Psi^{(1)} = \begin{bmatrix} 1 \\ 1 \\ 1 \end{bmatrix}, \quad \Psi^{(2)} = \begin{bmatrix} 2 \\ 5 \\ -9 \end{bmatrix}, \quad \Psi^{(3)} = \begin{bmatrix} 10 \\ 1 \\ -5 \end{bmatrix}.$$

Find the inertia matrix of the system.

2.4 Principal coordinates

When a system is vibrating freely in one of its principal modes all its generalised coordinates $q_1, q_2, \ldots, q_n$ may vary. An illustration of this fact is provided by the example at the end of § 2.3. A set of 'principal coordinates' for a system is a set of coordinates so chosen that when the system is vibrating in a principal mode, only one coordinate varies while the remainder remain zero. The general system will evidently have n principal coordinates and we shall denote them by $p_1, p_2, \ldots, p_n$. During free vibration in the sth mode, p_s will vary harmonically with frequency ω_s while the other p's will be zero. In the sth mode all the q's will be multiples of p_s.

In order to define the principal coordinates precisely it is necessary to introduce a scale factor and describe what is meant by the unit value of a principal coordinate. This can be explained best by reference to an example. It was shown in § 2.3 that, in the first mode of the system of fig. 2.2.1,

$$q_1 : q_2 : q_3 = 1 : 1{\cdot}802 : 2{\cdot}247. \tag{2.4.1}$$

In this mode the principal coordinates p_2 and p_3 are both zero and the q's are multiples of p_1. Let us choose p_1 so that when $p_1 = 1$ the coordinate $q_1 = 2$. For any other value of p_1 the coordinates q will have the values

$$q_1 = 2p_1; \quad q_2 = 3{\cdot}604p_1; \quad q_3 = 4{\cdot}494p_1. \tag{2.4.2}$$

It is clear that we may choose p_2 and p_3 in a similar way. For the sake of argument let us choose them so that in the second mode $q_1 = 1$ when $p_2 = 1$, and in the third mode $q_1 = 1$ when $p_3 = 1$. This means that in the second mode

$$q_1 = p_2; \quad q_2 = 0{\cdot}445p_2; \quad q_3 = -0{\cdot}802p_2, \tag{2.4.3}$$

while in the third mode

$$q_1 = p_3; \quad q_2 = -1{\cdot}247p_3; \quad q_3 = 0{\cdot}555p_3. \tag{2.4.4}$$

If the system is vibrating in its three modes simultaneously then

$$\left.\begin{aligned} q_1 &= 2p_1 + p_2 + p_3, \\ q_2 &= 3{\cdot}604p_1 + 0{\cdot}445p_2 - 1{\cdot}247p_3, \\ q_3 &= 4{\cdot}494p_1 - 0{\cdot}802p_2 + 0{\cdot}555p_3. \end{aligned}\right\} \tag{2.4.5}$$

This may be expressed in matrix notation by the equation

$$\begin{bmatrix} q_1 \\ q_2 \\ q_3 \end{bmatrix} = \begin{bmatrix} 2 & 1 & 1 \\ 3{\cdot}604 & 0{\cdot}445 & -1{\cdot}247 \\ 4{\cdot}494 & -0{\cdot}802 & 0{\cdot}555 \end{bmatrix} \begin{bmatrix} p_1 \\ p_2 \\ p_3 \end{bmatrix}. \tag{2.4.6}$$

This example indicates the way in which the generalised coordinates q are related to the principal coordinates p in the general system. Suppose that the frequency equation (2.3.5) has been solved and that the $\mathbf{\Psi}$ vectors corresponding to $\omega_1, \omega_2, \ldots, \omega_n$ have been found. Suppose further that unit values have been assigned to the principal coordinates so that in the first mode the coordinates q have the values

$$q_1 = \Psi_1^{(1)}; \quad q_2 = \Psi_2^{(1)}; \quad \ldots; \quad q_n = \Psi_n^{(1)}, \tag{2.4.7}$$

when $p_1 = 1$, and, generally, have the values

$$q_1 = \Psi_1^{(s)}; \quad q_2 = \Psi_2^{(s)}; \quad \ldots; \quad q_n = \Psi_n^{(s)}, \tag{2.4.8}$$

when $p_s = 1$ in the sth mode. Then when the system is vibrating in all its modes simultaneously the column vectors $\mathbf{q}$ and $\mathbf{p}$ will be related by the equation

$$\mathbf{q} = \mathbf{X}\mathbf{p}, \tag{2.4.9}$$

where $\mathbf{X}$ is the square matrix

$$\mathbf{X} = \begin{bmatrix} \Psi_1^{(1)} & \Psi_1^{(2)} & \ldots & \Psi_1^{(n)} \\ \Psi_2^{(1)} & \Psi_2^{(2)} & \ldots & \Psi_2^{(n)} \\ \cdots & \cdots & \cdots & \cdots \\ \Psi_n^{(1)} & \Psi_n^{(2)} & \ldots & \Psi_n^{(n)} \end{bmatrix}, \tag{2.4.10}$$

whose columns are the $\mathbf{\Psi}$ vectors.

Equation (2.4.9) expresses the q's, which are known, in terms of the p's, which are unknown, and is for this reason somewhat unsatisfactory. What is often required is an equation which will give the p's in terms of the q's. Such an equation may be obtained by inverting the matrix $\mathbf{X}$; for if $\boldsymbol{\xi} = \mathbf{X}^{-1}$, then

$$\mathbf{p} = \boldsymbol{\xi}\mathbf{q}. \tag{2.4.11}$$

We shall show in § 2.7 that this inverse can be obtained without recourse to the elaborate procedure outlined in § 1.6. As a preliminary to this we shall discuss the Orthogonality Principle and in so doing we shall discover some of the properties of the matrix $\mathbf{X}$.

EXAMPLES 2.4

1. The energies of a certain system are given by

$$2T = m\{4\dot{x}^2 + 7\dot{y}^2 + 23\dot{z}^2 + 6\dot{x}\dot{y} + 18\dot{x}\dot{z} + 18\dot{y}\dot{z}\},$$

$$2V = (mg/l)\{31x^2 + 106y^2 + 122z^2 + 96xy + 120xz + 204yz\}.$$

One of its natural frequencies is $\sqrt{(g/l)}$. Find the other two, and find the set of ratios $x:y:z$ in each of the principal modes.

Choose the modes of the system so that $y = 1$ when the system vibrates in any one mode and the corresponding principal coordinate has the value one. Find the equations giving x, y, z in terms of these principal coordinates, and invert them to find the principal coordinates in terms of x, y, z.

2. The principal coordinates for a certain system are chosen so that when $p_s = 1$ in the sth mode the coordinates q have the values

$$q_1 = \Psi_1^{(s)}, \quad q_2 = \Psi_2^{(s)}, \quad \ldots, \quad q_n = \Psi_n^{(s)}.$$

The equation linking the matrices q and p is then equation (2.4.9), namely

$$\mathbf{q} = \mathbf{X}\mathbf{p}.$$

The scaling of the principal coordinates is now changed so that when the new coordinate p_s^* has the value unity in the sth mode the coordinates q have the values

$$q_1 = \lambda_s \Psi_1^{(s)}, \quad q_2 = \lambda_s \Psi_2^{(s)}, \quad \ldots, \quad q_n = \lambda_s \Psi_n^{(s)}.$$

Find the new coordinates p_s^* in terms of the old.

In the system of Ex. 2.3.3 the coordinates p^* are chosen so that

$$\begin{aligned} q_2 &= 2 \quad \text{when} \quad p_1^* = 1 \quad \text{in the first mode,} \\ q_3 &= -9 \quad \text{when} \quad p_2^* = 1 \quad \text{in the second mode,} \\ q_1 &= 2 \quad \text{when} \quad p_3^* = 1 \quad \text{in the third mode.} \end{aligned}$$

Find the values of the p^*'s when the system has the displacement $\mathbf{q} = \{0 \quad 2 \quad 0\}$.

2.5 The Principle of Orthogonality

We have shown that a system with n degrees of freedom has n natural frequencies $\omega_1, \omega_2, \ldots, \omega_n$, which will here be assumed to be different. We have also shown that corresponding to each frequency ω_s there is a non-zero column matrix $\mathbf{\Psi}^{(s)}$ which satisfies equation (2.3.10), that is, which satisfies

$$(\mathbf{C} - \omega_s^2 \mathbf{A})\, \mathbf{\Psi}^{(s)} = \mathbf{0} \quad (s = 1, 2, \ldots, n). \tag{2.5.1}$$

The Principle of Orthogonality, considered in relation to matrix analysis, is a simple consequence of equation (2.5.1) and of the fact that the frequencies $\omega_1, \omega_2, \ldots, \omega_n$ are all different. We shall first state the principle in terms of the matrices $\mathbf{\Psi}$, we shall then give an alternative formulation in terms of the matrix $\mathbf{X}$ and finally we shall show that these formulations are equivalent to the usual statement of the principle in terms of the energy of the system.

Let ω_r and ω_s be two of the natural frequencies of the system and let $\mathbf{\Psi}^{(r)}$ and $\mathbf{\Psi}^{(s)}$ be the corresponding $\mathbf{\Psi}$ matrices. The matrix $\mathbf{\Psi}^{(s)}$ satisfies (2.5.1) while $\mathbf{\Psi}^{(r)}$ satisfies

$$(\mathbf{C} - \omega_r^2 \mathbf{A})\, \mathbf{\Psi}^{(r)} = \mathbf{0}. \tag{2.5.2}$$

Equation (2.5.1) can be premultiplied by $\mathbf{\Psi}'^{(r)}$ to give

$$\mathbf{\Psi}'^{(r)}(\mathbf{C} - \omega_s^2 \mathbf{A})\, \mathbf{\Psi}^{(s)} = 0. \tag{2.5.3}$$

The transpose of equation (2.5.2) is

$$\mathbf{\Psi}'^{(r)}(\mathbf{C} - \omega_r^2 \mathbf{A}) = \mathbf{0} \tag{2.5.4}$$

since both $\mathbf{A}$ and $\mathbf{C}$ are symmetric matrices. If equation (2.5.4) is now post-multiplied by $\mathbf{\Psi}^{(s)}$ it is found that

$$\mathbf{\Psi}'^{(r)}(\mathbf{C} - \omega_r^2 \mathbf{A})\, \mathbf{\Psi}^{(s)} = 0. \tag{2.5.5}$$

Subtraction of equation (2.5.5) from equation (2.5.3) gives

$$(\omega_r^2 - \omega_s^2)\, \mathbf{\Psi}'^{(r)} \mathbf{A} \mathbf{\Psi}^{(s)} = 0 \tag{2.5.6}$$

and since $\omega_r \neq \omega_s$ this implies that

$$\mathbf{\Psi}'^{(r)}\mathbf{A}\mathbf{\Psi}^{(s)} = 0 \quad (r \neq s). \tag{2.5.7}$$

This equation, when combined with (2.5.5), gives

$$\mathbf{\Psi}'^{(r)}\mathbf{C}\mathbf{\Psi}^{(s)} = 0 \quad (r \neq s). \tag{2.5.8}$$

Equations (2.5.7) and (2.5.8) are a statement in matrix form of the Principle of Orthogonality.†

Consider now the matrix product $\mathbf{X'AX}$. The element in the rth row and sth column may be written down by multiplying out the rth row of $\mathbf{X}'$ by the sth column of $\mathbf{AX}$. Equation (2.4.10) shows that the rth row of $\mathbf{X}'$ is simply $\mathbf{\Psi}'^{(r)}$ and that the sth column of $\mathbf{AX}$ is

$$\begin{bmatrix} a_{11}\Psi_1^{(s)} + a_{12}\Psi_2^{(s)} + \ldots + a_{1n}\Psi_n^{(s)} \\ a_{21}\Psi_1^{(s)} + a_{22}\Psi_2^{(s)} + \ldots + a_{2n}\Psi_n^{(s)} \\ \cdots\cdots\cdots\cdots\cdots\cdots \\ a_{n1}\Psi_1^{(s)} + a_{n2}\Psi_2^{(s)} + \ldots + a_{nn}\Psi_n^{(s)} \end{bmatrix} \tag{2.5.9}$$

which the reader will recognise as the column matrix $\mathbf{A\Psi}^{(s)}$. Therefore the element in the rth row and sth column of $\mathbf{X'AX}$ is $\mathbf{\Psi}'^{(r)}\mathbf{A}\mathbf{\Psi}^{(s)}$. But we have shown in equation (2.5.7) that $\mathbf{\Psi}'^{(r)}\mathbf{A}\mathbf{\Psi}^{(s)} = 0$ if $r \neq s$ so that

$$\mathbf{X'AX} = \mathbf{L}, \tag{2.5.10}$$

where $\mathbf{L}$ is a diagonal matrix. If this diagonal matrix is denoted by

$$\mathbf{L} = \begin{bmatrix} a_1 & 0 & \ldots & 0 \\ 0 & a_2 & \ldots & 0 \\ \cdots & \cdots & \cdots & \cdots \\ 0 & 0 & \ldots & a_n \end{bmatrix}, \tag{2.5.11}$$

then the quantities $a_1, a_2, \ldots, a_n$ will be given by the equations

$$a_r = \mathbf{\Psi}'^{(r)}\mathbf{A}\mathbf{\Psi}^{(r)}. \tag{2.5.12}$$

Since $\mathbf{A}$ is positive definite these quantities a_r will all be strictly positive.

By the same sort of reasoning it may be shown that

$$\mathbf{X'CX} = \mathbf{N}, \tag{2.5.13}$$

where $\mathbf{N}$ is the diagonal matrix

$$\mathbf{N} = \begin{bmatrix} c_1 & 0 & \ldots & 0 \\ 0 & c_2 & \ldots & 0 \\ \cdots & \cdots & \cdots & \cdots \\ 0 & 0 & \ldots & c_n \end{bmatrix}, \tag{2.5.14}$$

† The reader may verify that the $\mathbf{\Psi}$ matrices in the example at the end of §2.3 do indeed satisfy these relations.

each of whose elements is positive and whose rth element is

$$c_r = \mathbf{\Psi}'^{(r)}\mathbf{C}\mathbf{\Psi}^{(r)}. \tag{2.5.15}$$

In passing we note that equation (2.5.3) with $r = s$ shows that

$$c_r = \omega_r^2 a_r \quad (r = 1, 2, \ldots, n). \tag{2.5.16}$$

If the notation

$$\mathbf{\Omega} = \begin{bmatrix} \omega_1^2 & 0 & \ldots & 0 \\ 0 & \omega_2^2 & \ldots & 0 \\ \ldots & \ldots & \ldots & \ldots \\ 0 & 0 & \ldots & \omega_n^2 \end{bmatrix} \tag{2.5.17}$$

is introduced then the n equations (2.5.16) may be written as one matrix equation, viz.

$$\mathbf{N} = \mathbf{L\Omega}. \tag{2.5.18}$$

The Orthogonality Principle is usually expressed in the following form: 'The energies T and V of a conservative system, when it is distorted in two or more principal modes simultaneously, are equal to the sums of the energies that the system would have if it were distorted in each of the modes separately.' We shall now show that the two formulations of the principle are equivalent.

Suppose that the system is vibrating freely in the rth and sth modes simultaneously. The matrix q is of the form

$$\mathbf{q} = \mathbf{\Psi}^{(r)}\sin(\omega_r t+\theta_r) + \mathbf{\Psi}^{(s)}\sin(\omega_s t+\theta_s), \tag{2.5.19}$$

so that the kinetic energy T, given by equation (2.1.6), is

$$T = \tfrac{1}{2}[\omega_r \mathbf{\Psi}'^{(r)}\cos(\omega_r t+\theta_r) + \omega_s \mathbf{\Psi}'^{(s)}\cos(\omega_s t+\theta_s)]\,\mathbf{A} \\ \times [\omega_r \mathbf{\Psi}^{(r)}\cos(\omega_r t+\theta_r) + \omega_s \mathbf{\Psi}^{(s)}\cos(\omega_s t+\theta_s)] \tag{2.5.20}$$

$$= \tfrac{1}{2}\omega_r^2\cos^2(\omega_r t+\theta_r)\,\mathbf{\Psi}'^{(r)}\mathbf{A}\mathbf{\Psi}^{(r)} + \tfrac{1}{2}\omega_s^2\cos^2(\omega_s t+\theta_s)\,\mathbf{\Psi}'^{(s)}\mathbf{A}\mathbf{\Psi}^{(s)} \\ +\omega_r\omega_s\cos(\omega_r t+\theta_r)\cos(\omega_s t+\theta_s)\{\mathbf{\Psi}'^{(r)}\mathbf{A}\mathbf{\Psi}^{(s)} + \mathbf{\Psi}'^{(s)}\mathbf{A}\mathbf{\Psi}^{(r)}\}. \tag{2.5.21}$$

Equation (2.5.7) shows that the last two terms are both zero, so that

$$\left.\begin{aligned} T &= \tfrac{1}{2}\omega_r^2\cos^2(\omega_r t+\theta_r)\,\mathbf{\Psi}'^{(r)}\mathbf{A}\mathbf{\Psi}^{(r)} + \tfrac{1}{2}\omega_s^2\cos^2(\omega_s t+\theta_s)\,\mathbf{\Psi}'^{(s)}\mathbf{A}\mathbf{\Psi}^{(s)} \\ &= T^{(r)} + T^{(s)} \end{aligned}\right\} \tag{2.5.22}$$

proving the first part of the principle.

The second part of the principle, relating to the potential energies, follows similarly. For by equation (2.1.11)

$$V = \tfrac{1}{2}[\mathbf{\Psi}'^{(r)}\sin(\omega_r t+\theta_r) + \mathbf{\Psi}'^{(s)}\sin(\omega_s t+\theta_s)]\,\mathbf{C} \\ \times [\mathbf{\Psi}^{(r)}\sin(\omega_r t+\theta_r) + \mathbf{\Psi}^{(s)}\sin(\omega_s t+\theta_s)], \tag{2.5.23}$$

which when multiplied out is

$$V = \tfrac{1}{2}\sin^2(\omega_r t+\theta_r)\,\mathbf{\Psi}'^{(r)}\mathbf{C}\mathbf{\Psi}^{(r)} + \tfrac{1}{2}\sin^2(\omega_s t+\theta_s)\,\mathbf{\Psi}'^{(s)}\mathbf{C}\mathbf{\Psi}^{(s)} \\ +\sin(\omega_r t+\theta_r)\sin(\omega_s t+\theta_s)\{\mathbf{\Psi}'^{(r)}\mathbf{C}\mathbf{\Psi}^{(s)} + \mathbf{\Psi}'^{(s)}\mathbf{C}\mathbf{\Psi}^{(r)}\}. \tag{2.5.24}$$

Equation (2.5.8) shows that the last two terms are both zero, giving

$$V = \tfrac{1}{2}\sin^2(\omega_r t+\theta_r)\,\mathbf{\Psi}'^{(r)}\mathbf{C}\mathbf{\Psi}^{(r)} + \tfrac{1}{2}\sin^2(\omega_s t+\theta_s)\,\mathbf{\Psi}'^{(s)}\mathbf{C}\mathbf{\Psi}^{(s)} \tag{2.5.25}$$

$$= V^{(r)} + V^{(s)}. \tag{2.5.26}$$

EXAMPLES 2.5

1. Verify that the $\mathbf{\Psi}$ matrices in the example at the end of §2.3 satisfy the orthogonality relations.

2. It is required to find the inertia and stability matrices, the natural frequencies and the normal modes of a number of systems. The data given are as follows:

System (a)

$$\mathbf{A} = ml^2\begin{bmatrix} 2 & -1 & 3 \\ -1 & 1 & -2 \\ 3 & -2 & 16 \end{bmatrix}, \quad \mathbf{X} = \begin{bmatrix} 1 & 17 & 1 \\ 2 & 0 & 1 \\ -1 & -3 & 0 \end{bmatrix};$$

$$\omega_1 = \sqrt{(g/l)}, \quad \omega_2 = 1{\cdot}2\sqrt{(g/l)}, \quad \omega_3 = 5{\cdot}4\sqrt{(g/l)}.$$

System (b)

$$\mathbf{C} = mg/l\begin{bmatrix} 10 & ? & -17 \\ ? & 12 & -6 \\ -17 & -6 & 45 \end{bmatrix}, \quad \mathbf{\Psi}^{(1)} = \begin{bmatrix} 2 \\ 1 \\ 1 \end{bmatrix}, \quad \mathbf{\Psi}^{(3)} = \begin{bmatrix} -3 \\ 1 \\ 1 \end{bmatrix};$$

$$\omega_1 = \sqrt{(g/l)}, \quad \omega_2 = 2\sqrt{(g/l)}, \quad \omega_3 = 3\sqrt{(g/l)}.$$

System (c)

$$\mathbf{A} = I\begin{bmatrix} 3 & ? & ? \\ ? & 4 & ? \\ -3 & 8 & 18 \end{bmatrix}, \quad \mathbf{C} = k\begin{bmatrix} 34 & ? & ? \\ ? & 35 & ? \\ ? & ? & 97 \end{bmatrix},$$

$$\mathbf{\Psi}^{(1)} = \begin{bmatrix} 1 \\ -3 \\ 2 \end{bmatrix}, \quad \mathbf{\Psi}^{(2)} = \begin{bmatrix} 2 \\ -3 \\ 1 \end{bmatrix}.$$

In each case determine whether the data is consistent and, when it is, calculate the remaining quantities required; when it is not, specify the inconsistency.

3. Solve Ex. 2.3.3 by using equations (2.5.10), (2.5.13) and (2.5.18) and without solving any sets of linear equations.

2.6 The energy expressions in terms of principal coordinates

One of the advantages of working with principal coordinates is that the kinetic and potential energies then take particularly simple forms. To show that this is so we make use of the properties of the matrix $\mathbf{X}$ that we deduced in the previous section.

Equation (2.1.6) states that the kinetic energy of the system is

$$T = \tfrac{1}{2}\dot{\mathbf{q}}'\mathbf{A}\dot{\mathbf{q}}. \tag{2.6.1}$$

Since $\mathbf{X}$ is independent of time, equation (2.4.9) may be differentiated to give

$$\dot{\mathbf{q}} = \mathbf{X}\dot{\mathbf{p}}, \tag{2.6.2}$$ †

the transpose of which is

$$\dot{\mathbf{q}}' = \dot{\mathbf{p}}'\mathbf{X}'. \tag{2.6.3}$$

If equations (2.6.2) and (2.6.3) are substituted into (2.6.1) the resulting expression for T is

$$T = \tfrac{1}{2}\dot{\mathbf{p}}'\mathbf{X}'\mathbf{A}\mathbf{X}\dot{\mathbf{p}} = \tfrac{1}{2}\dot{\mathbf{p}}'\mathbf{L}\dot{\mathbf{p}} \tag{2.6.4}$$

on account of (2.5.10). When this last result is expanded it becomes

$$T = \tfrac{1}{2}(a_1\dot{p}_1^2 + a_2\dot{p}_2^2 + \ldots + a_n\dot{p}_n^2), \tag{2.6.5}$$

so that the kinetic energy expression now contains no product terms—only squares.

The expression for V may be found in a similar way. For

$$V = \tfrac{1}{2}\mathbf{q}'\mathbf{C}\mathbf{q}, \tag{2.6.6}$$

and thus

$$V = \tfrac{1}{2}\mathbf{p}'\mathbf{X}'\mathbf{C}\mathbf{X}\mathbf{p} = \tfrac{1}{2}\mathbf{p}'\mathbf{N}\mathbf{p} \tag{2.6.7}$$

$$= \tfrac{1}{2}(c_1 p_1^2 + c_2 p_2^2 + \ldots + c_n p_n^2), \tag{2.6.8}$$

so that all the product terms are again absent.

The advantage of having the expressions for T and V in this simple form will be seen when we deal with the receptances of the system in § 2.8. We conclude this section by showing that it is possible to get even simpler expressions for T and V.

In § 2.4 a method of affixing unit values to principal coordinates was described. This method entailed choosing the p's so that certain of the q's had specified values when $p_s = 1$ in the sth mode (see equation (2.4.8)). But there is another way of fixing unit values of the principal coordinates which does not involve reference to a particular generalised coordinate: this method we shall now describe.

The column matrices $\mathbf{\Psi}$ satisfy equations (2.5.7), (2.5.8), (2.5.12) and (2.5.15). The first two of these equations involve only the ratios of the elements $\Psi_s^{(r)}$ and cannot therefore be used to fix their magnitudes. But there are n equations of the type (2.5.12) each of which involves just one of the $\mathbf{\Psi}$ matrices. We may therefore choose the matrix $\mathbf{\Psi}^{(r)}$ so that

$$a_r = \mathbf{\Psi}'^{(r)}\mathbf{A}\mathbf{\Psi}^{(r)} = 1 \quad (r = 1, 2, \ldots, n), \tag{2.6.9}$$

which implies that

$$\mathbf{L} = \mathbf{I}, \tag{2.6.10}$$

where $\mathbf{I}$ denotes the unit matrix of order n. By combining these last two equations with equations (2.5.16) and (2.5.18) respectively we obtain

$$c_r = \omega_r^2 \quad (r = 1, 2, \ldots, n) \tag{2.6.11}$$

and

$$\mathbf{N} = \mathbf{\Omega}. \tag{2.6.12}$$

† This differentiation of the matrix equation as a whole is quite legitimate. For equation (2.4.9) gives n algebraic equations for the coordinates q_i and each of these equations may be differentiated. Equation (2.6.2) is merely the statement in matrix form of the n differentiated equations.

With these values for the matrices **L** and **N**, the expressions for T and V become

$$T = \tfrac{1}{2}\dot{\mathbf{p}}'\dot{\mathbf{p}} = \tfrac{1}{2}(\dot{p}_1^2+\dot{p}_2^2+\ldots+\dot{p}_n^2), \tag{2.6.13}$$

$$V = \tfrac{1}{2}\mathbf{p}'\mathbf{\Omega}\mathbf{p} = \tfrac{1}{2}(\omega_1^2 p_1^2+\omega_2^2 p_2^2+\ldots+\omega_n^2 p_n^2). \tag{2.6.14}$$

It should be pointed out, however, that this use of so-called 'normalised' principal coordinates has one major disadvantage. It is that although the constants a_r become equal to unity they still possess the dimensions of inertia; likewise, the coefficients c_r still retain the dimensions of stiffness and do not have the dimensions of (frequency)^2.

EXAMPLES 2.6

1. The potential energy of a system is

$$V = \tfrac{1}{2}mgl\{2\alpha^2+9\beta^2+3\gamma^2-2\alpha\beta+2\alpha\gamma+4\beta\gamma\}.$$

One of its natural frequencies is $2\surd(g/l)$ and the corresponding mode is $\alpha:\beta:\gamma = 2:5:-9$. If the normalised principal coordinate for the mode is denoted by p, find the value of α when the system vibrates in the mode and $p = \surd(ml^2)$.

2. If the data given in Ex. 2.5.2 (*b*) refer to coordinates x, y, z, find the normalised principal coordinates of the system. Is it possible to find them without knowing **A**?

3. If the principal coordinates referred to in the last exercise are denoted by p_1, p_2, p_3, find (*a*) the direct receptance at p_1; (*b*) the receptance between p_1 and p_2.

2.7 The inversion of the matrix X

In this section we return to a matter which was mentioned briefly in § 2.4. It was shown there that the column matrices **p** and **q** are related by the equation (2.4.9), viz.

$$\mathbf{q} = \mathbf{X}\mathbf{p}. \tag{2.7.1}$$

It was shown also that the solution of this equation for **p**—i.e. finding an expression for **p** in terms of **q**—may be effected by inverting the matrix **X**. For if $\boldsymbol{\xi} = \mathbf{X}^{-1}$ then

$$\mathbf{p} = \boldsymbol{\xi}\mathbf{q}. \tag{2.7.2}$$ †

In this section we shall assume that the matrix **X** is known. This means that the principal modes of the system have been found, and that unit values have been assigned to the principal coordinates in some way (see equation (2.4.8) or equation (2.6.9)). The problem is to find the principal coordinates **p** in terms of the generalised coordinates **q**.

The product **Xp** occurring in equation (2.7.1) is the column matrix

$$\begin{bmatrix} p_1\Psi_1^{(1)}+p_2\Psi_1^{(2)}+\ldots+p_n\Psi_1^{(n)} \\ p_1\Psi_2^{(1)}+p_2\Psi_2^{(2)}+\ldots+p_n\Psi_2^{(n)} \\ \cdots\cdots\cdots\cdots\cdots\cdots \\ p_1\Psi_n^{(1)}+p_2\Psi_n^{(2)}+\ldots+p_n\Psi_n^{(n)} \end{bmatrix}.$$

† In this section we shall actually find an explicit expression for $\mathbf{X}^{-1}$, and this will prove conclusively that **X** has an inverse. It may be noted, however, that it follows from equation (2.5.10) that

$$|\mathbf{X}'\mathbf{A}\mathbf{X}| = |\mathbf{X}'|.|\mathbf{A}|.|\mathbf{X}| = |\mathbf{L}|;$$

see Ex. 1.5.5. Since $|\mathbf{L}|$ and $|\mathbf{A}|$ are both non-zero, $|\mathbf{X}|$ is also non-zero; therefore **X** has an inverse.

This can be written as the sum of n column matrices, so that

$$\mathbf{q} = p_1 \begin{bmatrix} \Psi_1^{(1)} \\ \Psi_2^{(1)} \\ \vdots \\ \Psi_n^{(1)} \end{bmatrix} + p_2 \begin{bmatrix} \Psi_1^{(2)} \\ \Psi_2^{(2)} \\ \vdots \\ \Psi_n^{(2)} \end{bmatrix} + \dots + p_n \begin{bmatrix} \Psi_1^{(n)} \\ \Psi_2^{(n)} \\ \vdots \\ \Psi_n^{(n)} \end{bmatrix}, \tag{2.7.3}$$

or

$$\mathbf{q} = p_1 \mathbf{\Psi}^{(1)} + p_2 \mathbf{\Psi}^{(2)} + \dots + p_n \mathbf{\Psi}^{(n)}. \tag{2.7.4}$$

To find the coordinates p_r we premultiply this equation by $\mathbf{\Psi}'^{(r)}\mathbf{A}$ and use the Orthogonality Principle. The orthogonality relation (2.5.7) gives the result

$$p_r = \frac{\mathbf{\Psi}'^{(r)}\mathbf{A}\mathbf{q}}{\mathbf{\Psi}'^{(r)}\mathbf{A}\mathbf{\Psi}^{(r)}}. \tag{2.7.5†}$$

Consider the system of fig. 2.2.1 for which equations (2.4.5) give

$$\mathbf{\Psi}^{(1)} = \begin{bmatrix} 2{\cdot}000 \\ 3{\cdot}604 \\ 4{\cdot}494 \end{bmatrix}, \quad \mathbf{\Psi}^{(2)} = \begin{bmatrix} 1{\cdot}000 \\ 0{\cdot}445 \\ -0{\cdot}802 \end{bmatrix}, \quad \mathbf{\Psi}^{(3)} = \begin{bmatrix} 1{\cdot}000 \\ -1{\cdot}247 \\ 0{\cdot}555 \end{bmatrix}. \tag{2.7.6}$$

Suppose that the system is given the distortion

$$\mathbf{q} = \begin{bmatrix} 1 \\ 2 \\ 3 \end{bmatrix}. \tag{2.7.7}$$

Using the matrix $\mathbf{A}$ given in equation (2.3.12) we find that

$$p_1 = \frac{[2{\cdot}000 \quad 3{\cdot}604 \quad 4{\cdot}494] \begin{bmatrix} I & 0 & 0 \\ 0 & I & 0 \\ 0 & 0 & I \end{bmatrix} \begin{bmatrix} 1 \\ 2 \\ 3 \end{bmatrix}}{[2{\cdot}000 \quad 3{\cdot}604 \quad 4{\cdot}494] \begin{bmatrix} I & 0 & 0 \\ 0 & I & 0 \\ 0 & 0 & I \end{bmatrix} \begin{bmatrix} 2{\cdot}000 \\ 3{\cdot}604 \\ 4{\cdot}494 \end{bmatrix}}$$

$$= \frac{22{\cdot}690I}{37{\cdot}185I} = 0{\cdot}610. \tag{2.7.8}$$

In the same way it may be shown that

$$p_2 = -0{\cdot}279; \quad p_3 = 0{\cdot}060. \tag{2.7.9}$$

Alternatively we may premultiply equation (2.7.4) by $\mathbf{\Psi}'^{(r)}\mathbf{C}$ and use the orthogonality relation (2.5.8). This gives the result

$$p_r = \frac{\mathbf{\Psi}'^{(r)}\mathbf{C}\mathbf{q}}{\mathbf{\Psi}'^{(r)}\mathbf{C}\mathbf{\Psi}^{(r)}}, \tag{2.7.10}$$

† The reader who is familiar with Fourier analysis will notice the similarity between the way the coordinates p_r are obtained and the way the Fourier coefficients are found.

which will be found to give results which agree with those obtained from equation (2.7.5).

In the example above the Principle of Orthogonality provided a means of evaluating the coordinates p_1, p_2 and p_3 without inverting the matrix **X**. For this example equations (2.7.6) show that

$$\mathbf{X} = \begin{bmatrix} 2\cdot000 & 1\cdot000 & 1\cdot000 \\ 3\cdot604 & 0\cdot445 & -1\cdot247 \\ 4\cdot494 & -0\cdot802 & 0\cdot555 \end{bmatrix}. \tag{2.7.11}$$

It is a somewhat tedious business to invert this matrix (particularly by the method which is described in § 1.6). However, if the inversion is performed, it is found that

$$\mathbf{X}^{-1} = \boldsymbol{\xi} = -\frac{1}{14\cdot00}\begin{bmatrix} -0\cdot753 & -1\cdot357 & -1\cdot692 \\ -7\cdot604 & -3\cdot384 & 6\cdot098 \\ -4\cdot890 & 6\cdot094 & -2\cdot714 \end{bmatrix}. \tag{2.7.12}$$

Equation (2.7.2) states that the column matrix **p** is the product of this inverse and the matrix of (2.7.7). If this product is formed it is found to agree with the results (2.7.8) and (2.7.9).

We conclude this section by showing that instead of using the Orthogonality Principle to by-pass the inversion of **X** we may in fact use it to obtain an expression for the inverse. The inverse will be derived without recourse to the method of § 1.6.

The inverse $\boldsymbol{\xi}$ of **X** may be obtained either from equation (2.5.10) or from (2.5.13). Equation (2.5.10) is

$$\mathbf{L} = \mathbf{X}'\mathbf{A}\mathbf{X}, \tag{2.7.13}$$

and if this is postmultiplied by $\boldsymbol{\xi}$ it is found that

$$\mathbf{L}\boldsymbol{\xi} = \mathbf{X}'\mathbf{A}. \tag{2.7.14}$$

To show that we have here a means of evaluating the elements of $\boldsymbol{\xi}$ it is perhaps simplest to write these matrices out in full. We have, for the left-hand member,

$$\begin{bmatrix} a_1 & 0 & \dots & 0 \\ 0 & a_2 & \dots & 0 \\ \dots & \dots & \dots & \dots \\ 0 & 0 & \dots & a_n \end{bmatrix}\begin{bmatrix} \xi_{11} & \xi_{12} & \dots & \xi_{1n} \\ \xi_{21} & \xi_{22} & \dots & \xi_{2n} \\ \dots & \dots & \dots & \dots \\ \xi_{n1} & \xi_{n2} & \dots & \xi_{nn} \end{bmatrix},$$

which reduces to a square matrix having $a_r\xi_{rs}$ as the element in its rth row and sth column. The corresponding element in the square matrix $\mathbf{X}'\mathbf{A}$ is found by multiplying the elements of the rth row of $\mathbf{X}'$ by those of the sth column of **A**, that is, by forming the product

$$[\Psi_1^{(r)} \quad \Psi_2^{(r)} \quad \dots \quad \Psi_n^{(r)}]\begin{bmatrix} a_{1s} \\ a_{2s} \\ \vdots \\ a_{ns} \end{bmatrix}.$$

Therefore
$$a_r \xi_{rs} = a_{1s} \Psi_1^{(r)} + a_{2s} \Psi_2^{(r)} + \dots + a_{ns} \Psi_n^{(r)}. \tag{2.7.15}$$

But an expression for a_r may be found from equation (2.5.12), namely

$$a_r = a_{11}[\Psi_1^{(r)}]^2 + a_{22}[\Psi_2^{(r)}]^2 + \dots + a_{nn}[\Psi_n^{(r)}]^2 + 2a_{12}\Psi_1^{(r)}\Psi_2^{(r)} + \dots + 2a_{n-1,n}\Psi_{n-1}^{(r)}\Psi_n^{(r)}, \tag{2.7.16}$$

so that

$$\xi_{rs} = \frac{a_{1s}\Psi_1^{(r)} + a_{2s}\Psi_2^{(r)} + \dots + a_{ns}\Psi_n^{(r)}}{a_{11}[\Psi_1^{(r)}]^2 + a_{22}[\Psi_2^{(r)}]^2 + \dots + a_{nn}[\Psi_n^{(r)}]^2 + 2a_{12}\Psi_1^{(r)}\Psi_2^{(r)} + \dots + 2a_{n-1,n}\Psi_{n-1}^{(r)}\Psi_n^{(r)}}. \tag{2.7.17}$$

Equation (2.7.14) also gives an equation for the whole matrix $\boldsymbol{\xi}$; this is

$$\boldsymbol{\xi} = \mathbf{L}^{-1}\mathbf{X}'\mathbf{A}, \tag{2.7.18}$$

where the matrix $\mathbf{L}^{-1}$ is simply

$$\mathbf{L}^{-1} = \begin{bmatrix} a_1^{-1} & 0 & \dots & 0 \\ 0 & a_2^{-1} & \dots & 0 \\ \dots & \dots & \dots & \dots \\ 0 & 0 & \dots & a_n^{-1} \end{bmatrix}. \tag{2.7.19}$$

If we take the alternative course and start from equation (2.5.13), viz.

$$\mathbf{N} = \mathbf{X}'\mathbf{C}\mathbf{X}, \tag{2.7.20}$$

then postmultiplication by $\boldsymbol{\xi}$ gives

$$\mathbf{N}\boldsymbol{\xi} = \mathbf{X}'\mathbf{C}. \tag{2.7.21}$$

This gives an alternative expression for ξ_{rs} which is the same as (2.7.17) except that the a's are replaced by the c's. There is also an alternative equation for $\boldsymbol{\xi}$, namely

$$\boldsymbol{\xi} = \mathbf{N}^{-1}\mathbf{X}'\mathbf{C}. \tag{2.7.22}$$

EXAMPLES 2.7

1. Invert the matrix (2.7.11) by using the method described in the section. (The relevant matrices **A** and **C** are given in equation (2.3.12).)

2. If the data of Ex. 2.5.2 (c) refer to coordinates q_1, q_2, q_3, find the value of the principal coordinates when (a) $q_1 = 1$, $q_2 = -2$, $q_3 = 1$; (b) $q_1 = 2$, $q_2 = -3$, $q_3 = 1$.

3. Use equation (2.7.22) to solve Ex. 2.6.2.

2.8 The series form of receptances

Section 2.2 is concerned with the response of a system to a harmonic excitation $\boldsymbol{\Phi} \sin \omega t$. This response is expressed in matrix form in equation (2.2.12);

$$\mathbf{q} = \boldsymbol{\alpha}\boldsymbol{\Phi} \sin \omega t. \tag{2.8.1}$$

Here, $\boldsymbol{\alpha}$ is the receptance matrix for whose elements—the 'receptances'—we found an explicit expression in equation (2.2.15). In this section we are concerned with a way of expanding a receptance as a series of partial fractions. It is probably true to say that the potentialities of this series form in practical engineering work are far greater than those of the closed form mentioned before. In a practical problem the labour required to extract these partial fractions directly from the expressions given in § 2.2 would usually be prohibitive.† For this reason we shall employ an entirely different approach.

In § 2.2 we obtained the equation (2.2.8), namely

$$(\mathbf{C}-\omega^2\mathbf{A})\,\boldsymbol{\Psi} = \boldsymbol{\Phi}, \tag{2.8.2}$$

linking the amplitude $\boldsymbol{\Psi}$ of $\mathbf{q}$ with the amplitude $\boldsymbol{\Phi}$ of the column matrix $\mathbf{Q}$ representing the generalised forces. The solution of this equation, giving $\boldsymbol{\Psi}$ in terms of $\boldsymbol{\Phi}$ is

$$\boldsymbol{\Psi} = \boldsymbol{\alpha}\boldsymbol{\Phi}. \tag{2.8.3}$$

We shall now derive an expression for $\boldsymbol{\alpha}$ by making use of the principal coordinates of the system. Since $\mathbf{q}$ is related to $\mathbf{p}$, the vector of principal coordinates, by the equation (2.4.9), namely

$$\mathbf{q} = \mathbf{X}\mathbf{p}, \tag{2.8.4}$$

it follows that the corresponding amplitudes $\boldsymbol{\Psi}$ and $\boldsymbol{\Pi}$, say, are related by the equation

$$\boldsymbol{\Psi} = \mathbf{X}\boldsymbol{\Pi}. \tag{2.8.5}$$

Substituting this expression for $\boldsymbol{\Psi}$ into equation (2.8.2) and premultiplying the whole by $\mathbf{X}'$ we obtain

$$\mathbf{X}'(\mathbf{C}-\omega^2\mathbf{A})\,\mathbf{X}\boldsymbol{\Pi} = \mathbf{X}'\boldsymbol{\Phi}. \tag{2.8.6}$$

But by equations (2.5.10) and (2.5.13)

$$\mathbf{X}'\mathbf{A}\mathbf{X} = \mathbf{L}; \quad \mathbf{X}'\mathbf{C}\mathbf{X} = \mathbf{N}; \tag{2.8.7}$$

and therefore

$$(\mathbf{N}-\omega^2\mathbf{L})\,\boldsymbol{\Pi} = \mathbf{X}'\boldsymbol{\Phi}. \tag{2.8.8}$$

Let us pause for a moment to interpret this last equation. It is the equation for the principal coordinates corresponding to equation (2.8.2) for the generalised coordinates. This follows from the fact that for the principal coordinates, the matrices $\mathbf{A}$ and $\mathbf{C}$ are replaced by $\mathbf{L}$ and $\mathbf{N}$ respectively (see equations (2.6.4) and (2.6.7)). The matrix on the right of equation (2.8.2) is the matrix of the amplitudes of the forces Q_r. Therefore that on the right of equation (2.8.8) must be the matrix of amplitudes for the forces corresponding to the principal coordinates. If these forces are given by the equation

$$\mathbf{P} = \begin{bmatrix} P_1 \\ P_2 \\ \vdots \\ P_n \end{bmatrix} = \begin{bmatrix} \Xi_1 \\ \Xi_2 \\ \vdots \\ \Xi_n \end{bmatrix} \sin\omega t = \boldsymbol{\Xi}\sin\omega t, \tag{2.8.9}$$

then it follows that

$$\boldsymbol{\Xi} = \mathbf{X}'\boldsymbol{\Phi}. \tag{2.8.10}$$

If, further, the matrix $\mathbf{N}-\omega^2\mathbf{L}$ is written

$$\mathbf{N}-\omega^2\mathbf{L} = \boldsymbol{\zeta}, \tag{2.8.11}$$

† The reader may wish to investigate this for one of the simple receptances given in equation (2.2.25).

then
$$\boldsymbol{\zeta}\boldsymbol{\Pi} = \boldsymbol{\Xi}. \tag{2.8.12}$$

If ω is not a natural frequency, so that $\boldsymbol{\zeta}$ is non-singular, the solution of the last equation is
$$\boldsymbol{\Pi} = \boldsymbol{\zeta}^{-1}\boldsymbol{\Xi}. \tag{2.8.13}$$

This is the equation for the principal coordinates corresponding to equation (2.8.3) for the generalised coordinates. In other words, $\boldsymbol{\zeta}^{-1}$ is the receptance matrix for the principal coordinates. But the matrix $\boldsymbol{\zeta}$ is the diagonal matrix whose rth term is
$$c_r - \omega^2 a_r = a_r(\omega_r^2 - \omega^2). \tag{2.8.14}$$

The matrix $\boldsymbol{\zeta}^{-1}$ is therefore
$$\boldsymbol{\zeta}^{-1} = \begin{bmatrix} \alpha_1 & 0 & \dots & 0 \\ 0 & \alpha_2 & \dots & 0 \\ \dots & \dots & \dots & \dots \\ 0 & 0 & \dots & \alpha_n \end{bmatrix}, \tag{2.8.15}$$

where
$$\alpha_r = \frac{1}{a_r(\omega_r^2 - \omega^2)}, \tag{2.8.16}$$

as the reader may easily verify. The term α_r is therefore the (direct and only) receptance at the rth principal coordinate.

Returning to equation (2.8.8) we see that its solution is
$$\boldsymbol{\Pi} = \boldsymbol{\zeta}^{-1}\mathbf{X}'\boldsymbol{\Phi}, \tag{2.8.17}$$

which gives
$$\boldsymbol{\Psi} = \mathbf{X}\boldsymbol{\zeta}^{-1}\mathbf{X}'\boldsymbol{\Phi}. \tag{2.8.18}$$

Comparison of this equation with (2.8.3) now shows that
$$\boldsymbol{\alpha} = \mathbf{X}\boldsymbol{\zeta}^{-1}\mathbf{X}'. \tag{2.8.19}$$

This equation, which expresses the receptance matrix $\boldsymbol{\alpha}$ in terms of the known matrices, $\mathbf{X}$, $\boldsymbol{\zeta}^{-1}$ and $\mathbf{X}'$, may be rewritten in terms of the matrices $\boldsymbol{\Psi}^{(r)}$. By using the expressions for $\mathbf{X}$ and $\mathbf{X}'$ given by (2.4.10) it may be verified that
$$\boldsymbol{\alpha} = \alpha_1\boldsymbol{\Psi}^{(1)}\boldsymbol{\Psi}'^{(1)} + \alpha_2\boldsymbol{\Psi}^{(2)}\boldsymbol{\Psi}'^{(2)} + \dots + \alpha_n\boldsymbol{\Psi}^{(n)}\boldsymbol{\Psi}'^{(n)}, \tag{2.8.20}$$

and, on account of equation (2.8.16), this is
$$\boldsymbol{\alpha} = \frac{\boldsymbol{\Psi}^{(1)}\boldsymbol{\Psi}'^{(1)}}{a_1(\omega_1^2 - \omega^2)} + \frac{\boldsymbol{\Psi}^{(2)}\boldsymbol{\Psi}'^{(2)}}{a_2(\omega_2^2 - \omega^2)} + \dots + \frac{\boldsymbol{\Psi}^{(n)}\boldsymbol{\Psi}'^{(n)}}{a_n(\omega_n^2 - \omega^2)}. \tag{2.8.21}$$

Each term in this equation, being a column matrix postmultiplied by a row matrix, is a square matrix. The individual receptance α_{rs}—the term in the rth row and sth column of $\boldsymbol{\alpha}$—may be found by expanding the right-hand side of equation (2.8.21). It is found to be
$$\alpha_{rs} = \frac{\Psi_r^{(1)}\Psi_s^{(1)}}{a_1(\omega_1^2 - \omega^2)} + \frac{\Psi_r^{(2)}\Psi_s^{(2)}}{a_2(\omega_2^2 - \omega^2)} + \dots + \frac{\Psi_r^{(n)}\Psi_s^{(n)}}{a_n(\omega_n^2 - \omega^2)}. \tag{2.8.22}$$

This expression for α_{rs} may be written in an alternative way by using equation (2.4.9). That equation shows (as may be seen by reference to (2.4.10)) that

$$q_r = \sum_{t=1}^{n} \Psi_r^{(t)} p_t, \tag{2.8.23}$$

and therefore

$$\Psi_r^{(t)} = \frac{\partial q_r}{\partial p_t}. \tag{2.8.24}$$

The receptance α_{rs} is thus

$$\alpha_{rs} = \frac{\dfrac{\partial q_r}{\partial p_1}\dfrac{\partial q_s}{\partial p_1}}{a_1(\omega_1^2-\omega^2)} + \frac{\dfrac{\partial q_r}{\partial p_2}\dfrac{\partial q_s}{\partial p_2}}{a_2(\omega_2^2-\omega^2)} + \ldots + \frac{\dfrac{\partial q_r}{\partial p_n}\dfrac{\partial q_s}{\partial p_n}}{a_n(\omega_n^2-\omega^2)}. \tag{2.8.25}$$

This last result may be extended slightly so that it holds for any two independently chosen generalised coordinates x and y. Suppose that a harmonic generalised force $F \sin \omega t$ is applied to the system at the generalised coordinate y and that the resulting motion is required at the generalised coordinate x. Then

$$x = \alpha_{xy} F \sin \omega t, \tag{2.8.26}$$

where

$$\alpha_{xy} = \frac{\dfrac{\partial x}{\partial p_1}\dfrac{\partial y}{\partial p_1}}{a_1(\omega_1^2-\omega^2)} + \frac{\dfrac{\partial x}{\partial p_2}\dfrac{\partial y}{\partial p_2}}{a_2(\omega_2^2-\omega^2)} + \ldots + \frac{\dfrac{\partial x}{\partial p_n}\dfrac{\partial y}{\partial p_n}}{a_n(\omega_n^2-\omega^2)}. \tag{2.8.27}$$

This result, which is suggested by identifying q_r and q_s in (2.8.25) with x and y respectively, may be proved by combining equations (2.8.25) with (2.2.35).

Let us return to equation (2.8.21), which is of some interest. When combined with (2.8.3) it shows that

$$\mathbf{\Psi} = \frac{\mathbf{\Psi}^{(1)}\mathbf{\Psi}'^{(1)}\mathbf{\Phi}}{a_1(\omega_1^2-\omega^2)} + \frac{\mathbf{\Psi}^{(2)}\mathbf{\Psi}'^{(2)}\mathbf{\Phi}}{a_2(\omega_2^2-\omega^2)} + \ldots + \frac{\mathbf{\Psi}^{(n)}\mathbf{\Psi}'^{(n)}\mathbf{\Phi}}{a_n(\omega_n^2-\omega^2)}, \tag{2.8.28}$$

which may be written

$$\mathbf{\Psi} = \mu_1 \mathbf{\Psi}^{(1)} + \mu_2 \mathbf{\Psi}^{(2)} + \ldots + \mu_n \mathbf{\Psi}^{(n)}, \tag{2.8.29}$$

where the μ's are numbers given by

$$\mu_r = \frac{\mathbf{\Psi}'^{(r)}\mathbf{\Phi}}{a_r(\omega_r^2-\omega^2)}. \tag{2.8.30}$$

Equation (2.8.29) gives the intensities with which the principal modes occur in the total forced motion; μ_r is a measure of the intensity of the contribution in the rth principal mode. The expression for μ_r resembles that for the amplitude of the motion of a system which has a single degree of freedom and whose one natural frequency is ω_r.

EXAMPLES 2.8

1. Find a series expression for the receptance α_{12} corresponding to the data of Ex. 2.5.2 (*c*). Compare the result with that of Ex. 2.2.3.

2. Using the natural frequencies given in equation (2.3.15), express the receptance α_{11} given in equation (2.2.25) as a series of partial fractions.

3. A system has natural frequencies

$$\omega_1 = \sqrt{(g/l)},\quad \omega_2 = 2\sqrt{(g/l)},\quad \omega_3 = 5\sqrt{(g/l)},\quad \omega_4 = 6{\cdot}4\sqrt{(g/l)},$$

and when they are expressed in terms of coordinates q_1, q_2, q_3, q_4 its corresponding normalised principal modes are the columns of

$$\mathbf{X} = \sqrt{(1/m)}\begin{bmatrix} 1 & 2 & 7 & 1 \\ 2 & 3 & -2 & 6 \\ -1 & 1 & 1 & 2 \\ 4 & 5 & 3 & -1 \end{bmatrix}.$$

Find (*a*) the direct receptance at q_1; (*b*) the receptance between $q_1 - 2q_2 + q_3$ and $q_1 - q_4$.

4. (*a*) Derive equation (2.8.20) from (2.8.19), and (*b*) show that the value of the receptance matrix is independent of the scale factors with which the principal coordinates are chosen.

2.9 Forced motion; principal modes of excitation

This section is devoted to a brief discussion of an alternative way of tackling the problem of forced motion. We first look for sets of forces $Q_1, Q_2, \ldots, Q_n$ of a given frequency ω which produce a displacement in one of the principal modes of vibration of the system. Then, having found n such sets of forces we express an arbitrary set of forces in terms of them. Clearly this procedure is analogous to the way in which we earlier expressed an arbitrary set of displacements in terms of its components in the principal modes of vibration. When a system of forces produces a displacement only in the rth principal mode we shall say that the system is being driven by the rth 'principal mode of excitation'.

Suppose that a set of forces given by

$$\mathbf{Q} = \begin{bmatrix} Q_1 \\ Q_2 \\ \vdots \\ Q_n \end{bmatrix} = \begin{bmatrix} \Phi_1 \\ \Phi_2 \\ \vdots \\ \Phi_n \end{bmatrix} \sin \omega t = \mathbf{\Phi} \sin \omega t \tag{2.9.1}$$

produces a displacement in the rth principal mode. Then, since p_r is the only non-zero principal coordinate, equations (2.4.9) and (2.4.10) show that

$$\mathbf{q} = \mathbf{\Psi} \sin \omega t = \Pi_r \mathbf{\Psi}^{(r)} \sin \omega t, \tag{2.9.2}$$

where Π_r is a constant (the amplitude of p_r). Equation (2.2.8) gives

$$\mathbf{\Phi} = (\mathbf{C} - \omega^2 \mathbf{A})\,\mathbf{\Psi} = (\mathbf{C} - \omega^2 \mathbf{A})\,\Pi_r \mathbf{\Psi}^{(r)} = (\omega_r^2 - \omega^2)\,\Pi_r \mathbf{A}\mathbf{\Psi}^{(r)} \tag{2.9.3}$$

on account of (2.3.10).

Now let us select the intensity of the excitation by putting

$$\Pi_r = \frac{1}{\omega_r^2 - \omega^2}, \tag{2.9.4}$$

and let us denote the corresponding value of $\mathbf{\Phi}$ by $\mathbf{\Phi}^{(r)}$. Then

$$\mathbf{\Phi}^{(r)} = \mathbf{A\Psi}^{(r)} \tag{2.9.5}$$

and the force $\mathbf{\Phi}^{(r)} \sin \omega t$ produces the displacement

$$\mathbf{q} = \frac{\mathbf{\Psi}^{(r)} \sin \omega t}{\omega_r^2 - \omega^2}. \tag{2.9.6}$$

Just as we spoke of the column matrix $\mathbf{\Psi}^{(r)}$ as defining the rth principal mode of vibration so we may speak of $\mathbf{\Phi}^{(r)}$ as defining the rth principal mode of excitation.

To obtain an expression for an arbitrary matrix $\mathbf{\Phi}$ in terms of the matrices $\mathbf{\Phi}^{(r)}$ we start from equation (2.8.29), which states that

$$\mathbf{\Psi} = \mu_1 \mathbf{\Psi}^{(1)} + \mu_2 \mathbf{\Psi}^{(2)} + \ldots + \mu_n \mathbf{\Psi}^{(n)}, \tag{2.9.7}$$

where
$$\mu_r = \frac{\mathbf{\Psi}'^{(r)}\mathbf{\Phi}}{a_r(\omega_r^2 - \omega^2)}. \tag{2.9.8}$$

Premultiply equation (2.9.7) by $(\mathbf{C} - \omega^2\mathbf{A})$; this yields

$$(\mathbf{C} - \omega^2\mathbf{A})\,\mathbf{\Psi} = \mu_1(\mathbf{C} - \omega^2\mathbf{A})\,\mathbf{\Psi}^{(1)} + \mu_2(\mathbf{C} - \omega^2\mathbf{A})\,\mathbf{\Psi}^{(2)} + \ldots + \mu_n(\mathbf{C} - \omega^2\mathbf{A})\,\mathbf{\Psi}^{(n)}, \tag{2.9.9}$$

or
$$\mathbf{\Phi} = \mu_1(\omega_1^2 - \omega^2)\,\mathbf{\Phi}^{(1)} + \mu_2(\omega_2^2 - \omega^2)\,\mathbf{\Phi}^{(2)} + \ldots + \mu_n(\omega_n^2 - \omega^2)\,\mathbf{\Phi}^{(n)} \tag{2.9.10}$$

which is the required result. The significance of this equation may perhaps be seen best if μ_r is expressed in terms of $\mathbf{\Phi}^{(r)}$. Clearly

$$\mu_r = \frac{\mathbf{\Psi}'^{(r)}\{(\mathbf{C} - \omega^2\mathbf{A})\,\mathbf{\Psi}\}}{a_r(\omega_r^2 - \omega^2)} = \frac{\{\mathbf{\Psi}'^{(r)}(\mathbf{C} - \omega^2\mathbf{A})\}\,\mathbf{\Psi}}{a_r(\omega_r^2 - \omega^2)}$$
$$= \frac{(\omega_r^2 - \omega^2)\,\{\mathbf{\Psi}'^{(r)}\mathbf{A}\}\,\mathbf{\Psi}}{a_r(\omega_r^2 - \omega^2)} = \frac{\mathbf{\Phi}'^{(r)}\mathbf{\Psi}}{a_r}, \tag{2.9.11}$$

where, in the last step, we have used the transpose of equation (2.9.5).

Therefore

$$\mathbf{\Phi} = \frac{(\omega_1^2 - \omega^2)}{a_1}\,\mathbf{\Phi}^{(1)}\mathbf{\Phi}'^{(1)}\mathbf{\Psi} + \frac{(\omega_2^2 - \omega^2)}{a_2}\,\mathbf{\Phi}^{(2)}\mathbf{\Phi}'^{(2)}\mathbf{\Psi} + \ldots + \frac{(\omega_n^2 - \omega^2)}{a_n}\,\mathbf{\Phi}^{(n)}\mathbf{\Phi}'^{(n)}\mathbf{\Psi}, \tag{2.9.12}$$

which may be compared with equation (2.8.28). Equation (2.9.12) gives $\mathbf{\Phi}$ in terms of $\mathbf{\Psi}$ and must therefore be equivalent to equation (2.2.8). Therefore

$$\mathbf{Z} = \mathbf{C} - \omega^2\mathbf{A} = \frac{\omega_1^2 - \omega^2}{a_1}\,\mathbf{\Phi}^{(1)}\mathbf{\Phi}'^{(1)} + \frac{\omega_2^2 - \omega^2}{a_2}\,\mathbf{\Phi}^{(2)}\mathbf{\Phi}'^{(2)} + \ldots + \frac{\omega_n^2 - \omega^2}{a_n}\,\mathbf{\Phi}^{(n)}\mathbf{\Phi}'^{(n)}, \tag{2.9.13}$$

which may be compared with equation (2.8.21).

We shall now derive some results for the matrices $\mathbf{\Phi}^{(r)}$ similar to those already obtained for the matrices $\mathbf{\Psi}^{(r)}$. First, the Principle of Orthogonality may be expressed in terms of the matrices $\mathbf{\Phi}^{(r)}$. The part of the Principle referring to the kinetic energy of the system is expressed in the equation

$$\mathbf{\Psi}'^{(r)}\mathbf{A\Psi}^{(s)} = \begin{Bmatrix} a_r & \text{if} & r = s \\ 0 & \text{if} & r \neq s \end{Bmatrix} = a_r\,\delta_{rs}, \tag{2.9.14}$$

obtained by combining equations (2.5.7) and (2.5.12). By using equation (2.9.5) and its transpose we may write

$$\boldsymbol{\Psi}'^{(r)}\mathbf{A}\boldsymbol{\Psi}^{(s)} = (\boldsymbol{\Psi}'^{(r)}\mathbf{A})\,\mathbf{A}^{-1}(\mathbf{A}\boldsymbol{\Psi}^{(s)}) = \boldsymbol{\Phi}'^{(r)}\mathbf{A}^{-1}\boldsymbol{\Phi}^{(s)}, \tag{2.9.15}$$

so that an alternative expression for the first part of the Principle is

$$\boldsymbol{\Phi}'^{(r)}\mathbf{A}^{-1}\boldsymbol{\Phi}^{(s)} = a_r\delta_{rs}. \tag{2.9.16}$$

There is a similar expression for the second part of the Principle; it is

$$\boldsymbol{\Phi}'^{(r)}\mathbf{C}^{-1}\boldsymbol{\Phi}^{(s)} = \frac{c_r\delta_{rs}}{\omega_r^2\omega_s^2} = \frac{a_r}{\omega_r^2}\delta_{rs} \tag{2.9.17}$$

as the reader may prove for himself as an exercise.

Just as the matrix $\mathbf{X}$ was formed having the matrices $\boldsymbol{\Psi}^{(r)}$ as columns so we may form a matrix $\mathbf{Y}$ whose columns are the $\boldsymbol{\Phi}^{(r)}$ matrices. Equation (2.9.5) shows that $\mathbf{Y}$ is related to $\mathbf{X}$ by the equation

$$\mathbf{Y} = \mathbf{AX}. \tag{2.9.18}$$

Expressed in terms of $\mathbf{Y}$ the orthogonality relations (2.9.16) and (2.9.17) become

$$\mathbf{Y}'\mathbf{A}^{-1}\mathbf{Y} = \mathbf{L}; \quad \mathbf{Y}'\mathbf{C}^{-1}\mathbf{Y} = \mathbf{L}\boldsymbol{\Omega}^{-1}, \tag{2.9.19}$$

where, in the second equation, $\boldsymbol{\Omega}^{-1}$ is the inverse of the matrix given in (2.5.17).

The first of equations (2.9.19) may be written in the form

$$\mathbf{Y}'\mathbf{X} = \mathbf{L}, \tag{2.9.20}$$

which, on premultiplication by $\mathbf{L}^{-1}$, yields

$$\mathbf{L}^{-1}\mathbf{Y}'\mathbf{X} = \mathbf{I}. \tag{2.9.21}$$

This equation shows that

$$\mathbf{X}^{-1} = \mathbf{L}^{-1}\mathbf{Y}', \tag{2.9.22}$$

which, when the principal coordinates are normalised, becomes

$$\mathbf{X}^{-1} = \mathbf{Y}'. \tag{2.9.23}$$

Equation (2.9.22) may be used to find the equivalent of equation (2.9.13) for the matrices $\mathbf{Y}$ and $\mathbf{Y}'$. For the equation

$$\mathbf{X}'(\mathbf{C}-\omega^2\mathbf{A})\,\mathbf{X} = \mathbf{N}-\omega^2\mathbf{L}, \tag{2.9.24}$$

which follows from equations (2.5.10) and (2.5.13), gives

$$\mathbf{C}-\omega^2\mathbf{A} = (\mathbf{X}')^{-1}(\mathbf{N}-\omega^2\mathbf{L})\,\mathbf{X}^{-1} = \mathbf{YL}^{-1}(\mathbf{N}-\omega^2\mathbf{L})\,\mathbf{L}^{-1}\mathbf{Y}'. \tag{2.9.25}$$

This equation, when expanded, yields (2.9.13) because the matrix $\mathbf{L}^{-1}(\mathbf{N}-\omega^2\mathbf{L})\,\mathbf{L}^{-1}$ is the diagonal matrix with rth term

$$\frac{\omega_r^2-\omega^2}{a_r}.$$

Equation (2.9.25) may be written

$$\mathbf{Z} = \mathbf{C}-\omega^2\mathbf{A} = \mathbf{YL}^{-1}\boldsymbol{\zeta}\mathbf{L}^{-1}\mathbf{Y}', \tag{2.9.26}$$

and when the principal coordinates are normalised this becomes

$$\mathbf{Z} = \mathbf{Y}\boldsymbol{\zeta}\mathbf{Y}', \tag{2.9.27}$$

which may be compared with (2.8.19).

EXAMPLES 2.9

1. Find the principal modes of excitation of the system of Ex. 2.3.3. Is it possible to find them without knowing $\mathbf{A}$?

2. Prove (*a*) that if $\mathbf{A}$ and $\mathbf{C}$ are non-singular the principal modes of excitation satisfy

$$(\mathbf{C}^{-1} - \omega_r^{-2}\mathbf{A}^{-1})\,\boldsymbol{\Phi}^{(r)} = \mathbf{0} \quad (r = 1, 2, \ldots, n).$$

$$(b) \quad \mathbf{A} = \sum_{r=1}^{n} \frac{\boldsymbol{\Phi}^{(r)}\boldsymbol{\Phi}'^{(r)}}{a_r}; \quad \mathbf{C} = \sum_{r=1}^{n} \frac{\omega_r^2\boldsymbol{\Phi}^{(r)}\boldsymbol{\Phi}'^{(r)}}{a_r}.$$

Use the second of these three results to find the matrix $\mathbf{A}$ of Ex. 2.3.3.

3. Express the force distribution $\boldsymbol{\Phi} = F\{1 \quad 5 \quad 1\}$ in terms of the principal modes of excitation of the system of Ex. 2.5.2 (*c*).

2.10 Rayleigh's Principle

It was shown in § 2.3 that if a system is vibrating freely with frequency ω then the column matrix $\mathbf{q}$ is given by the equation

$$\mathbf{q} = \boldsymbol{\Psi}\sin(\omega t + \theta), \tag{2.10.1}$$

where $\boldsymbol{\Psi}$ satisfies the equation

$$\mathbf{C}\boldsymbol{\Psi} = \omega^2\mathbf{A}\boldsymbol{\Psi}. \tag{2.10.2}$$

From this it was deduced that for each of the n roots ω_r of (2.3.5) there will be a corresponding $\boldsymbol{\Psi}^{(r)}$ satisfying

$$\mathbf{C}\boldsymbol{\Psi}^{(r)} = \omega_r^2\mathbf{A}\boldsymbol{\Psi}^{(r)}. \tag{2.10.3}$$

Rayleigh's Principle is concerned with finding approximate values for the natural frequencies ω_r without solving the determinantal equation (2.3.5). In line with the rest of the chapter the discussion here will be largely mathematical; the physical implications of what follows have been mentioned elsewhere.†

Premultiply equation (2.10.3) by $\boldsymbol{\Psi}'^{(r)}$; the result is equation (2.5.16), viz.

$$\omega_r^2 = \frac{\boldsymbol{\Psi}'^{(r)}\mathbf{C}\boldsymbol{\Psi}^{(r)}}{\boldsymbol{\Psi}'^{(r)}\mathbf{A}\boldsymbol{\Psi}^{(r)}}. \tag{2.10.4}$$

Now consider an arbitrary non-zero column matrix $\boldsymbol{\Psi}$. The expression corresponding to that on the right-hand side of (2.10.4) is known as the 'Rayleigh Quotient' and we shall denote it‡ by the symbol ω_c^2 so that

$$\omega_c^2 = \frac{\boldsymbol{\Psi}'\mathbf{C}\boldsymbol{\Psi}}{\boldsymbol{\Psi}'\mathbf{A}\boldsymbol{\Psi}}. \tag{2.10.5}$$

† Bishop and Johnson, *The Mechanics of Vibration*, sect. 3.9. See also G. Temple and W. G. Bickley, *Rayleigh's Principle* (Oxford University Press, 1933).

‡ The use of the subscript c is of no consequence here. It stands for 'constrained' and is retained to preserve the notation that is used and explained in the book by Bishop and Johnson.

The value of ω_c^2 depends only on the ratios of the amplitudes $\Psi_1, \Psi_2, \ldots, \Psi_n$ and not on their absolute magnitudes. If the ratios are so adjusted as to form a principal mode, then clearly the value of ω_c will be the appropriate natural frequency. Rayleigh's Principle states that when $\mathbf{\Psi}$ is 'approximately' a principal mode, then ω_c will be 'very close' to the corresponding natural frequency. Expressed in a slightly more formal way it states that ω_c, considered as a function of the amplitudes $\Psi_1, \Psi_2, \ldots, \Psi_n$, is stationary in the principal modes. The second statement of the Principle implies that if a guess is made of the ratios of the amplitudes in some mode then, even if the guess is in appreciable error, the corresponding error in the natural frequency will still be quite small. It is for this reason that the Principle is so useful for obtaining approximate results.

This may be illustrated by referring, once more, to the system of fig. 2.2.1 for which

$$\mathbf{\Psi}^{(2)} = \begin{bmatrix} 1{\cdot}000 \\ 0{\cdot}445 \\ -0{\cdot}802 \end{bmatrix} \quad \text{and} \quad \omega_2^2 = 1{\cdot}555k/I. \tag{2.10.6}$$

Suppose that we take the matrix $\mathbf{\Psi}$ so as to be somewhere near $\mathbf{\Psi}^{(2)}$; for the sake of argument let

$$\mathbf{\Psi} = \begin{bmatrix} 1 \\ 0{\cdot}5 \\ -1 \end{bmatrix}. \tag{2.10.7}$$

By using the values of **A** and **C** given in (2.3.12) we find that

$$\omega_c^2 = \frac{\begin{bmatrix} 1 & 0{\cdot}5 & -1 \end{bmatrix} \begin{bmatrix} 2 & -1 & 0 \\ -1 & 2 & -1 \\ 0 & -1 & 1 \end{bmatrix} \begin{bmatrix} 1 \\ 0{\cdot}5 \\ -1 \end{bmatrix}}{\begin{bmatrix} 1 & 0{\cdot}5 & -1 \end{bmatrix} \begin{bmatrix} 1 & 0 & 0 \\ 0 & 1 & 0 \\ 0 & 0 & 1 \end{bmatrix} \begin{bmatrix} 1 \\ 0{\cdot}5 \\ -1 \end{bmatrix}} \times k/I. \tag{2.10.8}$$

This reduces to the value $\omega_c^2 = 1{\cdot}556k/I.$ (2.10.9)

which is extremely close to ω_2^2.

To establish Rayleigh's Principle algebraically, we note that the arbitrary amplitude matrix $\mathbf{\Psi}$ may be expressed in terms of the matrix of amplitudes of the principal coordinates by means of equation (2.8.5). Thus

$$\mathbf{\Psi} = \mathbf{X\Pi}, \tag{2.10.10}$$

where

$$\mathbf{\Pi} = \begin{bmatrix} \Pi_1 \\ \Pi_2 \\ \vdots \\ \Pi_n \end{bmatrix}, \tag{2.10.11}$$

and $\Pi_1, \Pi_2, \ldots, \Pi_n$ are the amplitudes of $p_1, p_2, \ldots, p_n$ in the harmonic motion. Equation (2.10.10) gives

$$\boldsymbol{\Psi}' = \boldsymbol{\Pi}'\mathbf{X}', \tag{2.10.12}$$

and therefore

$$\omega_c^2 = \frac{\boldsymbol{\Pi}'\mathbf{X}'\mathbf{C}\mathbf{X}\boldsymbol{\Pi}}{\boldsymbol{\Pi}'\mathbf{X}'\mathbf{A}\mathbf{X}\boldsymbol{\Pi}} = \frac{\boldsymbol{\Pi}'\mathbf{N}\boldsymbol{\Pi}}{\boldsymbol{\Pi}'\mathbf{L}\boldsymbol{\Pi}}, \tag{2.10.13}$$

because of equations (2.5.10) and (2.5.13). When this quotient is expanded by using the expressions for $\mathbf{L}$ and $\mathbf{N}$ given in (2.5.11) and (2.5.14) it becomes

$$\omega_c^2 = \frac{c_1^2\Pi_1^2 + c_2^2\Pi_2^2 + \ldots + c_n^2\Pi_n^2}{a_1^2\Pi_1^2 + a_2^2\Pi_2^2 + \ldots + a_n^2\Pi_n^2}. \tag{2.10.14}$$

It will simplify the presentation if we now specialise the treatment by assuming that the principal coordinates are normalised in the manner discussed in § 2.6. Then

$$\mathbf{L} = \mathbf{I}, \quad \mathbf{N} = \boldsymbol{\Omega} \tag{2.10.15}$$

and

$$\omega_c^2 = \frac{\boldsymbol{\Pi}'\boldsymbol{\Omega}\boldsymbol{\Pi}}{\boldsymbol{\Pi}'\boldsymbol{\Pi}} = \frac{\omega_1^2\Pi_1^2 + \omega_2^2\Pi_2^2 + \ldots + \omega_n^2\Pi_n^2}{\Pi_1^2 + \Pi_2^2 + \ldots + \Pi_n^2}. \tag{2.10.16}$$

Suppose that the motion $\boldsymbol{\Psi}$ is nearly in the jth mode, so that

$$\frac{\Pi_i}{\Pi_j} = \rho_i \ll 1 \quad (i \neq j). \tag{2.10.17}$$

The value of ω_c^2 is now expressible in the form

$$\omega_c^2 = \frac{\omega_1^2\rho_1^2 + \omega_2^2\rho_2^2 + \ldots + \omega_j^2 + \ldots + \omega_n^2\rho_n^2}{\rho_1^2 + \rho_2^2 + \ldots + 1 + \ldots + \rho_n^2}, \tag{2.10.18}$$

so that

$$\omega_c^2 - \omega_j^2 = \frac{(\omega_1^2 - \omega_j^2)\rho_1^2 + (\omega_2^2 - \omega_j^2)\rho_2^2 + \ldots + (\omega_n^2 - \omega_j^2)\rho_n^2}{\rho_1^2 + \rho_2^2 + \ldots + 1 + \ldots + \rho_n^2}. \tag{2.10.19}$$

This equation shows that the difference between ω_c^2 and ω_j^2 depends on the squares of the small quantities ρ, and, as we should expect, it is zero when all the ρ's are zero. Suppose that all the ρ's are zero except one, ρ_k, say, which has the value ϵ ($\ll 1$). Then

$$\omega_c^2 - \omega_j^2 = \frac{(\omega_k^2 - \omega_j^2)\epsilon^2}{1 + \epsilon^2} \doteqdot (\omega_k^2 - \omega_j^2)\epsilon^2, \tag{2.10.20}$$

so that

$$\omega_c \doteqdot \{\omega_j^2 + (\omega_k^2 - \omega_j^2)\epsilon^2\}^{\frac{1}{2}} \doteqdot \omega_j\left\{1 + \frac{1}{2}\left(\frac{\omega_k^2 - \omega_j^2}{\omega_j^2}\right)\epsilon^2\right\}, \tag{2.10.21}$$

and the error in ω_c is of the order ϵ^2. The result corresponding to this in the general case can easily be derived from equation (2.10.19) and is

$$\omega_c \doteqdot \omega_j\left\{1 + \frac{1}{2}\left(\frac{\omega_1^2 - \omega_j^2}{\omega_j^2}\right)\rho_1^2 + \frac{1}{2}\left(\frac{\omega_2^2 - \omega_j^2}{\omega_j^2}\right)\rho_2^2 + \ldots + \frac{1}{2}\left(\frac{\omega_n^2 - \omega_j^2}{\omega_j^2}\right)\rho_n^2\right\}. \tag{2.10.22}$$

This establishes Rayleigh's Principle in one of the ways we stated it. The statement that ω_c is stationary in the principal modes follows directly from (2.10.18). For

$$\frac{\partial\omega_c^2}{\partial\rho_i} = \frac{2\rho_i\{\omega_i^2(\rho_1^2 + \rho_2^2 + \ldots + \rho_n^2) - (\omega_1^2\rho_1^2 + \omega_2^2\rho_2^2 + \ldots + \omega_n^2\rho_n^2)\}}{(\rho_1^2 + \rho_2^2 + \ldots + 1 + \ldots + \rho_n^2)^2}, \tag{2.10.23}$$

and each ρ_i is zero in the jth principal mode.

By putting $j = 1$ and then $j = n$ in equation (2.10.19) we find that

$$\omega_1^2 \leqslant \omega_c^2 \leqslant \omega_n^2. \tag{2.10.24}$$

Thus if an attempt is made to estimate ω_1^2 by guessing the shape of the first principal mode then the result will be high (unless it happens to be exactly right). Equally, an estimate of ω_n^2 obtained by the method will be low.

EXAMPLES 2.10

1. Starting from the approximation $\boldsymbol{\Psi}^{(1)} = \{1 \quad 1 \quad 1\}$, find an approximation to the first natural frequency of the system of Ex. 2.1.2. Use this value of ω_1 in the *first two* of the Lagrangian equations for $\boldsymbol{\Psi}^{(1)}$ to improve the approximation to the mode and thence find a new approximation to ω_1. Finally, use this value of ω_1 in the same two equations to find a third approximation to $\boldsymbol{\Psi}^{(1)}$.

2. Show that for a system with two degrees of freedom the stationary values of the quotient

$$\frac{c_{11}x^2 + 2c_{12}x + c_{22}}{a_{11}x^2 + 2a_{12}x + a_{22}}$$

are the squares of the natural frequencies of the system. Find the natural frequencies and principal modes of the system for which

$$\mathbf{A} = m\begin{bmatrix} 2 & 1 \\ 1 & 4 \end{bmatrix}, \quad \mathbf{C} = mg/l\begin{bmatrix} 1 & 2 \\ 2 & 7 \end{bmatrix},$$

by finding the stationary values of the quotient.

2.11 Semi-definite systems—elimination of coordinates

This section is concerned with systems for which $\mathbf{A}$ and/or $\mathbf{C}$ are positive semi-definite. It was stated in § 2.1 that any positive semi-definite symmetric matrix is singular. In this section and the next the following simplifying assumption will be made.

Assumption 1. (*a*) If $\mathbf{A}$ is singular then at least one element of $|\mathbf{A}|$ has a non-zero cofactor.

(*b*) If $\mathbf{C}$ is singular then at least one element of $|\mathbf{C}|$ has a non-zero cofactor.

This will mean that any singular matrix that occurs will fall into the category of matrices considered in § 1.8. Consideration of the more general case in which this is not true will be reserved to § 4.6.

It was stated at the beginning of § 2.2, on forced motion, that the analysis of that section applies to positive semi-definite as well as to positive definite systems, and the reader may verify that at no point is it assumed that $\mathbf{A}$ or $\mathbf{C}$ is non-singular. The assumption that *is* made there is that $\mathbf{Z} = (\mathbf{C} - \omega^2\mathbf{A})$ is non-singular; provided that this is true the analysis holds. This means that it is unnecessary to discuss the forced motion of semi-definite systems as far as § 2.2 is concerned. Sections 2.8 and 2.9, on the series forms for receptances, will have to be modified—as we shall see—for semi-definite systems. However, since these sections make use of the principal modes of vibration of a system it will be necessary first to see what modifications must be made in the sections on free vibration.

There are two ways of dealing with the free vibration of a semi-definite system. The first is to eliminate certain coordinates and so reduce the problem to the study of the free vibration of a system which has fewer degrees of freedom, and which is positive definite. The other is to retain the semi-definiteness and see how much of the theory given in this chapter has to be modified. The first procedure will be discussed in this section and the second in § 2.12.

We shall introduce the first of these procedures—elimination—by studying the systems of figs 2.1.2–2.1.4. The equations governing the free vibration of the system of figs 2.1.3 may be obtained by using the expressions

$$\left.\begin{aligned} 2T &= I_1\dot{q}_1^2+I_2\dot{q}_2^2+I_3\dot{q}_3^2, \\ 2V &= k(q_1-q_2)^2+k(q_3-q_2)^2 \end{aligned}\right\} \tag{2.11.1}$$

in the Lagrangian equations (2.1.12) and putting $Q_1 = 0 = Q_2 = Q_3$. The equations are

$$\left.\begin{aligned} I_1\ddot{q}_1+kq_1-kq_2 &= 0, \\ I_2\ddot{q}_2-kq_1+2kq_2-kq_3 &= 0, \\ I_3\ddot{q}_3-kq_2+kq_3 &= 0. \end{aligned}\right\} \tag{2.11.2}$$

The sum of these three equations is

$$I_1\ddot{q}_1+I_2\ddot{q}_2+I_3\ddot{q}_3 = 0, \tag{2.11.3}$$

which may be integrated directly to give

$$I_1\dot{q}_1+I_2\dot{q}_2+I_3\dot{q}_3 = a, \tag{2.11.4}$$

$$I_1q_1+I_2q_2+I_3q_3 = at+b, \tag{2.11.5}$$

where a and b are constants. Equation (2.11.4) states merely that the angular momentum of the system is constant. Equation (2.11.5) shows that if the angular momentum is not *zero* then the displacements increase linearly with time—and therefore do not remain small. Therefore for any small vibration a must be zero. Since there is a choice to be made in fixing the plane from which the q's are measured b may be taken to be zero, so that the q's satisfy the equation

$$I_1q_1+I_2q_2+I_3q_3 = 0. \tag{2.11.6}$$

Any motion of the system can be made up by combining a small motion of the system with a rigid body rotation. For suppose the motion has angular momentum a, so that it satisfies (2.11.4) and (2.11.5). Let

$$q_1^{(1)} = \frac{at+b}{I_1+I_2+I_3} = q_2^{(1)} = q_3^{(1)}, \tag{2.11.7}$$

then $$q_1' = q_1-q_1^{(1)},\quad q_2' = q_2-q_2^{(1)},\quad q_3' = q_3-q_3^{(1)}$$

satisfy equations (2.11.2) and (2.11.6). In other words, q_1', q_2', q_3' correspond to a small motion of the system. It will be noticed that the motion given by equation

(2.11.7) is the first mode of the system; it corresponds to a zero natural frequency, $\omega_1 = 0$, and its existence was pointed out in § 2.1.

Having isolated the rigid-body mode we now concentrate our attention on the small vibration given by equations (2.11.2) and (2.11.6). Equation (2.11.6) may be used to express one of the q's, say q_3, in terms of the other two. If we substitute for q_3 in the second of equations (2.11.2) we obtain

$$\left.\begin{aligned} I_1\ddot{q}_1 + kq_1 - kq_2 &= 0, \\ I_2\ddot{q}_2 - k\left(1 - \frac{I_1}{I_3}\right)q_1 + k\left(2 + \frac{I_2}{I_3}\right)q_2 &= 0. \end{aligned}\right\} \tag{2.11.8}$$

These equations may be written as one matrix equation

$$\begin{bmatrix} I_1 & 0 \\ 0 & I_2 \end{bmatrix}\begin{bmatrix} \ddot{q}_1 \\ \ddot{q}_2 \end{bmatrix} + \begin{bmatrix} k & -k \\ -k\left(1 - \dfrac{I_1}{I_3}\right) & k\left(2 + \dfrac{I_2}{I_3}\right) \end{bmatrix}\begin{bmatrix} q_1 \\ q_2 \end{bmatrix} = 0, \tag{2.11.9}$$

but the second square matrix in this equation is not symmetric. If the equations of motion are left in this form then the theory of this chapter does not apply to them, for the theory holds only when **A** and **C** are both symmetric.

The way to get a symmetric matrix **C** in this case is to eliminate q_3 from the expressions for T and V, not from the equations of motion. For the Lagrange equations then ensure the symmetry of the resulting equations of motion. By substituting for q_3 in the expressions (2.11.1) we obtain

$$\left.\begin{aligned} 2T &= I_1\dot{q}_1^2 + I_2\dot{q}_2^2 + \frac{(I_1\dot{q}_1 + I_2\dot{q}_2)^2}{I_3}, \\ 2V &= k(q_2 - q_1)^2 + k\left(\frac{I_1q_1 + I_2q_2}{I_3} + q_2\right)^2, \end{aligned}\right\} \tag{2.11.10}$$

for which the matrices **A** and **C** are

$$\mathbf{A}_1 = \begin{bmatrix} I_1 + \dfrac{I_1^2}{I_3} & \dfrac{I_1I_2}{I_3} \\ \dfrac{I_1I_2}{I_3} & I_2 + \dfrac{I_2^2}{I_3} \end{bmatrix}; \quad \mathbf{C}_1 = \begin{bmatrix} k + k\dfrac{I_1^2}{I_3^2} & -k + k\dfrac{I_1(I_2+I_3)}{I_3^2} \\ -k + k\dfrac{I_1(I_2+I_3)}{I_3^2} & k + k\left(\dfrac{I_2+I_3}{I_3}\right)^2 \end{bmatrix}. \tag{2.11.11}$$

Since the expressions (2.11.10) are positive definite the analysis of this chapter may be applied to the system they describe. This analysis will show that the system has two natural frequencies and two corresponding principal modes. These two frequencies will clearly be natural frequencies of the original system, and the two modes will give rise to two modes for the original system.

The reader may verify that for the particular values

$$I_1/2 = I_2 = I_3 = I, \tag{2.11.12}$$

the natural frequencies calculated from $\mathbf{A}_1$ and $\mathbf{C}_1$ are

$$\omega_2 = 0{\cdot}8481\sqrt{(k/I)}, \quad \omega_3 = 1{\cdot}6675\sqrt{(k/I)}, \tag{2.11.13}$$

and the principal modes of the partial system are:

$$\left.\begin{aligned} q_1:q_2 &= 1:-0{\cdot}4385 \text{ corresponding to } \omega_2, \\ q_1:q_2 &= 1:-4{\cdot}5616 \text{ corresponding to } \omega_3. \end{aligned}\right\} \tag{2.11.14}$$

The other principal modes of the original system may be obtained by combining the last equation with equation (2.11.6), which may now be written as

$$q_3 = -2q_1 - q_2. \tag{2.11.15}$$

The principal modes of the whole system are therefore

$$\left.\begin{aligned} q_1:q_2:q_3 &= 1:\quad 1 \quad : \quad 1 \quad \text{ corresponding to } \omega_1, \\ q_1:q_2:q_3 &= 1:-0{\cdot}4385:-1{\cdot}5615 \text{ corresponding to } \omega_2, \\ q_1:q_2:q_3 &= 1:-4{\cdot}5616: \quad 2{\cdot}5616 \text{ corresponding to } \omega_3. \end{aligned}\right\} \tag{2.11.16}$$

We shall now consider a *general* system for which the matrix **C** is singular, but **A** is non-singular. The equation governing the free vibration is equation (2.2.4), namely

$$\mathbf{A}\ddot{\mathbf{q}} + \mathbf{C}\mathbf{q} = \mathbf{0}. \tag{2.11.17}$$

Since **C** is singular there is, as was pointed out in § 2.1, a rigid body mode corresponding to a zero natural frequency ω_1. This mode is specified by a column vector $\mathbf{\Psi}^{(1)}$ satisfying

$$\mathbf{C}\mathbf{\Psi}^{(1)} = \mathbf{0}. \tag{2.11.18}$$

Often the existence and shape of this mode will be obvious from kinematical considerations. Rarely, if ever, will it be necessary to solve equation (2.11.18) formally. Since **C** satisfies Assumption 1, this modal vector is unique apart from an arbitrary multiplying factor. The matrix **C** is symmetric, so that the transpose of equation (2.11.18) is

$$\mathbf{\Psi}'^{(1)}\mathbf{C} = \mathbf{0}. \tag{2.11.19}$$

If we premultiply equation (2.11.17) by $\mathbf{\Psi}'^{(1)}$ we obtain

$$\mathbf{\Psi}'^{(1)}\mathbf{A}\ddot{\mathbf{q}} + \mathbf{\Psi}'^{(1)}\mathbf{C}\mathbf{q} = 0, \tag{2.11.20}$$

so that

$$\mathbf{\Psi}'^{(1)}\mathbf{A}\ddot{\mathbf{q}} = 0. \tag{2.11.21}$$

This equation may be integrated directly to give

$$\mathbf{\Psi}'^{(1)}\mathbf{A}\mathbf{q} = at + b, \tag{2.11.22}$$

where a and b are constants. For any small motion a must be zero—or else the motion will increase linearly with time—and b may be taken to be zero.

We may now prove, just as before, that *any* motion of the system may be made up by combining a rigid body motion with a small motion. For if q satisfies (2.11.17) and (2.11.22), and if

$$\mathbf{q}^{(1)} = \frac{\mathbf{\Psi}^{(1)}(at+b)}{\mathbf{\Psi}'^{(1)}\mathbf{A}\mathbf{\Psi}^{(1)}}, \tag{2.11.23}$$

then $\mathbf{q}' = \mathbf{q} - \mathbf{q}^{(1)}$ is a small motion; that is, $\mathbf{q}'$ satisfies the equation (2.11.22) with $a = 0 = b$.

In any small motion, therefore, the q's satisfy the equation

$$\mathbf{\Psi}'^{(1)}\mathbf{Aq} = 0. \tag{2.11.24}$$

The interpretation of this equation is simply that the first principal coordinate p_1 is zero in every small motion (see equation (2.7.5)).

When equation (2.11.24) is written out in full it becomes

$$(a_{11}\Psi_1'^{(1)} + a_{21}\Psi_2'^{(1)} + \ldots + a_{n1}\Psi_n'^{(1)})\,q_1 + (a_{12}\Psi_1'^{(1)} + a_{22}\Psi_2'^{(1)} + \ldots + a_{n2}\Psi_n'^{(1)})\,q_2 + \ldots$$
$$+ (a_{1n}\Psi_1'^{(1)} + a_{2n}\Psi_2'^{(1)} + \ldots + a_{nn}\Psi_n'^{(1)})\,q_n = 0, \tag{2.11.25}$$

and at least one of the coefficients in this linear relation must be non-zero—for if they were all zero then $\mathbf{\Psi}'^{(1)}\mathbf{A}$ would be zero, which would imply that $\mathbf{A}$ was singular. We may therefore use this relation to express one of the coordinates, q_s say, in terms of the other $(n-1)$. Then, just as we did for the simple system of fig. 2.1.3 we substitute for $\dot{q}_s$ and q_s in the expressions for T and V and obtain the equations of motion from the new expressions. The new, partial, system which we obtain in this way will have $(n-1)$ degrees of freedom and have matrices $\mathbf{A}$ and $\mathbf{C}$ which are symmetric, non-singular and positive definite.

A particular case in which $\mathbf{C}$ is singular is when one of the coordinates, q_1 say, does not appear at all in the expression for V, but does appear in T. When this happens $c_{11}, c_{12}, \ldots, c_{1n}$ are all zero and the Lagrange equation for q_1 is simply

$$\frac{d}{dt}\left(\frac{\partial T}{\partial \dot{q}_1}\right) = 0, \tag{2.11.26}$$

i.e.

$$a_{11}\ddot{q}_1 + a_{12}\ddot{q}_2 + \ldots + a_{1n}\ddot{q}_n = 0, \tag{2.11.27}$$

which may be integrated to give

$$a_{11}q_1 + a_{12}q_2 + \ldots + a_{1n}q_n = 0, \tag{2.11.28}$$

for any small motion of the system. The rigid body mode $\mathbf{\Psi}^{(1)}$ of the system is now

$$q_1 : q_2 : q_3 : \ldots : q_n = 1:0:0:\ldots:0, \tag{2.11.29}$$

as will be clear from the relation (2.11.18). Coordinates which have this property are called 'ignorable'.† An example of a system with an ignorable coordinate is given in Ex. 2.11.1.

The second class of semi-definite systems consists of those which have a singular matrix $\mathbf{A}$ and a non-singular matrix $\mathbf{C}$. These systems are not so important as those for which $\mathbf{C}$ is singular, and if $\mathbf{A}$ is singular it is a sign that one or more of the coordinates are superfluous. They are discussed for the sake of completeness, and because there is often some doubt as to the conditions which must be satisfied by $\mathbf{A}$ and $\mathbf{C}$ if a system is to have the correct number of principal modes.

The system of fig. 2.1.2 falls into this category. The equations governing the free

† Whittaker, *Analytical Dynamics*, p. 56.

motion of this system may be obtained by putting $Q_1 = 0 = Q_2 = Q_3$ in equations (2.1.20). They are

$$\left.\begin{aligned} 2kq_1 - kq_2 &= 0, \\ I_2\ddot{q}_2 - kq_1 + 2kq_2 - kq_3 &= 0, \\ I_3\ddot{q}_3 - kq_2 + kq_3 &= 0. \end{aligned}\right\} \tag{2.11.30}$$

The first of these equations may be used to eliminate q_1 (or q_2) from the energy expressions. By substituting for q_1 in equation (2.1.17) and putting $I_1 = 0$ we obtain

$$2T = I_2\dot{q}_2^2 + I_3\dot{q}_3^2, \tag{2.11.31}$$

$$\begin{aligned} 2V &= \tfrac{1}{4}kq_2^2 + \tfrac{1}{4}kq_2^2 + k(q_3 - q_2)^2 \\ &= \tfrac{3}{2}kq_2^2 - 2kq_2q_3 + kq_3^2, \end{aligned} \tag{2.11.32}$$

which are the energy expressions for the system of fig. 2.1.2 when the coordinate q_1 is disregarded. The matrices **A** and **C** are now

$$\mathbf{A}_1 = \begin{bmatrix} I_2 & 0 \\ 0 & I_3 \end{bmatrix}, \quad \mathbf{C}_1 = \begin{bmatrix} \tfrac{3}{2}k & -k \\ -k & k \end{bmatrix}, \tag{2.11.33}$$

which are both symmetric, non-singular and positive definite.

The reader may verify that for the particular values

$$I_2 = \tfrac{1}{2}I_3 = I, \tag{2.11.34}$$

the natural frequencies of the system specified by $\mathbf{A}_1$ and $\mathbf{C}_1$ are

$$\omega_1 = 0{\cdot}3660\sqrt{(k/I)}, \quad \omega_2 = 1{\cdot}3660\sqrt{(k/I)} \tag{2.11.35}$$

and the principal modes are respectively

$$q_2 : q_3 :: 2 : 2{\cdot}7321 \quad \text{and} \quad q_2 : q_3 :: 2 : -0{\cdot}7321. \tag{2.11.36}$$

By combining the last equation with the first of equations (2.11.26) we obtain for the whole system the two modes

$$\left.\begin{aligned} &q_1 : q_2 : q_3 :: 1 : 2 : 2{\cdot}7321 \quad \text{corresponding to } \omega_1, \\ &q_1 : q_2 : q_3 :: 1 : 2 : -0{\cdot}7321 \text{ corresponding to } \omega_2. \end{aligned}\right\} \tag{2.11.37}$$

The discussion of the general system which has a singular matrix **A** proceeds along much the same lines as that for a system whose matrix **C** is singular. Thus there will be a single column matrix, which we shall denote by $\boldsymbol{\Psi}^{(n)}$, which will satisfy the equation

$$\mathbf{A}\boldsymbol{\Psi}^{(n)} = \mathbf{0} = \boldsymbol{\Psi}'^{(n)}\mathbf{A}. \tag{2.11.38}$$†

By pre-multiplying equation (2.11.17) by $\boldsymbol{\Psi}'^{(n)}$ we obtain

$$\boldsymbol{\Psi}'^{(n)}\mathbf{C}\mathbf{q} = 0, \tag{2.11.39}$$

which is a linear relation between the q's. One of the q's may therefore be eliminated from T and V and the resulting energy expressions will be those for a positive definite system having $(n-1)$ degrees of freedom.

† The reasons why it is convenient to call this matrix $\boldsymbol{\Psi}^{(n)}$ are discussed in §2.12.

The remaining class of semi-definite systems consists of those like that of fig. 2.1.4 for which both **A** and **C** are singular. The equations governing the free vibration of the system of fig. 2.1.4 are

$$\left.\begin{aligned} kq_1 - kq_2 &= 0, \\ I_2\ddot{q}_2 - kq_1 + 2kq_2 - kq_3 &= 0, \\ I_3\ddot{q}_3 - kq_2 + kq_3 &= 0. \end{aligned}\right\} \tag{2.11.40}$$

The first equation gives one linear relation between the coordinates, and by adding the three equations and integrating them we find that for any small motion the coordinates must also satisfy the equation

$$I_2q_2 + I_3q_3 = 0. \tag{2.11.41}$$

We may therefore eliminate two of the coordinates—q_1 and q_3 say—from the analysis, and if we do this we are left with a single equation for q_2, namely

$$I_2\ddot{q}_2 + k\left(1 + \frac{I_2}{I_3}\right)q_2 = 0. \tag{2.11.42}$$

The reader may verify that for the particular values given in (2.11.34) the non-zero natural frequency is

$$\omega_2 = 1{\cdot}2247\sqrt{(k/I)}, \tag{2.11.43}$$

and the two principal modes are

$$\left.\begin{aligned} q_1\!:\!q_2\!:\!q_3\!::\!1\!:\!1\!:\!1 \text{ corresponding to } \omega_1 &= 0, \\ q_1\!:\!q_2\!:\!q_3\!::\!1\!:\!1\!:\!-0{\cdot}5 \text{ corresponding to } \omega_2&. \end{aligned}\right\} \tag{2.11.44}$$

For the general system whose matrices **A** and **C** are both singular, but satisfy Assumption 1, there will be two linear relations between the cordinates—(2.11.24) and (2.11.39)—by means of which two of the coordinates may be eliminated. The resulting system will have $(n-2)$ degrees of freedom and will be positive definite. The orginal system will also have an $(n-1)$th mode corresponding to a rigid body motion.

EXAMPLES 2.11

1. The vibrating system shown consists of a ring P of mass $2m$ which may slide smoothly on the horizontal rod OA and to which is attached a light inextensible string PQR with masses m attached at Q and R ($PQ = QR = l$). Write down the kinetic and potential energies of the system for small vibrations about the initial position

$$x = 0 = \theta = \phi,$$

and show that for small vibrations the coordinates must satisfy the equation

$$4x + 2\theta + \phi = 0.$$

Give the physical interpretation of this equation.

2. Find the natural frequencies and principal modes of the system of Ex. 2.11.1.

3. Find the results (2.11.13) and (2.11.16) from equation (2.11.9), using the stability matrix which is not symmetric.

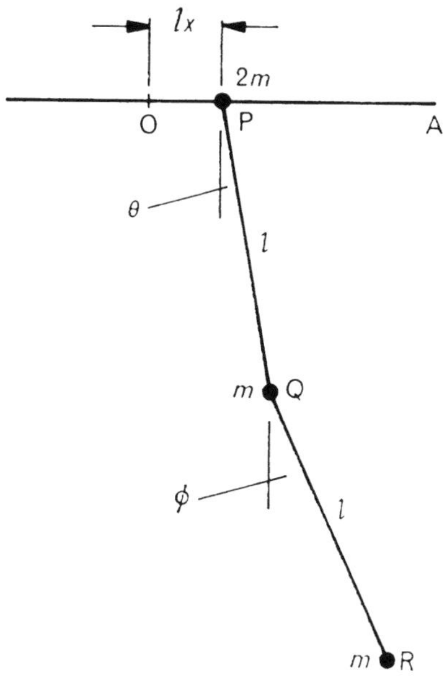

Ex. 2.11.1

4. The diagram shows three rigid masses $2M$, M and $2M$ connected by light springs of stiffness $2k$, k, k and λ. The left-hand spring has, attached to its free end, a light rigid frame by reference to which the displacements of the three masses are measured. When the system is at rest in its equilibrium position $q_1 = 0 = q_2 = q_3 = q_4$.

(*a*) Find the matrices **A** and **C** and show that both are singular.

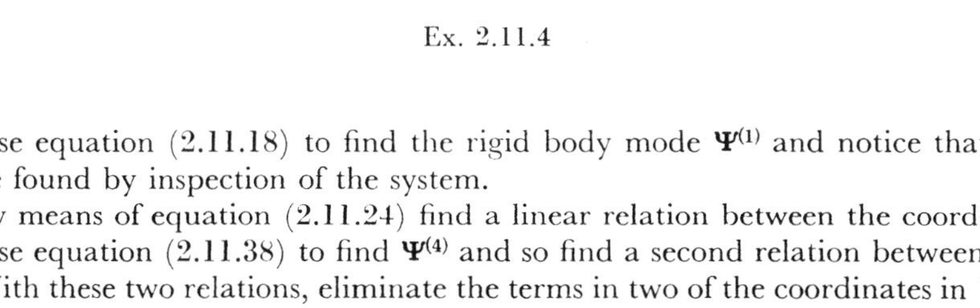

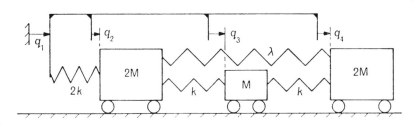

Ex. 2.11.4

(*b*) Use equation (2.11.18) to find the rigid body mode $\mathbf{\Psi}^{(1)}$ and notice that it may easily be found by inspection of the system.

(*c*) By means of equation (2.11.24) find a linear relation between the coordinates q.

(*d*) Use equation (2.11.38) to find $\mathbf{\Psi}^{(4)}$ and so find a second relation between the q's.

(*e*) With these two relations, eliminate the terms in two of the coordinates in T and V and so derive the inertia and stability matrices $\mathbf{A}_1$ and $\mathbf{C}_1$. Notice that both are positive definite.

(*f*) Find the remaining principal modes and natural frequencies from $\mathbf{A}_1$ and $\mathbf{C}_1$.

2.12 Degenerate systems—modal analysis

We shall now discuss the second of the procedures which may be adopted in connection with the free vibration of a semi-definite system. We shall retain the semi-definiteness and see how much of the theory has to be modified. Throughout this section we shall restrict the matrices **A** and **C** to satisfy:

Assumption 1. (*a*) If **A** is singular then at least one element of $|\mathbf{A}|$ has a non-zero cofactor.

(*b*) If **C** is singular at least one element of $|\mathbf{C}|$ has a non-zero cofactor.

Assumption 2. The non-zero natural frequencies of the system, i.e. the non-zero roots of $|\mathbf{C}-\omega^2\mathbf{A}| = 0$, are all different. The theory of this chapter, from § 2.3 onwards has been based on this assumption.

First we shall discuss the case in which **A** is non-singular and **C** is singular. We shall begin by studying the system of fig. 2.1.3, and for the sake of simplicity we shall take the particular values of the I's given in equation (2.11.12). We have already shown that the system has three natural frequencies—one of which is zero—and three corresponding principal modes. Using these modes as columns we may form the matrix

$$\mathbf{X} = \begin{bmatrix} 1 & 1 & 1 \\ 1 & -0{\cdot}4385 & -4{\cdot}5616 \\ 1 & -1{\cdot}5615 & 2{\cdot}5616 \end{bmatrix}, \tag{2.12.1}$$

according to equation (2.4.10). The determinant $|\mathbf{X}| = 12{\cdot}369$ so that **X** is non-singular. It may easily be verified that the modes satisfy the orthogonality relations (2.5.7) and (2.5.8). The constants a_r and c_r of equations (2.5.12) and (2.5.15) may be calculated and used to form the matrices **L** and **N**. These matrices are found to be

$$\mathbf{L} = I\begin{bmatrix} 4 & 0 & 0 \\ 0 & 4{\cdot}631 & 0 \\ 0 & 0 & 29{\cdot}38 \end{bmatrix}, \quad \mathbf{N} = k\begin{bmatrix} 0 & 0 & 0 \\ 0 & 3{\cdot}330 & 0 \\ 0 & 0 & 81{\cdot}67 \end{bmatrix}, \tag{2.12.2}$$

of which the first is non-singular, and the second singular. The matrix $\mathbf{\Omega}$ of equation (2.5.17) is also singular; it is

$$\mathbf{\Omega} = (k/I)\begin{bmatrix} 0 & 0 & 0 \\ 0 & 0{\cdot}7192 & 0 \\ 0 & 0 & 2{\cdot}7808 \end{bmatrix}, \tag{2.12.3}$$

and it may be verified that equation (2.5.18) still holds; that is

$$\mathbf{N} = \mathbf{L}\mathbf{\Omega}. \tag{2.12.4}$$

Since **L** is non-singular it may be used to find the inverse of **X** according to the equation

$$\boldsymbol{\xi} \equiv \mathbf{X}^{-1} = \mathbf{L}^{-1}\mathbf{X}'\mathbf{A}. \tag{2.12.5}$$

If we carry out this matrix multiplication we find that

$$\mathbf{X}^{-1} = \begin{bmatrix} 0{\cdot}25 & 0 & 0 \\ 0 & 0{\cdot}2159 & 0 \\ 0 & 0 & 0{\cdot}03404 \end{bmatrix} \begin{bmatrix} 1 & 1 & 1 \\ 1 & -0{\cdot}4385 & -1{\cdot}5615 \\ 1 & -4{\cdot}5616 & 2{\cdot}5616 \end{bmatrix} \begin{bmatrix} 2 & 0 & 0 \\ 0 & 1 & 0 \\ 0 & 0 & 1 \end{bmatrix}$$

$$= \begin{bmatrix} 0{\cdot}5 & 0{\cdot}25 & 0{\cdot}25 \\ 0{\cdot}4318 & -0{\cdot}0947 & -0{\cdot}3371 \\ 0{\cdot}0681 & -0{\cdot}1553 & 0{\cdot}0872 \end{bmatrix}, \tag{2.12.6}$$

and it may be verified that this is in fact the inverse of the matrix **X** in equation (2.12.1). Since **N** is singular the other equation for $\boldsymbol{\xi}$—equation (2.7.22)—is not valid.

Having found $\mathbf{X}^{-1}$ (that is, $\boldsymbol{\xi}$), we may find the matrix of principal coordinates by means of equation (2.7.2). This equation is now

$$\begin{bmatrix} p_1 \\ p_2 \\ p_3 \end{bmatrix} = \begin{bmatrix} 0{\cdot}5 & 0{\cdot}25 & 0{\cdot}25 \\ 0{\cdot}4318 & -0{\cdot}0947 & -0{\cdot}3371 \\ 0{\cdot}0681 & -0{\cdot}1553 & 0{\cdot}0872 \end{bmatrix} \begin{bmatrix} q_1 \\ q_2 \\ q_3 \end{bmatrix}, \tag{2.12.7}$$

which gives

$$\left.\begin{aligned} p_1 &= 0{\cdot}5q_1 + 0{\cdot}25q_2 + 0{\cdot}25q_3, \\ p_2 &= 0{\cdot}4318q_1 - 0{\cdot}0947q_2 - 0{\cdot}3371q_3, \\ p_3 &= 0{\cdot}0681q_1 - 0{\cdot}1553q_2 + 0{\cdot}0872q_3. \end{aligned}\right\} \tag{2.12.8}$$

The expressions for T and V may be written in terms of the principal coordinates by means of equations (2.6.4) and (2.6.7). For the present system we have

$$\left.\begin{aligned} 2T &= 4I\dot{p}_2^2 + 4{\cdot}631I\dot{p}_2^2 + 29{\cdot}31I\dot{p}_2^2, \\ 2V &= 3{\cdot}33kp_2^2 + 81{\cdot}67kp_3^2. \end{aligned}\right\} \tag{2.12.9}$$

The principal coordinates given in equation (2.12.8) are not normalised. The coordinates

$$\bar{p}_1 = \sqrt{(4I)}\,p_1, \quad \bar{p}_2 = \sqrt{(4{\cdot}631I)}\,p_2, \quad \bar{p}_3 = \sqrt{(29{\cdot}31I)}\,p_3 \tag{2.12.10}$$

are normalised, and if these are used the expressions for T and V become

$$\begin{aligned} T &= \tfrac{1}{2}(\dot{\bar{p}}_1^2 + \dot{\bar{p}}_2^2 + \dot{\bar{p}}_3^2) \\ V &= \tfrac{1}{2}[0{\cdot}7192(k/I)\,\bar{p}_2^2 + 2{\cdot}7808(k/I)\,\bar{p}_3^2] \end{aligned} \tag{2.12.11}$$

in agreement with equations (2.6.13) and (2.6.14).

So far we have confined our attention to the simple system of fig. 2.1.3. However, it should be clear that the situation will be qualitatively the same for any system satisfying Assumptions 1 and 2 whose matrix **A** is non-singular, but whose **C** is singular. The general system of this type will have a zero natural frequency, and the corresponding mode will refer to a rigid body displacement of the system. Just as for the system studied in detail, the matrices **X** and **L** will be non-singular, but the matrices **N** and **Ω** will be singular. The inverse of **X** may always be found by using equation (2.12.5). As in equation (2.12.9), one of the principal coordinates will be missing from the expression for V.

It may now be verified that these changes have very little effect on the §§ 2.8–2.10, and that the only equations which cease to be valid are equation (2.9.17) and the second of equations (2.9.19).

When **A** is singular, but **C** is non-singular, a large part of the theory given in this chapter is not valid. Thus equation (2.3.6) is now a polynomial equation of degree $(n-1)$, not n, in ω^2, so that the system has only $(n-1)$ natural frequencies and principal modes. This means that the matrix **X** is not square and therefore cannot have an inverse. To overcome these difficulties we approach the problem in a completely different way. We use equations (2.3.7) and (2.3.8), which are equivalent to equations (2.3.3) and (2.3.5) when the system is non-degenerate.

Equation (2.3.7) is

$$(\mathbf{A}-\lambda^2\mathbf{C})\,\mathbf{\Psi} = \mathbf{0}, \tag{2.12.12}$$

where $\lambda = 1/\omega$. Equation (2.3.8) is the condition for this equation to have a non-trivial solution, namely

$$|\mathbf{A}-\lambda^2\mathbf{C}| = 0. \tag{2.12.13}$$

For the system of fig. 2.1.2 the matrices **A** and **C** are those given in equations (2.1.18) and (2.1.19), and equation (2.12.13) is

$$\begin{vmatrix} -2k\lambda^2 & k\lambda^2 & 0 \\ k\lambda^2 & I_2-2k\lambda^2 & k\lambda^2 \\ 0 & k\lambda^2 & I_3-k\lambda^2 \end{vmatrix} = 0, \tag{2.12.14}$$

or

$$k^3\lambda^6-(2I_2+3I_3)\,k^2\lambda^4+2I_2I_3k\lambda^2 = 0. \tag{2.12.15}$$

This equation has $\lambda^2 = 0$ as one root, and when the I's have the particular values given in equation (2.11.34) the other two roots are given by

$$\lambda_1 = 2{\cdot}7321\surd(I/k), \quad \lambda_2 = 0{\cdot}7321\surd(I/k). \tag{2.12.16}$$

If we label the roots in this way and call the zero root λ_3 then the reciprocals of λ_1 and λ_2 will give the natural frequencies ω_1 and ω_2 shown in equation (2.11.35), and λ_3 will give an infinite natural frequency.

The significance of this infinite natural frequency may be understood by referring back to the original system of fig. 2.1.1, from which the present system was obtained by putting $I_1 = 0$. If in the original system we take the values of I_2 and I_3 given in equation (2.11.34) and take

$$I_1 = I\epsilon, \tag{2.12.17}$$

where ϵ is small, then the natural frequencies are found to be

$$\left.\begin{aligned} \omega_1 &= \surd(k/I)\,(0{\cdot}3660-0{\cdot}0097\epsilon), \\ \omega_2 &= \surd(k/I)\,(1{\cdot}3660-0{\cdot}1347\epsilon), \\ \omega_3 &= \surd(2k/I\epsilon)\,(1+0{\cdot}125\epsilon). \end{aligned}\right\} \tag{2.12.18}$$

In these expressions the terms involving ϵ^2 and higher powers of ϵ have been neglected. If we now let ϵ tend to zero then ω_1 and ω_2 will tend to the values given in equation (2.11.35) while ω_3 will tend to infinity. This indicates that the degenerate system should properly be treated as the limit, as I_1 tends to zero, of the system of fig. 2.1.1. The analysis of the degenerate system will give a first approximation to the behaviour of the original system when I_1 is small.

The solutions of equation (2.12.12) corresponding to the three roots of equation (2.12.13) give the principal modes of the system. They are

$$\left.\begin{aligned} q_1:q_2:q_3 &= 1:2:2\cdot7321 && \text{corresponding to } \lambda_1,\\ q_1:q_2:q_3 &= 1:2:-0\cdot7321 && \text{corresponding to } \lambda_2,\\ q_1:q_2:q_3 &= 1:0:\quad 0 && \text{corresponding to } \lambda_3,\end{aligned}\right\} \tag{2.12.19}$$

and the first two modes are the same as those given in equation (2.11.37). The justification for calling the last of these a 'principal mode' is that the third mode of the original system is

$$q_1:q_2:q_3 = 1:-0\cdot5\epsilon-0\cdot375\epsilon^2:0\cdot125\epsilon^2, \tag{2.12.20}$$

and the mode in question is the limit of this as ϵ tends to zero.

Using these modes as columns we may form the matrix

$$\mathbf{X} = \begin{bmatrix} 1 & 1 & 1 \\ 2 & 2 & 0 \\ 2\cdot7321 & -0\cdot7321 & 0 \end{bmatrix}. \tag{2.12.21}$$

This matrix has determinant $|\mathbf{X}| = -6\cdot9284$ and so is non-singular. Again it may be verified that the modes satisfy the orthogonality relations (2.5.7) and (2.5.8). The matrices $\mathbf{L}$ and $\mathbf{N}$ may be calculated and are found to be

$$\mathbf{L} = I\begin{bmatrix} 18\cdot928 & 0 & 0 \\ 0 & 5\cdot072 & 0 \\ 0 & 0 & 0 \end{bmatrix}, \quad \mathbf{N} = k\begin{bmatrix} 2\cdot536 & 0 & 0 \\ 0 & 9\cdot464 & 0 \\ 0 & 0 & 2 \end{bmatrix}, \tag{2.12.22}$$

of which the first is singular and the second non-singular. Since ω_3 is infinite we cannot form the matrix $\mathbf{\Omega}$. Instead we introduce the matrix, $\mathbf{\Omega}_0$ say, given by

$$\mathbf{\Omega}_0 = \begin{bmatrix} \lambda_1^2 & 0 & \dots & 0 \\ 0 & \lambda_2^2 & \dots & 0 \\ \dots & \dots & \dots & \dots \\ 0 & 0 & \dots & \lambda_n^2 \end{bmatrix}. \tag{2.12.23}$$

For a positive definite system, $\mathbf{\Omega}_0$ is the inverse of $\mathbf{\Omega}$, but for a system which has a singular matrix $\mathbf{A}$, the matrix $\mathbf{\Omega}_0$ is singular. In the system under consideration

$$\mathbf{\Omega}_0 = (I/k)\begin{bmatrix} 7\cdot464 & 0 & 0 \\ 0 & 0\cdot536 & 0 \\ 0 & 0 & 0 \end{bmatrix}. \tag{2.12.24}$$

The counterpart of equation (2.5.18) when $\mathbf{A}$ is singular is

$$\mathbf{L} = \mathbf{N}\boldsymbol{\Omega}_0. \tag{2.12.25}$$

Previously, we found the inverse of $\mathbf{X}$ by using equation (2.12.5), that is, equation (2.7.18). Since $\mathbf{L}$ is now singular this equation is not valid. Equation (2.7.22) is valid, however, and this gives

$$\mathbf{X}^{-1} \equiv \boldsymbol{\xi} = \mathbf{N}^{-1}\mathbf{X}'\mathbf{C}. \tag{2.12.26}$$

In the present problem this matrix product gives

$$\mathbf{X}^{-1} = \begin{bmatrix} 0{\cdot}3943 & 0 & 0 \\ 0 & 0{\cdot}1057 & 0 \\ 0 & 0 & 0{\cdot}5 \end{bmatrix}\begin{bmatrix} 1 & 2 & 2{\cdot}7321 \\ 1 & 2 & -0{\cdot}7321 \\ 1 & 0 & 0 \end{bmatrix}\begin{bmatrix} 2 & -1 & 0 \\ -1 & 2 & -1 \\ 0 & -1 & 1 \end{bmatrix} \tag{2.12.27}$$

$$= \begin{bmatrix} 0 & 0{\cdot}1057 & 0{\cdot}2887 \\ 0 & 0{\cdot}3943 & -0{\cdot}2887 \\ 1 & -0{\cdot}5 & 0 \end{bmatrix}, \tag{2.12.28}$$

and again it may be verified that this is the inverse of the matrix $\mathbf{X}$ in equation (2.12.21).

The principal coordinates may now be found using equation (2.7.2). They are

$$\left.\begin{aligned} p_1 &= 0{\cdot}1057q_2 + 0{\cdot}2887q_3, \\ p_2 &= 0{\cdot}3943q_2 - 0{\cdot}2887q_3, \\ p_3 &= q_1 - 0{\cdot}5q_2. \end{aligned}\right\} \tag{2.12.29}$$

When the energy equations are written in terms of these coordinates they become

$$\left.\begin{aligned} 2T &= 18{\cdot}928I\dot{p}_1^2 + 5{\cdot}072I\dot{p}_2^2, \\ 2V &= 2{\cdot}536kp_1^2 + 9{\cdot}464kp_2^2 + 2kp_3^2. \end{aligned}\right\} \tag{2.12.30}$$

There is now no term in $\dot{p}_3^2$ in T, and this means that the principal coordinates cannot be normalised to make T have the form

$$\tfrac{1}{2}(\dot{\bar{p}}_1^2 + \dot{\bar{p}}_2^2 + \dot{\bar{p}}_3^2),$$

this is the way we previously normalised the coordinates. One way of normalising the coordinates is to make

$$V = \tfrac{1}{2}(\bar{p}_1^2 + \bar{p}_2^2 + \bar{p}_3^2). \tag{2.12.31}$$

The coordinates required for this are

$$\bar{p}_1 = \sqrt{(2{\cdot}536k)}\,p_1, \quad \bar{p}_2 = \sqrt{(9{\cdot}464k)}\,p_2, \quad \bar{p}_3 = \sqrt{(2k)}\,p_3. \tag{2.12.32}$$

When T is written in terms of these coordinates it becomes

$$T = \tfrac{1}{2}[7{\cdot}464(I/k)\,\dot{\bar{p}}_1^2 + 0{\cdot}536(I/k]\,\dot{\bar{p}}_2^2), \tag{2.12.33}$$

which will be recognised as

$$T = \tfrac{1}{2}(\lambda_1^2\dot{\bar{p}}_1^2 + \lambda_2^2\dot{\bar{p}}_2^2). \tag{2.12.34}$$

Any system satisfying Assumptions 1 and 2 whose matrix **A** is singular, and whose matrix **C** is non-singular, may be analysed in a similar way. The equation (2.12.13) for any such system will have a zero root, λ_n, and $(n-1)$ non-zero roots. The system will have n modes which will satisfy the orthogonality principle. The matrix **X** formed from these modes will be non-singular, and its inverse will be given by equation (2.12.26). The system will have n principal coordinates. All these coordinates will appear in the expression for V, but p_n will be absent from T. The principal coordinates may always be normalised to make

$$\left.\begin{aligned} T &= \tfrac{1}{2}(\lambda_1^2\dot{p}_1^2+\lambda_2^2\dot{p}_2^2+\ldots+\lambda_{n-1}^2\dot{p}_{n-1}^2),\\ V &= \tfrac{1}{2}(p_1^2+p_2^2+\ldots+p_n^2). \end{aligned}\right\} \tag{2.12.35}$$

Since a_n is zero and **L** is singular, many of the equations in §§ 2.7–2.10 are not valid. The first such equation is equation (2.7.5) when $r=n$. This equation can be avoided as p_n may be found by using equation (2.7.10) instead. We have already noted that equation (2.7.18) is not valid. In § 2.8 various modifications have to be made because equation (2.8.14) is not valid when $r=n$. Equation (2.8.14) is

$$c_r-\omega^2 a_r = a_r(\omega_r^2-\omega^2), \tag{2.12.36}$$

and when $r=n$ the left-hand side is just c_n. Therefore, whenever $a_n(\omega_n^2-\omega^2)$ occurs it must be replaced by c_n. The last term in equation (2.8.21), for instance, is now

$$\frac{\mathbf{\Psi}^{(n)}\mathbf{\Psi}'^{(n)}}{c_n} \quad \text{instead of} \quad \frac{\mathbf{\Psi}^{(n)}\mathbf{\Psi}'^{(n)}}{a_n(\omega_n^2-\omega^2)}.$$

Modifications also have to be made in § 2.9, on principal modes of excitation, but these will be left to the reader.

The changes which have to be made when ω_n is infinite spoil the symmetry of the expressions. Symmetry may be preserved by working entirely in terms of λ and λ_r, instead of ω and ω_r. This may be illustrated by referring to equation (2.8.14). When both **A** and **C** are positive definite the equation may be written

$$c_r-\frac{1}{\lambda^2}a_r = c_r\left(1-\frac{\lambda_r^2}{\lambda^2}\right), \tag{2.12.37}$$

and this is still valid *for all values of* r when **A** is singular. If we use this expression then we may write the receptance $\boldsymbol{\alpha}$, given by equation (2.8.21), in the form

$$\boldsymbol{\alpha} = \frac{\mathbf{\Psi}^{(1)}\mathbf{\Psi}'^{(1)}}{c_1\left(1-\dfrac{\lambda_1^2}{\lambda^2}\right)}+\frac{\mathbf{\Psi}^{(2)}\mathbf{\Psi}'^{(2)}}{c_2\left(1-\dfrac{\lambda_2^2}{\lambda^2}\right)}+\ldots+\frac{\mathbf{\Psi}^{(n)}\mathbf{\Psi}'^{(n)}}{c_n\left(1-\dfrac{\lambda_n^2}{\lambda^2}\right)}. \tag{2.12.38}$$

The last term reduces to that given above because $\lambda_n=0$.

When **A** is singular, Rayleigh's Principle may be put into a valid form by using the formulation in terms of λ and λ_r. Thus equation (2.10.3) may be replaced by

$$\lambda_r^2\mathbf{C}\mathbf{\Psi}^{(r)} = \mathbf{A}\mathbf{\Psi}^{(r)}, \tag{2.12.39}$$

and equation (2.10.5) by

$$\lambda_c^2 = \frac{\mathbf{\Psi}'\mathbf{A}\mathbf{\Psi}}{\mathbf{\Psi}'\mathbf{C}\mathbf{\Psi}}. \tag{2.12.40}$$

The reader may verify that when principal coordinates are introduced, and normalised in the sense of equation (2.12.35) then equation (2.12.40) becomes

$$\lambda_c^2 = \frac{\lambda_1^2 \Pi_1^2 + \lambda_2^2 \Pi_2^2 + \ldots + \lambda_n^2 \Pi_n^2}{\Pi_1^2 + \Pi_2^2 + \ldots + \Pi_n^2}. \tag{2.12.41}$$

This may be compared with equation (2.10.16).

Finally, when both **A** and **C** are singular it is not possible to apply either the modal analysis in terms of ω or that in terms of λ. In this case it is necessary to eliminate one or two of the coordinates by the procedure outlined in § 2.11. The resulting system will then have $(n-1)$ or $(n-2)$ degrees of freedom and will fall into one of the categories which have been analysed above.

EXAMPLES 2.12

1. For a particular system

$$\mathbf{A} = M\begin{bmatrix} 2 & 0 & 0 \\ 0 & 1 & 0 \\ 0 & 0 & 2 \end{bmatrix}, \quad \mathbf{X} = \begin{bmatrix} 1 & 1 & 1 \\ 1 & 0 & -4 \\ 1 & -1 & 1 \end{bmatrix},$$

and

$$\omega_1^2 = 0, \quad \omega_2^2 = \frac{k+2\lambda}{2M}, \quad \omega_3^2 = \frac{5k}{2M}.$$

Find the matrices **L**, **N**, $\boldsymbol{\xi}$ (without inverting **X** directly) and **C**, and the alternative matrix **X** whose columns are the *normalised* principal modes of the system.

2. For a certain system,

$$\mathbf{A} = M\begin{bmatrix} 20 & 10 & 10 \\ 10 & 6 & 4 \\ 10 & 4 & 6 \end{bmatrix}, \quad \mathbf{C} = k\begin{bmatrix} 50 & 25 & 25 \\ 25 & 16 & 11 \\ 25 & 11 & 14 \end{bmatrix}.$$

Find its natural frequencies and principal modes.

3. Let $\lambda = k$ in the system of Ex. 2.11.4, for which **A** and **C** are both singular. Using the rigid body mode corresponding to $\omega_1^2 = 0$, eliminate q_3 in the manner of §2.11. Hence find the remaining non-zero roots λ_2, λ_3, giving the corresponding principal modes together with that for $\lambda_4 = 0$. What is the physical interpretation of the latter?

CHAPTER 3

LINEAR EQUATIONS

Oh! Then your lordship is of opinion that married happiness is *not* inconsistent with discrepancy in rank?

H.M.S. Pinafore

In Chapter 1 we gave an account of elementary matrix theory and this was sufficient for the discussion of simple dynamical systems presented in Chapter 2. The reader who wishes to follow up this introductory theory by a systematic study of matrix methods has now three possibilities from which to choose.

The reader who is new to the subject, or whose interests are essentially practical, may wish to pass on immediately to Chapter 5. By this means he will escape what is necessarily a more mathematical portion of the book without missing any of the physically more important matters of dynamics.

Secondly, the reader who wants a fairly complete treatment of dynamical problems, but whose mathematical background is limited, should study §§ 3.1, 3.2, the first part of § 3.7 and the whole of Chapter 4. In this way he will become familiar with the matrix analysis of systems having repeated natural frequencies. These systems do not occur *exactly* in reality, but they are important because they may be approximated very closely by real systems. Their analysis requires a little more care, but this is rewarded by added insight into the nature of vibration in general.

Finally, the reader who requires a thorough background in matrix methods—for the purposes of research, for instance—should become familiar with this chapter and the next (as well as with the rest of the book if need be).

3.1 Introduction

In § 1.8 we considered a system (1.8.1) of n homogeneous equations in n unknowns $x_1, x_2, \ldots, x_n$. We proved that a necessary condition for the equations to have a non-trivial solution is that the determinant of coefficients, $|\mathbf{A}|$, should be zero. We stated, without proof, that, when not all the cofactors A_{rs} are zero, there is essentially only one solution, and we gave a construction for it.

We shall widen the discussion in this chapter by considering the equation

$$\mathbf{Ax} = \mathbf{0}, \tag{3.1.1}$$

in which $\mathbf{A}$ is of order m by n, and is no longer restricted to be square. That is, we shall contemplate the m equations

$$\left.\begin{aligned} a_{11}x_1 + a_{12}x_2 + \ldots + a_{1n}x_n &= 0,\\ a_{21}x_1 + a_{22}x_2 + \ldots + a_{2n}x_n &= 0,\\ \ldots\ldots\ldots\ldots\ldots\ldots&\ldots\ldots\\ a_{m1}x_1 + a_{m2}x_2 + \ldots + a_{mn}x_n &= 0,\end{aligned}\right\} \tag{3.1.2}$$

in the n unknowns $x_1, x_2, \ldots, x_n$. The quantities a_{rs} are merely numerical coefficients and are not intended to be taken as elements of an inertia matrix. In our consideration of equations (3.1.2) we shall not place any restriction on the coefficients. We shall then be able to treat the equations considered in § 1.8 as the special case in which $m = n$, and we shall be able to relax the condition that not all the cofactors are zero.

There are two fundamental questions which we may ask concerning any set of equations of the type (3.1.2). First 'Do the equations have a solution apart from

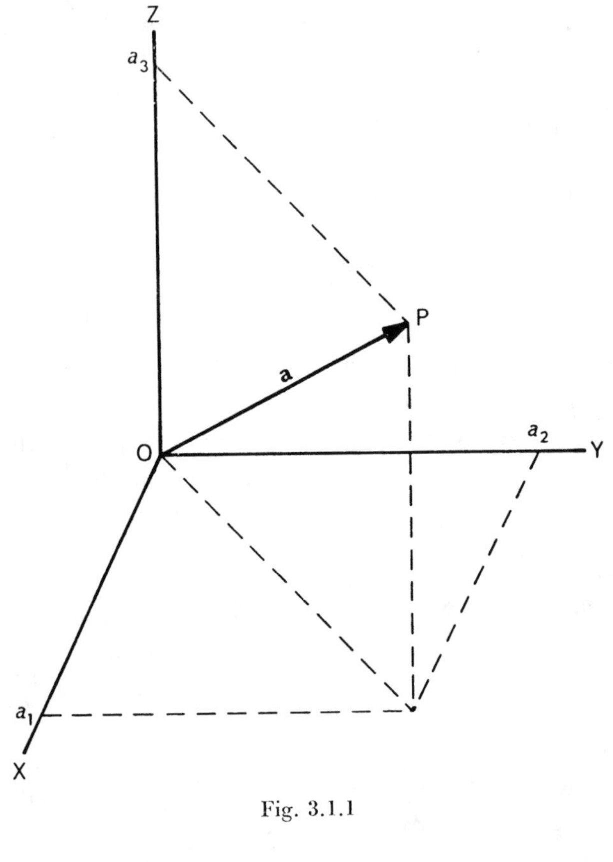

Fig. 3.1.1

the trivial one in which all the x's are zero?' To answer this we shall have to introduce the ideas of linear dependence and linear independence. The second question is this—'If the equations do have a non-trivial solution, do they have more than one; in fact, how many essentially different solutions do they have?' To answer this we must introduce the idea of the rank of a matrix.

The terms 'linear dependence' and 'linear independence' mentioned above may be viewed from an algebraic or a geometric point of view. The former is the one we shall take up in the next section, but it may be helpful at this point to indicate the geometric significance of the terms, and this we may do by showing the connection between column matrices and vectors.

The vector **a** in 3-dimensional space is defined by its three components a_1, a_2, a_3. It may be represented, as in fig. 3.1.1, by the directed line joining the origin O of a rectangular Cartesian frame to the point P whose coordinates are a_1, a_2, a_3. We may find the sum of two vectors from the parallelogram law. This states that if the vectors **a** and **b** are represented by the directed lines OP and OQ, as in fig. 3.1.2, then the vector $\mathbf{a}+\mathbf{b}$ is represented by the diagonal OR of the parallelogram $OPRQ$. Using this law we may prove that if **a** and **b** have components a_1, a_2, a_3 and b_1, b_2, b_3, then $\mathbf{a}+\mathbf{b}$ has components $a_1+b_1, a_2+b_2, a_3+b_3$. This means that vectors are added

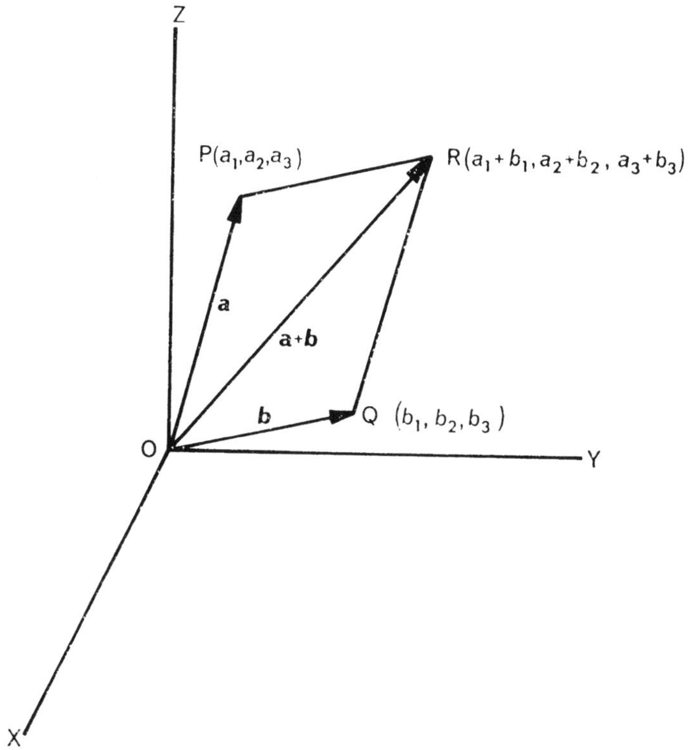

Fig. 3.1.2

in just the same way as column (or row) matrices. In other words, if the vectors **a** and **b** are represented by the column matrices

$$\mathbf{a} = \begin{bmatrix} a_1 \\ a_2 \\ a_3 \end{bmatrix}, \quad \mathbf{b} = \begin{bmatrix} b_1 \\ b_2 \\ b_3 \end{bmatrix}, \tag{3.1.3}$$

then the vector $\mathbf{a}+\mathbf{b}$ will be represented by the column matrix $\mathbf{a}+\mathbf{b}$. It is because of this correspondence between vectors and column matrices—and there is clearly a similar correspondence between vectors and row matrices—that it is permissible to use the terms 'column vector' and 'row vector' for column and row matrices.

Of course, in our use of these terms we have extended them to cover matrices with n components instead of the original three.

Two vectors **a** and **b** are 'proportional' if their components are proportional. If they are represented by lines from the origin then, when they are proportional, they will be represented by segments of the same line. Thus when they are proportional there will be a number λ such that

$$b_1 = \lambda a_1, \quad b_2 = \lambda a_2, \quad b_3 = \lambda a_3, \tag{3.1.4}$$

or, in the matrix notation of equation (3.1.3),

$$\mathbf{b} = \lambda \mathbf{a}. \tag{3.1.5}$$

By putting $\lambda = -x_1/x_2$ we can write the last equation in the alternative, and preferable, form

$$x_1 \mathbf{a} + x_2 \mathbf{b} = \mathbf{0}. \tag{3.1.6}$$

If the (column, row, or ordinary) vectors **a** and **b** are connected by a relation of the type (3.1.6) for some values of x_1 and x_2 *which are not both zero* they are said to be 'linearly dependent'. The linear dependence of two vectors is slightly more general than proportionality. Proportionality corresponds to equation (3.1.6) when *neither* x_1 *nor* x_2 *is zero*; linear dependence holds between any vector and the zero vector (if $\mathbf{a} = 0$ then equation (3.1.6) can be satisfied by taking $x_1 \neq 0$, $x_2 = 0$). If two vectors are not linearly dependent they are said to be 'linearly independent'. Any two vectors which do not have the same direction are linearly independent.

The relation for three vectors corresponding to equation (3.1.6) is

$$x_1 \mathbf{a} + x_2 \mathbf{b} + x_3 \mathbf{c} = \mathbf{0} \quad \text{or} \quad x_1 \mathbf{a} + x_2 \mathbf{b} + x_3 \mathbf{c} = \mathbf{0}, \tag{3.1.7}$$

where the second equation is a matrix equation, and in which we again demand that *not all of* x_1, x_2, x_3 shall be zero. This is the condition for the three vectors to lie in one plane, but again there are various degenerate cases. Thus if **a** and **b** are in the same line then **a**, **b** and any other vector **c** will lie in the same plane and so be linearly dependent. If **a**, **b** and **c** are not in the same plane then no relation of the type (3.1.7) will hold between them and they will be linearly independent.

EXAMPLES 3.1

1. Using the parallelogram law of addition show that if **a** and **b** are represented by the directed lines OP and OQ then $\mathbf{a}-\mathbf{b}$ is represented by a line equal and parallel to (and in the same sense as) the diagonal QP. Hence or otherwise show that if **a**, **b** and **c** are equal and parallel to the sides BC, CA and AB of the triangle ABC then

$$\mathbf{a}+\mathbf{b}+\mathbf{c} = \mathbf{0}.$$

[This result is equivalent to the 'triangle of forces'; **c** is the equilibrant of **a** and **b**.]

2. Using the result of Ex. 3.1.1 prove that if $\mathbf{a}^{(1)}, \mathbf{a}^{(2)}, \ldots, \mathbf{a}^{(n)}$ are equal and parallel to the sides $A_1A_2, A_2A_3, \ldots, A_{n-1}A_n, A_nA_1$ of the polygon $A_1A_2, \ldots, A_nA_1$ then

$$\mathbf{a}^{(1)}+\mathbf{a}^{(2)}+\ldots+\mathbf{a}^{(n)} = \mathbf{0}.$$

[This is equivalent to the 'polygon of forces'; $\mathbf{a}^{(n)}$ is the equilibrant of $\mathbf{a}^{(1)}, \mathbf{a}^{(2)}, \ldots, \mathbf{a}^{n-(1)}$.]

3.2 Linear dependence

The geometric significance of linear dependence for vectors in three-dimensional space has been briefly discussed in the last section. We shall now proceed with an algebraic treatment of the subject. We shall deal with column vectors, but the only reason for this is that we have to choose rows or columns, one or the other.

A set of column matrices, or column vectors, is said to form a 'linearly dependent set' if there is a linear relation between them. Thus the three column matrices

$$\mathbf{A}^{(1)} = \begin{bmatrix} 3 \\ -2 \\ -2 \end{bmatrix}, \quad \mathbf{A}^{(2)} = \begin{bmatrix} -1 \\ 4 \\ -1 \end{bmatrix}, \quad \mathbf{A}^{(3)} = \begin{bmatrix} 2 \\ 2 \\ -3 \end{bmatrix} \qquad (3.2.1)$$

are linearly dependent because

$$\mathbf{A}^{(1)} = -\mathbf{A}^{(2)} + \mathbf{A}^{(3)}, \qquad (3.2.2)$$

or, as we shall prefer to write such a relation,

$$\mathbf{A}^{(1)} + \mathbf{A}^{(2)} - \mathbf{A}^{(3)} = \mathbf{0}. \qquad (3.2.3)$$

We shall now give a formal definition of linear dependence. The column matrices

$$\mathbf{A}^{(1)} = \begin{bmatrix} a_{11} \\ a_{21} \\ \vdots \\ a_{m1} \end{bmatrix}, \quad \mathbf{A}^{(2)} = \begin{bmatrix} a_{12} \\ a_{22} \\ \vdots \\ a_{m2} \end{bmatrix}, \quad \dots, \quad \mathbf{A}^{(n)} = \begin{bmatrix} a_{1n} \\ a_{2n} \\ \vdots \\ a_{mn} \end{bmatrix} \qquad (3.2.4)$$

are said to be linearly dependent if numbers $x_1, x_2, \dots, x_n$ can be found such that

$$x_1\mathbf{A}^{(1)} + x_2\mathbf{A}^{(2)} + \dots + x_n\mathbf{A}^{(n)} = \mathbf{0}. \qquad (3.2.5)$$

But, since we can always satisfy this equation by taking all of the x's equal to zero, we add the restriction that $x_1, x_2, \dots, x_n$ shall not all be zero.

If the column matrices $\mathbf{A}^{(1)}, \mathbf{A}^{(2)}, \dots, \mathbf{A}^{(n)}$ are not linearly dependent they are said to be 'linearly independent'. One way of showing the linear independence of a set of column matrices is to show that an equation like (3.2.5) can be satisfied only if all the numbers $x_1, x_2, \dots, x_n$ are zero, that is only if

$$x_1 = 0 = x_2 = \dots = x_n. \qquad (3.2.6)$$

Equation (3.2.5) implies m algebraic equations, one for each row of the column matrices. Moreover, these m equations are exactly the m equations (3.1.2). It follows that the linear dependence of the matrices $\mathbf{A}^{(1)}, \mathbf{A}^{(2)}, \dots, \mathbf{A}^{(n)}$ is equivalent to the existence of a non-trivial solution of equation (3.1.2)—each implies the other. A corollary of this which we shall find useful may be stated as follows: 'The column matrices $\mathbf{A}^{(1)}, \mathbf{A}^{(2)}, \dots, \mathbf{A}^{(n)}$ are linearly dependent if, and only if, equations (3.1.2) have a non-trivial solution.'

The reader should note that we can write not only equations (3.1.2), but also (3.2.5), in the matrix form

$$\mathbf{Ax} = \mathbf{0} \qquad (3.2.7)$$

with
$$\mathbf{A} = \begin{bmatrix} a_{11} & a_{12} & \dots & a_{1n} \\ a_{21} & a_{22} & \dots & a_{2n} \\ \dots & \dots & \dots & \dots \\ a_{m1} & a_{m2} & \dots & a_{mn} \end{bmatrix}, \quad \mathbf{x} = \begin{bmatrix} x_1 \\ x_2 \\ \vdots \\ x_n \end{bmatrix}. \tag{3.2.8}$$

With these preliminary results on linear dependence we are in a position to tackle the first question posed in § 3.1—that concerning the existence of a solution of equations (3.1.2). In doing this, we shall consider the three possibilities $m = n$, $m < n$ and $m > n$, and shall take them in that order.

3.2.1 Number of equations equal to the number of unknowns: $m = n$

We shall prove the following theorem which may well be called the fundamental theorem on linear dependence.

Theorem 3.2.1 (*a*). A necessary and sufficient condition for the n column matrices $\mathbf{A}^{(1)}, \mathbf{A}^{(2)}, \dots, \mathbf{A}^{(n)}$ (each having n components) to be linearly dependent is that $|\mathbf{A}|$ shall be zero.

Proof of necessity. We have to prove that if $\mathbf{A}^{(1)}, \mathbf{A}^{(2)}, \dots, \mathbf{A}^{(n)}$ are linearly dependent then $|\mathbf{A}| = 0$. If the column matrices are linearly dependent then there is a non-trivial solution $\mathbf{x}$ of equation (3.2.7). The matrix $\mathbf{A}$ in that equation is now square. Suppose, if possible, that $|\mathbf{A}|$ is not zero; then $\mathbf{A}$ has an inverse $\mathbf{A}^{-1}$. Premultiply the equation by this inverse. The result is

$$\mathbf{A}^{-1}\mathbf{A}\mathbf{x} = \mathbf{I}\mathbf{x} = \mathbf{x} = \mathbf{0}, \tag{3.2.9}$$

which contradicts the assumption that $\mathbf{x}$ is non-trivial. Therefore $\mathbf{A}$ cannot have an inverse, and $|\mathbf{A}|$ must be zero. This is the argument that we used in § 1.8.

Proof of sufficiency. We have to prove that if $|\mathbf{A}| = 0$ then $\mathbf{A}^{(1)}, \mathbf{A}^{(2)}, \dots, \mathbf{A}^{(n)}$ are linearly dependent. We shall do this by proving that if the column matrices are linearly independent then $|\mathbf{A}|$ is not zero. This is equivalent to the result we have to prove.

Suppose that $\mathbf{A}^{(1)}, \mathbf{A}^{(2)}, \dots, \mathbf{A}^{(n)}$ are linearly independent. $\mathbf{A}^{(1)}$ must have at least one non-zero element; for if it contained only zeros we could satisfy (3.2.5) by putting $x_1 = 1, x_2 = 0 = \dots = x_n$, so that the column matrices would be linearly dependent, contrary to hypothesis. Suppose then that a_{i1}, the element in the ith row of $\mathbf{A}^{(1)}$ is non-zero. From $\mathbf{A}^{(2)}, \mathbf{A}^{(3)}, \dots, \mathbf{A}^{(n)}$ subtract such multiples of $\mathbf{A}^{(1)}$ as will make the resultant column matrices $\mathbf{B}^{(2)}, \mathbf{B}^{(3)}, \dots, \mathbf{B}^{(n)}$ each have a zero in row i. Then

$$\mathbf{B}^{(r)} = \mathbf{A}^{(r)} - \frac{a_{ir}}{a_{i1}}\mathbf{A}^{(1)} \quad (r = 2, 3, \dots, n). \tag{3.2.10}$$

By rule (*i*) of § 1.5, if a multiple of one column of a determinant is subtracted from another, the value of the determinant remains the same. Therefore the determinant of the matrices $\mathbf{A}^{(1)}, \mathbf{B}^{(2)}, \mathbf{B}^{(3)}, \dots, \mathbf{B}^{(n)}$ is equal to $|\mathbf{A}|$.

We may illustrate this procedure by taking the matrices

$$\mathbf{A}^{(1)} = \begin{bmatrix} 1 \\ 2 \\ 3 \end{bmatrix}, \quad \mathbf{A}^{(2)} = \begin{bmatrix} 1 \\ -4 \\ 1 \end{bmatrix}, \quad \mathbf{A}^{(3)} = \begin{bmatrix} 2 \\ -3 \\ 1 \end{bmatrix}. \tag{3.2.11}$$

Their determinant is

$$|\mathbf{A}| = \begin{vmatrix} 1 & 1 & 2 \\ 2 & -4 & -3 \\ 3 & 1 & 1 \end{vmatrix} = 16. \tag{3.2.12}$$

Taking the second row (i.e. taking $i = 2$) we now have

$$\mathbf{A}^{(1)} = \begin{bmatrix} 1 \\ 2 \\ 3 \end{bmatrix}, \quad \mathbf{B}^{(2)} = \begin{bmatrix} 3 \\ 0 \\ 7 \end{bmatrix}, \quad \mathbf{B}^{(3)} = \begin{bmatrix} 7/2 \\ 0 \\ 11/2 \end{bmatrix}, \tag{3.2.13}$$

and the new determinant is

$$\begin{vmatrix} 1 & 3 & 7/2 \\ 2 & 0 & 0 \\ 3 & 7 & 11/2 \end{vmatrix} = 16 = |\mathbf{A}|. \tag{3.2.14}$$

Returning to the general case we note that $\mathbf{A}^{(1)}, \mathbf{B}^{(2)}, \ldots, \mathbf{B}^{(n)}$ must be linearly independent. For suppose that

$$y_1\mathbf{A}^{(1)} + y_2\mathbf{B}^{(2)} + \ldots + y_n\mathbf{B}^{(n)} = \mathbf{0}, \tag{3.2.15}$$

then

$$\left\{y_1 - \frac{a_{i2}y_2 + a_{i3}y_3 + \ldots + a_{in}y_n}{a_{i1}}\right\}\mathbf{A}^{(1)} + y_2\mathbf{A}^{(2)} + \ldots + y_n\mathbf{A}^{(n)} = \mathbf{0}. \tag{3.2.16}$$

Since the $\mathbf{A}$'s are linearly independent each of the coefficients in this equation must be zero. This means that

$$y_1 = 0 = y_2 = \ldots = y_n, \tag{3.2.17}$$

and therefore $\mathbf{A}^{(1)}, \mathbf{B}^{(2)}, \ldots, \mathbf{B}^{(n)}$ are linearly independent. Since $\mathbf{A}^{(1)}, \mathbf{B}^{(2)}, \ldots, \mathbf{B}^{(n)}$ are linearly independent, $\mathbf{B}^{(2)}$ must have at least one non-zero element—for otherwise we could satisfy (3.2.15) by putting

$$y_1 = 0, \quad y_2 = 1, \quad y_3 = 0 = y_4 = \ldots = y_n. \tag{3.2.18}$$

Suppose that the element b_{j2} in the jth row is non-zero. Subtract such multiples of $\mathbf{B}^{(2)}$ from $\mathbf{B}^{(3)}, \mathbf{B}^{(4)}, \ldots, \mathbf{B}^{(n)}$ as will make the resulting matrices $\mathbf{C}^{(3)}, \mathbf{C}^{(4)}, \ldots, \mathbf{C}^{(n)}$ have zero in the jth row. Then

$$b_{jr} = a_{jr} - \frac{a_{ir}}{a_{i1}}a_{j1} \quad (r = 2, 3, \ldots, n), \tag{3.2.19}$$

and

$$\mathbf{C}^{(r)} = \mathbf{B}^{(r)} - \frac{b_{jr}}{b_{j2}}\mathbf{B}^{(2)} \quad (r = 3, 4, \ldots, n). \tag{3.2.20}$$

In the above example, for instance, we may use the element in row 3 of $\mathbf{B}^{(2)}$ for this purpose. We then arrive at the set

$$\mathbf{A}^{(1)} = \begin{bmatrix} 1 \\ 2 \\ 3 \end{bmatrix}, \quad \mathbf{B}^{(2)} = \begin{bmatrix} 3 \\ 0 \\ 7 \end{bmatrix}, \quad \mathbf{C}^{(3)} = \begin{bmatrix} 8/7 \\ 0 \\ 0 \end{bmatrix}, \tag{3.2.21}$$

and it may easily be verified that the determinant which has these matrices as columns still has the value 16.

Returning to the general case we note that this process may be continued and that, at each stage, the column matrices are linearly independent, and their determinant has the same value, namely $|\mathbf{A}|$. The n numbers $i, j, k, \ldots, t$ which are arrived at in this way are all different and are therefore the numbers $1, 2, 3, \ldots, n$ in some order. The set of n column matrices $\mathbf{A}^{(1)}, \mathbf{B}^{(2)}, \mathbf{C}^{(3)}, \ldots, \mathbf{Z}^{(n)}$ are such that $\mathbf{A}^{(1)}$ is the only one with a non-zero term in row i; $\mathbf{B}^{(2)}$, but no subsequent matrix, has a non-zero term in row j; $\mathbf{C}^{(3)}$, but no subsequent matrix, has a non-zero term in yet another row, and so on. The form of the determinant is therefore

$$\begin{vmatrix} a_{11} & b_{12} & c_{13} & d_{14} & \cdots \\ \cdots & \cdots & \cdots & \cdots & \cdots \\ a_{j1} & b_{j2} & 0 & 0 & \cdots \\ \cdots & \cdots & \cdots & \cdots & \cdots \\ a_{k1} & b_{k2} & c_{k3} & 0 & \cdots \\ \cdots & \cdots & \cdots & \cdots & \cdots \\ a_{i1} & 0 & 0 & 0 & \cdots \\ a_{n1} & b_{n2} & c_{n3} & d_{n4} & \cdots \end{vmatrix}. \tag{3.2.22}$$

We may find the value of this determinant by suitably rearranging the rows; this rearranging will either leave the value of the determinant the same, or merely change its sign. Rearrange the rows so that they occur in the order $i, j, k, \ldots, t$. The new determinant will be

$$\begin{vmatrix} a_{i1} & 0 & 0 & 0 & \cdots & 0 \\ a_{j1} & b_{j2} & 0 & 0 & \cdots & 0 \\ a_{k1} & b_{k2} & c_{k3} & 0 & \cdots & 0 \\ \cdots & \cdots & \cdots & \cdots & \cdots & \cdots \\ a_{t1} & b_{t2} & c_{t3} & d_{t4} & \cdots & z_{tn} \end{vmatrix}. \tag{3.2.23}$$

In this determinant all the elements above the diagonal are zero and all those on the diagonal are non-zero. By rule (*d*) of § 1.5 the determinant is just the product of the diagonal terms, so that the value of the original determinant is

$$|\mathbf{A}| = \pm a_{i1} b_{j2} c_{k3} \cdots z_{tn}, \tag{3.2.24}^\dagger$$

which is not zero. This proves the required result.

† The sign to be taken is positive or negative according as the determinant (3.2.23) is obtained from the determinant (3.2.22) by an even or an odd number of interchanges of two rows.

We may illustrate the last step by evaluating the determinant for the matrices (3.2.21) in the form

$$|\mathbf{A}| = \begin{vmatrix} 1 & 3 & 8/7 \\ 2 & 0 & 0 \\ 3 & 7 & 0 \end{vmatrix} = +\begin{vmatrix} 2 & 0 & 0 \\ 3 & 7 & 0 \\ 1 & 3 & 8/7 \end{vmatrix} = 16. \tag{3.2.25}$$

It follows from the argument at the beginning of this section that the theorem which we have proved is equivalent to

Theorem 3.2.1 (*b*). A necessary and sufficient condition for the n equations (3.1.2) to have a non-trivial solution is that the determinant $|\mathbf{A}|$ shall be zero.

This provides a complete answer to the first of our two questions concerning equations (3.1.2) when $m = n$.

3.2.2 Number of equations less than the number of unknowns: $m < n$

The corresponding theorems for the case $m < n$ follow as simple corollaries of Theorems 3.2.1. Consider the $n \times n$ determinant

$$\begin{vmatrix} a_{11} & a_{12} & \dots & a_{1n} \\ a_{21} & a_{22} & \dots & a_{2n} \\ \dots & \dots & \dots & \dots \\ a_{m1} & a_{m2} & \dots & a_{mn} \\ 0 & 0 & \dots & 0 \\ \dots & \dots & \dots & \dots \\ 0 & 0 & \dots & 0 \end{vmatrix},$$

which is formed by adding $n-m$ rows of noughts to the original array. This determinant is clearly zero so that, by Theorem 3.2.1 (*a*), its columns must be linearly dependent. In particular, the first m members of its columns, that is, those elements which form the original column matrices, must be linearly dependent. This proves the theorem:

Theorem 3.2.2 (*a*). If $m < n$, that is, if the number of rows is less than the number of vectors, then any set of n column vectors is a linearly dependent set. This may be stated as a theorem concerning the solutions of equation (3.1.2), namely:

Theorem 3.2.2 (*b*). If $m < n$, that is, if the number of equations is less than the number of unknowns, then the equations (3.1.2) always have a non-trivial solution.

To illustrate these theorems we may assert that the three column matrices

$$\mathbf{A}^{(1)} = \begin{bmatrix} 3 \\ 4 \end{bmatrix}, \quad \mathbf{A}^{(2)} = \begin{bmatrix} 2 \\ -1 \end{bmatrix}, \quad \mathbf{A}^{(3)} = \begin{bmatrix} -1 \\ 3 \end{bmatrix} \tag{3.2.26}$$

must be such that we can find numbers x_1, x_2, x_3 which are not all zero and which satisfy

$$x_1\mathbf{A}^{(1)} + x_2\mathbf{A}^{(2)} + x_3\mathbf{A}^{(3)} = \mathbf{0}. \tag{3.2.27}$$

This is equivalent to the statement that the equations

$$\left.\begin{aligned} 3x_1+2x_2-x_3 &= 0, \\ 4x_1-x_2+3x_3 &= 0 \end{aligned}\right\} \tag{3.2.28}$$

must have a non-trivial solution. The values

$$x_1 = 5, \quad x_2 = -13, \quad x_3 = -11 \tag{3.2.29}$$

satisfy these equations.

3.2.3 Number of equations greater than the number of unknowns: $m > n$

When $m > n$, the matrix $\mathbf{A}$ of equation (3.2.7) has more rows than columns. Under these circumstances the set of n column matrices $\mathbf{A}^{(1)}, \mathbf{A}^{(2)}, \ldots, \mathbf{A}^{(n)}$ (each having m elements) may or may not be linearly dependent. For instance the matrices

$$\mathbf{A}^{(1)} = \begin{bmatrix} 1 \\ 2 \\ 3 \end{bmatrix}, \quad \mathbf{A}^{(2)} = \begin{bmatrix} -3 \\ -6 \\ -9 \end{bmatrix} \tag{3.2.30}$$

are linearly dependent since

$$3\mathbf{A}^{(1)}+\mathbf{A}^{(2)} = \mathbf{0}. \tag{3.2.31}$$

On the other hand, the matrices

$$\mathbf{A}^{(1)} = \begin{bmatrix} 1 \\ 2 \\ 3 \end{bmatrix}, \quad \mathbf{A}^{(2)} = \begin{bmatrix} -3 \\ -3 \\ -9 \end{bmatrix} \tag{3.2.32}$$

form a linearly independent set.

We shall prove the following theorem:

Theorem 3.2.3 (*a*). A necessary and sufficient condition for the n ($<m$) column matrices $\mathbf{A}^{(1)}, \mathbf{A}^{(2)}, \ldots, \mathbf{A}^{(n)}$ to be linearly dependent is that all the $n \times n$ determinants obtained by selecting n rows from $\mathbf{A}$ shall be zero.

The condition is necessary. For if the n column matrices are linearly dependent then the column matrices formed by taking the first n rows of the matrices must be linearly dependent. But by Theorem 3.2.1 (*a*) this means that the uppermost $n \times n$ determinant of $\mathbf{A}$ is zero. Similarly, each set of n column matrices obtained by selecting elements from a certain n rows must be a linearly dependent set and so the corresponding determinant must be zero.

We may also prove that the condition is sufficient. However, we can prove this in almost the same way as we proved the sufficiency of Theorem 3.2.1 (*a*), and for this reason we shall not give the details. We prove that if the column matrices are linearly independent then, by the process of 'triangulation' which we described (equations (3.2.10) to (3.2.23)), it must be possible to find a non-zero determinant of order n in the matrix $\mathbf{A}$. We leave the details of the proof to the reader who may also apply the theorem to the sets (3.2.30) and (3.2.32).

At the beginning of this section we showed that the column vectors

$$\mathbf{A}^{(1)}, \quad \mathbf{A}^{(2)}, \quad \ldots, \quad \mathbf{A}^{(n)}$$

are linearly dependent if and only if the equations (3.1.2) have a non-trivial solution. Combining this with Theorem 3.2.3 (*a*) we obtain

Theorem 3.2.3 (*b*). When the number of equations is greater than the number of unknowns, that is, $m > n$, then a necessary and sufficient condition for the equations (3.1.2) to have a non-trivial solution is that all the $n \times n$ determinants obtained by selecting n rows from $\mathbf{A}$ shall be zero. We may illustrate the theorem by considering the equations corresponding to (3.2.30) and (3.2.32). The first set,

$$x_1 - 3x_2 = 0; \quad 2x_1 - 6x_2 = 0; \quad 3x_1 - 9x_2 = 0 \tag{3.2.33}$$

have a non-trivial solution, whereas the second set,

$$x_1 - 3x_2 = 0; \quad 2x_1 - 3x_2 = 0; \quad 3x_1 - 9x_2 = 0 \tag{3.2.34}$$

have not.

We have now answered the first of the two questions which we raised in § 3.1. We have found the conditions under which the equations (3.1.2) have a non-trivial solution. Before we can answer the second question we must consider further the notions of linear dependence and linear independence.

EXAMPLES 3.2

1. Test the following sets of vectors for linear dependence

(*a*) $\mathbf{A}^{(1)} = \begin{bmatrix} 0{\cdot}162 \\ -1{\cdot}219 \\ 3{\cdot}217 \end{bmatrix}, \quad \mathbf{A}^{(2)} = \begin{bmatrix} 1{\cdot}963 \\ 2{\cdot}145 \\ -7{\cdot}904 \end{bmatrix}, \quad \mathbf{A}^{(3)} = \begin{bmatrix} 1{\cdot}904 \\ -0{\cdot}061 \\ 4{\cdot}915 \end{bmatrix}, \quad \mathbf{A}^{(4)} = \begin{bmatrix} 1{\cdot}065 \\ 1{\cdot}992 \\ -3{\cdot}215 \end{bmatrix}.$

(*b*) $\mathbf{A}^{(1)} = \begin{bmatrix} 1 \\ 6 \\ 2 \\ -4 \end{bmatrix}, \quad \mathbf{A}^{(2)} = \begin{bmatrix} -5 \\ 4 \\ 3 \\ 17 \end{bmatrix}, \quad \mathbf{A}^{(3)} = \begin{bmatrix} -7 \\ 26 \\ 12 \\ 22 \end{bmatrix}.$

(*c*) $\mathbf{A}^{(1)} = \begin{bmatrix} 1 \\ 4 \\ 2 \end{bmatrix}, \quad \mathbf{A}^{(2)} = \begin{bmatrix} 1 \\ 9 \\ 21 \end{bmatrix}, \quad \mathbf{A}^{(3)} = \begin{bmatrix} -4 \\ 1 \\ 17 \end{bmatrix}.$

2. Without expanding it, evaluate the determinant

$$\begin{vmatrix} 1 & 2 & 5 & 4 & 6 \\ 9 & 1 & 4 & 2 & -4 \\ -2 & 6 & -1 & -11 & -16 \\ 3 & 2 & 6 & 31 & 32 \\ 4 & 1 & 7 & 63 & 65 \end{vmatrix}.$$

3. Discuss whether the following sets of equations have non-zero solutions:

(*a*)
$$0{\cdot}162x+1{\cdot}963y+1{\cdot}904z+7{\cdot}241t = 0;$$
$$-1{\cdot}219x+2{\cdot}145y-0{\cdot}061z+2{\cdot}159t = 0;$$
$$3{\cdot}217x-7{\cdot}904y+4{\cdot}915z-4{\cdot}197t = 0.$$

(*b*)
$$x+5y-2z = 0, \quad 4x+2y-7z = 0, \quad 5x-21y-4z = 0.$$

(*c*)
$$x+2y-17z = 0, \quad 2x+5y-11z = 0.$$
$$4x+3y+62z = 0, \quad 5x+12y-39z = 0.$$

4. Fill in the details of the proof of sufficiency of Theorem 3.2.3 (*a*).

5. For what values of a do the equations
$$x+3y-2z = 0, \quad ax+5y+3z = 0, \quad 3x+11y-az = 0$$
have a solution?

3.3 The minors of a matrix; rank

The determinants which occur in the statement of Theorem 3.2.3 (*a*) are called 'minors' of the matrix $\mathbf{A}$; more precisely, they are called 'minors of order n'. Since the concept of a minor is more general and is of importance, we need a formal definition.

Consider a matrix $\mathbf{A}$ of order $m \times n$ and delete all the elements of $\mathbf{A}$ except those in a certain r rows and r columns. If $r > 1$, the remaining elements form a square matrix of order r and the determinant of this matrix is called a minor of $\mathbf{A}$ of order r. Thus the determinant

$$\begin{vmatrix} -1 & 3 \\ 4 & 7 \end{vmatrix} \text{ is a minor of } \begin{bmatrix} 2 & -1 & 5 & 3 \\ 1 & 2 & -4 & 5 \\ 9 & 4 & 0 & 7 \end{bmatrix} \tag{3.3.1}$$

of order 2. Each element of a matrix is a minor of order 1.

The notion of a minor is, in some ways, an extension to rectangular matrices of the idea of a cofactor introduced for determinants. A cofactor in an $n \times n$ determinant $|\mathbf{A}|$ is a minor of order $(n-1)$ of the matrix $\mathbf{A}$, with a sign attached.

Any minor of order $r+1$ (where $r \geqslant 1$), being a determinant, can be expanded (e.g. along its first row) and so expressed as a sum of multiples of minors of order r. Hence if *every* minor of order r is zero, then every minor of order $(r+1)$ must also be zero. For example in the matrix

$$\mathbf{A} = \begin{bmatrix} 3 & 1 & 2 \\ -6 & -2 & -4 \\ 9 & 3 & 6 \end{bmatrix} \tag{3.3.2}$$

every minor of order 2 is zero and, therefore, the only minor of order 3, the determinant $|\mathbf{A}|$, is zero.

The converse of this last property is not true. That is, if every minor of order $(r+1)$ is zero, then it does *not* follow that every minor of order r is zero also. For example, in the matrix above every minor of order 2 is zero whereas *no* minor of order 1 is.

We are now in a position to define the rank of a matrix. Roughly speaking the rank of a matrix is a measure of its degeneracy—or, rather, of its lack of degeneracy. Consider first the matrix **A** of equation (3.3.2). The minor of order 3, (viz. $|\mathbf{A}|$) is zero while *every* minor of order 2 (and there are 9 of them) is zero also. There is at least one non-zero minor of order 1—in fact none are zero—and we shall thus refer to this matrix as being 'of rank 1'.

The matrix (3.3.1) is of order 3×4. The minor

$$\begin{vmatrix} 2 & -1 & 5 \\ 1 & 2 & -4 \\ 9 & 4 & 0 \end{vmatrix} \tag{3.3.3}$$

has the value -2 so that at least one minor of order 3 is non-zero. This matrix therefore has rank 3.

The rank of a matrix **A** may be defined formally as follows. If every single element of **A** is zero its rank is zero. If **A** is not a null matrix there will be at least one set of minors of the same order which are not all zero. If every minor of order $(r+1)$ is zero while at least one minor of order r is non-zero, then **A** is said to be of rank r.

The rank of a matrix must obey certain rules, three of the more elementary ones being as follows.

(i) The rank of a rectangular matrix $\mathbf{A}(m, n)$ is less than or equal to the smaller of m and n; that is, in symbols,

$$r \leqslant \min(m, n). \tag{3.3.4}$$

(ii) The rank of a non-singular matrix of order n is n, while the rank of a singular matrix is less than n.

(iii) The rank of $\mathbf{A}'$ is equal to the rank of **A**.

The first two of these properties follow from the definition of a minor and of rank. It is left as an example for the reader to establish the third.

In § 3.2 we proved three theorems relating to the solution of equations (3.1.2). Using the concept of rank we may now combine these theorems to make one theorem which covers the three possibilities; $m = n$, $m < n$, $m > n$. We shall now state this theorem and the reader should check carefully that it covers the previous three (i.e. Theorems 3.2.1 (*b*), 3.2.2 (*b*) and 3.2.3 (*b*)).

Theorem 3.3.1. A necessary and sufficient condition for the m equations (3.1.2) in the n unknowns $x_1, x_2, \ldots, x_n$ to have a non-trivial solution is that the rank of **A** shall be less than n.

EXAMPLES 3.3

1. Determine the ranks of the following matrices

(*a*) $[1 \quad 0 \quad 0 \quad 0 \quad 0]$, (*b*) $\begin{bmatrix} 4 & 1 & 6 \\ 3 & 2 & 17 \\ 0 & 5 & 50 \end{bmatrix}$, (*c*) $\begin{bmatrix} 1 & 9 & 6 & 3 & -17 \\ 1 & 0 & 4 & 91 & 106 \\ 0 & 2 & 3 & 269 & 123 \end{bmatrix}$,

(*d*) $\begin{bmatrix} 1 & 4 & 2 & 1 & 17 \\ 2 & 9 & 1 & 15 & 105 \\ 0 & 1 & -3 & 13 & 71 \end{bmatrix}$.

2. State whether it is possible to have a matrix

(*a*) $\mathbf{A}(3, 5)$ of rank 2; (*b*) $\mathbf{A}(2, 5)$ of rank 3;
(*c*) $\mathbf{A}(1, 7)$ of rank 4; (*d*) $\mathbf{A}(4, 2)$ of rank 1.

When it is possible give one example of such a matrix.

3. Find the values of a and b for which

$$\begin{bmatrix} a & 4 & 2 & b \\ 2 & 8 & 4a & 10 \\ -3a & -12 & -3a & -15 \end{bmatrix}$$

has rank (*a*) 1, (*b*) 2, (*c*) 3.

3.4 Rank and linear dependence

So far we have discussed the linear dependence and linear independence of sets of column vectors. Thus equation (3.2.5) concerns the matrices (3.2.4) and we noted that it is equivalent to the n algebraic equations (3.1.2). However, we may also discuss the linear dependence of row vectors. The row vectors

$$\left.\begin{aligned} \mathbf{A}'_{(1)} &= [a_{11} \quad a_{12} \quad \dots \quad a_{1n}], \\ \mathbf{A}'_{(2)} &= [a_{21} \quad a_{22} \quad \dots \quad a_{2n}], \\ &\dots\dots\dots\dots\dots\dots \\ \mathbf{A}'_{(m)} &= [a_{m1} \quad a_{m2} \quad \dots \quad a_{mn}], \end{aligned}\right\} \tag{3.4.1}$$

are linearly dependent if we can find numbers $x_1, x_2, \dots, x_m$, not all zero, such that

$$x_1\mathbf{A}'_{(1)} + x_2\mathbf{A}'_{(2)} + \dots + x_m\mathbf{A}'_{(m)} = \mathbf{0}. \tag{3.4.2}$$

Equation (3.4.2) is equivalent to the n equations

$$\left.\begin{aligned} a_{11}x_1 + a_{21}x_2 + \dots + a_{m1}x_m &= 0, \\ a_{12}x_1 + a_{22}x_2 + \dots + a_{m2}x_m &= 0, \\ &\dots\dots\dots\dots \\ a_{1n}x_1 + a_{2n}x_2 + \dots + a_{mn}x_m &= 0, \end{aligned}\right\} \tag{3.4.3}$$

in the m unknowns $x_1, x_2, \ldots, x_m$. We may write these equations as the single matrix equation

$$\mathbf{A}'\mathbf{x} = \mathbf{0} \quad \text{or} \quad \mathbf{x}'\mathbf{A} = \mathbf{0}, \tag{3.4.4}$$

where $\mathbf{A}'$ is the transpose of the matrix $\mathbf{A}$ given in equation (3.2.8), and $\mathbf{x}$ is the column vector

$$\mathbf{x} = \begin{bmatrix} x_1 \\ x_2 \\ \vdots \\ x_m \end{bmatrix}. \tag{3.4.5}$$

Equations (3.4.3) are of the same type as equations (3.1.2). By comparing the two sets we see that the question of the linear dependence or independence of the row vectors $\mathbf{A}'_{(1)}, \mathbf{A}'_{(2)}, \ldots, \mathbf{A}'_{(m)}$ is exactly equivalent to that for the column vectors

$$\mathbf{A}_{(1)} = \begin{bmatrix} a_{11} \\ a_{12} \\ \vdots \\ a_{1n} \end{bmatrix}, \quad \mathbf{A}_{(2)} = \begin{bmatrix} a_{21} \\ a_{22} \\ \vdots \\ a_{2n} \end{bmatrix}, \quad \ldots, \quad \mathbf{A}_{(m)} = \begin{bmatrix} a_{m1} \\ a_{m2} \\ \vdots \\ a_{mn} \end{bmatrix}, \tag{3.4.6}$$

obtained by transposing the row vectors. This means that each theorem concerning the linear dependence or independence of column vectors may be interpreted to give a corresponding theorem for row vectors. Thus Theorem 3.2.1 (*a*) leads to

Theorem 3.4.1. A necessary and sufficient condition for the n rows $\mathbf{A}'_{(1)}, \mathbf{A}'_{(2)}, \ldots, \mathbf{A}'_{(n)}$ of the square matrix $\mathbf{A}$ to be linearly dependent is that $|\mathbf{A}|$ shall be zero.

To obtain the theorem corresponding to Theorem 3.2.2 (*a*) we have merely to interchange m and n. The theorem is

Theorem 3.4.2. If $n < m$, that is, if the number of columns is less than the number of vectors, then any set of m row vectors is a linearly dependent set.

The derivation of the third theorem—corresponding to Theorem 3.2.3 (*a*)—requires a little more care. The row vectors $\mathbf{A}'_{(1)}, \mathbf{A}'_{(2)}, \ldots, \mathbf{A}'_{(m)}$ are linearly dependent if and only if the column vectors $\mathbf{A}_{(1)}, \mathbf{A}_{(2)}, \ldots, \mathbf{A}_{(m)}$ are. To apply Theorem 3.2.3 (*a*) to these column vectors we notice that they are the m columns of the matrix $\mathbf{A}'$. The condition for linear dependence is therefore that all the $m \times m$ determinants obtained by selecting m rows from $\mathbf{A}'$ shall be zero. But each of these determinants may be obtained by selecting m columns from $\mathbf{A}$. This leads to

Theorem 3.4.3. A necessary and sufficient condition for the $m\,(<n)$ row matrices $\mathbf{A}'_{(1)}, \mathbf{A}'_{(2)}, \ldots, \mathbf{A}'_{(m)}$, that is, the rows of $\mathbf{A}$, to be linearly dependent is that all the $m \times m$ determinants obtained by selecting m columns from $\mathbf{A}$ shall be zero.

In the last section, by using the concept of the rank of a matrix, we stated one composite theorem concerning the solutions of equations (3.1.2) which applied equally to the three cases $m = n$, $m < n$, $m > n$. We may also use this concept to give a composite theorem concerning the linear dependence of sets of column vectors. We leave it to the reader to check that the following theorem covers our previous theorems 3.2.1 (*a*), 3.2.2 (*a*), 3.2.3 (*a*).

Theorem 3.4.4. A necessary and sufficient condition for the n column matrices $\mathbf{A}^{(1)}, \mathbf{A}^{(2)}, \ldots, \mathbf{A}^{(n)}$, that is, the columns of $\mathbf{A}$, to be linearly dependent, is that the rank of $\mathbf{A}$ shall be less than n.

We have shown in this section that any theorem relating to column matrices has a counterpart for row matrices. The counterpart of the last theorem is

Theorem 3.4.5. A necessary and sufficient condition for the m row matrices $\mathbf{A}'_{(1)}, \mathbf{A}'_{(2)}, \ldots, \mathbf{A}'_{(m)}$, that is, the rows of $\mathbf{A}$, to be linearly dependent, is that the rank of $\mathbf{A}$ shall be less than m.

If we have a set of column or row vectors and we know that they are linearly dependent, then it is a relevant question (and also an important question, as we shall see later) to ask 'How many linearly independent vectors are there in the set?'. Perhaps we can explain the meaning of this statement best by means of an example.

The rows of the 3×4 matrix

$$\mathbf{A} = \begin{bmatrix} 1 & 2 & 3 & 4 \\ -2 & -4 & -6 & -8 \\ 4 & 3 & 2 & 1 \end{bmatrix} \tag{3.4.7}$$

are linearly dependent. In fact a set of multipliers of the type referred to in equation (3.4.2) is

$$x_1 = 2, \quad x_2 = 1, \quad x_3 = 0. \tag{3.4.8}$$

If we consider the first two rows *only* then we see that they also are linearly dependent, and appropriate multipliers are

$$x_1 = 2, \quad x_2 = 1. \tag{3.4.9}$$

If we consider rows 1 and 3 *only* we see that they are linearly independent because it is impossible to find non-zero multipliers y_1 and y_3 such that

$$y_1 \times \mathbf{row}\,(1) + y_3 \times \mathbf{row}\,(3) = \mathbf{0}. \tag{3.4.10}$$

Similarly, rows 2 and 3 are linearly independent. Thus we can say: 'The three rows of the matrix are linearly dependent, but there is a subset consisting of two rows, viz. rows 1 and 3 (or rows 2 and 3) which is a linearly independent set.' This is the meaning which we shall attach to the statement that the matrix $\mathbf{A}$ of equation (3.4.7) has two linearly independent rows.

This definition may be generalised so that it applies to an arbitrary set of row or column vectors. The general statement is: If in a certain set of vectors every subset consisting of $(r+1)$ vectors is linearly dependent, but there is at least one linearly independent subset consisting of r vectors, then the set is said to have r linearly independent vectors.

Using this definition let us see how many linearly independent *columns* there are in the matrix of (3.4.7). By inspection we see that

$$2 \times \mathbf{col}\,(2) = \mathbf{col}\,(1) + \mathbf{col}\,(3), \tag{3.4.11}$$

$$\mathbf{col}\,(1) - \mathbf{col}\,(2) - \mathbf{col}\,(3) + \mathbf{col}\,(4) = \mathbf{0}. \tag{3.4.12}$$

The first of these relations shows that columns 1, 2, 3 are linearly dependent. The first combined with the second shows that all the other subsets consisting of three columns, viz. 1, 2, 4; 1, 3, 4 and 2, 3, 4 are linearly dependent. Therefore each of the 4 sets of 3 columns is linearly dependent. Since columns 1 and 2 are linearly independent there are just 2 linearly independent columns.

Before proving the general theorems concerning the number of linearly independent vectors in a set we notice that in this example the rank of the matrix **A** is 2.

The general theorems are:

Theorem 3.4.6. If the matrix **A** has rank r then it has just r linearly independent columns and just r linearly independent rows.

Theorem 3.4.7. If the matrix **A** has just r linearly independent columns (rows) then its rank is r and it has just r linearly independent rows (columns).

In order to prove these theorems we first establish two lemmas:

Lemma 1. If the matrix **A** has rank r then it has *at least* r linearly independent rows and *at least* r linearly independent columns.

Proof. If **A** has rank r then it has at least one non-zero minor of order r. Since this minor is non-zero, Theorem 3.2.1 (a) shows that its columns are linearly independent, while Theorem 3.4.1 shows that its rows are linearly independent. Therefore the corresponding r columns and r rows of **A** of which they form parts will also be linearly independent.

Lemma 2. If **A** has r linearly independent columns (rows) then its rank is at least equal to r.

Proof. Suppose that the first r columns of **A** are linearly independent. (If they are not, then some other set of r columns will be, and the proof will be essentially the same as the following with only slight modifications.) Consider the matrix

$$\mathbf{B} = \begin{bmatrix} a_{11} & a_{12} & \cdots & a_{1r} \\ a_{21} & a_{22} & \cdots & a_{2r} \\ \cdots & \cdots & \cdots & \cdots \\ a_{m1} & a_{m2} & \cdots & a_{mr} \end{bmatrix}. \tag{3.4.13}$$

At least one of the determinants obtained by selecting r rows† from **B** must be non-zero. This follows from Theorem 3.2.1 (a) if $r = m$, or from Theorem 3.2.3 (a) if $r < m$. But therefore **A** has at least one non-zero minor of order r so that its rank is at least equal to r.

We may now prove Theorem 3.4.6. Lemma 1 shows that **A** has at least r linearly independent rows and columns. Lemma 2 shows that it cannot have more than r, for if it had $r+1$ its rank would be at least equal to $r+1$.

† r, the rank of **A**, must always be less than or equal to m.

To prove Theorem 3.4.7 we use the lemmas in the reverse order. If $\mathbf{A}$ has just r linearly independent columns then, by Lemma 2, its rank is at least equal to r. Its rank cannot be greater for then, by Lemma 1, it would have more than r linearly independent columns. Therefore its rank is r and it has r linearly independent rows.

An important corollary of these theorems is:

Theorem 3.4.8. If $\mathbf{A}$ is of rank r then it is possible to choose r columns (rows), namely r linearly independent columns (rows), and express every other column (row) as a linear combination of them. It is not possible to do this with less than r columns (rows). The proof of the theorem, which is obvious, is left to the reader.

The problem of finding the rank of a large matrix is difficult, and at this stage we can do no more than give a few hints. It is completely impracticable to hunt for the largest non-zero minor. Rather one should use the sort of procedure which was advocated in § 1.5 for the evaluation of determinants. We may use this procedure because the rank of $\mathbf{A}$ remains the same if:

(*a*) All the elements of a row (column) are multiplied by a constant factor.

(*b*) Two rows (columns) are interchanged.

(*c*) A linear combination of some of the rows (columns) is added to one row (column).

None of these operations changes the number of linearly independent rows (columns) of $\mathbf{A}$, and we may combine them to form a 'triangulation' process.

Suppose, for example, that

$$\mathbf{A} = \begin{bmatrix} 1 & 2 & 1 & 1 \\ 1 & -1 & 1 & 1 \\ 3 & 0 & 3 & 3 \\ 1 & -1 & -1 & -2 \end{bmatrix}, \tag{3.4.14}$$

then by performing operations (*a*), (*b*) and (*c*) on the rows of $\mathbf{A}$ we may reduce it as follows:

$$\mathbf{A} \rightarrow \begin{bmatrix} 1 & 2 & 1 & 1 \\ 0 & -3 & 0 & 0 \\ 0 & -6 & 0 & 0 \\ 0 & -3 & -2 & -3 \end{bmatrix} \rightarrow \begin{bmatrix} 1 & 2 & 1 & 1 \\ 0 & 1 & 0 & 0 \\ 0 & 0 & 0 & 0 \\ 0 & 0 & -2 & -3 \end{bmatrix} \rightarrow \begin{bmatrix} 1 & 2 & 1 & 1 \\ 0 & 1 & 0 & 0 \\ 0 & 0 & -2 & -3 \\ 0 & 0 & 0 & 0 \end{bmatrix}. \tag{3.4.15}$$

The last matrix clearly has just 3 linearly independent rows so that $\mathbf{A}$ has rank 3.

EXAMPLES 3.4

1. For what values of a are the row vectors

$$\mathbf{A}_{(1)} = [1 \quad a \quad 9], \quad \mathbf{A}_{(2)} = [4 \quad 2 \quad 17], \quad \mathbf{A}_{(3)} = [a \quad -1 \quad -10]$$

linearly dependent?

2. Determine the ranks of the following matrices by triangulation:

$$(a)\ \begin{bmatrix} 3 & 2 & -1 & 0 & 1 \\ 1 & 4 & 0 & 0 & -1 \\ 1 & 3 & -4 & 2 & 1 \\ 4 & 8 & 7 & -4 & -4 \\ 4 & 6 & -1 & 0 & 0 \end{bmatrix}, \qquad (b)\ \begin{bmatrix} 1 & 2 & -2 & 0 & 3 & 15 \\ -1 & 0 & 4 & 2 & -3 & 1 \\ 0 & 6 & 6 & 1 & -4 & -12 \\ 2 & 8 & 0 & 1 & -1 & 3 \\ 3 & 6 & -6 & -1 & 2 & 2 \end{bmatrix},$$

$$(c)\ \begin{bmatrix} 5 & 1 & 4 & 2 \\ 3 & -1 & 0 & -4 \\ 1 & 6 & 2 & -2 \\ 2 & -6 & -8 & -20 \end{bmatrix}.$$

3. Determine the number of linearly independent vectors in the following sets

(a) $[1\ \ 4\ \ 2\ \ 7]$, $[2\ \ -9\ \ 6\ \ 4]$, $[-1\ \ 30\ \ -6\ \ 13]$, $[11\ \ -7\ \ 28\ \ 47]$.

$$(b)\ \begin{bmatrix} 1 \\ 6 \\ 11 \end{bmatrix}, \begin{bmatrix} 2 \\ 4 \\ 3 \end{bmatrix}, \begin{bmatrix} 1 \\ 9 \\ 6 \end{bmatrix}, \begin{bmatrix} 7 \\ -1 \\ 2 \end{bmatrix}.$$

3.5 Linearly independent solutions of homogeneous equations

We now come to the second of the two questions raised in § 3.1, namely: 'How many essentially different solutions are there of equations (3.1.2)?' The equations are

$$\left.\begin{aligned} a_{11}x_1 + a_{12}x_2 + \ldots + a_{1n}x_n &= 0, \\ a_{21}x_1 + a_{22}x_2 + \ldots + a_{2n}x_n &= 0, \\ \ldots\ldots\ldots\ldots\ldots\ldots\ldots\ldots\ldots \\ a_{m1}x_1 + a_{m2}x_2 + \ldots + a_{mn}x_n &= 0. \end{aligned}\right\} \tag{3.5.1}$$

Each solution of these equations gives rise to a column matrix $\mathbf{x}$ and the meaning we shall give to our question is: 'How many linearly independent column matrices can one form from the solutions of the equations?' We may write equations (3.5.1) in the form

$$\mathbf{Ax} = \mathbf{0}, \tag{3.5.2}$$

and clearly the solutions depend on the structure of $\mathbf{A}$; in fact, the number of solutions depends on the rank of $\mathbf{A}$, as we shall show.

For the sake of illustration, consider the equations

$$\left.\begin{aligned} x_1 + 2x_2 + x_3 + x_4 &= 0, \\ x_1 - x_2 + x_3 + x_4 &= 0, \\ 3x_1 \qquad + 3x_3 + 3x_4 &= 0, \\ x_1 - x_2 - x_3 - 2x_4 &= 0. \end{aligned}\right\} \tag{3.5.3}$$

The matrix of these equations is that given in equation (3.4.14) and we have already shown that its rank is 3. Equation (3.4.15), from which we deduced this rank, shows that for the matrix **A** {**row** (3) − 3 **row** (1) − 2[**row** (2) − **row** (1)]} produces a row of noughts in **row** (3); that is,

$$\textbf{row}\,(3) = \textbf{row}\,(1) + 2\times\textbf{row}\,(2). \tag{3.5.4}$$

This equation shows that for equations (3.5.3)

$$\textbf{equation}\,(3) = \textbf{equation}\,(1) + 2\times\textbf{equation}\,(2), \tag{3.5.5}$$

which means that if values of x_1, x_2, x_3 and x_4 are found which satisfy equations (1) and (2), then equation (3) will automatically be satisfied. We may therefore disregard equation (3) and consider only equations (1), (2) and (4). We shall describe this situation by saying that equations (1), (2) and (3) are linearly dependent, while equations (1), (2) and (4) are linearly independent.

Since **A** in equation (3.4.14) has rank 3 it has just 3 linearly independent columns. Equation (3.4.15) shows that the first three are linearly independent. If therefore equations (1), (2) and (4) are written in the form

$$\left.\begin{aligned} x_1 + 2x_2 + x_3 &= -x_4,\\ x_1 - x_2 + x_3 &= -x_4,\\ x_1 - x_2 - x_3 &= 2x_4,\end{aligned}\right\} \tag{3.5.6}$$

or

$$\begin{bmatrix} 1 & 2 & 1\\ 1 & -1 & 1\\ 1 & -1 & -1\end{bmatrix}\begin{bmatrix} x_1\\ x_2\\ x_3\end{bmatrix} = \begin{bmatrix} -x_4\\ -x_4\\ 2x_4\end{bmatrix}, \tag{3.5.7}$$

then, since the matrix on the left-hand side is non-singular, we have a set of equations of the type considered in equation (1.6.2). For a given value of x_4 there is a unique set of values of x_1, x_2, x_3. That is to say we may find the required ratios between the x's by assigning some convenient value to one of them, namely x_4.

Actually the solution can be obtained far more easily by going back to equation (3.4.15). By operating only on *rows* we reduced **A** to the matrix

$$\begin{bmatrix} 1 & 2 & 1 & 1\\ 0 & 1 & 0 & 0\\ 0 & 0 & -2 & -3\\ 0 & 0 & 0 & 0\end{bmatrix}.$$

But this means that equations (3.5.3) are equivalent to

$$\left.\begin{aligned} x_1 + 2x_2 + x_3 + x_4 &= 0,\\ x_2 &= 0,\\ -2x_3 - 3x_4 &= 0,\end{aligned}\right\} \tag{3.5.8}$$

from which we obtain $$x_1 : x_2 : x_3 : x_4 = 1 : 0 : -3 : 2. \tag{3.5.9}$$

In the general case we may use Theorems 3.4.6 and 3.4.8 and apply them to the equations of the system (3.5.1) instead of to the rows of the matrix **A**. The first of these theorems shows that if **A** has rank r then, of the m equations, only r are linearly independent. The second shows that it is possible to express each of the other $m-r$ equations as linear combinations of the r linearly independent ones. Without loss of generality we may assume that the first r equations are linearly independent. (If they are not then we may rearrange the equations so that the first r equations of the new set are linearly independent.) We need only consider the r equations with which we are left, namely

$$\left.\begin{array}{l} a_{11}x_1+a_{12}x_2+\ldots+a_{1n}x_n = 0, \\ a_{21}x_1+a_{22}x_2+\ldots+a_{2n}x_n = 0, \\ \cdots\cdots\cdots\cdots\cdots\cdots\cdots\cdots \\ a_{r1}x_1+a_{r2}x_2+\ldots+a_{rn}x_n = 0. \end{array}\right\} \tag{3.5.10}$$

The coefficients of these equations form a matrix of order $r \times n$, and equation (3.3.4) states that $r \leqslant n$. The matrix has r linearly independent rows and therefore, by Theorem 3.4.7 it has rank r and has r linearly independent columns. Without loss of generality we may assume that the first r columns are linearly independent. (If they are not then we may renumber the unknowns so that the first r of the rearranged columns are linearly independent.) If we write equations (3.5.10) in the form

$$\left.\begin{array}{l} a_{11}x_1+a_{12}x_2+\ldots+a_{1r}x_r = -a_{1,r+1}x_{r+1}-\ldots-a_{1n}x_n, \\ a_{21}x_1+a_{22}x_2+\ldots+a_{2r}x_r = -a_{2,r+1}x_{r+1}-\ldots-a_{2n}x_n, \\ \cdots\cdots\cdots\cdots\cdots\cdots\cdots\cdots\cdots\cdots\cdots\cdots \\ a_{r1}x_1+a_{r2}x_2+\ldots+a_{rr}x_r = -a_{r,r+1}x_{r+1}-\ldots-a_{rn}x_n, \end{array}\right\} \tag{3.5.11}$$

then the determinant of coefficients on the left-hand side is non-zero. Now the equations are of the type considered in equation (1.6.2) and we may obtain the solutions by assigning arbitrary values to $x_{r+1}, x_{r+2}, \ldots, x_n$ and then solving the equations for $x_1, x_2, \ldots, x_r$. Once the values of $x_{r+1}, x_{r+2}, \ldots, x_n$ have been assigned, the values of $x_1, x_2, \ldots, x_r$ are uniquely determined.

In order to find how many linearly independent solutions equations (3.5.11) have we proceed as follows. If $r = n$ they have only the trivial solution—that is the case we dealt with in Chapter 1. If $r < n$ we put

$$x_{r+1} = 1, \quad x_{r+2} = 0 = x_{r+3} = \ldots = x_n, \tag{3.5.12}$$

and find the unique set of values for $x_1, x_2, \ldots, x_r$. This will give, as a solution, the column vector

$$\mathbf{z}^{(1)} = \{(\text{these values for } x_1 \text{ to } x_r), 1, 0, 0 \ldots 0\}. \tag{3.5.13}$$

We now give $x_{r+1}, x_{r+2}, \ldots, x_n$ the following values in turn

$$\left.\begin{array}{lllll} x_{r+1} = 0, & x_{r+2} = 1, & x_{r+3} = 0, & \ldots, & x_n = 0; \\ x_{r+1} = 0, & x_{r+2} = 0, & x_{r+3} = 1, & \ldots, & x_n = 0; \\ \cdots\cdots & \cdots\cdots & \cdots\cdots & \cdots & \cdots\cdots \\ x_{r+1} = 0, & x_{r+2} = 0, & x_{r+3} = 0, & \ldots, & x_n = 1; \end{array}\right\} \tag{3.5.14}$$

and obtain solutions $\mathbf{z}^{(2)}, \mathbf{z}^{(3)}, \ldots, \mathbf{z}^{(n-r)}$. The column matrices $\mathbf{z}^{(1)}, \mathbf{z}^{(2)}, \ldots, \mathbf{z}^{(n-r)}$ are linearly independent since the number 1 occurs in a different position among the last $n-r$ elements in each.

Since the equations (3.5.11) are homogeneous, any linear combination of solutions is also a solution. In particular, any linear combination of $\mathbf{z}^{(1)}, \mathbf{z}^{(2)}, \ldots, \mathbf{z}^{(n-r)}$ is a solution. We shall now prove the converse property, namely, that any solution of the equations is a linear combination of $\mathbf{z}^{(1)}, \mathbf{z}^{(2)}, \ldots, \mathbf{z}^{(n-r)}$. To prove this we suppose that the column vector $\mathbf{x} = \{x_1, x_2, \ldots, x_n\}$ is a solution. By subtracting from $\mathbf{x}$ suitable multiples of $\mathbf{z}^{(1)}, \mathbf{z}^{(2)}, \ldots, \mathbf{z}^{(n-r)}$ we construct another solution $\mathbf{w}$ which has its last $n-r$ components all zero. This solution $\mathbf{w}$ is given by

$$\mathbf{w} = \mathbf{x} - x_{r+1}\mathbf{z}^{(1)} - x_{r+2}\mathbf{z}^{(2)} - \ldots - x_n\mathbf{z}^{(n-r)}, \tag{3.5.15}$$

and we shall prove that $\mathbf{w}$ is identically zero; this will give the required result. To see that $\mathbf{w}$ is zero we notice that it is a solution of equations (3.5.11) for which

$$w_{r+1} = 0 = w_{r+2} = \ldots = w_n. \tag{3.5.16}$$

By substituting these values into equation (3.5.11) we find that the equations for $w_1, w_2, \ldots, w_r$ are

$$\left.\begin{aligned} a_{11}w_1 + a_{12}w_2 + \ldots + a_{1r}w_r &= 0, \\ a_{21}w_1 + a_{22}w_2 + \ldots + a_{2r}w_r &= 0, \\ \ldots\ldots\ldots\ldots\ldots\ldots\ldots\ldots & \\ a_{r1}w_1 + a_{r2}w_2 + \ldots + a_{rr}w_r &= 0, \end{aligned}\right\} \tag{3.5.17}$$

and because the determinant of coefficients for these equations is non-zero the only solution which they possess is

$$w_1 = 0 = w_2 = \ldots = w_r, \tag{3.5.18}$$

so that $\mathbf{w}$ is identically zero. Equation (3.5.15) now shows that

$$\mathbf{x} = x_{r+1}\mathbf{z}^{(1)} + x_{r+2}\mathbf{z}^{(2)} + \ldots + x_n\mathbf{z}^{(n-r)}, \tag{3.5.19}$$

which is the required result.

We have now obtained a complete answer to the second of our questions and we may state the result as a theorem:

Theorem 3.5.1. If the rank of the matrix $\mathbf{A}$ in equation (3.5.2) is r, then equations (3.5.1) have just $n-r$ linearly independent solutions.

This theorem provides an indirect proof of a result which was used in § 1.8. There it was stated that when not all the cofactors of the square matrix $\mathbf{A}$ are zero the set of ratios

$$A_{j1} : A_{j2} : \ldots : A_{jn}$$

is the same for all values of j. This follows from the theorem applied to the case $m = n$ ($\mathbf{A}$ is square) and $r = n-1$ (at least one cofactor, or minor of order $n-1$, is non-zero). The theorem states that in this case there is just one linearly independent solution. Therefore the set of ratios, which provides a solution for all values of j, must be the same for all j.

Theorem 3.5.1 has another important corollary, which is

Theorem 3.5.2. If **U** and **V** are non-singular then the matrices **VA**, **AU** and **VAU** all have the same rank.

Proof. For the theorem to be meaningful the matrices must be such that the products may be formed. If **A** is of order $m \times n$ then **U** and **V** must be of orders n and m respectively; the products will all be of order $m \times n$.

Suppose **A** has rank r then the equation

$$\mathbf{Ax} = \mathbf{0} \tag{3.5.20}$$

has $n-r$ linearly independent solutions. Since **V** is non-singular this equation holds if, and only if,

$$\mathbf{VAx} = \mathbf{0}. \tag{3.5.21}$$

Therefore this equation also has $n-r$ solutions and the rank of **VA** is r. Moreover, if equation (3.5.20) holds then

$$\mathbf{AU}.\mathbf{U}^{-1}\mathbf{x} = \mathbf{0}, \tag{3.5.22}$$

so that $\mathbf{y} = \mathbf{U}^{-1}\mathbf{x}$ is a solution of

$$\mathbf{AUy} = \mathbf{0}. \tag{3.5.23}$$

Conversely, if **y** satisfies equation (3.5.23) then $\mathbf{x} = \mathbf{Uy}$ satisfies equation (3.5.20). Therefore the two equations have the same number of solutions so that **AU** has rank r. Finally equation (3.5.23) holds if and only if

$$\mathbf{VAUy} = \mathbf{0}, \tag{3.5.24}$$

so that **VAU** also has rank r.

EXAMPLES 3.5

1. Give a complete solution, where one exists, of the following sets of equations:

(*a*) $x+2y-3z = 0, \quad 2x+5y-7z = 0, \quad 4x+y-2z = 0,$

(*b*) $4x+3y-7z+3t = 0, \quad x+2y+8z-16t = 0.$

$5x-38z+54t = 0, \quad 5y+39z-67t = 0.$

(*c*) $5x-16y+3z-4t+2u = 0,$

$2x+17y-3z+11t-u = 0,$

$x-50y+9z-26t+4u = 0.$

2. Discuss how the number of solutions of the following equations varies with a and b, and give complete solutions for each of the various possibilities:

$$\begin{aligned} ax+2y-z+4t &= 0,\\ x+2y-az+bt &= 0,\\ 2ax+by-2z+8t &= 0. \end{aligned}$$

3. The operations (*a*), (*b*) and (*c*) mentioned at the end of §3.4 may be expressed in terms of multiplication by non-singular matrices. Theorem 3.5.2 ensures that the ranks of the matrices obtained will be the same as that of the original matrix.

Let **A** be a matrix of order $m \times n$. Find:

(*a*) a matrix **U** such that **UA** is the same as **A** except that the first row is multiplied by k;

(*b*) a matrix **V** such that **AV** is the same as **A** except that column r is the sum of columns 1 and r of **A**;

(*c*) how one should multiply **A** to obtain a matrix which has columns 1 and 2 interchanged.

3.6 The solutions of non-homogeneous equations

This section is concerned with the solutions of the non-homogeneous matrix equation

$$\mathbf{Ax} = \mathbf{y}, \tag{3.6.1}$$

which arises in various contexts, and especially in problems of forced vibration. In § 1.6 we showed that when **A** is a square non-singular matrix the equation has the solution

$$\mathbf{x} = \mathbf{A}^{-1}\mathbf{y}. \tag{3.6.2}$$

This is then the only solution of the equation, for if $\bar{\mathbf{x}}$ is a solution then

$$\mathbf{A}(\bar{\mathbf{x}} - \mathbf{A}^{-1}\mathbf{y}) = \mathbf{A}\bar{\mathbf{x}} - \mathbf{y} = \mathbf{y} - \mathbf{y} = \mathbf{0}, \tag{3.6.3}$$

and since **A** is non-singular this implies that $\bar{\mathbf{x}} = \mathbf{A}^{-1}\mathbf{y}$. In order to discuss the properties of equation (3.6.1) more fully, it is convenient, once more, to widen the investigation and allow **A** to be a rectangular matrix of order $m \times n$. Equation (3.6.1) then represents

$$\left.\begin{array}{l} a_{11}x_1 + a_{12}x_2 + \ldots + a_{1n}x_n = y_1, \\ a_{21}x_1 + a_{22}x_2 + \ldots + a_{2n}x_n = y_2, \\ \ldots\ldots\ldots\ldots\ldots\ldots\ldots\ldots\ldots\ldots \\ a_{m1}x_1 + a_{m2}x_2 + \ldots + a_{mn}x_n = y_m, \end{array}\right\} \tag{3.6.4}$$

so that **x** is now a column vector of order n and **y** is one of order m.

Once more it is useful to ask two questions about the equations. The first is 'When do the equations have a solution?' and the second is 'When they do, do they have more than one; in fact how many linearly independent solutions do they have?' The answer to the first may be stated in the form of a theorem, the proof of which will lead to the answer to the second question.

Theorem 3.6.1. A necessary and sufficient condition for equations (3.6.4) to have a solution is that the matrices

$$\mathbf{A} = \begin{bmatrix} a_{11} & a_{12} & \ldots & a_{1n} \\ a_{21} & a_{22} & \ldots & a_{2n} \\ \ldots & \ldots & \ldots & \ldots \\ a_{m1} & a_{m2} & \ldots & a_{mn} \end{bmatrix}, \quad \mathbf{B} = \begin{bmatrix} a_{11} & a_{12} & \ldots & a_{1n} & y_1 \\ a_{21} & a_{22} & \ldots & a_{2n} & y_2 \\ \ldots & \ldots & \ldots & \ldots & \ldots \\ a_{m1} & a_{m2} & \ldots & a_{mn} & y_m \end{bmatrix} \tag{3.6.5}$$

shall be of the same rank.

Before proving this theorem, we may first observe that when **A** is a square non-singular matrix of order n then **A** and **B** both have rank n, which fits in with the result which is stated in equation (3.6.2).

Let the ranks of $\mathbf{A}$ and $\mathbf{B}$ be r and r' respectively. Then $\mathbf{A}$ has just r linearly independent columns and since these belong to $\mathbf{B}$ also, $\mathbf{B}$ must have at least r linearly independent columns. Therefore $r' \geqslant r$.

We shall now show that the condition $r = r'$ is necessary; that is, we shall prove that if the equations have a solution then $r = r'$. In the notation of equation (3.2.4), equation (3.6.4) is

$$\mathbf{y} = x_1\mathbf{A}^{(1)} + x_2\mathbf{A}^{(2)} + \ldots + x_n\mathbf{A}^{(n)}. \tag{3.6.6}$$

This means that if the equations have a solution then the last column of $\mathbf{B}$ is a linear combination of the first n columns, the columns of $\mathbf{A}$. Therefore $\mathbf{B}$ has the same number of linearly independent columns as $\mathbf{A}$, and $r' = r$.

We now have to show that if $r' = r$ then the equations have a solution. Since $\mathbf{A}$ has rank r it has just r linearly independent rows. As before, we may assume without loss of generality that these are the first r rows. This means that the first r rows of $\mathbf{B}$ are linearly independent and, since $r' = r$, $\mathbf{B}$ has no more linearly independent rows. This has two corollaries. First, by Theorem 3.4.8 applied to equations instead of rows, the last $m-r$ equations of the system (3.6.4) may be expressed as linear combinations of the first r equations. Secondly, there is a non-zero minor among the first r rows of $\mathbf{A}$, and again there is no loss in generality in assuming it to be the minor in the top left-hand corner. If, therefore, we write the first r of equations (3.6.4) in the form

$$\left.\begin{aligned} a_{11}x_1 + a_{12}x_2 + \ldots + a_{1r}x_r &= y_1 - a_{1,r+1}x_{r+1} - \ldots - a_{1n}x_n, \\ a_{21}x_1 + a_{22}x_2 + \ldots + a_{2r}x_r &= y_2 - a_{2,r+1}x_{r+1} - \ldots - a_{2n}x_n, \\ &\cdots\cdots\cdots\cdots \\ a_{r1}x_1 + a_{r2}x_2 + \ldots + a_{rr}x_r &= y_r - a_{r,r+1}x_{r+1} - \ldots - a_{rn}x_n, \end{aligned}\right\} \tag{3.6.7}$$

then the determinant of coefficients on the left-hand side is non-zero, so that the equations have a solution. We have therefore proved that the condition $r' = r$ is both necessary and sufficient for the equations to have a solution.

We shall now show that if $\mathbf{A}$ and $\mathbf{B}$ are both of rank r then equations (3.6.4) have just $n-r+1$ linearly independent solutions. Let $\mathbf{x}^{(0)}$ be a solution of the equations and let $\mathbf{x}$ be any other solution; then $\mathbf{x}-\mathbf{x}^{(0)}$ is a solution of the homogeneous equations. For

$$\mathbf{A}(\mathbf{x}-\mathbf{x}^{(0)}) = \mathbf{y}-\mathbf{y} = \mathbf{0}. \tag{3.6.8}$$

Therefore $\mathbf{x}-\mathbf{x}^{(0)}$ is a linear combination of the column matrices $\mathbf{z}^{(1)}, \mathbf{z}^{(2)}, \ldots, \mathbf{z}^{(n-r)}$ defined in the previous section, i.e.

$$\mathbf{x} = \mathbf{x}^{(0)} + \lambda_1\mathbf{z}^{(1)} + \lambda_2\mathbf{z}^{(2)} + \ldots + \lambda_{n-r}\mathbf{z}^{(n-r)}. \tag{3.6.9}$$

This equation shows that the general solution of the non-homogeneous equations may be written as the sum of a particular solution and the general solution of the homogeneous equations. In particular it shows that the non-homogeneous equations have just $n-r+1$ linearly independent solutions.

For an example consider the equations for which

$$\mathbf{B} = \begin{bmatrix} 1 & 2 & -3 & 7 & 3 & 4 & 7 \\ 3 & 2 & 4 & 9 & 5 & 1 & 2 \\ 6 & 4 & 3 & 2 & 2 & 7 & 1 \\ 5 & 4 & 2 & 9 & 5 & 6 & 5 \\ 1 & 0 & 1 & -7 & -3 & 1 & -4 \end{bmatrix}. \tag{3.6.10}$$

By operating on the rows of **B** we may reduce it as follows

$$\mathbf{B} \to \begin{bmatrix} 1 & 2 & -3 & 7 & 3 & 4 & 7 \\ 0 & -4 & 13 & -12 & -4 & -11 & -19 \\ 0 & -8 & 21 & -40 & -16 & -17 & -41 \\ 0 & -6 & 17 & -26 & -10 & -14 & -30 \\ 0 & -2 & 4 & -14 & -6 & -3 & -11 \end{bmatrix} \to \begin{bmatrix} 1 & 2 & -3 & 7 & 3 & 4 & 7 \\ 0 & -2 & 4 & -14 & -6 & -3 & -11 \\ 0 & 0 & 5 & 16 & 8 & -5 & 3 \\ 0 & 0 & 5 & 16 & 8 & -5 & 3 \\ 0 & 0 & 5 & 16 & 8 & -5 & 3 \end{bmatrix}, \tag{3.6.11}$$

which shows that rank $(\mathbf{A}) = 3 = \text{rank}(\mathbf{B})$. Therefore the equations have a solution, and the equations may be written

$$\left.\begin{aligned} x_1 + 2x_2 - 3x_3 &= \quad 7 - 7x_4 - 3x_5 - 4x_6, \\ -2x_2 + 4x_3 &= -11 + 14x_4 + 6x_5 + 3x_6, \\ 5x_3 &= \quad 3 - 16x_4 - 8x_5 + 5x_6. \end{aligned}\right\} \tag{3.6.12}$$

By putting $x_4 = 0 = x_5 = x_6$ we obtain

$$\mathbf{x}^{(0)} = \tfrac{1}{10}\{-46 \quad 67 \quad 6 \quad 0 \quad 0 \quad 0\} \tag{3.6.13}$$

as a particular solution. There are three linearly independent solutions of the homogeneous equations and a convenient set is

$$\left.\begin{aligned} \mathbf{z}^{(1)} &= \{\ 51 \quad -67 \quad -16 \quad 5 \quad 0 \quad 0\}, \\ \mathbf{z}^{(2)} &= \{\ 23 \quad -31 \quad -8 \quad 0 \quad 5 \quad 0\}, \\ \mathbf{z}^{(3)} &= \{-4 \quad 1 \quad 2 \quad 0 \quad 0 \quad 2\}. \end{aligned}\right\} \tag{3.6.14}$$

Any solution of the equations may be represented in the form

$$\begin{aligned} \mathbf{x} = \tfrac{1}{10}\{-46 \quad 67 \quad 6 \quad 0 \quad 0 \quad 0\} + \lambda_1\{51 \quad -67 \quad -16 \quad 5 \quad 0 \quad 0\} \\ + \lambda_2\{23 \quad -31 \quad -8 \quad 0 \quad 5 \quad 0\} + \lambda_3\{-4 \quad 1 \quad 2 \quad 0 \quad 0 \quad 2\}. \end{aligned} \tag{3.6.15}$$

It should be mentioned, however, that this formula for the solutions is not unique. Other formulae may be obtained by choosing different particular solutions and different solutions of the homogeneous equations.

Equation (3.6.15) shows that by taking

$$\left.\begin{aligned} \lambda_1 = 0 \quad \lambda_2 = 0 \quad \lambda_3 = 0, \\ \lambda_1 = 1 \quad \lambda_2 = 0 \quad \lambda_3 = 0, \\ \lambda_1 = 0 \quad \lambda_2 = 1 \quad \lambda_3 = 0, \\ \lambda_1 = 0 \quad \lambda_2 = 0 \quad \lambda_3 = 1, \end{aligned}\right\} \tag{3.6.16}$$

in turn we obtain 4 linearly independent solutions of the non-homogeneous equations.

The discussion of equation (3.6.1) will be taken up again at the end of the next section.

EXAMPLES 3.6

1. Find the solutions, if there are any, of the following sets of equations:

$$(a)\quad 2x-3y+7z = 2,\quad 3x+9y-11z = 4,\quad 5x-2y+5z = 1;$$
$$(b)\quad 5x+2y-11z = 3,\quad 4x-3y+13z = 2,\quad x+5y-24z = 1;$$
$$(c)\quad x-y+2z = 3,\quad 5x+2y-6z = 2,\quad 2x+5y-12z = 1.$$

2. Find a formula which gives all the solutions of the equations:

$$x+2y-3z+4t+u = 9,\qquad 6x-18y+15z+11t+2u = -1,$$
$$3x-4y+2z+5t-7u = 14,\qquad x-8y+8z+t+4u = -12.$$

3. How many linearly independent solutions do the equations

$$x_1+3x_2-4x_3+6x_4-4x_5+7x_6 = 9,\qquad 2x_1-5x_2-3x_3+11x_4+13x_5+15x_6 = 11$$
$$4x_1+x_2-11x_3+23x_4-2x_5+x_6 = 13,\qquad 7x_1-x_2-18x_3+40x_4+7x_5+23x_6 = 33$$

possess? Find a set which contains the greatest number of solutions.

3.7 Orthogonality

The similarities between column vectors and ordinary vectors in three-dimensional space were discussed in § 3.1. This section is devoted to a discussion of some other properties of column vectors. These properties are concerned, for the most part, with the idea of orthogonality, which is an extension of the concept of perpendicularity in ordinary three-dimensional space. An extension of the concept of orthogonality, as it will be discussed *here*, occurs in connection with vibrating systems, and will be taken up in § 4.1. This section, therefore, as well as presenting matters of importance in themselves, also serves as an introduction to that section. The reader will observe in fact that there is a close connection between the n-dimensional space treated here and a dynamical system with n degrees of freedom.

If a_1, a_2, a_3 are the components of the vector $\mathbf{a}$ in some rectangular Cartesian frame (see fig. 3.1.1) then the magnitude of $\mathbf{a}$ is

$$|\mathbf{a}| = \{a_1^2+a_2^2+a_3^2\}^{\frac{1}{2}}. \qquad (3.7.1)$$

If $\mathbf{b}$ is another vector, with components b_1, b_2, b_3, then the scalar product of $\mathbf{a}$ and $\mathbf{b}$ is

$$(\mathbf{a}.\mathbf{b}) = a_1b_1+a_2b_2+a_3b_3. \qquad (3.7.2)$$

The geometrical meaning of the scalar product is given by the equation

$$(\mathbf{a}.\mathbf{b}) = |\mathbf{a}|.|\mathbf{b}|.\cos\theta, \qquad (3.7.3)$$

where θ is the angle between $\mathbf{a}$ and $\mathbf{b}$. From this equation it follows that the scalar product of $\mathbf{a}$ with itself is

$$(\mathbf{a}.\mathbf{a}) = |\mathbf{a}|^2, \qquad (3.7.4)$$

and the condition for **a** and **b** to be perpendicular is that

$$(\mathbf{a}.\mathbf{b}) = 0, \tag{3.7.5}$$

that is

$$a_1 b_1 + a_2 b_2 + a_3 b_3 = 0. \tag{3.7.6}$$

We shall now extend these definitions to column vectors of order n. Let **u** be the column vector

$$\mathbf{u} = \{u_1 \quad u_2 \quad \dots \quad u_n\}. \tag{3.7.7}$$†

The 'norm' of **u** is defined as

$$|\mathbf{u}| = \{u_1^2 + u_2^2 + \dots + u_n^2\}^{\frac{1}{2}} = \{\mathbf{u}'\mathbf{u}\}^{\frac{1}{2}}. \tag{3.7.8}$$

If $|\mathbf{u}| = 1$, **u** is said to be 'normalised', or a 'unit vector'. The 'scalar product' of two column vectors **u** and **v** is defined to be

$$\mathbf{u}'\mathbf{v} = \mathbf{v}'\mathbf{u} = u_1 v_1 + u_2 v_2 + \dots + u_n v_n, \tag{3.7.9}$$

and **u** and **v** are said to be 'orthogonal' if

$$\mathbf{u}'\mathbf{v} = 0.$$

Thus the vectors

$$\mathbf{u} = \{1 \quad 0 \quad -1 \quad 5\}, \quad \mathbf{v} = \{0 \quad 2 \quad 5 \quad 1\} \tag{3.7.10}$$

are orthogonal.

We shall now discuss a problem involving orthogonality which arises in various contexts. The problem has a simple counterpart in ordinary space, and we shall first discuss this. The problem in ordinary space is that we are given a vector **v** and we have to write it as the sum of two vectors, one in the direction of a given vector **u**, and the other perpendicular to **u**. The required vectors are shown in fig. 3.7.1. The first, **OA**, is the component of **v** in the direction of **u**, and the second, **OB**, which we shall denote by **z**, is the projection of **v** on the plane perpendicular to **u**. Thus

$$\mathbf{v} = \mathbf{OC} = \mathbf{OA} + \mathbf{AC} = \mathbf{OA} + \mathbf{OB}, \tag{3.7.11}$$

and

$$\mathbf{OA} = |\mathbf{OC}| \cos\theta . \frac{\mathbf{OA}}{|\mathbf{OA}|}$$

$$= |\mathbf{v}| \cos\theta . \frac{\mathbf{u}}{|\mathbf{u}|} = \frac{(\mathbf{u}.\mathbf{v})\,\mathbf{u}}{|\mathbf{u}|^2}, \tag{3.7.12}$$

so that

$$\mathbf{z} = \mathbf{OB} = \mathbf{OC} - \mathbf{OA} = \mathbf{v} - \frac{(\mathbf{u}.\mathbf{v})\,\mathbf{u}}{|\mathbf{u}|^2}, \tag{3.7.13}$$

and the required representation is

$$\mathbf{v} = \frac{(\mathbf{u}.\mathbf{v})\,\mathbf{u}}{|\mathbf{u}|^2} + \mathbf{z}. \tag{3.7.14}$$

We note that the problem is equivalent to finding a vector **z** which is a linear combination of **u** and **v** and which is orthogonal to **u**.

To solve the counterpart of this problem for column vectors we adopt this alternative standpoint and construct a vector **z**, given by

$$\mathbf{z} = \mathbf{v} + \lambda \mathbf{u}, \tag{3.7.15}$$

† The aggregate of all possible column vectors of order n is sometimes said to define 'the Euclidean n-dimensional vector space, E_n'.

which is orthogonal to $\mathbf{u}$. Suppose that

$$\mathbf{u} = \{1 \quad 0 \quad -1 \quad 5\}, \quad \mathbf{v} = \{1 \quad -1 \quad 4 \quad 2\}, \tag{3.7.16}$$

then

$$\mathbf{z} = \{1+\lambda \quad -1 \quad 4-\lambda \quad 2+5\lambda\}, \tag{3.7.17}$$

and we have to choose λ to make $\mathbf{z}$ orthogonal to $\mathbf{u}$. This will occur if

$$1+\lambda-(4-\lambda)+5(2+5\lambda) = 0, \tag{3.7.18}$$

that is, if

$$\lambda = \frac{-7}{27}. \tag{3.7.19}$$

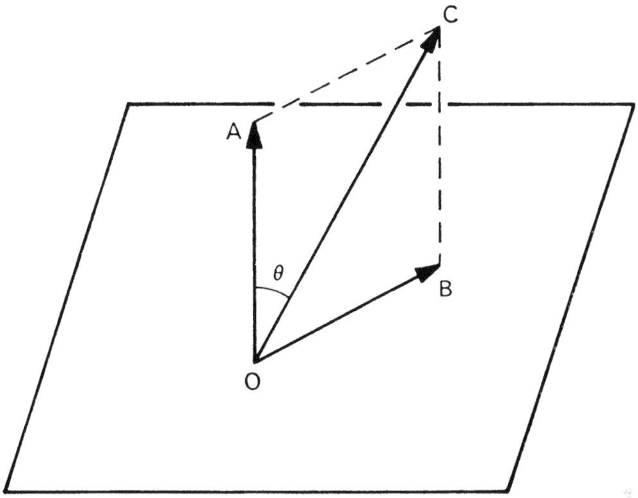

Fig. 3.7.1

In general, the multiplier λ in equation (3.7.15) is given by

$$\mathbf{u}'\mathbf{z} = 0 = \mathbf{u}'\mathbf{v}+\lambda\mathbf{u}'\mathbf{u}, \tag{3.7.20}$$

so that

$$\mathbf{z} = \mathbf{v}-\frac{(\mathbf{u}'\mathbf{v})\,\mathbf{u}}{|\mathbf{u}|^2}, \tag{3.7.21}$$

in analogy with equation (3.7.13).

We notice that $|\mathbf{u}|^2$, the denominator of λ is always non-zero because it is the sum of the squares of the elements of $\mathbf{u}$. Also the vector $\mathbf{z}$ will be zero if and only if $\mathbf{u}$ and $\mathbf{v}$ are linearly dependent. If $\mathbf{u}$ is a unit vector equation (3.7.21) becomes, as a special case,

$$\mathbf{z} = \mathbf{v}-(\mathbf{u}'\mathbf{v})\,\mathbf{u}, \tag{3.7.22}$$

and we may rearrange this equation to give the analogue of equation (3.7.14), namely,

$$\mathbf{v} = (\mathbf{u}'\mathbf{v})\,\mathbf{u}+\mathbf{z}. \tag{3.7.23}$$

The column vectors $\mathbf{u}^{(1)}, \mathbf{u}^{(2)}, \ldots, \mathbf{u}^{(m)}$ are said to be 'mutually orthogonal' if

$$\mathbf{u}'^{(r)}\mathbf{u}^{(s)} = 0 \quad (r \neq s). \tag{3.7.24}$$

If they are also unit vectors they are said to be 'orthonormal'. In this case

$$\mathbf{u}'^{(r)}\mathbf{u}^{(s)} = \delta_{rs}. \tag{3.7.25}$$

For example, the vectors

$$\mathbf{e}^{(1)} = \begin{bmatrix} 1 \\ 0 \\ 0 \\ \vdots \\ 0 \end{bmatrix}, \quad \mathbf{e}^{(2)} = \begin{bmatrix} 0 \\ 1 \\ 0 \\ \vdots \\ 0 \end{bmatrix}, \quad \ldots, \quad \mathbf{e}^{(n)} = \begin{bmatrix} 0 \\ 0 \\ 0 \\ \vdots \\ 1 \end{bmatrix} \tag{3.7.26}$$

are orthonormal.

A set of orthonomal vectors may be formed from any set of mutually orthogonal vectors. For example, the vectors

$$\mathbf{u}^{(1)} = \begin{bmatrix} 1 \\ 2 \\ 1 \\ -1 \end{bmatrix}, \quad \mathbf{u}^{(2)} = \begin{bmatrix} 2 \\ 1 \\ -3 \\ 1 \end{bmatrix}, \quad \mathbf{u}^{(3)} = \begin{bmatrix} 1 \\ -1 \\ 0 \\ -1 \end{bmatrix} \tag{3.7.27}$$

are mutually orthogonal and the vectors

$$\mathbf{v}^{(1)} = \frac{\mathbf{u}^{(1)}}{\sqrt{7}}, \quad \mathbf{v}^{(2)} = \frac{\mathbf{u}^{(2)}}{\sqrt{15}}, \quad \mathbf{v}^{(3)} = \frac{\mathbf{u}^{(3)}}{\sqrt{3}}, \tag{3.7.28}$$

are orthonormal.

If the vectors $\mathbf{u}^{(1)}, \mathbf{u}^{(2)}, \ldots, \mathbf{u}^{(m)}$ are orthonormal (or just mutually orthogonal, provided that they are all non-zero) they are linearly independent. For suppose

$$\lambda_1 \mathbf{u}^{(1)} + \lambda_2 \mathbf{u}^{(2)} + \ldots + \lambda_m \mathbf{u}^{(m)} = \mathbf{0}, \tag{3.7.29}$$

then

$$\mathbf{u}'^{(1)}(\lambda_1 \mathbf{u}^{(1)} + \lambda_2 \mathbf{u}^{(2)} + \ldots + \lambda_m \mathbf{u}^{(m)}) = 0, \tag{3.7.30}$$

and on account of (3.7.25) this reduces to $\lambda_1 = 0$. Therefore λ_1, and similarly $\lambda_2, \lambda_3, \ldots, \lambda_m$, are all zero, and the vectors are linearly independent.

Results concerning the orthogonality of column vectors have counterparts in matrix multiplication. For let $\mathbf{V}$ be a square matrix with columns $\mathbf{v}^{(1)}, \mathbf{v}^{(2)}, \ldots, \mathbf{v}^{(n)}$. Then the element in row r and column s of the product $\mathbf{V}'\mathbf{V}$ is $\mathbf{v}'^{(r)}\mathbf{v}^{(s)}$ (see equation (1.3.8)). If therefore the columns are orthonormal then

$$\mathbf{v}'^{(r)}\mathbf{v}^{(s)} = \delta_{rs}, \tag{3.7.31}$$

which means that

$$\mathbf{V}'\mathbf{V} = \mathbf{I}. \tag{3.7.32}$$

Such a matrix $\mathbf{V}$ is called an 'orthogonal' matrix.

A matrix having the form

$$\begin{bmatrix} a_{11} & a_{12} & \ldots & a_{1r} & 0 & 0 & \ldots & 0 \\ a_{21} & a_{22} & \ldots & a_{2r} & 0 & 0 & \ldots & 0 \\ \cdots & \cdots & \cdots & \cdots & \cdots & \cdots & \cdots & \cdots \\ a_{r1} & a_{r2} & \ldots & a_{rr} & 0 & 0 & \ldots & 0 \\ 0 & 0 & \ldots & 0 & b_{11} & b_{12} & \ldots & b_{1,n-r} \\ 0 & 0 & \ldots & 0 & b_{21} & b_{22} & \ldots & b_{2,n-r} \\ \cdots & \cdots & \cdots & \cdots & \cdots & \cdots & \cdots & \cdots \\ 0 & 0 & \ldots & 0 & b_{n-r,1} & b_{n-r,2} & \ldots & b_{n-r,n-r} \end{bmatrix} \tag{3.7.33}$$

is a simple example of a 'partitioned' matrix. Such matrices occur in connection with orthogonality, and the symbolism which will be used to represent them is

$$\begin{bmatrix} \mathbf{A}_r & \mathbf{0} \\ \mathbf{0} & \mathbf{B}_{n-r} \end{bmatrix}. \tag{3.7.34}$$

Suppose, for instance, that the first column of the square matrix $\mathbf{V}$ is orthogonal to the remaining $n-1$, then

$$\mathbf{v}'^{(1)}\mathbf{v}^{(s)} = 0 = \mathbf{v}'^{(s)}\mathbf{v}^{(1)} \quad \text{if} \quad s \neq 1, \tag{3.7.35}$$

so that the product $\mathbf{V}'\mathbf{V}$ will have zeros in the first row and first column except for the leading diagonal position. That is, with the 'partitioned' notation

$$\mathbf{V}'\mathbf{V} = \begin{bmatrix} \mathbf{A}_1 & \mathbf{0} \\ \mathbf{0} & \mathbf{B}_{n-1} \end{bmatrix}, \tag{3.7.36}$$

where $\mathbf{A}_1 = \mathbf{v}'^{(1)}\mathbf{v}^{(1)}$. Again, if $\mathbf{v}^{(1)}$ and $\mathbf{v}^{(2)}$ are orthonormal and each is orthogonal to the remaining $n-2$ columns then, in the partitioned symbolism,

$$\mathbf{V}'\mathbf{V} = \begin{bmatrix} \mathbf{I}_2 & \mathbf{0} \\ \mathbf{0} & \mathbf{B}_{n-2} \end{bmatrix}; \quad \mathbf{I}_2 = \begin{bmatrix} 1 & 0 \\ 0 & 1 \end{bmatrix}. \tag{3.7.37}$$

Two important properties of partitioned matrices are shown in the equations

$$\begin{bmatrix} \mathbf{A}_r & \mathbf{0} \\ \mathbf{0} & \mathbf{B}_{n-r} \end{bmatrix} \begin{bmatrix} \mathbf{C}_r & \mathbf{0} \\ \mathbf{0} & \mathbf{D}_{n-r} \end{bmatrix} = \begin{bmatrix} \mathbf{A}_r\mathbf{C}_r & \mathbf{0} \\ \mathbf{0} & \mathbf{B}_{n-r}\mathbf{D}_{n-r} \end{bmatrix}, \tag{3.7.38}$$

$$\begin{vmatrix} \mathbf{A}_r & \mathbf{0} \\ \mathbf{0} & \mathbf{B}_{n-r} \end{vmatrix} = |\mathbf{A}_r| \cdot |\mathbf{B}_{n-r}|, \tag{3.7.39}$$

which may easily be verified by forming the quantities concerned. For equation (3.7.38) to be meaningful it is essential that the partitioning of the two matrices should be the same, for otherwise the products $\mathbf{A}_r\mathbf{C}_r$, $\mathbf{B}_{n-r}\mathbf{D}_{n-r}$ cannot be formed.

The reader may, if he wishes, omit the remainder of this section on a first reading. It is not required in Chapter 4.

It has been proved that orthonormal vectors are always linearly independent. The converse is clearly not true; linearly independent vectors are not necessarily orthonormal. However, from a set of m linearly independent vectors it is possible to choose m linear combinations of them which are orthonormal. The process is called the Gram–Schmidt orthogonalisation procedure, and it is an extension of the procedure described in equations (3.7.15)–(3.7.22).

To illustrate this procedure we shall now show how we may obtain three orthonormal vectors $\mathbf{u}^{(1)}$, $\mathbf{u}^{(2)}$, $\mathbf{u}^{(3)}$ from the vectors

$$\mathbf{v}^{(1)} = \begin{bmatrix} 1 \\ 0 \\ 2 \end{bmatrix}; \quad \mathbf{v}^{(2)} = \begin{bmatrix} 3 \\ 1 \\ 0 \end{bmatrix}; \quad \mathbf{v}^{(3)} = \begin{bmatrix} 5 \\ 1 \\ 2 \end{bmatrix}. \tag{3.7.40}$$

For the first vector we just take any one of the vectors and normalise it. Thus we may take $\mathbf{v}^{(1)}$ and put

$$\mathbf{u}^{(1)} = \frac{\mathbf{v}^{(1)}}{|\mathbf{v}^{(1)}|} = \frac{1}{\sqrt{5}}\begin{bmatrix}1\\0\\2\end{bmatrix}. \tag{3.7.41}$$

For the second vector we first find a linear combination, $\mathbf{z}^{(2)}$ say, of $\mathbf{u}^{(1)}$ and $\mathbf{v}^{(2)}$ which is orthogonal to $\mathbf{u}^{(1)}$. We may find $\mathbf{z}^{(2)}$ by putting

$$\mathbf{u} = \mathbf{u}^{(1)}, \quad \mathbf{v} = \mathbf{v}^{(2)} \tag{3.7.42}$$

in equation (3.7.22); it is

$$\mathbf{z}^{(2)} = \mathbf{v}^{(2)} - (\mathbf{u}'^{(1)}\mathbf{v}^{(2)})\,\mathbf{u}^{(1)} \tag{3.7.43}$$

$$= \begin{bmatrix}3\\1\\0\end{bmatrix} - \frac{3}{\sqrt{5}}\cdot\frac{1}{\sqrt{5}}\begin{bmatrix}1\\0\\2\end{bmatrix} = \tfrac{1}{5}\begin{bmatrix}12\\5\\-6\end{bmatrix}. \tag{3.7.44}$$

If we now normalise $\mathbf{z}^{(2)}$ and put

$$\mathbf{u}^{(2)} = \frac{\mathbf{z}^{(2)}}{|\mathbf{z}^{(2)}|} = \frac{1}{\sqrt{205}}\begin{bmatrix}12\\5\\-6\end{bmatrix}, \tag{3.7.45}$$

then we shall have

$$\mathbf{u}'^{(1)}\mathbf{u}^{(1)} = 1 = \mathbf{u}'^{(2)}\mathbf{u}^{(2)}, \quad \mathbf{u}'^{(1)}\mathbf{u}^{(2)} = 0 = \mathbf{u}'^{(2)}\mathbf{u}^{(1)}, \tag{3.7.46}$$

so that $\mathbf{u}^{(1)}$ and $\mathbf{u}^{(2)}$ will be orthonormal. We now have to find a third unit vector $\mathbf{u}^{(3)}$ which is orthogonal to both $\mathbf{u}^{(1)}$ and $\mathbf{u}^{(2)}$, and we may do this by first finding a linear combination $\mathbf{z}^{(3)}$ of $\mathbf{u}^{(1)}$, $\mathbf{u}^{(2)}$ and $\mathbf{v}^{(3)}$ which is orthogonal to both $\mathbf{u}^{(1)}$ and $\mathbf{u}^{(2)}$. (This vector $\mathbf{z}^{(3)}$ will clearly be a linear combination of the original vectors $\mathbf{v}^{(1)}$, $\mathbf{v}^{(2)}$ and $\mathbf{v}^{(3)}$.) Thus, if we put

$$\mathbf{z}^{(3)} = \mathbf{v}^{(3)} + \lambda\mathbf{u}^{(1)} + \mu\mathbf{u}^{(2)}, \tag{3.7.47}$$

then the two equations

$$\mathbf{u}'^{(1)}\mathbf{z}^{(3)} = 0 = \mathbf{u}'^{(2)}\mathbf{z}^{(3)} \tag{3.7.48}$$

will be two equations for the unknown multipliers λ and μ. The equations are

$$\left.\begin{aligned}\mathbf{u}'^{(1)}\mathbf{z}^{(3)} &= 0 = \mathbf{u}'^{(1)}\mathbf{v}^{(3)} + \lambda\mathbf{u}'^{(1)}\mathbf{u}^{(1)} + \mu\mathbf{u}'^{(1)}\mathbf{u}^{(2)},\\ \mathbf{u}'^{(2)}\mathbf{z}^{(3)} &= 0 = \mathbf{u}'^{(2)}\mathbf{v}^{(3)} + \lambda\mathbf{u}'^{(2)}\mathbf{u}^{(1)} + \mu\mathbf{u}'^{(2)}\mathbf{u}^{(2)},\end{aligned}\right\} \tag{3.7.49}$$

and because of the equations (3.7.46) these reduce to

$$\mathbf{u}'^{(1)}\mathbf{v}^{(3)} + \lambda = 0 = \mathbf{u}'^{(2)}\mathbf{v}^{(3)} + \mu. \tag{3.7.50}$$

Substituting these values of λ and μ into equation (3.7.47) we obtain

$$\mathbf{z}^{(3)} = \mathbf{v}^{(3)} - (\mathbf{u}'^{(1)}\mathbf{v}^{(3)})\,\mathbf{u}^{(1)} - (\mathbf{u}'^{(2)}\mathbf{v}^{(3)})\,\mathbf{u}^{(2)}. \tag{3.7.51}$$

In our example $\mathbf{z}^{(3)}$ is

$$\mathbf{z}^{(3)} = \begin{bmatrix} 5 \\ 1 \\ 2 \end{bmatrix} - \frac{9}{5}\begin{bmatrix} 1 \\ 0 \\ 2 \end{bmatrix} - \frac{53}{205}\begin{bmatrix} 12 \\ 5 \\ -6 \end{bmatrix} = \frac{2}{41}\begin{bmatrix} 2 \\ -6 \\ -1 \end{bmatrix} \tag{3.7.52}$$

and

$$\mathbf{u}^{(3)} = \frac{\mathbf{z}^{(3)}}{|\mathbf{z}^{(3)}|} = \frac{1}{\sqrt{41}}\begin{bmatrix} 2 \\ -6 \\ -1 \end{bmatrix}. \tag{3.7.53}$$

If we have m linearly independent vectors $\mathbf{v}^{(1)}, \mathbf{v}^{(2)}, \ldots, \mathbf{v}^{(m)}$ then by applying the process described above we may find m orthonormal vectors $\mathbf{u}^{(1)}, \mathbf{u}^{(2)}, \ldots, \mathbf{u}^{(m)}$. The equations for $\mathbf{z}^{(r)}$ and $\mathbf{u}^{(r)}$ will be

$$\mathbf{z}^{(r)} = \mathbf{v}^{(r)} - (\mathbf{u}'^{(1)}\mathbf{v}^{(r)})\,\mathbf{u}^{(1)} - \ldots - (\mathbf{u}'^{(r-1)}\mathbf{v}^{(r)})\,\mathbf{u}^{(r-1)}, \tag{3.7.54}$$

and

$$\mathbf{u}^{(r)} = \mathbf{z}^{(r)}/|\mathbf{z}^{(r)}|. \tag{3.7.55}$$

If we apply the process to a linearly dependent set of vectors $\mathbf{v}^{(1)}, \mathbf{v}^{(2)}, \ldots, \mathbf{v}^{(m)}$ we shall obtain as many orthonormal vectors as there are linearly independent vectors among the $\mathbf{v}$'s. We may illustrate this last statement by applying the process to the vectors $\mathbf{v}^{(1)}, \mathbf{v}^{(2)}, \mathbf{v}^{(3)}$ in equation (3.7.40) together with

$$\mathbf{v}^{(4)} = \{7 \quad 1 \quad 2\}. \tag{3.7.56}$$

For the vector $\mathbf{z}^{(4)}$ we obtain

$$\mathbf{z}^{(4)} = \begin{bmatrix} 7 \\ 1 \\ 2 \end{bmatrix} - \frac{11}{5}\begin{bmatrix} 1 \\ 0 \\ 2 \end{bmatrix} - \frac{77}{205}\begin{bmatrix} 12 \\ 5 \\ -6 \end{bmatrix} - \frac{6}{41}\begin{bmatrix} 2 \\ -6 \\ -1 \end{bmatrix} = 0. \tag{3.7.57}$$

We find the same result, $\mathbf{z}^{(4)} = \mathbf{0}$, whatever vector we take for $\mathbf{v}^{(4)}$. This agrees with Theorem 3.2.2 (*a*) which shows that any 4 vectors with 3 components are linearly dependent.

We have proved that from any set of m linearly independent vectors we may form m orthonormal vectors. We shall now use this fact to prove an important theorem concerning the solutions of equation (3.6.1) for the case where $\mathbf{A}$ is square.

Theorem 3.7.1. *Either* the equation

$$\mathbf{Ax} = \mathbf{y} \tag{3.7.58}$$

has a unique solution for any column vector $\mathbf{y}$ *or* the associated homogeneous equation

$$\mathbf{A}'\mathbf{x} = \mathbf{0} \tag{3.7.59}$$

has $n-r$ linearly independent solutions, where r is the rank of $\mathbf{A}$. In the latter event the non-homogeneous equation has a solution if and only if $\mathbf{y}$ is orthogonal to all the solutions of the homogeneous equation (3.7.59).

If $\mathbf{A}$ is non-singular the equation has a unique solution for all vectors $\mathbf{y}$, and it is given by equation (3.6.2). If $\mathbf{A}$ is singular and has rank r $(r < n)$ the matrix $\mathbf{A}'$

will also have rank r. The equation (3.7.59) will then have $n-r$ linearly independent solutions. Let $\mathbf{u}^{(1)}, \mathbf{u}^{(2)}, \ldots, \mathbf{u}^{(n-r)}$ be $n-r$ mutually orthogonal solutions of the equation. The condition stated in the theorem is *necessary*. For if equation (3.7.58) has a solution then

$$\mathbf{y}'\mathbf{u}^{(s)} = \mathbf{x}'\mathbf{A}'\mathbf{u}^{(s)} = \mathbf{x}'.\mathbf{0} = 0, \tag{3.7.60}$$

so that $\mathbf{y}$ must be orthogonal to all the solutions of the homogeneous equation (3.7.59).

To prove that the condition is *sufficient* we proceed as follows. The matrix $\mathbf{A}$, being of rank r, has just r linearly independent columns. Without loss of generality we may suppose that the first r columns of $\mathbf{A}$ (viz. $\mathbf{A}^{(1)}, \mathbf{A}^{(2)}, \ldots, \mathbf{A}^{(r)}$) are linearly independent. (For if they are not we may re-number the unknowns so that the first r columns of the new matrix of coefficients are linearly independent.) The term in row t of the product

$$\mathbf{A}'\mathbf{u}^{(s)} = \mathbf{0} \tag{3.7.61}$$

is $\mathbf{A}'^{(t)}\mathbf{u}^{(s)}$, and therefore each vector $\mathbf{u}^{(s)}$ is orthogonal to each column of $\mathbf{A}$. This means that the n vectors $\mathbf{A}^{(1)}, \mathbf{A}^{(2)}, \ldots, \mathbf{A}^{(r)}, \mathbf{u}^{(1)}, \mathbf{u}^{(2)}, \ldots, \mathbf{u}^{(n-r)}$ are linearly independent. For if

$$\lambda_1\mathbf{A}^{(1)} + \lambda_2\mathbf{A}^{(2)} + \ldots + \lambda_r\mathbf{A}^{(r)} + \mu_1\mathbf{u}^{(1)} + \mu_2\mathbf{u}^{(2)} + \ldots + \mu_{n-r}\mathbf{u}^{(n-r)} = \mathbf{0}, \tag{3.7.62}$$

then premultiplication by $\mathbf{u}'^{(s)}$ gives

$$\mu_s\mathbf{u}'^{(s)}\mathbf{u}^{(s)} = 0, \tag{3.7.63}$$

so that

$$\mu_1 = 0 = \mu_2 = \ldots = \mu_{n-r}, \tag{3.7.64}$$

and

$$\lambda_1\mathbf{A}^{(1)} + \lambda_2\mathbf{A}^{(2)} + \ldots + \lambda_r\mathbf{A}^{(r)} = \mathbf{0}. \tag{3.7.65}$$

But the $\mathbf{A}$'s are linearly independent so that

$$\lambda_1 = 0 = \lambda_2 = \ldots = \lambda_r. \tag{3.7.66}$$

Since the n vectors are linearly independent each vector $\mathbf{y}$ may be represented (uniquely) in the form

$$\mathbf{y} = \theta_1\mathbf{A}^{(1)} + \theta_2\mathbf{A}^{(2)} + \ldots + \theta_r\mathbf{A}^{(r)} + \phi_1\mathbf{u}^{(1)} + \phi_2\mathbf{u}^{(2)} + \ldots + \phi_{n-r}\mathbf{u}^{(n-r)}. \tag{3.7.67}$$

But if

$$\mathbf{y}'\mathbf{u}^{(s)} = 0 \quad (s = 1, 2, \ldots, n-r), \tag{3.7.68}$$

then

$$\phi_1 = 0 = \phi_2 = \ldots = \phi_{n-r} \tag{3.7.69}$$

and

$$\mathbf{y} = \theta_1\mathbf{A}^{(1)} + \theta_2\mathbf{A}^{(2)} + \ldots + \theta_r\mathbf{A}^{(r)}, \tag{3.7.70}$$

which means that

$$\mathbf{x}^{(0)} = \{\theta_1 \quad \theta_2 \quad \ldots \quad \theta_r \quad 0 \quad 0 \quad \ldots \quad 0\} \tag{3.7.71}$$

is a solution of equation (3.7.58) and this must be true for any vector $\mathbf{y}$ provided that it is orthogonal to all the solutions $\mathbf{u}^{(s)}$ of equation (3.7.59). Other solutions of the equation may be obtained by adding to $\mathbf{x}^{(0)}$ linear combinations of the solution of the homogeneous equation

$$\mathbf{A}\mathbf{x} = \mathbf{0}. \tag{3.7.72}$$

This matter was discussed in § 3.6; the full solution of the equation is given by equation (3.6.9).

EXAMPLES 3.7

1. Find all the values of t for which the following three vectors are mutually orthogonal

$$\begin{bmatrix} t \\ 2 \\ 1 \end{bmatrix}, \quad \begin{bmatrix} 3t \\ -2t \\ 1 \end{bmatrix}, \quad \begin{bmatrix} 3t+3 \\ 3t \\ -13t+1 \end{bmatrix}.$$

2. Give a formula for vectors $\{x_1 \quad x_2 \quad x_3\}$ orthogonal to both $\{-1 \quad 4 \quad 3\}$ and $\{7 \quad -1 \quad 2\}$. Find a unit vector which is orthogonal to the given vectors.

3. Obtain a set of orthogonal vectors from the following:

$$\mathbf{v}^{(1)} = \begin{bmatrix} -1 \\ 3 \\ 2 \end{bmatrix}, \quad \mathbf{v}^{(2)} = \begin{bmatrix} 5 \\ -1 \\ 6 \end{bmatrix}, \quad \mathbf{v}^{(3)} = \begin{bmatrix} 2 \\ 1 \\ 4 \end{bmatrix}, \quad \mathbf{v}^{(4)} = \begin{bmatrix} 3 \\ 5 \\ 10 \end{bmatrix}.$$

CHAPTER 4

FURTHER DEVELOPMENT OF THE THEORY OF CONSERVATIVE SYSTEMS

Let's give three cheers for the sailor's bride
Who casts all thought of rank aside... *H.M.S. Pinafore*

This chapter is a continuation of Chapter 2 and, among other things, it contains a discussion of some of the finer mathematical points which arose in the discussion of the free vibration of a system in § 2.3.

4.1 Orthogonality for dynamical systems

In discussing the purely geometrical properties of the n-dimensional vector space in § 3.7 we noted that there was a resemblance between such a space and a dynamical system with n degrees of freedom. In this section we shall continue this discussion.

The configuration of a dynamical system at any moment is specified by the values of the n generalised displacements $q_1, q_2, ..., q_n$, and these will be some functions of the time t. This means that we may associate the column vector $\mathbf{q} = \{q_1, q_2, ..., q_n\}$ with the dynamical system at any moment.†

For a dynamical system the quantity $\mathbf{q}'\mathbf{q}$ encountered in § 3.7 is not of particular interest. The important quantities are the kinetic and potential energies of the system, given by

$$T = \tfrac{1}{2}\dot{\mathbf{q}}'\mathbf{A}\dot{\mathbf{q}}, \quad V = \tfrac{1}{2}\mathbf{q}'\mathbf{C}\mathbf{q}. \tag{4.1.1}$$

We recollect also that when dealing with the principal modes of a system we found that they satisfied the relations

$$\mathbf{\Psi}'^{(r)}\mathbf{A}\mathbf{\Psi}^{(s)} = 0, \quad \mathbf{\Psi}'^{(r)}\mathbf{C}\mathbf{\Psi}^{(s)} = 0, \tag{4.1.2}$$

and that these relations were the expression of the so-called Orthogonality Principle. These considerations suggest that in dealing with n-dimensional vector space we should study the quantities $\mathbf{u}'\mathbf{A}\mathbf{u}$ and $\mathbf{u}'\mathbf{C}\mathbf{u}$ instead of $\mathbf{u}'\mathbf{u}$, and replace the relation $\mathbf{u}'\mathbf{v} = 0$ by relations of the type $\mathbf{u}'\mathbf{A}\mathbf{v} = 0$ and $\mathbf{u}'\mathbf{C}\mathbf{v} = 0$. We shall now do this, and our procedure will be first to introduce some formal definitions; only afterwards shall we discuss the physical interpretations which can be given to them.

Let $\mathbf{A}$ denote any symmetric positive definite matrix—not necessarily the inertia matrix. The 'norm' of a column vector $\mathbf{u}$ relative to $\mathbf{A}$ is defined as

$$|\mathbf{u}|_{\mathbf{A}} = \{\mathbf{u}'\mathbf{A}\mathbf{u}\}^{\frac{1}{2}}. \tag{4.1.3}$$

If $|\mathbf{u}|_{\mathbf{A}} = 1$, then $\mathbf{u}$ is said to be 'normalised relative to $\mathbf{A}$'.

† The motion of the system may thus be thought of as corresponding to the motion of the end-point of the vector $\mathbf{q}$ through an n-dimensional space.

The column vector $\mathbf{u}$ is said to be orthogonal to $\mathbf{v}$ relative to $\mathbf{A}$ if

$$\mathbf{u}'\mathbf{A}\mathbf{v} = 0. \tag{4.1.4}$$

Since $\mathbf{A}$ is symmetric by hypothesis, this relation is symmetrical; for its transpose is

$$\mathbf{v}'\mathbf{A}\mathbf{u} = 0, \tag{4.1.5}$$

which states that $\mathbf{v}$ is orthogonal to $\mathbf{u}$ relative to $\mathbf{A}$. The order of writing $\mathbf{u}$ and $\mathbf{v}$ is thus immaterial and when equation (4.1.4) holds we may say that $\mathbf{u}$ and $\mathbf{v}$ are '$\mathbf{A}$-orthogonal'. If we want to retain a geometric point of view we may think of the vectors $\mathbf{u}$ and $\mathbf{v}$ as being in a space whose 'geometry' is defined by the matrix $\mathbf{A}$. However, this is not necessary, and we can deal with the matter algebraically.

The physical interpretations which we can give to these ideas are quite straightforward. If $\mathbf{A}$ is identified with the inertia matrix of a system, and $\mathbf{u}$ with the vector $\dot{\mathbf{q}}$ then the norm of $\dot{\mathbf{q}}$ is related to the kinetic energy of the system by the equation

$$T = \tfrac{1}{2}\,|\dot{\mathbf{q}}|^2_{\mathbf{A}}. \tag{4.1.6}$$

When $\mathbf{A}$ is the inertia matrix the interpretation of $\mathbf{A}$-orthogonality is as follows. If we write $\dot{\mathbf{q}}$ as the sum of two $\mathbf{A}$-orthogonal parts $\dot{\mathbf{u}}$ and $\dot{\mathbf{v}}$, i.e. if

$$\dot{\mathbf{q}} = \dot{\mathbf{u}}+\dot{\mathbf{v}} \quad \text{and} \quad \dot{\mathbf{u}}'\mathbf{A}\dot{\mathbf{v}} = 0, \tag{4.1.7}$$

then there will be no coupling between $\dot{\mathbf{u}}$ and $\dot{\mathbf{v}}$ in the expression for the kinetic energy. For

$$T = \tfrac{1}{2}\dot{\mathbf{q}}'\mathbf{A}\dot{\mathbf{q}} = \tfrac{1}{2}(\dot{\mathbf{u}}'+\dot{\mathbf{v}}')\,\mathbf{A}(\dot{\mathbf{u}}+\dot{\mathbf{v}}) = \tfrac{1}{2}\dot{\mathbf{u}}'\mathbf{A}\dot{\mathbf{u}}+\tfrac{1}{2}\dot{\mathbf{v}}'\mathbf{A}\dot{\mathbf{v}}. \tag{4.1.8}$$

We may express this result in two ways—either

$$|\dot{\mathbf{q}}|^2_{\mathbf{A}} = |\dot{\mathbf{u}}|^2_{\mathbf{A}}+|\dot{\mathbf{v}}|^2_{\mathbf{A}}, \tag{4.1.9}$$

or, what is equivalent,

$$T(\dot{\mathbf{q}}) = T(\dot{\mathbf{u}}) + T(\dot{\mathbf{v}}). \tag{4.1.10}$$

Equation (4.1.9) is the analogue of Pythagoras's theorem for the space, and equation (4.1.10) is a generalised form of the first part of the Orthogonality Principle. The previous result, equation (2.5.22), may be obtained from it by putting

$$\mathbf{u} = \mathbf{\Psi}^{(r)}\sin(\omega_r t+\theta_r), \quad \mathbf{v} = \mathbf{\Psi}^{(s)}\sin(\omega_s t+\theta_s). \tag{4.1.11}$$

When the matrix $\mathbf{A}$ in equations (4.1.3) and (4.1.4) is identified with the stability matrix $\mathbf{C}$ and $\mathbf{u}$ with the vector $\mathbf{q}$, the results are similar. First

$$V = \tfrac{1}{2}\,|\mathbf{q}|^2_{\mathbf{C}} \tag{4.1.12}$$

and secondly, if

$$\mathbf{q} = \mathbf{u}+\mathbf{v} \quad \text{and} \quad \mathbf{u}'\mathbf{C}\mathbf{v} = 0, \tag{4.1.13}$$

then

$$|\mathbf{q}|^2_{\mathbf{C}} = |\mathbf{u}|^2_{\mathbf{C}}+|\mathbf{v}|^2_{\mathbf{C}}, \tag{4.1.14}$$

and

$$V(\mathbf{q}) = V(\mathbf{u}) + V(\mathbf{v}). \tag{4.1.15}$$

Again, equation (4.1.14) is the form of Pythagoras's theorem and equation (4.1.15) is the generalisation of the second part of the Orthogonality Principle.

We notice in passing that because the principal modes of a system satisfy the equation

$$\mathbf{C}\mathbf{\Psi}^{(s)} = \omega_s^2\mathbf{A}\mathbf{\Psi}^{(s)}, \tag{4.1.16}$$

A-orthogonality of two vectors implies **C**-orthogonality when one of the vectors is a principal mode. We may therefore talk about a motion of a system being 'orthogonal' to the sth mode, meaning **A**-orthogonal and **C**-orthogonal. In symbols, if the displacement $\mathbf{q}$ is such that

$$\mathbf{q}'\mathbf{A}\boldsymbol{\Psi}^{(s)} = 0, \tag{4.1.17}$$

then $\mathbf{q}$ also satisfies

$$\mathbf{q}'\mathbf{C}\boldsymbol{\Psi}^{(s)} = 0. \tag{4.1.18}$$

The converse (**C**-orthogonality implies **A**-orthogonality) is true, provided that $\boldsymbol{\Psi}^{(s)}$ is not a rigid body mode, that is, provided that $\omega_s \neq 0$.

We notice also that equation (4.1.17) is equivalent to the statement that the sth mode is absent from $\mathbf{q}$, that is, that

$$p_s = 0. \tag{4.1.19}$$

This follows from the equations

$$\mathbf{q} = p_1 \boldsymbol{\Psi}^{(1)} + p_2 \boldsymbol{\Psi}^{(2)} + \ldots + p_n \boldsymbol{\Psi}^{(n)}, \tag{4.1.20}$$

and

$$0 = \mathbf{q}'\mathbf{A}\boldsymbol{\Psi}^{(s)} = (p_1 \boldsymbol{\Psi}'^{(1)} + p_2 \boldsymbol{\Psi}'^{(2)} + \ldots + p_n \boldsymbol{\Psi}'^{(n)})\mathbf{A}\boldsymbol{\Psi}^{(s)}$$

$$= p_s(\boldsymbol{\Psi}'^{(s)}\mathbf{A}\boldsymbol{\Psi}^{(s)}). \tag{4.1.21}$$

The concepts of mutual orthogonality and orthonormality have counterparts in a dynamical system. The formal concepts are as follows. The vectors $\mathbf{u}^{(1)}, \mathbf{u}^{(2)}, \ldots, \mathbf{u}^{(m)}$ are said to be mutually **A**-orthogonal if

$$\mathbf{u}'^{(r)}\mathbf{A}\mathbf{u}^{(s)} = 0 \quad (r \neq s). \tag{4.1.22}$$

If the vectors are also normalised they are said to be orthonormal relative to **A**. In this case

$$\mathbf{u}'^{(r)}\mathbf{A}\mathbf{u}^{(s)} = \delta_{rs}. \tag{4.1.23}$$

It may be proved that if a number of non-zero vectors are mutually **A**-orthogonal they are linearly independent. Suppose for instance that

$$\lambda_1 \mathbf{u}^{(1)} + \lambda_2 \mathbf{u}^{(2)} + \ldots + \lambda_m \mathbf{u}^{(m)} = \mathbf{0}, \tag{4.1.24}$$

then pre-multiplication by $\mathbf{u}'^{(1)}\mathbf{A}$ gives

$$\lambda_1 \mathbf{u}'^{(1)}\mathbf{A}\mathbf{u}^{(1)} = 0, \tag{4.1.25}$$

and since **A** is positive definite this means that $\lambda_1 = 0$. Similarly,

$$\lambda_2 = 0 = \lambda_3 = \ldots = \lambda_m,$$

which establishes the linear independence.

It is because the principal modes of a system are mutually **A**-orthogonal and mutually **C**-orthogonal that the expressions for the kinetic and potential energies are just sums of squares when the principal coordinates are used. This matter was discussed in § 2.6; there it was shown that if the displacement vector is expressed by means of equation (4.1.20) and if the arbitrary multiplying factors occurring in the modes are chosen so that the modes are normalised relative to **A**, that is so that

$$\boldsymbol{\Psi}'^{(r)}\mathbf{A}\boldsymbol{\Psi}^{(s)} = \delta_{rs}, \tag{4.1.26}$$

then
$$\left.\begin{aligned} 2T &= p_1^2+p_2^2+\ldots+p_n^2, \\ 2V &= \omega_1^2 p_1^2+\omega_2^2 p_2^2+\ldots+\omega_n^2 p_n^2. \end{aligned}\right\} \tag{4.1.27}$$

EXAMPLES 4.1

1. If $\mathbf{A} = \begin{bmatrix} 2 & -1 & 3 \\ -1 & 1 & -2 \\ 3 & -2 & 16 \end{bmatrix}$ and $\mathbf{u} = \begin{bmatrix} 1 \\ 1 \\ 3 \end{bmatrix}$

find two linearly independent vectors of the form $\mathbf{v} = \{1 \quad x \quad y\}$ which are **A**-orthogonal to **u**.

2. Normalise the modes

$$\mathbf{\Psi}^{(1)} = \begin{bmatrix} 4 \\ 1 \\ -3 \end{bmatrix}, \quad \mathbf{\Psi}^{(2)} = \begin{bmatrix} -4 \\ 1 \\ 1 \end{bmatrix}, \quad \mathbf{\Psi}^{(3)} = \begin{bmatrix} 2 \\ 1 \\ -1 \end{bmatrix}$$

relative to the matrix

$$\mathbf{A} = m\begin{bmatrix} 4 & 3 & 9 \\ 3 & 7 & 9 \\ 9 & 9 & 23 \end{bmatrix}.$$

3. A certain dynamical system has the inertia matrix

$$\mathbf{A} = I\begin{bmatrix} 3 & -1 & -3 \\ -1 & 4 & 8 \\ -3 & 8 & 18 \end{bmatrix}$$

and two of its modes are

$$\mathbf{\Psi}^{(1)} = \{1 \quad -3 \quad 2\}, \quad \mathbf{\Psi}^{(2)} = \{2 \quad -3 \quad 1\}.$$

Find the normalised principal modes of the system.

4.2 Change of coordinates

To specify the motion of a system with n degrees of freedom we need n independent coordinates $q_1, q_2, \ldots, q_n$ which must be such that, between them, they take into account all the possible motions of the system. These coordinates can give the displacements (or rotations) of n physical points of the system or, more generally, they can specify the motions of the system in n independent modes.† The n modes which are chosen are the analogues for the system of the coordinate axes in an

† We use the word 'mode' in the sense that a mode defines a set of ratios between the generalised displacements at n points of the system. When the system vibrates in one mode the response at the coordinates $q_1, q_2, \ldots, q_n$ will all be in phase. These modes need not necessarily be principal modes of the system.

n-dimensional space. Any displacement $\mathbf{q}$ of the system may be expressed in terms of them by means of the equation

$$\mathbf{q} = \begin{bmatrix} q_1 \\ q_2 \\ \vdots \\ q_n \end{bmatrix} = q_1 \begin{bmatrix} 1 \\ 0 \\ \vdots \\ 0 \end{bmatrix} + q_2 \begin{bmatrix} 0 \\ 1 \\ \vdots \\ 0 \end{bmatrix} + \ldots + q_n \begin{bmatrix} 0 \\ 0 \\ \vdots \\ 1 \end{bmatrix}$$

$$= q_1 \mathbf{e}^{(1)} + q_2 \mathbf{e}^{(2)} + \ldots + q_n \mathbf{e}^{(n)}, \tag{4.2.1}$$

in which $\mathbf{e}^{(1)}, \mathbf{e}^{(2)}, \ldots, \mathbf{e}^{(n)}$ denote the columns of the unit matrix $\mathbf{I}_n$.

Choosing another set of coordinates $\bar{q}_1, \bar{q}_2, \ldots, \bar{q}_n$ is equivalent to choosing a different set of modes. Any set of n modes may be chosen so long as the column vectors associated with them, $\mathbf{z}^{(1)}, \mathbf{z}^{(2)}, \ldots, \mathbf{z}^{(n)}$, say, are linearly independent. Then

$$\mathbf{q} = \bar{q}_1 \mathbf{z}^{(1)} + \bar{q}_2 \mathbf{z}^{(2)} + \ldots + \bar{q}_n \mathbf{z}^{(n)}. \tag{4.2.2}$$

This equation may be written in the alternative form

$$\mathbf{q} = \mathbf{Z}\bar{\mathbf{q}}, \tag{4.2.3}$$

where $\mathbf{Z}$ is the non-singular square matrix which has the $\mathbf{z}$'s as columns.

To find the inertia and stability matrices, $\bar{\mathbf{A}}$ and $\bar{\mathbf{C}}$ respectively, for the new coordinates, in terms of $\mathbf{A}$ and $\mathbf{C}$ we notice that

$$\left.\begin{aligned} 2T &= \dot{\mathbf{q}}'\mathbf{A}\dot{\mathbf{q}} = \dot{\bar{\mathbf{q}}}'\mathbf{Z}'\mathbf{A}\mathbf{Z}\dot{\bar{\mathbf{q}}}, \\ 2V &= \mathbf{q}'\mathbf{C}\mathbf{q} = \bar{\mathbf{q}}'\mathbf{Z}'\mathbf{C}\mathbf{Z}\bar{\mathbf{q}}, \end{aligned}\right\} \tag{4.2.4}$$

so that

$$\bar{\mathbf{A}} = \mathbf{Z}'\mathbf{A}\mathbf{Z}; \quad \bar{\mathbf{C}} = \mathbf{Z}'\mathbf{C}\mathbf{Z}. \tag{4.2.5}$$

The individual elements of $\bar{\mathbf{A}}$ and $\bar{\mathbf{C}}$—that is, the inertia and stability coefficients corresponding to the new coordinates—are given by

$$\bar{a}_{rs} = \mathbf{z}'^{(r)}\mathbf{A}\mathbf{z}^{(s)}, \quad \bar{c}_{rs} = \mathbf{z}'^{(r)}\mathbf{C}\mathbf{z}^{(s)}. \tag{4.2.6}$$

It is physically obvious that the natural frequencies of a system will be the same whatever coordinates are used to describe its motion. To prove this mathematically we notice that, when we use the coordinates q, the equation for the natural frequencies is

$$|\mathbf{C} - \omega^2\mathbf{A}| = 0, \tag{4.2.7}$$

and when we use the coordinates $\bar{q}$ it is

$$|\bar{\mathbf{C}} - \omega^2\bar{\mathbf{A}}| = 0. \tag{4.2.8}$$

Equation (4.2.5) shows that

$$\bar{\mathbf{C}} - \omega^2\bar{\mathbf{A}} = \mathbf{Z}'(\mathbf{C} - \omega^2\mathbf{A})\,\mathbf{Z}, \tag{4.2.9}$$

so that

$$|\bar{\mathbf{C}} - \omega^2\bar{\mathbf{A}}| = |\mathbf{Z}'|\,.\,|\mathbf{C} - \omega^2\mathbf{A}|\,.\,|\mathbf{Z}|. \tag{4.2.10}$$

But $\mathbf{Z}$ is non-singular, so that $|\mathbf{Z}|$ and $|\mathbf{Z}'|$ are both non-zero. Therefore the values of ω for which equation (4.2.8) holds are the same as those for which equation (4.2.7) holds. Therefore the natural frequencies are the same.

It is also physically obvious that the principal modes of a system will remain the same (physically) whatever coordinates are used. However, when we use one set of coordinates q, a given mode will be associated with a certain column vector $\mathbf{\Psi}$ which will give the set of ratios $q_1 : q_2 : \ldots : q_n$ for the mode. When we change to the coordinates $\bar{q}$ the same mode will be associated with another vector $\bar{\mathbf{\Psi}}$ giving the set of ratios $\bar{q}_1 : \bar{q}_2 : \ldots : \bar{q}_n$. To find the relation between $\mathbf{\Psi}$ and $\bar{\mathbf{\Psi}}$ we notice that if they correspond to a natural frequency ω_s of the system they satisfy the equations

$$(\mathbf{C} - \omega_s^2 \mathbf{A})\, \mathbf{\Psi} = \mathbf{0} \tag{4.2.11}$$

and

$$(\bar{\mathbf{C}} - \omega_s^2 \bar{\mathbf{A}})\, \bar{\mathbf{\Psi}} = \mathbf{0} \tag{4.2.12}$$

respectively. Using equation (4.2.5) we may write the second equation

$$\mathbf{Z}'(\mathbf{C} - \omega_s^2 \mathbf{A})\, \mathbf{Z}\bar{\mathbf{\Psi}} = \mathbf{0}. \tag{4.2.13}$$

Because $\mathbf{Z}'$ is non-singular this equation implies that

$$(\mathbf{C} - \omega_s^2 \mathbf{A})\, \mathbf{Z}\bar{\mathbf{\Psi}} = \mathbf{0}, \tag{4.2.14}$$

and by comparing this with equation (4.2.11) we see that

$$\mathbf{\Psi} = \mathbf{Z}\bar{\mathbf{\Psi}}. \tag{4.2.15}$$

In § 4.4 we shall deal with systems for which two or more of the natural frequencies coincide. It will emerge that, to one of these multiple natural frequencies there will correspond a number of modes. Clearly we should expect that the number of modes corresponding to any given multiple frequency would be the same whatever coordinates are used. This we may easily demonstrate by considering equations (4.2.11) and (4.2.12) further. Every solution $\mathbf{\Psi}$ of equation (4.2.11) will give rise to a solution $\bar{\mathbf{\Psi}} = \mathbf{Z}^{-1}\mathbf{\Psi}$ of equation (4.2.12), and, conversely, every solution $\bar{\mathbf{\Psi}}$ of the second will give a solution $\mathbf{\Psi} = \mathbf{Z}\bar{\mathbf{\Psi}}$ of the first. Therefore the two equations will have the same number of linearly independent solutions.

EXAMPLES 4.2

1. The system to which the figures relate consists of a light string of length $4l$ stretched to a tension S. Three equal masses m are attached to the string at equal intervals l. Find expressions for the kinetic and potential energies of small transverse vibration of the system first in terms of the coordinates q of diagram (a) and then in terms of the coordinates $\bar{q}$ of diagram (b), and verify equations (4.2.5). (The figures are on the following page.)

4.3 Vibration under constraint

Suppose that a vibrating system with n degrees of freedom is subjected to a constraint which can be expressed as a relation

$$f(q_1, q_2, \ldots, q_n) = 0 \tag{4.3.1}$$

between the coordinates. We shall now discuss the effects which such a constraint has on the system.

Any constraint may be expanded by means of the extended Taylor's theorem to give a relation of the form

$$d_1 q_1 + d_2 q_2 + \ldots + d_n q_n = 0, \tag{4.3.2}$$

where

$$d_r = \left.\frac{\partial f}{\partial q_r}\right|_{q_1=0=q_2=\ldots=q_n} \tag{4.3.3}$$

There will be no constant term because the condition is to be satisfied in the position of equilibrium, in which all the coordinates vanish; and higher powers of the q's may be neglected because only small vibrations are permitted.

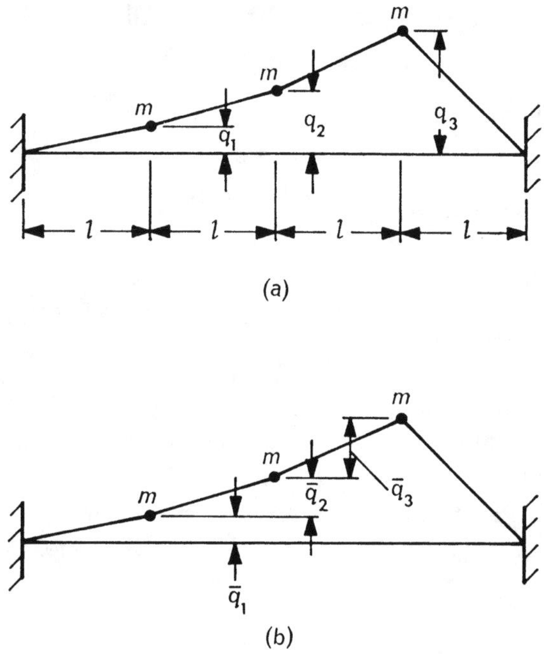

Ex. 4.2.1

It is convenient to work in terms of the normalised principal coordinates $p_1, p_2, \ldots, p_n$ of the unconstrained system, and then the constraint (4.3.2) becomes

$$b_1 p_1 + b_2 p_2 + \ldots + b_n p_n = 0, \tag{4.3.4}$$

and the equations for T and V are

$$\left.\begin{aligned} 2T &= \dot{p}_1^2 + \dot{p}_2^2 + \ldots + \dot{p}_n^2, \\ 2V &= \omega_1^2 p_1^2 + \omega_2^2 p_2^2 + \ldots + \omega_n^2 p_n^2. \end{aligned}\right\} \tag{4.3.5}$$

The simplest constraint is

$$p_s = 0. \tag{4.3.6}$$

It was shown in §4.1 (see equations (4.1.17)–(4.1.21)) that when a system is subjected to this constraint its motion is orthogonal to the sth principal mode, or, in other words, there is no motion in this mode. The constraint may be expressed in terms of the q's by using equation (4.1.17); it is

$$\mathbf{\Psi}'^{(s)}\mathbf{Aq} = 0. \tag{4.3.7}$$

If we introduce the notation

$$\mathbf{\Phi}^{(s)} = \mathbf{A\Psi}^{(s)}, \tag{4.3.8}$$

then we may express the constraint as

$$\mathbf{\Phi}'^{(s)}\mathbf{q} = 0 = \Phi_1^{(s)}q_1 + \Phi_2^{(s)}q_2 + \ldots + \Phi_n^{(s)}q_n. \tag{4.3.9}$$

Clearly the natural frequencies of the system subject to this constraint will be

$$\omega_1, \omega_2, \ldots, \omega_{s-1}, \omega_{s+1}, \ldots, \omega_n.$$

Before we consider more general constraints we note that we may demand that the system be orthogonal to a number of principal modes. For instance, it may be constrained to be orthogonal to the first two modes, and then

$$\mathbf{\Psi}'^{(1)}\mathbf{Aq} = 0 = \mathbf{\Psi}'^{(2)}\mathbf{Aq}, \tag{4.3.10}$$

or, in other words,

$$p_1 = 0 = p_2. \tag{4.3.11}$$

If it is subjected to these constraints then its natural frequencies will be $\omega_3, \omega_4, \ldots, \omega_n$.

We must now consider a system subjected to the general constraint shown in equation (4.3.4). We shall show that the constrained system has $n-1$ natural frequencies and that they alternate regularly with the natural frequencies of the unconstrained system.

Using equation (4.3.4) we may eliminate one of the coordinates p_r; thus if $b_n \neq 0$ we may eliminate p_n. Then

$$\left.\begin{aligned} 2T &= \dot{p}_1^2 + \dot{p}_2^2 + \ldots + \dot{p}_{n-1}^2 + \frac{1}{b_n^2}(b_1\dot{p}_1 + b_2\dot{p}_2 + \ldots + b_{n-1}\dot{p}_{n-1})^2, \\ 2V &= \omega_1^2 p_1^2 + \omega_2^2 p_2^2 + \ldots + \omega_{n-1}^2 p_{n-1}^2 + \frac{\omega_n^2}{b_n^2}(b_1 p_1 + b_2 p_2 + \ldots + b_{n-1}p_{n-1})^2. \end{aligned}\right\} \tag{4.3.12}$$

The Lagrangian equations for $p_1, p_2, \ldots, p_{n-1}$ are then

$$\begin{aligned} \ddot{p}_r + \omega_r^2 p_r + b_r\biggl\{\frac{1}{b_n^2}(b_1\ddot{p}_1 + b_2\ddot{p}_2 + \ldots + b_{n-1}\ddot{p}_{n-1}) \\ + \frac{\omega_n^2}{b_n^2}(b_1 p_1 + b_2 p_2 + \ldots + b_{n-1}p_{n-1})\biggr\} = 0 \quad (r = 1, 2, \ldots, n-1). \end{aligned} \tag{4.3.13}$$

When the constrained system is vibrating freely in one of its principal modes the coordinates $p_1, p_2, \ldots, p_{n-1}$ will have the form

$$p_1 = \Pi_1 \sin\omega t, \quad p_2 = \Pi_2 \sin\omega t, \quad \ldots, \quad p_{n-1} = \Pi_{n-1}\sin\omega t. \tag{4.3.14}$$

If we substitute these values into the equations (4.3.13) we shall obtain $n-1$ homogeneous equations for the amplitudes $\Pi_1, \Pi_2, \ldots, \Pi_{n-1}$. The condition for these $n-1$ equations to have a non-zero solution is that the determinant of coefficients should vanish. This gives an equation for ω whose roots are the natural frequencies of the constrained system. But this is not the best way to treat the matter since the coordinate p_n is treated in a different way from the others, and the determinantal equation for the natural frequencies is unnecessarily complicated.

It is easier to introduce a further unknown, μ, and write equation (4.3.13) as

$$\ddot{p}_r + \omega_r^2 p_r + \mu b_r = 0 \quad (r = 1, 2, \ldots, n-1), \tag{4.3.15}$$

where

$$\mu = \frac{1}{b_n^2}(b_1\ddot{p}_1 + b_2\ddot{p}_2 + \ldots + b_{n-1}\ddot{p}_{n-1}) + \frac{\omega_n^2}{b_n^2}(b_1 p_1 + b_2 p_2 + \ldots + b_{n-1}p_{n-1})$$

$$= \frac{-b_n\ddot{p}_n}{b_n^2} - \frac{\omega_n^2 b_n p_n}{b_n^2}. \tag{4.3.16}$$

This last equation for μ is the equation obtained by putting $r = n$ in equation (4.3.15). Therefore there are now the n equations

$$\ddot{p}_r + \omega_r^2 p_r + \mu b_r = 0 \quad (r = 1, 2, \ldots, n), \tag{4.3.17}$$

in which all the coordinates appear on an equal footing, and the $n+1$ unknowns $p_1, p_2, \ldots, p_n, \mu$ have to be determined from these n equations together with equation (4.3.4).

These equations may be solved by using equation (4.3.14) and writing

$$\mu = \nu \sin \omega t, \tag{4.3.18}$$

so that

$$\Pi_r(\omega_r^2 - \omega^2) + \nu b_r = 0 \quad (r = 1, 2, \ldots, n). \tag{4.3.19}$$

Substituting the values of $\Pi_1, \Pi_2, \ldots, \Pi_n$ given by these equations in the equation

$$b_1\Pi_1 + b_2\Pi_2 + \ldots + b_n\Pi_n = 0 \tag{4.3.20}$$

we have

$$\frac{b_1^2}{\omega_1^2 - \omega^2} + \frac{b_2^2}{\omega_2^2 - \omega^2} + \ldots + \frac{b_n^2}{\omega_n^2 - \omega^2} = 0. \tag{4.3.21}$$

This equation in ω^2 has $n-1$ roots and they are the squares of the natural frequencies $\tilde{\omega}_1^2, \tilde{\omega}_2^2, \ldots, \tilde{\omega}_{n-1}^2$, say, of the constrained system.

To show that the quantities $\tilde{\omega}_1^2, \tilde{\omega}_2^2, \ldots, \tilde{\omega}_{n-1}^2$ alternate with $\omega_1^2, \omega_2^2, \ldots, \omega_n^2$ we multiply the equation by $(\omega_1^2 - \omega^2)(\omega_2^2 - \omega^2)\ldots(\omega_n^2 - \omega^2)$ and write

$$F(\omega^2) \equiv b_1^2(\omega_2^2 - \omega^2)(\omega_3^2 - \omega^2)\ldots(\omega_n^2 - \omega^2) + b_2^2(\omega_1^2 - \omega^2)(\omega_3^2 - \omega^2)$$
$$\ldots(\omega_n^2 - \omega^2) + \ldots + b_n^2(\omega_1^2 - \omega^2)(\omega_2^2 - \omega^2)\ldots(\omega_{n-1}^2 - \omega^2) = 0. \tag{4.3.22}$$

If

$$\omega_1 < \omega_2 < \ldots < \omega_n, \tag{4.3.23}$$

then $F(\omega_1^2)$ is positive, $F(\omega_2^2)$ is negative, $F(\omega_3^2)$ is positive, and so on alternately, so that

$$\omega_1 \leqslant \tilde{\omega}_1 \leqslant \omega_2 \leqslant \tilde{\omega}_2 \leqslant \ldots \leqslant \omega_{n-1} \leqslant \tilde{\omega}_{n-1} \leqslant \omega_n. \tag{4.3.24}$$

We notice, moreover, that $\tilde{\omega}_1$ can be equal to ω_1 if and only if b_1 is zero, and generally

$$\tilde{\omega}_r = \omega_r \quad \text{if and only if} \quad b_r = 0. \tag{4.3.25}$$

On the other hand, $\tilde{\omega}_1$ can be equal to ω_2 if and only if

$$b_2 = 0 = b_3 = \ldots = b_n, \tag{4.3.26}$$

and then

$$\tilde{\omega}_2 = \omega_3, \quad \tilde{\omega}_3 = \omega_4, \quad \ldots, \quad \tilde{\omega}_{n-1} = \omega_n. \tag{4.3.27}$$

This is a particular case of the simple constraint that was considered first.

EXAMPLES 4.3

1. Calculate, by Newton's method or otherwise, the roots of the equation

$$\frac{1}{(g/l)-\omega^2}+\frac{4}{(3g/l)-\omega^2}+\frac{16}{(25g/l)-\omega^2}+\frac{9}{(30g/l)-\omega^2}=0$$

correct to three decimal places.

2. The system described in Ex. 2.11.1 is constrained to vibrate in a certain way. The constraint is supplied by a framework of light rigid rods which is so constructed that the system may swing about P with $\theta:\phi = 1:k$. By eliminating θ or ϕ find the natural frequencies of the constrained system. For what values of k does the non-zero natural frequency take stationary values and what are the modes corresponding to these values of k?

3. Find the coefficients b_1, b_2 and b_3 of equation (4.3.4) corresponding to the constraint described in Ex. 4.3.2.

4.4 A rigorous treatment of natural frequencies and principal modes

In this section we shall give a rigorous treatment of certain matters which arose in § 2.3 when we were defining the principal modes of a system. We shall prove the following three results which we may conveniently enunciate as theorems. The first is

Theorem 4.4.1. When the natural frequencies of a system are all different the principal mode corresponding to each natural frequency is determined uniquely apart from an arbitrary multiplying factor.

This may be considered to be a special case of

Theorem 4.4.2. If ω_s is a simple (non-repeated) natural frequency then the principal mode $\mathbf{\Psi}^{(s)}$ is uniquely determined apart from an arbitrary multiplying factor. This theorem applies to any system whether the natural frequencies are all different or not; it holds provided that the particular frequency ω_s is simple.

Finally, we shall prove the general result, namely

Theorem 4.4.3. If ω_s is an r-fold natural frequency, that is, if r of the roots of the frequency equation are equal to ω_s, then there are exactly r principal modes corresponding to it. Moreover, these r modes may be chosen to be mutually **A**-orthogonal and **C**-orthogonal.

The conclusions which we can draw from these theorems are quite straightforward. A system with n degrees of freedom has n principal modes whether the natural frequencies are all different or not; and the modes may be chosen to satisfy the orthogonality relations (2.5.7) and (2.5.8). This means, as the reader may easily verify, that the modal analysis given in Chapter 2 applies without any modification to systems which have repeated natural frequencies. The theorems apply even when the system has rigid-body modes (corresponding to $\omega_s = 0$) so that **C** may be singular. It will be assumed that **A** is non-singular, but it is shown how this restriction may be lifted.

Since the formal proof of these results, and especially the last one, can be rather complicated† we shall content ourselves with a (rigorous) *description* of the method employed in the proof.

The three theorems may be proved by using two simple lemmas:

Lemma 1. If ω_s is a natural frequency of the system, that is, if it satisfies the equation

$$|\mathbf{C}-\omega_s^2\mathbf{A}| = 0, \tag{4.4.1}$$

then it is possible to find at least one column vector $\mathbf{\Psi}^{(s)}$ satisfying

$$(\mathbf{C}-\omega_s^2\mathbf{A})\,\mathbf{\Psi}^{(s)} = \mathbf{0}. \tag{4.4.2}$$

Lemma 2. Once such a column vector $\mathbf{\Psi}^{(s)}$ has been found it is possible to find a new set of coordinates $\bar{\mathbf{q}}$ given by

$$\mathbf{q} = \mathbf{Z}\bar{\mathbf{q}}, \tag{4.4.3}$$

for which the inertia and stability matrices are

$$\bar{\mathbf{A}} = \mathbf{Z}'\mathbf{A}\mathbf{Z} = \begin{bmatrix} 1 & \mathbf{0} \\ \mathbf{0} & \mathbf{A}_{n-1} \end{bmatrix}, \quad \bar{\mathbf{C}} = \mathbf{Z}'\mathbf{C}\mathbf{Z} = \begin{bmatrix} \omega_s^2 & \mathbf{0} \\ \mathbf{0} & \mathbf{C}_{n-1} \end{bmatrix}. \tag{4.4.4}$$

The first lemma is simply a special case of Theorem 3.2.1 (b) which states that a necessary and sufficient condition for equation (4.4.2) to have a solution is that ω_s should satisfy equation (4.4.1). It is perhaps of interest to note that previously (in § 2.3) we referred to § 1.8 when passing from equation (4.4.2) to (4.4.1), but actually all that we proved in § 1.8 is that *if* equation (4.4.2) has a non-zero solution *then* equation (4.4.1) must hold (i.e. that the condition is necessary). Only when the determinant $|\mathbf{C}-\omega_s^2\mathbf{A}|$ of coefficients has at least one non-zero co-factor (see § 1.5) does § 1.8 guarantee that the equation has a solution when equation (4.4.1) holds (i.e. that the condition is sufficient).

The proof of the second lemma is simply a matter of showing how the new coordinates may be found. The values of the new inertia and stability coefficients are given by equations (4.2.6), namely

$$\bar{a}_{rs} = \mathbf{z}'^{(r)}\mathbf{A}\mathbf{z}^{(s)}, \quad \bar{c}_{rs} = \mathbf{z}'^{(r)}\mathbf{C}\mathbf{z}^{(s)}. \tag{4.4.5}$$

Therefore, if the new matrices are to have the required form the following equations must hold:

$$\mathbf{z}'^{(1)}\mathbf{A}\mathbf{z}^{(1)} = 1, \quad \mathbf{z}'^{(1)}\mathbf{C}\mathbf{z}^{(1)} = \omega_s^2; \tag{4.4.6}$$

$$\mathbf{z}'^{(1)}\mathbf{A}\mathbf{z}^{(r)} = 0 = \mathbf{z}'^{(1)}\mathbf{C}\mathbf{z}^{(r)} \quad (r = 2, 3, \ldots, n). \tag{4.4.7}$$

The first two equations may be satisfied by putting

$$\mathbf{z}^{(1)} = \frac{\mathbf{\Psi}^{(s)}}{\{\mathbf{\Psi}'^{(s)}\mathbf{A}\mathbf{\Psi}^{(s)}\}^{\frac{1}{2}}}, \tag{4.4.8}$$

and the remainder may be satisfied by taking $\mathbf{z}^{(2)}, \mathbf{z}^{(3)}, \ldots, \mathbf{z}^{(n)}$ to be any $n-1$ vectors orthogonal (i.e. **A**-orthogonal and **C**-orthogonal) to $\mathbf{z}^{(1)}$. We shall now show how the vectors $\mathbf{z}^{(2)}, \mathbf{z}^{(3)}, \ldots, \mathbf{z}^{(n)}$ may be found.‡

† G. M. L. Gladwell, 'Vibrating systems with equal natural frequencies', *J. Mech. Engng Sci.* **3** (1961), pp. 174–81.

‡ On a first reading the reader may prefer to take the **z**'s as given: he may then omit the following paragraph in brackets.

[First, it is sufficient to determine $\mathbf{z}^{(2)}, \mathbf{z}^{(3)}, \ldots, \mathbf{z}^{(n)}$ so that they are $\mathbf{A}$-orthogonal to $\mathbf{z}^{(1)}$, for, then, because $\mathbf{z}^{(1)}$ is a principal mode, they will also be $\mathbf{C}$-orthogonal to $\mathbf{z}^{(1)}$ (see equations (4.1.17), (4.1.18)). Now we use the analogue for $\mathbf{A}$-orthogonality of the method described in § 3.7. If $\mathbf{u}$ and $\mathbf{v}$ are any two vectors the vector

$$\mathbf{z} = \mathbf{v} + \lambda \mathbf{u} \tag{4.4.9}$$

will be $\mathbf{A}$-orthogonal to $\mathbf{u}$ provided that

$$\mathbf{u}'\mathbf{A}\mathbf{z} = 0 = \mathbf{u}'\mathbf{A}\mathbf{v} + \lambda \mathbf{u}'\mathbf{A}\mathbf{u}, \tag{4.4.10}$$

i.e.

$$\lambda = -\mathbf{u}'\mathbf{A}\mathbf{v}/(\mathbf{u}'\mathbf{A}\mathbf{u}). \tag{4.4.11}$$

The recipe for determining the $\mathbf{z}$'s is therefore as follows:

(1) Choose $n-1$ vectors $\mathbf{v}^{(2)}, \mathbf{v}^{(3)}, \ldots, \mathbf{v}^{(n)}$. If the vectors $\mathbf{z}^{(1)}, \mathbf{z}^{(2)}, \ldots, \mathbf{z}^{(n)}$ are to be linearly independent then the vectors $\mathbf{z}^{(1)}, \mathbf{v}^{(2)}, \ldots, \mathbf{v}^{(n)}$ must be also: but apart from this the $\mathbf{v}$'s may be chosen arbitrarily.

(2) To obtain $\mathbf{z}^{(r)}$ from $\mathbf{v} = \mathbf{v}^{(r)}$ subtract the multiple of $\mathbf{u} = \mathbf{z}^{(1)}$ given by equation (4.4.11). Thus

$$\mathbf{z}^{(r)} = \mathbf{v}^{(r)} + \lambda_r \mathbf{z}^{(1)}, \tag{4.4.12}$$

where

$$\lambda_r = -\frac{\mathbf{z}'^{(1)}\mathbf{A}\mathbf{v}^{(r)}}{\mathbf{z}'^{(1)}\mathbf{A}\mathbf{z}^{(1)}} = -\mathbf{z}'^{(1)}\mathbf{A}\mathbf{v}^{(r)}. \tag{4.4.13}$$

(3) It may easily be shown that if the first element, $z_1^{(1)}$, of $\mathbf{z}^{(1)}$ is non-zero then a suitable set of vectors $\mathbf{v}^{(2)}, \ldots, \mathbf{v}^{(n)}$ are the columns $\mathbf{e}^{(2)}, \mathbf{e}^{(3)}, \ldots, \mathbf{e}^{(n)}$ of the unit matrix $\mathbf{I}_n$. If $z_1^{(1)} = 0$ but some other element $z_t^{(1)}$ is non-zero, then $\mathbf{e}^{(1)}, \mathbf{e}^{(2)}, \ldots, \mathbf{e}^{(t-1)}, \mathbf{e}^{(t+1)}, \ldots, \mathbf{e}^{(n)}$ are a suitable set.]

We shall now use the lemmas to prove Theorem 4.4.2. In the new coordinates $\bar{q}_1, \bar{q}_2, \ldots, \bar{q}_n$, the frequency equation is

$$|\overline{\mathbf{C}} - \omega^2 \overline{\mathbf{A}}| = 0, \tag{4.4.14}$$

and by combining the two parts of equation (4.4.4) we see that

$$\overline{\mathbf{C}} - \omega^2 \overline{\mathbf{A}} = \begin{bmatrix} \omega_s^2 - \omega^2 & \mathbf{0} \\ \mathbf{0} & \mathbf{C}_{n-1} - \omega^2 \mathbf{A}_{n-1} \end{bmatrix}. \tag{4.4.15}$$

To obtain the determinant of this partitioned matrix we use equation (3.7.39) which gives

$$|\overline{\mathbf{C}} - \omega^2 \overline{\mathbf{A}}| = (\omega_s^2 - \omega^2)\,|\mathbf{C}_{n-1} - \omega^2 \mathbf{A}_{n-1}|. \tag{4.4.16}$$

Therefore the equation for the natural frequencies is

$$(\omega_s^2 - \omega^2)\,|\mathbf{C}_{n-1} - \omega^2 \mathbf{A}_{n-1}| = 0. \tag{4.4.17}$$

But, by hypothesis, ω_s is a simple (non-repeated) frequency and therefore

$$|\mathbf{C}_{n-1} - \omega_s^2 \mathbf{A}_{n-1}| \neq 0. \tag{4.4.18}$$

In the new coordinates, the equation governing the mode (or modes) corresponding to ω_s is (see equations (4.2.11)–(4.2.15) and the discussion following)

$$(\overline{\mathbf{C}} - \omega_s^2 \overline{\mathbf{A}})\,\overline{\mathbf{\Psi}} = \mathbf{0}, \tag{4.4.19}$$

and equation (4.4.15) shows that this is

$$\begin{bmatrix} 0 & \mathbf{0} \\ \mathbf{0} & \mathbf{C}_{n-1}-\omega_s^2\mathbf{A}_{n-1} \end{bmatrix} \begin{bmatrix} \bar{\Psi}_1 \\ \bar{\Psi}_2 \\ \vdots \\ \bar{\Psi}_n \end{bmatrix} = \mathbf{0}. \tag{4.4.20}$$

We shall now show that these equations have one and only one solution.† When multiplied out, these equations for the n components of $\bar{\mathbf{\Psi}}$ consist of $n-1$ equations for the $n-1$ components $\bar{\Psi}_2, \bar{\Psi}_3, \ldots, \bar{\Psi}_n$; all the coefficients involving $\bar{\Psi}_1$ are zero. Moreover, the determinant of the coefficients occurring in the equations for the $n-1$ $\bar{\Psi}$'s is shown in equation (4.4.18) to be non-zero. Therefore, the only solution of the $n-1$ equations is the trivial one in which

$$\bar{\Psi}_2 = 0 = \bar{\Psi}_3 = \ldots = \bar{\Psi}_n. \tag{4.4.21}$$

Therefore there is only one mode corresponding to ω_s. In the new coordinates it is

$$\bar{\mathbf{\Psi}} = \{c, 0, 0, \ldots, 0\}, \tag{4.4.22}$$

where c is an arbitrary constant, and in the old coordinates it is (see equation (4.2.15))

$$\begin{aligned} \mathbf{\Psi} &= \mathbf{Z}\bar{\mathbf{\Psi}} = \mathbf{Z}\{c, 0, \ldots, 0\} \\ &= c\mathbf{z}^{(1)} = c\mathbf{\Psi}^{(s)}/\{\mathbf{\Psi}'^{(s)}\mathbf{A}\mathbf{\Psi}^{(s)}\}^{\frac{1}{2}}. \end{aligned} \tag{4.4.23}$$

This proves Theorem 4.4.2.

If all the natural frequencies are simple (non-repeated), then the above argument can be applied to each frequency to yield Theorem 4.4.1.

We shall now prove Theorem 4.4.3. If we introduce the notation

$$\mathbf{q}^{(n-1)} = \{\bar{q}_2, \bar{q}_3, \ldots, \bar{q}_n\} \tag{4.4.24}$$

and use equation (4.4.4) we may express the kinetic and potential energies of the system in the form

$$\left.\begin{aligned} 2T &= \dot{\bar{q}}_1^2 + \dot{\mathbf{q}}'^{(n-1)}\mathbf{A}_{n-1}\dot{\mathbf{q}}^{(n-1)}, \\ 2V &= \omega_s^2\bar{q}_1^2 + \mathbf{q}'^{(n-1)}\mathbf{C}_{n-1}\mathbf{q}^{(n-1)}. \end{aligned}\right\} \tag{4.4.25}$$

Suppose now that we constrain the system to be orthogonal to the mode $\mathbf{\Psi}^{(s)}$ which has been found, then, since this is equivalent to the constraint

$$\bar{q}_1 = 0, \tag{4.4.26}$$

the kinetic and potential energies of the constrained system will be T and V where

$$\left.\begin{aligned} 2T &= \dot{\mathbf{q}}'^{(n-1)}\mathbf{A}_{n-1}\dot{\mathbf{q}}^{(n-1)}, \\ 2V &= \mathbf{q}'^{(n-1)}\mathbf{C}_{n-1}\mathbf{q}^{(n-1)}. \end{aligned}\right\} \tag{4.4.27}$$

Thus $\mathbf{A}_{n-1}$ and $\mathbf{C}_{n-1}$ are the inertia and stability matrices when the motion of the

† Since, as equation (4.4.18) shows, the matrix of coefficients has a non-zero cofactor and is therefore of rank $n-1$, the result can be deduced from Theorem 3.5.1; however, the argument which will now be given does not depend on ideas of rank.

constrained system is expressed in terms of the new coordinates. The equations for the frequencies and modes of this sytem are

$$|\mathbf{C}_{n-1} - \omega^2 \mathbf{A}_{n-1}| = 0 \tag{4.4.28}$$

and

$$(\mathbf{C}_{n-1} - \omega^2 \mathbf{A}_{n-1}) \boldsymbol{\Psi} = \mathbf{0}. \tag{4.4.29}$$

The theorem can now be proved by a process of 'induction', as follows. Start with coordinates q and inertia and stability matrices $\mathbf{A}$ and $\mathbf{C}$, and assume that $\mathbf{A}$ is non-singular. Find one natural frequency of the system and a corresponding mode (Lemma 1 ensures that this can always be done). Change the coordinates, using Lemma 2, and deal in exactly the same way with the system constrained to be orthogonal to the mode which has been found. The constrained system will be exactly similar to the original system except that it will have one degree of freedom less, its inertia and stability matrices will be $\mathbf{A}_{n-1}$ and $\mathbf{C}_{n-1}$, and equation (4.4.4) shows that $\mathbf{A}_{n-1}$ will be non-singular. Now find a natural frequency and principal mode of the constrained system (and again Lemma 1 applied to the constrained system ensures that this can always be done). The frequency will be a natural frequency of the original system and the mode will be a principal mode of the original system corresponding to this frequency (though it will be expressed in terms of the new coordinates).† Because it is a mode of the constrained system this mode will automatically be orthogonal to the first mode which was found, and this will be so whether the first two frequencies are different or not. Now we apply Lemma 2 to the constrained system, change the coordinates a second time and repeat the process with the system constrained to be orthogonal to the first two modes, and so on. The process can be continued until n natural frequencies and n mutually orthogonal modes have been found. This proves the theorem, for, if ω_s is an r-fold natural frequency, it will occur exactly r times in the process, and each time it occurs it will be allotted a mode which is $\mathbf{A}$-orthogonal and $\mathbf{C}$-orthogonal to all the previous modes.‡

If the matrix $\mathbf{C}$ is singular the foregoing argument still holds. Theorem 4.4.3 then states that if $\omega^2 = 0$ is an r-fold root of the frequency equation then there will be r mutually orthogonal rigid-body modes.

If the matrix $\mathbf{A}$ is singular and is of rank r there will be $n-r$ modes,

$$\boldsymbol{\Psi}^{(r+1)}, \quad \boldsymbol{\Psi}^{(r+2)}, \quad \ldots, \quad \boldsymbol{\Psi}^{(n)},$$

say, satisfying the equation

$$\mathbf{A}\boldsymbol{\Psi}^{(s)} = \mathbf{0} \quad (s = r+1, r+2, \ldots, n), \tag{4.4.30}$$

and therefore corresponding to an 'infinite' natural frequency. Free vibration of the system is governed by the equation

$$\mathbf{A}\ddot{\mathbf{q}} + \mathbf{C}\mathbf{q} = \mathbf{0} \tag{4.4.31}$$

† If the mode found from equation (4.4.29) is $\{\Psi_2, \Psi_3, \ldots, \Psi_n\}$ then in the original coordinates it will be $\mathbf{Z}\{0, \Psi_2, \Psi_3, \ldots, \Psi_n\}$ (see equation (4.2.15)).

‡ We note that the principal modes will be found in terms of the current set of coordinates which are being used, and to find the modes in terms of the original coordinates we have to take account of all the successive coordinate changes.

and premultiplication of this equation by $\mathbf{\Psi}'^{(s)}$ shows that

$$\mathbf{\Psi}'^{(s)}\mathbf{C}\mathbf{q} = 0 \quad (s = r+1, r+2, \ldots, n). \tag{4.4.32}$$

There are thus $n-r$ linear relations between the coordinates from which $n-r$ of them can be eliminated. When these are eliminated the resulting inertia matrix will be non-singular.

The work of this section may be used to prove another result which was taken for granted in Chapter 2, namely, 'If $\mathbf{C}$ is positive semi-definite it is necessarily singular'. This may be proved as follows. It has been shown that if $\mathbf{A}$ is non-singular it is always possible to find n mutually orthogonal principal modes $\mathbf{\Psi}^{(1)}, \mathbf{\Psi}^{(2)}, \ldots, \mathbf{\Psi}^{(n)}$. It is therefore always possible to introduce normalised principal coordinates p, as in § 2.6, such that

$$\mathbf{q} = \mathbf{X}\mathbf{p}, \tag{4.4.33}$$

and then the potential energy of the system will be given by

$$2V = \omega_1^2 p_1^2 + \omega_2^2 p_2^2 + \ldots + \omega_n^2 p_n^2. \tag{4.4.34}$$

If $\mathbf{C}$ is semi-definite then V can be zero for non-trivial values of $p_1, p_2, \ldots, p_n$; but this can happen only if one of the ω's is zero, which can happen only if $\mathbf{C}$ is singular.

EXAMPLES 4.4

1. One principal mode of the system in diagram (*a*) of Ex. 4.2.1 is given by

$$q_1 : q_2 : q_3 = 1 : 0 : -1.$$

Find:

(*a*) the corresponding normalised mode;
(*b*) the corresponding natural frequency;
(*c*) a set of new coordinates satisfying the conditions of Lemma 2;
(*d*) the matrices $\mathbf{A}_2$, $\mathbf{C}_2$ for the constrained system;
(*e*) the remaining natural frequencies and normalised principal modes.

2. The system shown consists of a mass m connected to the mid-points of the sides of a square of side $2a$ by equal springs of natural length a and stiffness k. Find the natural frequency of the system and enumerate the kinds of free vibration that the mass may execute at this frequency. (The figure is on p. 145.)

3. The figure shows a system consisting of a mass M suspended by an inextensible string of length l from the mass m of Ex. 4.4.2. Assuming that the mass m moves only in the horizontal plane, find expressions for the energies of the system when the pendulum executes small vibrations about the vertical position. (The figure is on p. 146.)

4. Determine the natural frequencies and a set of mutually orthogonal principal modes of the system of Ex. 4.4.3 if, in a consistent set of units,

$$M = 6, \quad m = 3, \quad k = 8, \quad l = 8, \quad g = 32.$$

4.5 Further discussion of vibration under constraint

In § 4.3 we considered the effect of imposing a constraint on a system whose principal coordinates and natural frequencies are known. In this section we shall consider the problem of constraints more generally. We shall begin by discussing systems subject to a single constraint.

The Lagrangian equations

$$\frac{d}{dt}\left(\frac{\partial T}{\partial \dot{q}_r}\right)+\frac{\partial V}{\partial q_r} = Q_r \quad (r = 1, 2, ..., n) \tag{4.5.1}$$

are the equations governing the vibration of a system whose motion is specified by *n independent* generalised coordinates $q_1, q_2, ..., q_n$. When a vibrating system is subjected to the constraint

$$d_1 q_1 + d_2 q_2 + ... + d_n q_n = 0, \tag{4.5.2}$$

the coordinates $q_1, q_2, ..., q_n$ are no longer independent and it is necessary to discuss the changes that must be made to the Lagrangian equations if they are to be applicable to constrained motion.

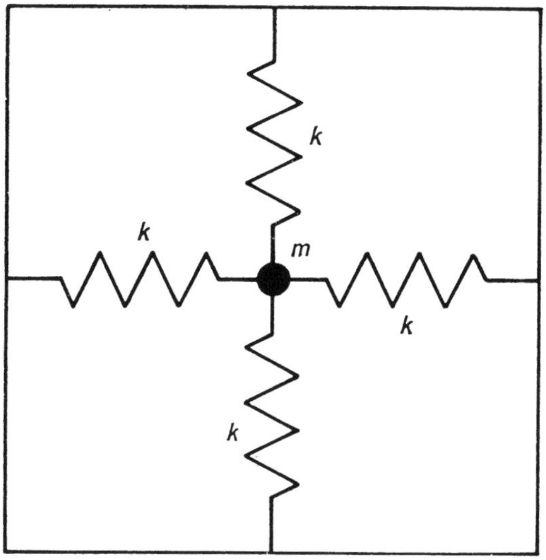

Ex. 4.4.2

The problem may be *by-passed* by eliminating one of the original coordinates from the equation of constraint. The remaining $n-1$ coordinates will then be independent and may be used in the Lagrange equations. This is the method that was used first in § 4.3. It is subject to the objection that one coordinate is treated in a different way from the rest, and the equations of motion are unnecessarily complicated.

The modifications that have to be made to the Lagrange equations may be found by looking at the problem in a slightly different way, and regarding the constrained system as the unconstrained system acted on by certain external forces, namely those forces which have to be exerted in order to compel the system to obey the constraint. This formulation has the advantage that in it the coordinates $q_1, q_2, ..., q_n$ may be regarded as independent; the constraints now appear as the effect of additional forces, and not as relations between the coordinates. Because the coordinates are independent, Lagrange's equations may be used, and the equations of motion of the constrained systems may be obtained by including the effects of

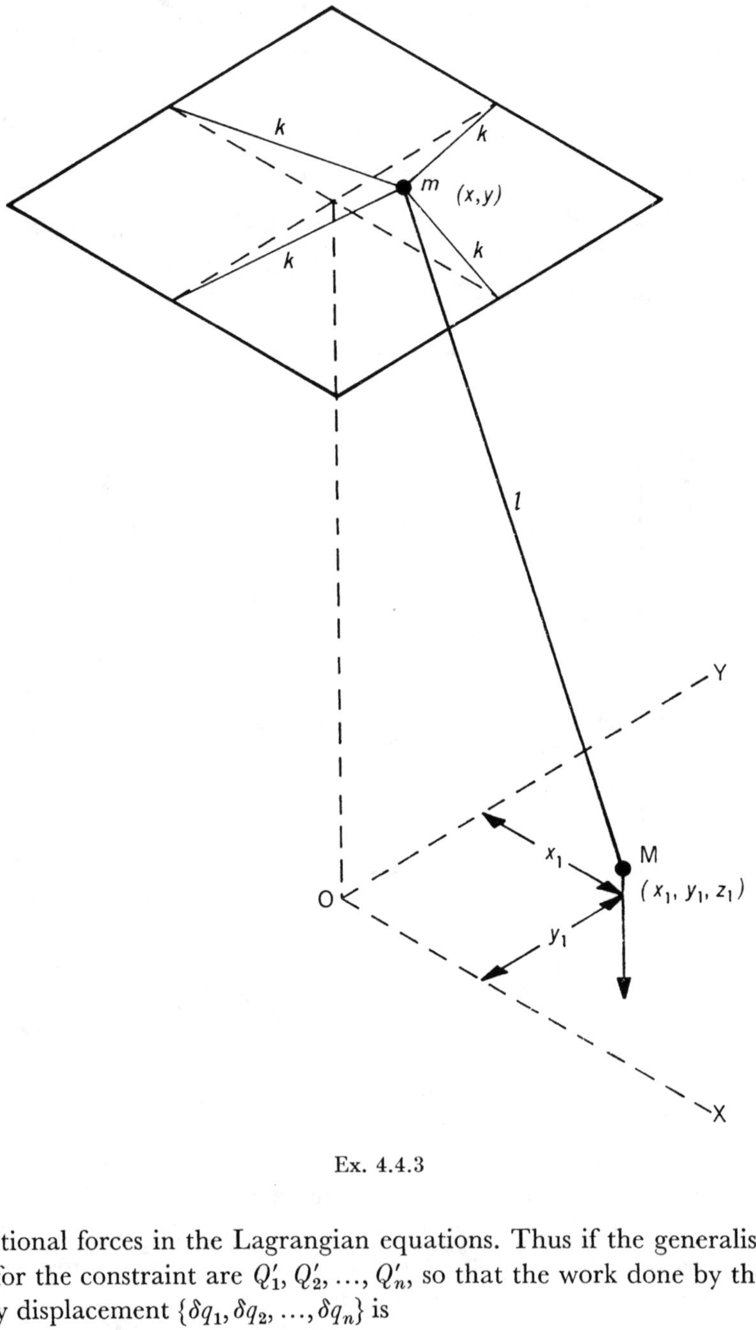

Ex. 4.4.3

the additional forces in the Lagrangian equations. Thus if the generalised forces needed for the constraint are $Q_1', Q_2', \ldots, Q_n'$, so that the work done by them in an arbitrary displacement $\{\delta q_1, \delta q_2, \ldots, \delta q_n\}$ is

$$W' = Q_1'\delta q_1 + Q_2'\delta q_2 + \ldots + Q_n'\delta q_n, \tag{4.5.3}$$

then the equations of motion for the constrained system will be

$$\frac{d}{dt}\left(\frac{\partial T}{\partial \dot{q}_r}\right)+\frac{\partial V}{\partial q_r} = Q_r+Q'_r. \tag{4.5.4}$$

The values of the additional external forces may be found by using the fact that they do no work in any displacement that is consistent with the constraint. Their purpose is to counteract any displacement which would break the constraint. This means that

$$W' = 0 \quad \text{whenever} \quad d_1 q_1 + d_2 q_2 + \ldots + d_n q_n = 0, \tag{4.5.5}$$

and this implies that

$$Q'_1 = \lambda d_1, \quad Q'_2 = \lambda d_2, \quad \ldots, \quad Q'_n = \lambda d_n, \tag{4.5.6}$$

where λ is unknown. Therefore, using the fact that when the constraint is expressed in the form (4.3.1) the constants d_r are given by

$$d_r = \left.\frac{\partial f}{\partial q_r}\right|_{q_1=0=q_2=\ldots=q_n}, \tag{4.5.7}$$

we may write the equations of motion as

$$\frac{d}{dt}\left(\frac{\partial T}{\partial \dot{q}_r}\right)+\frac{\partial V}{\partial q_r}-\lambda\left.\frac{\partial f}{\partial q_r}\right|_0 = Q_r \quad (r = 1, 2, \ldots, n). \tag{4.5.8}$$

We notice that if the q's are the principal coordinates for the system, so that T and V are given by equations (4.3.5), equation (4.5.8) yields the same set of equations as before, namely (4.3.17), for the free vibration of the constrained system.

The n equations governing the vibration of a constrained system may be combined to give a single matrix equation similar to equations (2.1.15), namely

$$\mathbf{A}\ddot{\mathbf{q}}+\mathbf{C}\mathbf{q}-\lambda\mathbf{d} = \mathbf{Q}, \tag{4.5.9}$$

where $\mathbf{d}$ is the column vector

$$\mathbf{d} = \{d_1, d_2, \ldots, d_n\}. \tag{4.5.10}$$

This equation, together with the equation of constraint, is sufficient for the determination of the $n+1$ unknowns $q_1, q_2, \ldots, q_n$ and λ.

To solve the equations governing free vibration we put

$$\mathbf{q} = \boldsymbol{\Psi}\sin\omega t, \quad \lambda = \nu\sin\omega t, \tag{4.5.11}$$

where $\boldsymbol{\Psi}$ is a column vector and ν is a constant scalar. Then the equations become

$$\left.\begin{aligned}(\mathbf{C}-\omega^2\mathbf{A})\,\boldsymbol{\Psi}-\nu\mathbf{d} &= \mathbf{0},\\ \text{and} \qquad \mathbf{d}'\boldsymbol{\Psi} &= 0.\end{aligned}\right\} \tag{4.5.12}$$

We may write these equations in the partitioned form

$$\begin{bmatrix}\mathbf{C}-\omega^2\mathbf{A} & -\mathbf{d}\\ \mathbf{d}' & 0\end{bmatrix}\begin{bmatrix}\boldsymbol{\Psi}\\ \nu\end{bmatrix} = \mathbf{0}, \tag{4.5.13}$$

and then we see that the $n+1$ equations have a non-zero solution provided that

$$\begin{vmatrix} c_{11}-\omega^2 a_{11} & c_{12}-\omega^2 a_{12} & \dots & c_{1n}-\omega^2 a_{1n} & d_1 \\ c_{21}-\omega^2 a_{21} & c_{22}-\omega^2 a_{22} & \dots & c_{2n}-\omega^2 a_{2n} & d_2 \\ \dots & \dots & \dots & \dots & \dots \\ c_{n1}-\omega^2 a_{n1} & c_{n2}-\omega^2 a_{n2} & \dots & c_{nn}-\omega^2 a_{nn} & d_n \\ d_1 & d_2 & \dots & d_n & 0 \end{vmatrix} = 0. \tag{4.5.14}$$

This is an equation of degree $n-1$ in ω^2 whose roots are the natural frequencies of the constrained system.

The Lagrangian equations of motion for a system subjected to a number of constraints may now be obtained without any difficulty. Suppose that the system is subjected to m ($<n$) independent constraints

$$f^{(i)}(q_1, q_2, \dots, q_n) = 0 \quad (i = 1, 2, \dots, m). \tag{4.5.15}$$

Each constraint may be expanded and written as

$$d_1^{(i)} q_1 + d_2^{(i)} q_2 + \dots + d_n^{(i)} q_n = 0 \quad (i = 1, 2, \dots, m), \tag{4.5.16}$$

or

$$d'^{(i)} q = 0, \tag{4.5.17}$$

where

$$d_r^{(i)} = \left.\frac{\partial f^{(i)}}{\partial q_r}\right|_{q_1=0=q_2=\dots=q_n}. \tag{4.5.18}$$

The external forces which must be applied to compel the system to obey the constraints must now be such that they do no work in any displacement consistent with the m constraints (4.5.17). The reader may verify that this implies that

$$Q'_r = \lambda_1 d_r^{(1)} + \lambda_2 d_r^{(2)} + \dots + \lambda_m d_r^{(m)} \quad (r = 1, 2, \dots, n), \tag{4.5.19}$$

where $\lambda_1, \lambda_2, \dots, \lambda_m$ are unknown multipliers. Substituting these values into the equations of motion (4.5.4), and using equation (4.5.18), we obtain

$$\frac{d}{dt}\left(\frac{\partial T}{\partial \dot{q}_r}\right) + \frac{\partial V}{\partial q_r} - \sum_{i=1}^{m} \lambda_i \left.\frac{\partial f^{(i)}}{\partial q_r}\right|_0 = Q_r \quad (r = 1, 2, \dots, n). \tag{4.5.20}$$

These n equations, together with the m equations of constraint, are sufficient for the determination of the $n+m$ unknowns $q_1, q_2, \dots, q_n, \lambda_1, \lambda_2, \dots, \lambda_m$.

There are a number of variational theorems concerned with the imposition of constraints on a system. The simplest is

Theorem 4.5.1. Let the natural frequencies of a system be

$$\omega_1 \leqslant \omega_2 \leqslant \dots \leqslant \omega_n.$$

Then the greatest value that can be attained by $\tilde{\omega}_1(d)$, the least natural frequency of the system subject to the constraint

$$\mathbf{d'q} = 0 = d_1 q_1 + d_2 q_2 + \dots + d_n q_n \tag{4.5.21}$$

is ω_2. This maximum value is attained when the system is constrained to be orthogonal to the first mode, that is, when

$$\mathbf{\Psi}'^{(1)}\mathbf{Aq} = 0, \tag{4.5.22}$$

This theorem follows directly from equation (4.3.24). There are similar theorems which are concerned with the imposition of a number of constraints, and with the effect of constraints on higher natural frequencies of a system. The reader will find a full account of these theorems elsewhere.†

EXAMPLES 4.5

1. Deduce equation (4.5.6) from equation (4.5.5) and prove the result stated in equation (4.5.19).

2. The natural frequencies of the pendulum system shown are

$$\omega_1 = 0{\cdot}6454\surd(g/l), \quad \omega_2 = 1{\cdot}5146\surd(g/l), \quad \omega_3 = 2{\cdot}5078\surd(g/l).$$

Find the natural frequencies of the system subject to the constraints

$$(a)\ \theta = 0, \quad (b)\ \theta = \psi, \quad (c)\ \theta = -\psi,$$

and verify that in each case they lie between those of the unconstrained system, in conformity with equation (4.3.24).

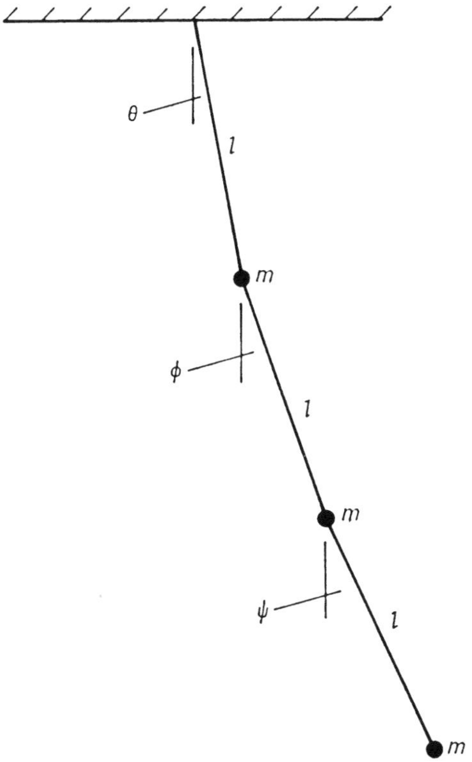

Ex. 4.5.2

3. (*a*) Find the natural frequency of the system of Ex. 4.5.2 when it vibrates under the two constraints

$$\theta = \phi, \quad \phi = \psi,$$

using a frequency equation of the general type (4.5.14).

(*b*) Find a determinantal form of the frequency equation for a system whose free vibration is governed by equations (4.5.20), with $Q_r = 0$.

† Whittaker, *Analytical Dynamics*, ch. 7; and S. H. Gould, *Variational Methods for Eigenvalue Problems* (University of Toronto Press, 1957), chs. 1, 2.

CHAPTER 5

DAMPED FORCED VIBRATION

. . . toffee in moderation is a capital thing. *Patience*

Every real vibrating system is subject to damping in one form or another, but so far we have taken no account of it. This is because the theory of undamped vibration, which is much less complicated than that of damped vibration, is quite sufficient for most purposes. This chapter is devoted to a discussion of the *nature* of the forced, damped vibration of systems having a finite number of degrees of freedom.

5.1 Complex numbers as elements of matrices

Later in this chapter it will be shown that the introduction of damping into a system produces phase-differences between the motions at the coordinates and the excitation. It will be shown that a convenient way of allowing for this is to use matrices having complex elements. All the matrices that we have dealt with so far have had real elements, but there is no reason, in matrix theory, why the elements should not be complex, provided that the mathematical rules for manipulating complex numbers are observed. Using these rules we may show that the whole of the analysis of Chapter 1 applies, without any change, to complex matrices. In this section, an important property of complex matrices will be examined.

Consider a column matrix

$$\mathbf{u} = \mathbf{r} + i\mathbf{s} = \begin{bmatrix} \rho_1 \\ \rho_2 \\ \vdots \\ \rho_n \end{bmatrix} + i \begin{bmatrix} \sigma_1 \\ \sigma_2 \\ \vdots \\ \sigma_n \end{bmatrix} = \begin{bmatrix} \rho_1 + i\sigma_1 \\ \rho_2 + i\sigma_2 \\ \vdots \quad \vdots \\ \rho_n + i\sigma_n \end{bmatrix}. \tag{5.1.1}$$

The 'transposed-conjugate' of this is defined as the matrix $\mathbf{u}^*$ which is formed by transposing $\mathbf{u}$ and replacing each element by its complex conjugate. That is

$$\mathbf{u}^* = [\rho_1 - i\sigma_1, \rho_2 - i\sigma_2, \ldots, \rho_n - i\sigma_n] = (\mathbf{r} - i\mathbf{s})'. \tag{5.1.2}$$

Generally, if $\mathbf{H}$ is any complex matrix and $\bar{\mathbf{H}}$ denotes the matrix formed from the complex conjugates of the elements of $\mathbf{H}$, then the transposed-conjugate $\mathbf{H}^*$ is defined by the equation

$$\mathbf{H}^* = (\bar{\mathbf{H}})', \tag{5.1.3}$$

so that the element in the ith row and jth column of $\mathbf{H}^*$ is $\bar{h}_{ji}$. Clearly, if $\mathbf{H}$ is real then the transposed-conjugate is identical with the transpose, i.e.

$$\mathbf{H}^* = \mathbf{H}'. \tag{5.1.4}$$

Thus the inertia and stiffness matrices possess the property that

$$\mathbf{A}^* = \mathbf{A}; \quad \mathbf{C}^* = \mathbf{C} \tag{5.1.5}$$

because they are both real and symmetric.

It was shown earlier that if $\mathbf{x}$ is a column matrix and $\mathbf{Y}$ is a square matrix then the product $\mathbf{x}'\mathbf{Y}\mathbf{x}$ is a scalar quantity. It is of interest now to examine the effect of making the column matrix complex using, instead of the transposed matrix, the transposed-conjugate. The product $\mathbf{u}^*\mathbf{C}\mathbf{u}$, for instance, gives

$$\begin{aligned}(\mathbf{r}-i\mathbf{s})'\mathbf{C}(\mathbf{r}+i\mathbf{s}) &= \mathbf{r}'\mathbf{C}\mathbf{r}+i\mathbf{r}'\mathbf{C}\mathbf{s}-i\mathbf{s}'\mathbf{C}\mathbf{r}+\mathbf{s}'\mathbf{C}\mathbf{s}\\ &= \mathbf{r}'\mathbf{C}\mathbf{r}+\mathbf{s}'\mathbf{C}\mathbf{s}\end{aligned} \tag{5.1.6}$$

since $\mathbf{C}$ (the stiffness matrix) is symmetric. Now $\mathbf{C}$ is positive semi-definite so that both members on the right-hand side of equation (5.1.4) are non-negative. It follows that

$$\mathbf{u}^*\mathbf{C}\mathbf{u} \geqslant 0. \tag{5.1.7}$$

The usefulness of this result may be demonstrated in a simple proof that free vibration of an undamped system in a principal mode is associated with a real (as opposed to a complex) natural frequency. In other words, we may show that the ω_r^2 in the equation

$$(\mathbf{C}-\omega_r^2\mathbf{A})\,\mathbf{\Psi}^{(r)} = \mathbf{0} \quad (r = 1, 2, \ldots, n) \tag{5.1.8}$$

are all real and non-negative.

Suppose that $\mathbf{\Psi}^{(r)}$ contains some complex elements. Premultiplication of equation (5.1.8) by $\mathbf{\Psi}^{*(r)}$ shows that

$$\omega_r^2 = \frac{\mathbf{\Psi}^{*(r)}\mathbf{C}\mathbf{\Psi}^{(r)}}{\mathbf{\Psi}^{*(r)}\mathbf{A}\mathbf{\Psi}^{(r)}}. \tag{5.1.9}$$

Now both the numerator and denominator of this expression are real and non-negative so that

$$\omega_r^2 \geqslant 0 \quad (r = 1, 2, \ldots, n). \tag{5.1.10}$$

It follows that the principal modes $\mathbf{\Psi}^{(r)}$ are all real. For the mode $\mathbf{\Psi}^{(r)}$ is determined by the ratios of cofactors in the determinant $|\mathbf{C}-\omega_r^2\mathbf{A}|$, every element of which is real.

EXAMPLES 5.1

1. If

$$\mathbf{H} = \begin{bmatrix} e^{i\theta} & e^{-i\theta} \\ e^{i\theta} & e^{-i\theta} \end{bmatrix}$$

find the value of $\mathbf{H}\mathbf{H}^*$, $\mathbf{H}^*\mathbf{H}$ and $\mathbf{H}'\mathbf{H}^*$, expressing the results in terms of trigonometric functions.

2. (*a*) Complete the entries in the following table of special types of matrices:

Type of matrix	Matrix relation	Nature of elements
Skew symmetric	$\mathbf{A} = -\mathbf{A}'$	$a_{ii} = 0$; $a_{ij} = -a_{ji}$
Real	$\mathbf{A} = \bar{\mathbf{A}}$	—
Imaginary	—	a_{ij} all pure imaginary
Hermitian	$\mathbf{A} = \mathbf{A}^*$	—
Skew Hermitian	$\mathbf{A} = -\mathbf{A}^*$	—
Orthogonal	—	$\sum_{k=1}^{n} a_{ik}a_{jk} = \sum_{k=1}^{n} a_{ki}a_{kj} = \delta_{ij}$
Unitary	—	$\sum_{k=1}^{n} a_{ik}\bar{a}_{jk} = \sum_{k=1}^{n} \bar{a}_{ki}a_{kj} = \delta_{ij}$

(*b*) Show that the real part of an Hermitian matrix is symmetric while the imaginary part is skew symmetric.

(*c*) Show that a real orthogonal matrix is unitary.

3. Show that if $\mathbf{F} = \mathbf{ABC}$, then

$$(a)\ \overline{\mathbf{F}} = \overline{\mathbf{A}}\overline{\mathbf{B}}\overline{\mathbf{C}} \quad \text{and} \quad (b)\ \mathbf{F}^* = \mathbf{C}^*\mathbf{B}^*\mathbf{A}^*.$$

5.2 Systems with one degree of freedom

There are two simple mathematical models for the damping in a system; the damping may be viscous or hysteretic. In the first the energy dissipation per cycle is proportional to the forcing frequency, while in hysteretic damping it is independent of frequency. Mathematically, the two types of damping are very similar. We shall therefore give a brief comparison of their effects on the forced vibration of a system with one degree of freedom and then discuss the effect of one of them on the forced vibration of a system with many degrees of freedom. We shall choose viscous damping; the reader will find a discussion of the effects of hysteretic damping on a system with many degrees of freedom, together with a full treatment of the whole subject, elsewhere.† Some of the properties of systems with hysteretic damping form the substance of examples.

5.2.1 Viscous damping

Consider the typical system with one degree of freedom shown in fig. 5.2.1. The equation of motion is

$$M\ddot{y} + b\dot{y} + ky = F \sin \omega t. \tag{5.2.1}$$

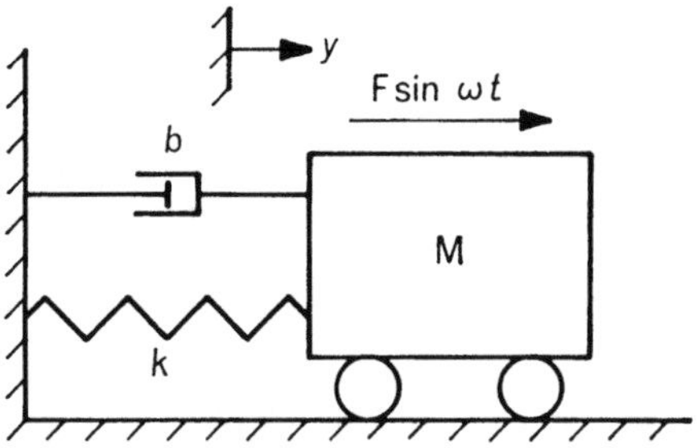

Fig. 5.2.1

The trial solution

$$y = Y \sin (\omega t - \zeta) \tag{5.2.2}$$

satisfies the equation provided that

$$Y = \frac{F}{\sqrt{[(k-\omega^2 M)^2 + \omega^2 b^2]}} \tag{5.2.3}$$

and

$$\tan \zeta = \frac{\omega b}{k - \omega^2 M}. \tag{5.2.4}$$

† R. E. D. Bishop and G. M. L. Gladwell, 'An investigation into the theory of resonance testing', *Phil. Trans.* A, **255** (1963), pp. 241–80.

The reader who is not familiar with these results may easily establish them for himself.

The above results may be expressed in dimensionless form by introducing the notation

$$\omega_1 = \sqrt{\frac{k}{M}}, \quad \beta = \frac{\omega}{\omega_1}, \quad \mu = \frac{b\omega_1}{k}. \tag{5.2.5†}$$

In this notation we have

$$Y = \frac{F}{k}\frac{1}{\sqrt{[(1-\beta^2)^2+\mu^2\beta^2]}} \tag{5.2.6}$$

and

$$\tan\zeta = \frac{\mu\beta}{1-\beta^2}. \tag{5.2.7}$$

Fig. 5.2.2 shows the dimensionless quantities kY/F and ζ plotted against β for several values of the dimensionless damping coefficient μ.

5.2.2 Hysteretic damping

The equation of motion for a hysteretically damped system is

$$M\ddot{y} + \frac{h}{\omega}\dot{y} + ky = F\sin\omega t. \tag{5.2.8}$$

The trial solution

$$y = Y\sin(\omega t - \eta) \tag{5.2.9}$$

leads to the result

$$Y = \frac{F}{\sqrt{[(k-\omega^2 M)^2+h^2]}}, \tag{5.2.10}$$

where

$$\tan\eta = \frac{h}{k-\omega^2 M}. \tag{5.2.11}$$

These results may be expressed in dimensionless form by using the notation

$$\mu = h/k. \tag{5.2.12}$$

in addition to that of equation (5.2.5). Then

$$Y = \frac{F}{k}\frac{1}{\sqrt{[(1-\beta^2)^2+\mu^2]}} \tag{5.2.13}$$

and

$$\tan\eta = \frac{\mu}{1-\beta^2}. \tag{5.2.14}$$

A comparison of equations (5.2.13) and (5.2.14), with equations (5.2.6) and (5.2.7), shows that when the system is vibrating at the natural frequency $\omega_1 (\beta = 1)$, hysteretic and viscous damping corresponding to the same value of μ have exactly the same effect. The hysteretic damping μ may therefore be called the 'equivalent hysteretic damping' to the viscous damping μ. Fig. 5.2.3 shows the dimensionless quantities kY/F and η plotted against β for the same values of μ as were used in fig. 5.2.2.

† This dimensionless measure μ for viscous damping is equal to twice the damping factor $\nu = b/b_{crit}$ of the system. The damping $\mu = 0{\cdot}04$ therefore corresponds to 2% critical damping.

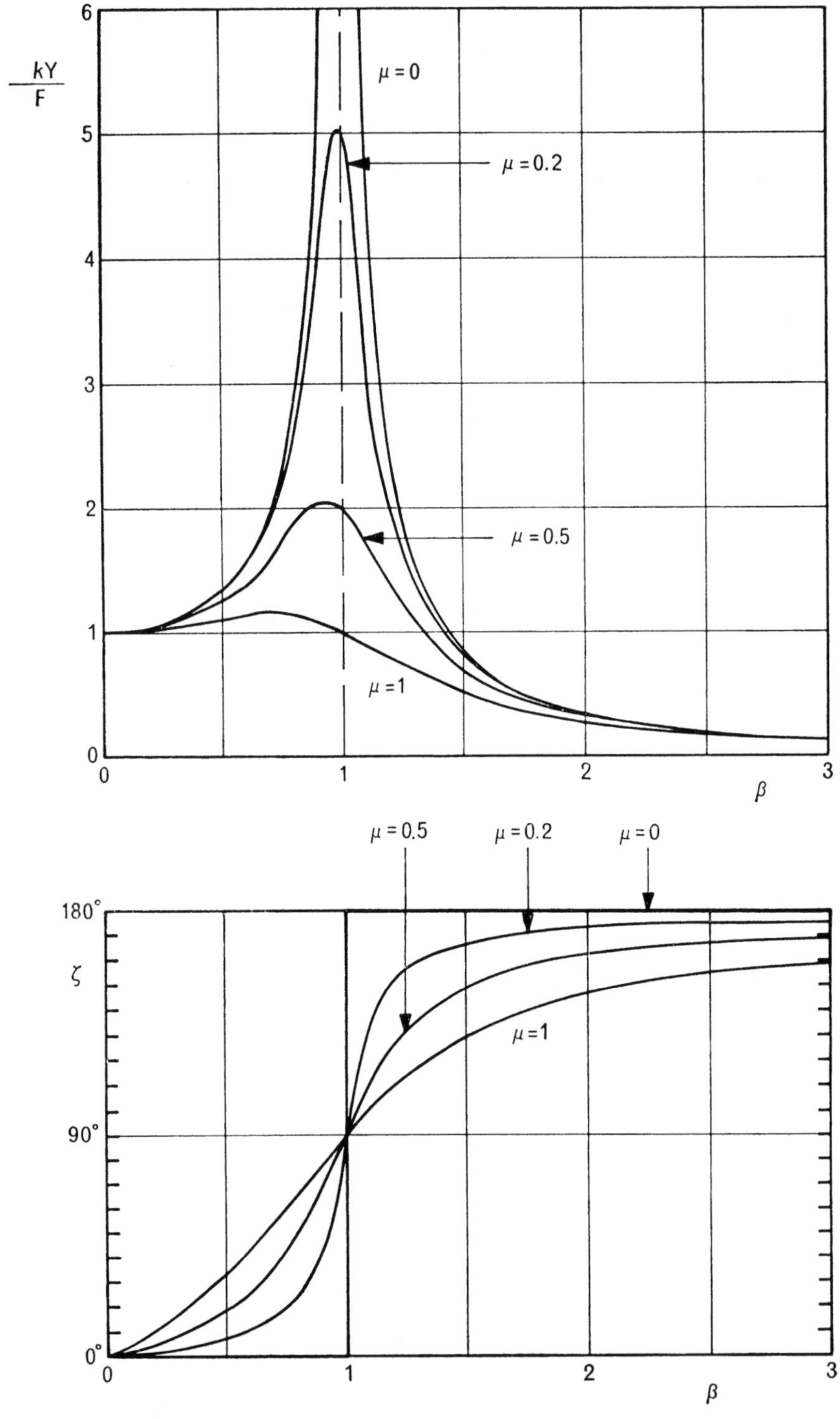

Fig. 5.2.2

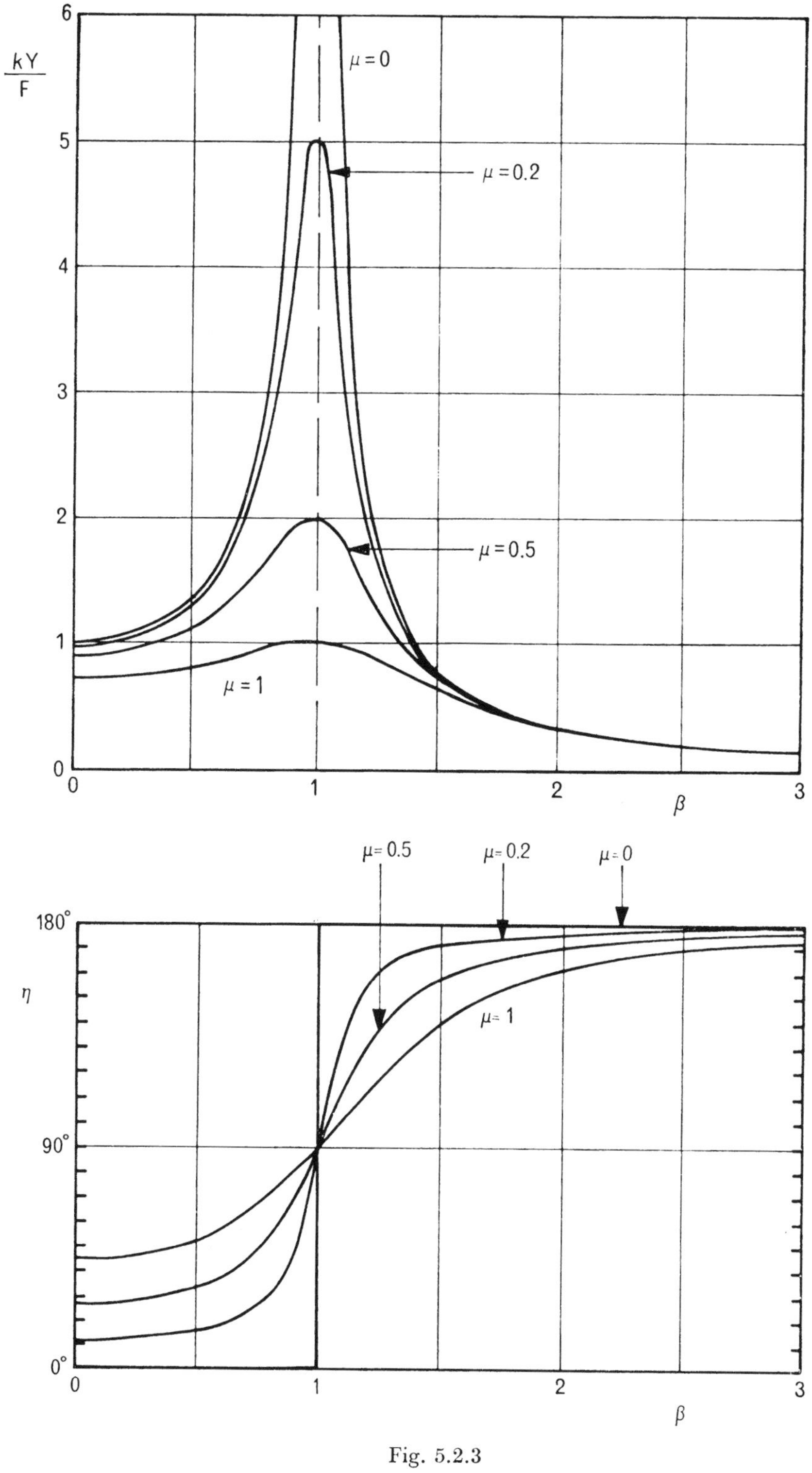

Fig. 5.2.3

EXAMPLES 5.2

1. Explain carefully why the use of an hysteretic damping idealisation is restricted to forced oscillation and does not apply without further definition to free vibration.

2. (*a*) Show that the result in equations (5.2.10) and (5.2.11) may be expressed in the form

$$y = \frac{F \sin \eta \,.\, \sin(\omega t - \eta)}{h}.$$

(*b*) Hence, or otherwise, show that if the line of length F in the figure represents the amplitude of the applied force $F \sin \omega t$, then the amplitude and phase of the response are represented by the line of length $(F/h) \sin \eta$. Show further that the locus of the end of the response vector is a circle, and find its radius.

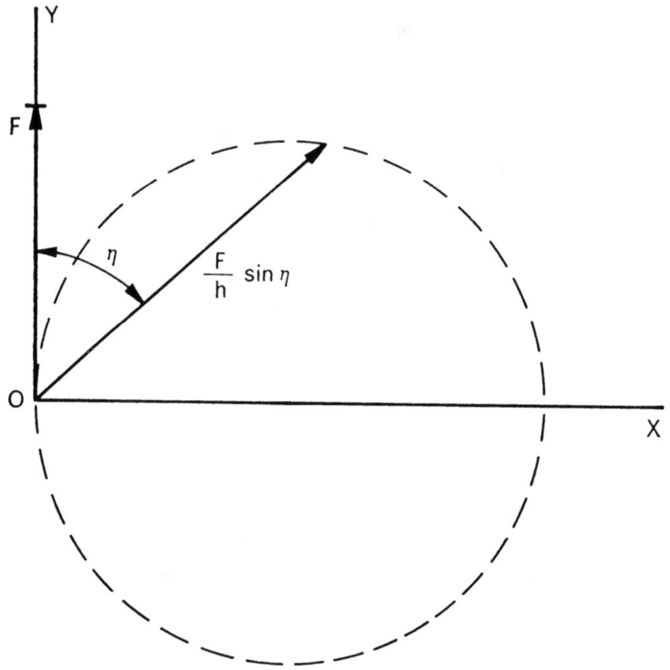

Ex. 5.2.2

3. (*a*) Sketch a three-dimensional curve of the locus of the end of the response vector in Ex. 5.2.2, measuring ω in a direction perpendicular to the plane of the figure for that example.

(*b*) Make a sketch of the same sort, relating to viscous damping.

5.3 Systems with many degrees of freedom

The effects of viscous and hysteretic damping on the forced vibration of a system with one degree of freedom were discussed in § 5.2. In this section we shall discuss the forced vibration of a system with many degrees of freedom subject to viscous damping.

In Chapter 2 it was shown how the equations of motion of an undamped system may be found by introducing the expressions $\frac{1}{2}\dot{\mathbf{q}}'\mathbf{A}\dot{\mathbf{q}}$ and $\frac{1}{2}\mathbf{q}'\mathbf{C}\mathbf{q}$ for the kinetic and potential energies, and using the Lagrangian equations (2.1.12). Lord Rayleigh† showed that one may allow for viscous damping forces by introducing a third function, the 'dissipation function' D into the Lagrangian equations.‡ The dissipation function is defined as half the instantaneous rate of energy dissipation and it may be shown to be expressible in the form

$$D = \tfrac{1}{2}[b_{11}\dot{q}_1^2 + b_{22}\dot{q}_2^2 + \ldots + b_{nn}\dot{q}_n^2 + 2b_{12}\dot{q}_1\dot{q}_2 + \ldots + 2b_{n-1,n}\dot{q}_{n-1}\dot{q}_n], \tag{5.3.1}$$

where the b's are constants and

$$b_{ij} = b_{ji}. \tag{5.3.2}$$

The matrix expression for D is therefore

$$D = \tfrac{1}{2}\dot{\mathbf{q}}'\mathbf{B}\dot{\mathbf{q}}, \tag{5.3.3}$$

where $\mathbf{B}$ is a symmetric square matrix called the 'damping matrix'. For the systems that are dealt with in this book D is always non-negative, so that $\mathbf{B}$ is positive semi-definite (see § 2.1).

The Lagrangian equations now take the form

$$\frac{d}{dt}\left(\frac{\partial T}{\partial \dot{q}_r}\right) + \frac{\partial D}{\partial \dot{q}_r} + \frac{\partial V}{\partial q_r} = Q_r \quad (r = 1, 2, \ldots, n), \tag{5.3.4}$$

so that the damping terms are found by forming the derivatives $\partial D/\partial \dot{q}_r$. The n equations of motion found in this way (cf. equation (2.1.15)) may be abbreviated to the single matrix equation

$$\mathbf{A}\ddot{\mathbf{q}} + \mathbf{B}\dot{\mathbf{q}} + \mathbf{C}\mathbf{q} = \mathbf{\Phi}\sin\omega t. \tag{5.3.5}$$

Since we shall here be concerned with harmonic excitation, Q has been replaced by the column matrix $\mathbf{\Phi}\sin\omega t$.

We notice in passing that the equation governing the forced motion of a system subject to hysteretic damping is obtained by assuming that all the elements of $\mathbf{B}$ are proportional to $1/\omega$. That is

$$\mathbf{B} = \mathbf{D}/\omega, \tag{5.3.6}$$

where $\mathbf{D}$ is a square matrix, not to be confused with the dissipation function in equation (5.3.3). The equation is

$$\mathbf{A}\ddot{\mathbf{q}} + (\mathbf{D}/\omega)\,\dot{\mathbf{q}} + \mathbf{C}\mathbf{q} = \mathbf{\Phi}\sin\omega t. \tag{5.3.7}$$

For an example of a system subject to viscous damping, consider the system shown in fig. 5.3.1. Damping is provided by viscous dashpots whose coefficients are b_1 and b_2. For this system

$$\left.\begin{aligned} T &= \tfrac{1}{2}[M\dot{q}_1^2 + m(\dot{q}_1 + l\dot{q}_2)^2], \\ D &= \tfrac{1}{2}[b_1\dot{q}_1^2 + b_2\dot{q}_2^2], \\ V &= \tfrac{1}{2}[kq_1^2 + mglq_2^2], \end{aligned}\right\} \tag{5.3.8}$$

so that

$$\mathbf{A} = \begin{bmatrix} M+m & ml \\ ml & ml^2 \end{bmatrix}, \quad \mathbf{B} = \begin{bmatrix} b_1 & 0 \\ 0 & b_2 \end{bmatrix}, \quad \mathbf{C} = \begin{bmatrix} k & 0 \\ 0 & mgl \end{bmatrix}, \tag{5.3.9}$$

† *The Theory of Sound* (Macmillan, 2nd ed., 1894), vol. II, p. 102.
‡ Bishop and Johnson, *The Mechanics of Vibration*, § 8.4.

while the excitation indicated in the figure is such that

$$\mathbf{Q} = \begin{bmatrix} F \\ 0 \end{bmatrix} \sin \omega t. \tag{5.3.10}$$

In general the solution of equation (5.3.5) for given $\mathbf{\Phi}$ and ω is rather complicated. Though they are all harmonic with the impressed frequency ω, the responses at each of the coordinates q_r due to an arbitrary force $\mathbf{\Phi} \sin \omega t$ are not in phase with each other or with the exciting force. However, for certain forces $\mathbf{\Phi}$, the responses

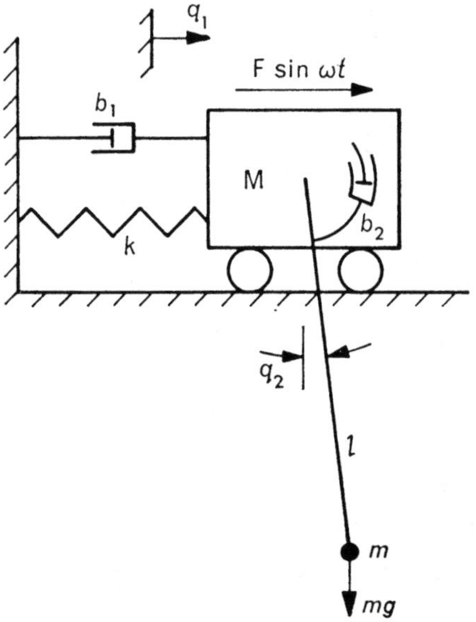

Fig. 5.3.1

at each of the coordinates are all in phase with each other, though still out of phase with the force. We shall now find these forces and the corresponding responses. Let

$$\mathbf{q} = \begin{bmatrix} q_1 \\ q_2 \\ \vdots \\ q_n \end{bmatrix} = \begin{bmatrix} \kappa_1 \\ \kappa_2 \\ \vdots \\ \kappa_n \end{bmatrix} \sin(\omega t - \theta) = \boldsymbol{\kappa} \sin(\omega t - \theta), \tag{5.3.11}$$

so that there is a common phase difference θ between the response and the excitation. If this expression is introduced into equation (5.3.5) the resulting equation is

$$\sin(\omega t - \theta) . (\mathbf{C} - \omega^2 \mathbf{A}) \boldsymbol{\kappa} + \omega \cos(\omega t - \theta) . \mathbf{B}\boldsymbol{\kappa} = \mathbf{\Phi} \sin \omega t \tag{5.3.12}$$

and, when the sine and cosine terms are separated out, the equation becomes

$$\cos\theta . (\mathbf{C} - \omega^2 \mathbf{A}) \boldsymbol{\kappa} + \omega \sin\theta . \mathbf{B}\boldsymbol{\kappa} = \mathbf{\Phi}, \tag{5.3.13}$$

$$\sin\theta . (\mathbf{C} - \omega^2 \mathbf{A}) \boldsymbol{\kappa} - \omega \cos\theta . \mathbf{B}\boldsymbol{\kappa} = \mathbf{0}. \tag{5.3.14}$$

Since ω is given, these equations contain three unknowns, $\mathbf{\Phi}$, $\boldsymbol{\kappa}$ and θ. Equation

(5.3.13) expresses $\mathbf{\Phi}$ in terms of $\boldsymbol{\varkappa}$ and θ, while (5.3.14) links $\boldsymbol{\varkappa}$ and θ. If $\cos\theta$ is not zero, equation (5.3.14) may be divided by $\cos\theta$ to give

$$[\tan\theta.(\mathbf{C}-\omega^2\mathbf{A})-\omega\mathbf{B}]\boldsymbol{\varkappa} = \mathbf{0}. \tag{5.3.15}$$

This equation, being a homogeneous equation for the column matrix $\boldsymbol{\varkappa}$, possesses a non-trivial solution only if

$$|\tan\theta.(\mathbf{C}-\omega^2\mathbf{A})-\omega\mathbf{B}| = 0. \tag{5.3.16}$$

This equation is an algebraic equation of order n in $\tan\theta$ and therefore has n solutions $\tan\theta_r$ $(r = 1, 2, \ldots, n)$. For each solution $\tan\theta_r$ there is a column matrix $\boldsymbol{\varkappa}^{(r)}$ satisfying the equation

$$[\tan\theta_r.(\mathbf{C}-\omega^2\mathbf{A})-\omega\mathbf{B}]\boldsymbol{\varkappa}^{(r)} = \mathbf{0}. \tag{5.3.17}$$

The roots $\tan\theta_r$ and the corresponding matrices $\boldsymbol{\varkappa}^{(r)}$ are all real. For premultiplication of (5.3.17) by the transposed-conjugate $\boldsymbol{\varkappa}^{*(r)}$ of $\boldsymbol{\varkappa}^{(r)}$ (see § 5.1) gives

$$\tan\theta_r = \frac{\omega\boldsymbol{\varkappa}^{*(r)}\mathbf{B}\boldsymbol{\varkappa}^{(r)}}{\boldsymbol{\varkappa}^{*(r)}(\mathbf{C}-\omega^2\mathbf{A})\boldsymbol{\varkappa}^{(r)}}. \tag{5.3.18}$$

Since $\mathbf{A}$, $\mathbf{B}$ and $\mathbf{C}$ are all real symmetric matrices this equation gives $\tan\theta_r$ as the ratio of two real quantities. Therefore $\tan\theta_r$, and hence $\boldsymbol{\varkappa}^{(r)}$ are both real, so that the transposed-conjugate matrix $\boldsymbol{\varkappa}^{*(r)}$ is the same as the ordinary transpose $\boldsymbol{\varkappa}'^{(r)}$ of $\boldsymbol{\varkappa}^{(r)}$.

Thus the original trial solution (5.3.11) has led to the discovery of n modes of distortion corresponding to the given value of the frequency ω. These modes are quite different from the principal modes of the undamped system in that both the phase angle θ_r and the shape of the mode vary with the frequency.

Equations (5.3.16) and (5.3.17) show that the modes $\boldsymbol{\varkappa}^{(r)}$ depend only on the 'shape' of the damping matrix $\mathbf{B}$, and not on its intensity. Suppose that every element of the matrix $\mathbf{B}$ is increased in the ratio $k:1$. Equation (5.3.16) shows that the roots $\tan\theta_r$ will all be increased in the same ratio so that equation (5.3.17), which is the equation determining $\boldsymbol{\varkappa}^{(r)}$, will be multiplied throughout by the factor k, and therefore $\boldsymbol{\varkappa}^{(r)}$ will be unchanged. If the shape of the damping matrix is kept the same while the intensity of the damping is progressively decreased, the modes will remain the same; but when $\mathbf{B} = \mathbf{0}$ the modes are the principal modes of the system. Therefore the change from the damped to the undamped state is discontinuous, and undamped motion may be fundamentally different from damped motion—a conclusion which is perhaps rather surprising.

The matrix $\boldsymbol{\varkappa}^{(r)}$ given by (5.3.17) is undefined to the extent of an arbitrary factor. Just as with the principal modes (see § 2.4) this scale factor may be chosen so that one of the coordinates q_i has a specified value when $\sin(\omega t-\theta_r) = 1$ in the mode concerned.

The way in which the $\boldsymbol{\varkappa}$'s and θ's are labelled and the scale factor is assigned to $\boldsymbol{\varkappa}^{(r)}$ deserves to be considered further. To do this we shall return to equation (5.3.14). In deriving equation (5.3.15) from (5.3.14) it was assumed that $\cos\theta \neq 0$. If, however, $\cos\theta = 0$, equation (5.3.14) becomes

$$(\mathbf{C}-\omega^2\mathbf{A})\boldsymbol{\varkappa} = \mathbf{0}, \tag{5.3.19}$$

and the condition for $\boldsymbol{\varkappa}$ to be non-trivial is that

$$|\mathbf{C}-\omega^2\mathbf{A}| = 0. \tag{5.3.20}$$

This means that ω must be a natural frequency of the (undamped) system. If then $\omega = \omega_s$, the $\boldsymbol{\varkappa}$ mode corresponding to this value of ω and the solution $\cos\theta = 0$, may be identified with the sth principal mode $\boldsymbol{\Psi}^{(s)}$. If the θ solution and the corresponding value of $\boldsymbol{\varkappa}$ are labelled θ_s and $\boldsymbol{\varkappa}^{(s)}$ respectively then one may write

$$\theta_s = \tfrac{1}{2}\pi, \quad \boldsymbol{\varkappa}^{(s)} = \boldsymbol{\Psi}^{(s)} \quad \text{when} \quad \omega = \omega_s. \tag{5.3.21}$$

For this value of ω there will also be $(n-1)$ other $\boldsymbol{\varkappa}^{(r)}$ modes corresponding to the remaining $(n-1)$ roots $\tan\theta_r$ of equation (5.3.16).

Equation (5.3.21) may be used to give a consistent way of labelling the θ's and $\boldsymbol{\varkappa}$'s for values of ω other than the natural frequencies. Each of the roots θ of equation (5.3.16) is a continuous function of ω. We may therefore write $\theta = \theta(\omega)$. Equation (5.3.18) shows that for small values of ω each root $\tan\theta_r$ is small and positive so that each $\theta_r(\omega)$ is a small positive angle. As ω becomes larger and approaches ω_1, one of the roots $\theta(\omega)$ will approach the value $\frac{1}{2}\pi$; label this root $\theta_1(\omega)$. For values of ω greater than ω_1, the denominator of the expression for $\tan\theta_1$ will be negative while the numerator will still be positive; $\theta_1(\omega)$ will thus be greater than $\frac{1}{2}\pi$. As ω is increased indefinitely $\theta_1(\omega)$ will approach the value π. The remaining $(n-1)$ roots $\theta_s(\omega)$ may be labelled in a similar way; $\theta_s(\omega)$ is that root which has the value $\frac{1}{2}\pi$ when $\omega = \omega_s$. This way of labelling θ_s may be carried over to the modes—at any frequency ω, the shape of the sth mode is given by the solution $\boldsymbol{\varkappa}^{(s)}(\omega)$ of

$$[\tan\theta_s.(\mathbf{C}-\omega^2\mathbf{A})-\omega\mathbf{B}]\,\boldsymbol{\varkappa}^{(s)} = \mathbf{0}. \tag{5.3.22}$$

The scale factor for $\boldsymbol{\varkappa}^{(s)}(\omega)$ may also be chosen consistently. Instead of choosing one for each value of ω separately one may choose a scale factor for $\boldsymbol{\Psi}^{(s)}$ and then choose that for $\boldsymbol{\varkappa}^{(s)}(\omega)$ so that each component of $\boldsymbol{\varkappa}^{(s)}(\omega)$ is a continuous function of ω equal to the corresponding component of $\boldsymbol{\Psi}^{(s)}$ when $\omega = \omega_s$. In particular, if there is a component of $\boldsymbol{\varkappa}^{(s)}(\omega)$ which is non-zero for all values of ω, then the mode may be scaled so that that component is unity for all values of ω. When the modes are scaled in either of these ways the transition to the undamped system ($\mathbf{B} \equiv \mathbf{0}$) means

$$\theta_r(\omega) = 0, \quad \boldsymbol{\varkappa}^{(r)}(\omega) = \boldsymbol{\Psi}^{(r)}, \tag{5.3.23}$$

though, as has already been shown, the transition to the undamped case is not continuous.

We shall now illustrate some of the matters discussed above by referring to the system of fig. 5.3.1. Suppose that

$$M = 2\,\text{lb}, \quad m = 1\,\text{lb}, \quad l = 1\,\text{ft}, \quad k = 256\,\text{pdl}\,\text{ft}^{-1}, \quad g = 32\,\text{ft}\,\text{sec}^{-2}, \tag{5.3.24}$$

and that $F\sin\omega t$ is an applied force in pdl. The values of b_1 and b_2 will be taken in the form

$$b_1 = 5b\,\text{pdl}\,\text{ft}^{-1}\,\text{sec}^{-1}, \quad b_2 = b\,\text{pdl}\,\text{ft}\,\text{rad}^{-1}\,\text{sec}^{-1}, \tag{5.3.25}$$

where b is an assignable positive constant; this will simplify the presentation a little. With these values

$$\mathbf{A} = \begin{bmatrix} 3 & 1 \\ 1 & 1 \end{bmatrix}, \quad \mathbf{B} = b\begin{bmatrix} 5 & 0 \\ 0 & 1 \end{bmatrix}, \quad \mathbf{C} = \begin{bmatrix} 256 & 0 \\ 0 & 32 \end{bmatrix} \tag{5.3.26}$$

and the determinantal equation (5.3.16) in $\tan\theta$ reduces to the quadratic

$$(8192-352\omega^2+2\omega^4)\tan^2\theta - 2(208-4\omega^2)\,\omega b\tan\theta + 5\omega^2 b^2 = 0. \tag{5.3.27}$$

The natural frequencies of the system are the roots of the equation when $b = 0$, that is of

$$8192 - 352\omega^2 + 2\omega^4 = 0, \tag{5.3.28}$$

and are $\omega_1 = 5{\cdot}25$ and $\omega_2 = 12{\cdot}18$ rad/sec. When ω is equal to a natural frequency then the first term in (5.3.27) is zero and one of the angles θ_r has the value $\frac{1}{2}\pi$ while the other has the value given by

$$\tan\theta = \frac{5\omega b}{2(208-4\omega^2)}. \tag{5.3.29}$$

By making the substitution

$$\sigma = \frac{\omega b}{\tan\theta} \tag{5.3.30}$$

we may transform equation (5.3.27) into

$$5\sigma^2 - 2(208-4\omega^2)\,\sigma + 8192 - 352\omega^2 + 2\omega^4 = 0, \tag{5.3.31}$$

whose roots are

$$\sigma = \frac{\omega b}{\tan\theta} = \frac{208 - 4\omega^2 \pm \sqrt{[(208-4\omega^2)^2 - 5(8192-352\omega^2+2\omega^4)]}}{5}. \tag{5.3.32}$$

When $\omega = \omega_1$, this equation gives

$$\frac{\omega b}{\tan\theta} = \frac{208-4\omega_1^2 \pm (208-4\omega_1^2)}{5}, \tag{5.3.33}$$

so that the negative sign corresponds to θ_1 while the positive corresponds to θ_2.

Fig. 5.3.2 shows the values of θ_1 and θ_2 for a range of values of ω. The three values of b which have been chosen correspond roughly to 'light', 'medium' and 'heavy' damping. By comparing fig. 5.3.2 with fig. 5.2.2 (b) we see that each of the angles θ_r depends on ω in qualitatively the same way as ζ. Equation (5.3.32) shows how $\tan\theta$ behaves for small and for large values of ω. In fact

$$\frac{\tan\theta}{\omega b} \to \tfrac{1}{32} \quad \text{or} \quad \tfrac{5}{256} \quad \text{as} \quad \omega \to 0, \tag{5.3.34}$$

and

$$\frac{\omega\tan\theta}{b} \to \frac{-4 \pm \sqrt{6}}{2} \quad \text{as} \quad \omega \to \infty. \tag{5.3.35}$$

The modal shapes are given by equation (5.3.17), which reduces to

$$\begin{bmatrix} (256-3\omega^2)\tan\theta_r - 5b\omega & -\omega^2\tan\theta_r \\ -\omega^2\tan\theta_r & (32-\omega^2)\tan\theta_r - b\omega \end{bmatrix}\begin{bmatrix} \kappa_1^{(r)} \\ \kappa_2^{(r)} \end{bmatrix} = 0 \quad (r = 1, 2), \tag{5.3.36}$$

where, for instance, $\kappa_1^{(2)}$ means the deflection at q_1 in the second mode. The solution is

$$\kappa_1^{(r)}:\kappa_2^{(r)} = \omega^2 \tan\theta_r : (256-3\omega^2)\tan\theta_r - 5b\omega$$
$$= \omega^2 : 256 - 3\omega^2 - \frac{5b\omega}{\tan\theta_r}. \tag{5.3.37}$$

By substituting the value of σ from (5.3.32) we obtain

$$\kappa_1^{(r)}:\kappa_2^{(r)} = \omega^2 : 48+\omega^2 \pm \sqrt{[(48+\omega^2)^2+5\omega^4]}, \tag{5.3.38}$$

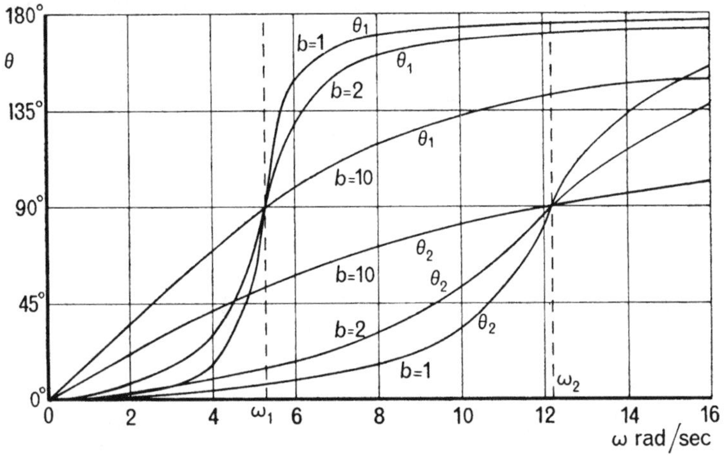

Fig. 5.3.2

where the positive sign now corresponds to $r = 1$ and the negative to $r = 2$. This equation shows that for $\omega = 0$

$$\left.\begin{aligned} \kappa_1^{(1)}:\kappa_2^{(1)} &= 0:(\text{finite}),\\ \kappa_1^{(2)}:\kappa_2^{(2)} &= (\text{finite}):0, \end{aligned}\right\} \tag{5.3.39}$$

while none of the components is zero for any other value of ω. It also shows that as $\omega \to \infty$

$$\kappa_1^{(r)}:\kappa_2^{(r)} \to 1:1\pm\sqrt{6}, \tag{5.3.40}$$

the positive sign again relating to the first mode.

In the first mode the second component, $\kappa_2^{(1)}$, is never zero. The mode may therefore be scaled so that $\kappa_2^{(1)} = 1$ for all values of ω. If the first *principal* mode is scaled by making the second component unity it is found to be

$$\mathbf{\Psi}^{(1)} = \begin{bmatrix} 0\cdot160 \\ 1 \end{bmatrix}. \tag{5.3.41}$$

Similarly, in the second mode the first component is never zero and may therefore be taken to be unity for all values of ω. With this scaling the second principal mode is

$$\mathbf{\Psi}^{(2)} = \begin{bmatrix} 1 \\ -1\cdot275 \end{bmatrix}. \tag{5.3.42}$$

Fig. 5.3.3 shows the graphs of the two modes.

We have as yet paid no attention to equation (5.3.13). But it is now clear that for each value of ω and each mode $\boldsymbol{\varkappa}^{(r)}(\omega)$ there is a corresponding forcing matrix $\boldsymbol{\Sigma}^{(r)}\sin\omega t$ given by

$$\cos\theta_r.(\mathbf{C}-\omega^2\mathbf{A})\,\boldsymbol{\varkappa}^{(r)}+\omega\sin\theta_r.\mathbf{B}\boldsymbol{\varkappa}^{(r)} = \boldsymbol{\Sigma}^{(r)}. \qquad (5.3.43)$$

This equation shows that when $\omega = \omega_r$ (i.e. $\theta_r = \frac{1}{2}\pi$) the forcing term is given by

$$\boldsymbol{\Sigma}^{(r)} = \omega_r\mathbf{B}\boldsymbol{\varkappa}^{(r)} = \omega_r\mathbf{B}\boldsymbol{\Psi}^{(r)}. \qquad (5.3.44)$$

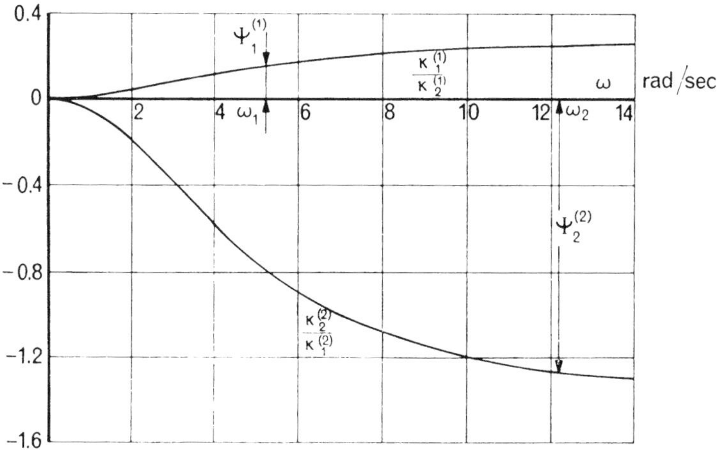

Fig. 5.3.3

On the other hand, when $\mathbf{B} = \mathbf{0}$ it is

$$\boldsymbol{\Sigma}^{(r)} = (\mathbf{C}-\omega^2\mathbf{A})\,\boldsymbol{\Psi}^{(r)} = (\omega_r^2-\omega^2)\,\mathbf{A}\boldsymbol{\Psi}^{(r)} = (\omega_r^2-\omega^2)\,\boldsymbol{\Phi}^{(r)}, \qquad (5.3.45)$$

where $\boldsymbol{\Phi}^{(r)}$ is the principal mode of excitation defined in equation (2.9.5).

The conclusions of this section may be summarised as follows:

(*a*) For each frequency of excitation there are n modes of forced vibration, each corresponding to certain sets of forces.

(*b*) The shapes of the modes and the corresponding sets of forces are given by real column matrices $\boldsymbol{\varkappa}^{(r)}$ and $\boldsymbol{\Sigma}^{(r)}$ respectively, and depend on ω, **A**, **B** and **C**.

(*c*) The $\boldsymbol{\varkappa}^{(r)}$ modes depend on the 'shape' of the matrix **B**, and not on the intensity of the damping.

(*d*) In each mode the responses at the coordinates are all in phase, but lag behind the excitation by an angle θ_r. This lag also depends on ω, **A**, **B** and **C**.

The existence of these modes appears to have been pointed out first by Fraeijs de Veubeke.† We shall follow his nomenclature and call the angles θ_r 'characteristic phase-differences'. We shall refer to the modes $\boldsymbol{\varkappa}^{(r)}$ as 'forced modes', and to the sets of amplitudes $\boldsymbol{\Sigma}^{(r)}$ as 'forced modes of excitation'. Some of the applications of the theory presented in this section have been pointed out in a paper,‡ in which references to other works will be found.

† B. M. Fraeijs de Veubeke, 'Déphasages Caractéristiques et Vibrations Forcées d'un Système Amorti', *Bull. Acad. Belg. Sci.* ser. 5, **34** (1948), pp. 626–41.

‡ Bishop and Gladwell, *Phil. Trans.* A, **255**. As mentioned previously, the comparable theory for hysteretic damping is developed in the paper.

EXAMPLES 5.3

1. Show that, in general, it is not possible to excite a system in one of its principal modes using a distribution of forces which are all in phase, except at the appropriate resonant frequency.

2. (*a*) Referred to coordinates $\mathbf{q}$ the inertia, stability and viscous damping matrices of a system are $\mathbf{A}$, $\mathbf{C}$ and $\mathbf{B}$ respectively, and the matrix of principal modes is $\mathbf{X}$. Show that, referred to principal coordinates $\mathbf{p}$, where

$$\mathbf{q} = \mathbf{Xp},$$

the corresponding matrices are

$$\mathbf{L} = \mathbf{X'AX}, \quad \mathbf{N} = \mathbf{X'CX} \quad \text{and} \quad \mathbf{M} = \mathbf{X'BX},$$

and find the equation governing the forced motion of the system referred to these coordinates.

(*b*) Show that if $\mathbf{M}$ is a diagonal matrix the excitation

$$\mathbf{q} = \Pi_r \mathbf{\Psi}^{(r)} \sin \omega t$$

in the rth principal mode may be excited at any frequency by an in-phase force distribution. Find its magnitude and express the result in terms of $\beta_r = \omega/\omega_r$ and $\mu_r = \omega b_r/c_r$, where b_r is the rth element in the diagonal matrix $\mathbf{M}$.

3. Show that, for a system of the particular type mentioned in Ex. 5.3.2, the damped modes $\boldsymbol{\varkappa}^{(r)}$ are the same as the principal modes $\mathbf{\Psi}^{(r)}$ and are not frequency-dependent.

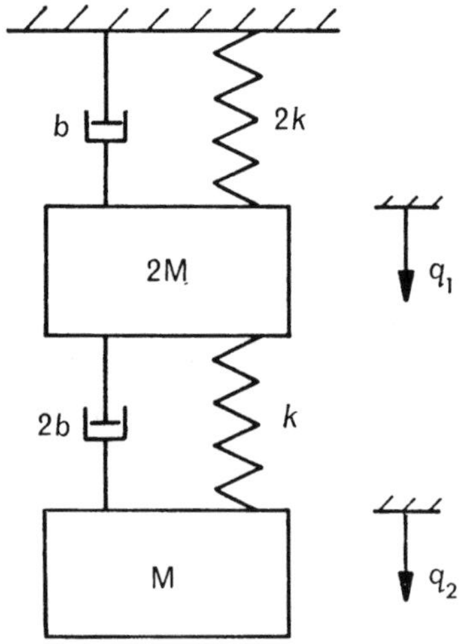

Ex. 5.3.4

4. Find the characteristic phase angles θ_1 and θ_2 and also the vectors

$$\boldsymbol{\varkappa}^{(1)}, \quad \boldsymbol{\varkappa}^{(2)}, \quad \mathbf{\Sigma}^{(1)} \quad \text{and} \quad \mathbf{\Sigma}^{(2)},$$

for the system shown, in which $b = \sqrt{(kM)}$ and where $\omega = \sqrt{(k/M)}$. Scale the vectors so that

$$\varkappa_1^{(1)} = 1 = \varkappa_1^{(2)}.$$

5.4 Orthogonality properties—modal expansions

It was shown in Chapter 2 that the orthogonality properties of principal modes are of considerable importance. In this section we shall discuss certain analogous properties of forced modes. Just as in Chapter 2 (§§ 2.5 and 2.9) we shall express the orthogonality properties in two equivalent forms—one involving the forced modes $\boldsymbol{\varkappa}^{(r)}$ only and one involving the forced modes of excitation $\boldsymbol{\Sigma}^{(r)}$ also.

Premultiply equations (5.3.17) and (5.3.22) by $\boldsymbol{\varkappa}'^{(s)}$ and $\boldsymbol{\varkappa}'^{(r)}$ respectively; the resulting equations are

$$\tan\theta_r . \boldsymbol{\varkappa}'^{(s)}(\mathbf{C}-\omega^2\mathbf{A})\,\boldsymbol{\varkappa}^{(r)} - \omega . \boldsymbol{\varkappa}'^{(s)}\mathbf{B}\boldsymbol{\varkappa}^{(r)} = 0, \tag{5.4.1}$$

$$\tan\theta_s . \boldsymbol{\varkappa}'^{(r)}(\mathbf{C}-\omega^2\mathbf{A})\,\boldsymbol{\varkappa}^{(s)} - \omega . \boldsymbol{\varkappa}'^{(r)}\mathbf{B}\boldsymbol{\varkappa}^{(s)} = 0. \tag{5.4.2}$$

Remembering that **A**, **B** and **C** are all symmetric matrices we may transpose the second equation to obtain

$$\tan\theta_s . \boldsymbol{\varkappa}'^{(s)}(\mathbf{C}-\omega^2\mathbf{A})\,\boldsymbol{\varkappa}^{(r)} - \omega . \boldsymbol{\varkappa}'^{(s)}\mathbf{B}\boldsymbol{\varkappa}^{(r)} = 0. \tag{5.4.3}$$

By subtracting this equation from equation (5.4.1) we obtain

$$\left.\begin{aligned} \boldsymbol{\varkappa}'^{(s)}(\mathbf{C}-\omega^2\mathbf{A})\,\boldsymbol{\varkappa}^{(r)} &= 0,\\ \boldsymbol{\varkappa}'^{(s)}\mathbf{B}\boldsymbol{\varkappa}^{(r)} &= 0, \end{aligned}\right\} \tag{5.4.4}$$

provided that $\theta_r \neq \theta_s$. The forced modes are orthogonal in the sense that equations (5.4.4) are valid. The condition $\theta_r \neq \theta_s$ will be relaxed in this section since $\theta_r = \theta_s$ implies that the determinantal equation (5.3.16) has repeated roots—a special case of little physical significance. Thus the orthogonality conditions (5.4.4) hold good provided, simply, that $r \neq s$. To obtain the other expression for the orthogonality properties we combine equations (5.4.4) with equation (5.3.43). This leads to the equations

$$\boldsymbol{\varkappa}'^{(r)}\boldsymbol{\Sigma}^{(s)} = 0 = \boldsymbol{\varkappa}'^{(s)}\boldsymbol{\Sigma}^{(r)} \quad (r \neq s). \tag{5.4.5}$$

In considering the equation governing the damped forced vibration of a system in the last section we took the trial solution (5.3.11), and this led to the discovery of n forced modes. In the rth of these modes the displacement vector $\mathbf{q}$ has the form

$$\mathbf{q} = \mathbf{q}^{(r)} = \boldsymbol{\varkappa}^{(r)}\sin(\omega t - \theta_r) \tag{5.4.6}$$

and is the response to the generalised force vector

$$\boldsymbol{\Sigma}^{(r)}\sin\omega t.$$

The vectors $\boldsymbol{\varkappa}^{(r)}$ and $\boldsymbol{\Sigma}^{(r)}$ are linked by equation (5.3.43). Equation (5.3.12) is linear and therefore any combination of solutions is also a solution. This means that if $\lambda_1, \lambda_2, \ldots, \lambda_n$ are numerical multipliers then the response to the force

$$\boldsymbol{\Phi}\sin\omega t = [\lambda_1\boldsymbol{\Sigma}^{(1)} + \lambda_2\boldsymbol{\Sigma}^{(2)} + \ldots + \lambda_n\boldsymbol{\Sigma}^{(n)}]\sin\omega t \tag{5.4.7}$$

is given by

$$\mathbf{q} = \lambda_1\boldsymbol{\varkappa}^{(1)}\sin(\omega t-\theta_1) + \lambda_2\boldsymbol{\varkappa}^{(2)}\sin(\omega t-\theta_2) + \ldots + \lambda_n\boldsymbol{\varkappa}^{(n)}\sin(\omega t-\theta_n). \tag{5.4.8}$$

We shall now show that

(i) any set of forces $\mathbf{\Phi}\sin\omega t$ may be expressed in the form (5.4.7), and that

(ii) the appropriate constants λ_r for a given force vector $\mathbf{\Phi}$ may be found by using the orthogonality relation (5.4.5).

To prove the first of these statements it is sufficient to prove that the column vectors $\mathbf{\Sigma}^{(1)}, \mathbf{\Sigma}^{(2)}, \ldots, \mathbf{\Sigma}^{(n)}$ are linearly independent.† For then the determinant of coefficients in the equations for $\lambda_1, \lambda_2, \ldots, \lambda_n$ will be non-zero and the λ's may be found by Cramer's rule, as is shown in § 1.6. For the $\mathbf{\Sigma}$'s to be linearly independent the equation

$$x_1\mathbf{\Sigma}^{(1)}+x_2\mathbf{\Sigma}^{(2)}+\ldots+x_n\mathbf{\Sigma}^{(n)} = \mathbf{0}, \tag{5.4.9}$$

in which the x's are numerical multipliers, must imply that

$$x_1 = 0 = x_2 = \ldots = x_n. \tag{5.4.10}$$

Suppose that equation (5.4.9) holds. Premultiply the equation by $\boldsymbol{\kappa}'^{(r)}$ and use the first of equations (5.4.5); the result is

$$x_r\boldsymbol{\kappa}'^{(r)}\mathbf{\Sigma}^{(r)} = 0. \tag{5.4.11}$$

Equations (5.3.17) and (5.3.43) show, between them, that

$$\sin\theta_r.\mathbf{\Sigma}^{(r)} = \omega\mathbf{B}\boldsymbol{\kappa}^{(r)}, \tag{5.4.12}$$

$$\cos\theta_r.\mathbf{\Sigma}^{(r)} = (\mathbf{C}-\omega^2\mathbf{A})\,\boldsymbol{\kappa}^{(r)}, \tag{5.4.13}$$

and therefore
$$\{\boldsymbol{\kappa}'^{(r)}\mathbf{\Sigma}^{(r)}\}^2 = \{\omega\boldsymbol{\kappa}'^{(r)}\mathbf{B}\boldsymbol{\kappa}^{(r)}\}^2+\{\boldsymbol{\kappa}'^{(r)}(\mathbf{C}-\omega^2\mathbf{A})\,\boldsymbol{\kappa}^{(r)}\}^2 > 0 \tag{5.4.14}$$

which, when combined with equation (5.4.11), shows that

$$x_r = 0. \tag{5.4.15}$$

The vectors $\mathbf{\Sigma}^{(1)}, \mathbf{\Sigma}^{(2)}, \ldots, \mathbf{\Sigma}^{(n)}$ are linearly independent. The constants $\lambda_1, \lambda_2, \ldots, \lambda_n$ in (5.4.7) may now be found by premultiplying the equation by $\boldsymbol{\kappa}'^{(r)}$ and using equation (5.4.5). By this means, it is found that

$$\boldsymbol{\kappa}'^{(r)}\mathbf{\Phi} = \lambda_r\boldsymbol{\kappa}'^{(r)}\mathbf{\Sigma}^{(r)}, \tag{5.4.16}$$

or

$$\lambda_r = \frac{\boldsymbol{\kappa}'^{(r)}\mathbf{\Phi}}{\boldsymbol{\kappa}'^{(r)}\mathbf{\Sigma}^{(r)}}. \tag{5.4.17}$$

These expressions for the λ's may now be substituted into equation (5.4.8) to give the response $\mathbf{q}$ due to the forces $\mathbf{\Phi}\sin\omega t$. It is

$$\mathbf{q} = \frac{\boldsymbol{\kappa}^{(1)}\boldsymbol{\kappa}'^{(1)}\mathbf{\Phi}}{\boldsymbol{\kappa}'^{(1)}\mathbf{\Sigma}^{(1)}}\sin(\omega t-\theta_1)+\frac{\boldsymbol{\kappa}^{(2)}\boldsymbol{\kappa}'^{(2)}\mathbf{\Phi}}{\boldsymbol{\kappa}'^{(2)}\mathbf{\Sigma}^{(2)}}\sin(\omega t-\theta_2)+\ldots+\frac{\boldsymbol{\kappa}^{(n)}\boldsymbol{\kappa}'^{(n)}\mathbf{\Phi}}{\boldsymbol{\kappa}'^{(n)}\mathbf{\Sigma}^{(n)}}\sin(\omega t-\theta_n). \tag{5.4.18}$$

Clearly this may be written in the form

$$\mathbf{q} = \left\{\frac{\boldsymbol{\kappa}^{(1)}\boldsymbol{\kappa}'^{(1)}}{\boldsymbol{\kappa}'^{(1)}\mathbf{\Sigma}^{(1)}}\sin(\omega t-\theta_1)+\frac{\boldsymbol{\kappa}^{(2)}\boldsymbol{\kappa}'^{(2)}}{\boldsymbol{\kappa}'^{(2)}\mathbf{\Sigma}^{(2)}}\sin(\omega t-\theta_2)+\ldots+\frac{\boldsymbol{\kappa}^{(n)}\boldsymbol{\kappa}'^{(n)}}{\boldsymbol{\kappa}'^{(n)}\mathbf{\Sigma}^{(n)}}\sin(\omega t-\theta_n)\right\}\mathbf{\Phi}, \tag{5.4.19}$$

† The reader who has not done so may wish to familiarise himself with the preliminary remarks in § 3.2.

in which it will be seen that each of the terms $\boldsymbol{\varkappa}^{(r)}\boldsymbol{\varkappa}'^{(r)}$ is a square matrix. The denominator $\boldsymbol{\varkappa}'^{(r)}\boldsymbol{\Sigma}^{(r)}$ may be expressed in terms of the $\boldsymbol{\varkappa}$'s in various ways by using equations (5.3.43) or (5.4.12) to (5.4.14). Thus the first two of these equations give, respectively,

$$\left.\begin{aligned}\boldsymbol{\varkappa}'^{(r)}\boldsymbol{\Sigma}^{(r)} &= \cos\theta_r.\boldsymbol{\varkappa}'^{(r)}(\mathbf{C}-\omega^2\mathbf{A})\,\boldsymbol{\varkappa}^{(r)}+\omega\sin\theta_r.\boldsymbol{\varkappa}'^{(r)}\mathbf{B}\boldsymbol{\varkappa}^{(r)},\\ \boldsymbol{\varkappa}'^{(r)}\boldsymbol{\Sigma}^{(r)} &= \omega\operatorname{cosec}\theta_r.\boldsymbol{\varkappa}'^{(r)}\mathbf{B}\boldsymbol{\varkappa}^{(r)}.\end{aligned}\right\} \tag{5.4.20}$$

It is necessary to point out that it is *not* possible to expand an arbitrary column matrix $\mathbf{q}$ having components

$$q_1 = A_1\sin(\omega t-\alpha_1),\quad q_2 = A_2\sin(\omega t-\alpha_2)\quad \ldots,\quad q_n = A_n\sin(\omega t-\alpha_n), \tag{5.4.21}$$

in a series of the type (5.4.8). The reason for this is that such a set of displacements would, in general, be the response to a set of forces of the type

$$Q_1 = \Phi_1\sin(\omega t-\phi_1),\quad Q_2 = \Phi_2\sin(\omega t-\phi_2),\quad \ldots,\quad Q_n = \Phi_n\sin(\omega t-\phi_n), \tag{5.4.22}$$

whereas any response of type (5.4.8) results from a set of forces which are all in phase.

We conclude this section by comparing the behaviour of the system of fig. 5.3.1 with that of a system with a single degree of freedom. In particular we shall compare the amplitude multipliers λ_1 and λ_2 with the amplitude Y occurring in equation (5.2.3).

The amplitude multipliers λ_1 and λ_2 for the system of fig. 5.3.1 may best be calculated from equations (5.4.17) and (5.4.20). These give

$$\lambda_r = \frac{\boldsymbol{\varkappa}'^{(r)}\boldsymbol{\Phi}\sin\theta_r}{\omega\boldsymbol{\varkappa}'^{(r)}\mathbf{B}\boldsymbol{\varkappa}^{(r)}}$$

$$= \frac{[\kappa_1^{(r)}\ \ \kappa_2^{(r)}]\begin{bmatrix}F\\0\end{bmatrix}\sin\theta_r}{\omega[\kappa_1^{(r)}\ \ \kappa_2^{(r)}]\begin{bmatrix}5b & 0\\ 0 & b\end{bmatrix}\begin{bmatrix}\kappa_1^{(r)}\\ \kappa_2^{(r)}\end{bmatrix}} \tag{5.4.23}$$

$$= \frac{F\kappa_1^{(r)}\sin\theta_r}{\omega b\{5(\kappa_1^{(r)})^2+(\kappa_2^{(r)})^2\}}. \tag{5.4.24}$$

If the modes are scaled in the manner suggested in § 5.3 then the values of λ_1 and λ_2 are given by

$$\frac{\lambda_1}{F} = \frac{\kappa_1^{(1)}\sin\theta_1}{\omega b\{5(\kappa_1^{(1)})^2+1\}};\quad \frac{\lambda_2}{F} = \frac{\sin\theta_2}{\omega b\{5+(\kappa_2^{(2)})^2\}}. \tag{5.4.25}$$

The variation of λ_1 and λ_2 with ω and b is shown in fig. 5.4.1, and by comparing this with fig. 5.2.2 (*a*) we see that λ_1 and λ_2 both depend on ω in qualitatively the same way as Y in equation (5.2.3). The essential difference between the two systems is that in the one case Y specifies the amplitude of the motion completely, while in the other λ_r gives only a partial specification of the amplitude in the rth forced mode.

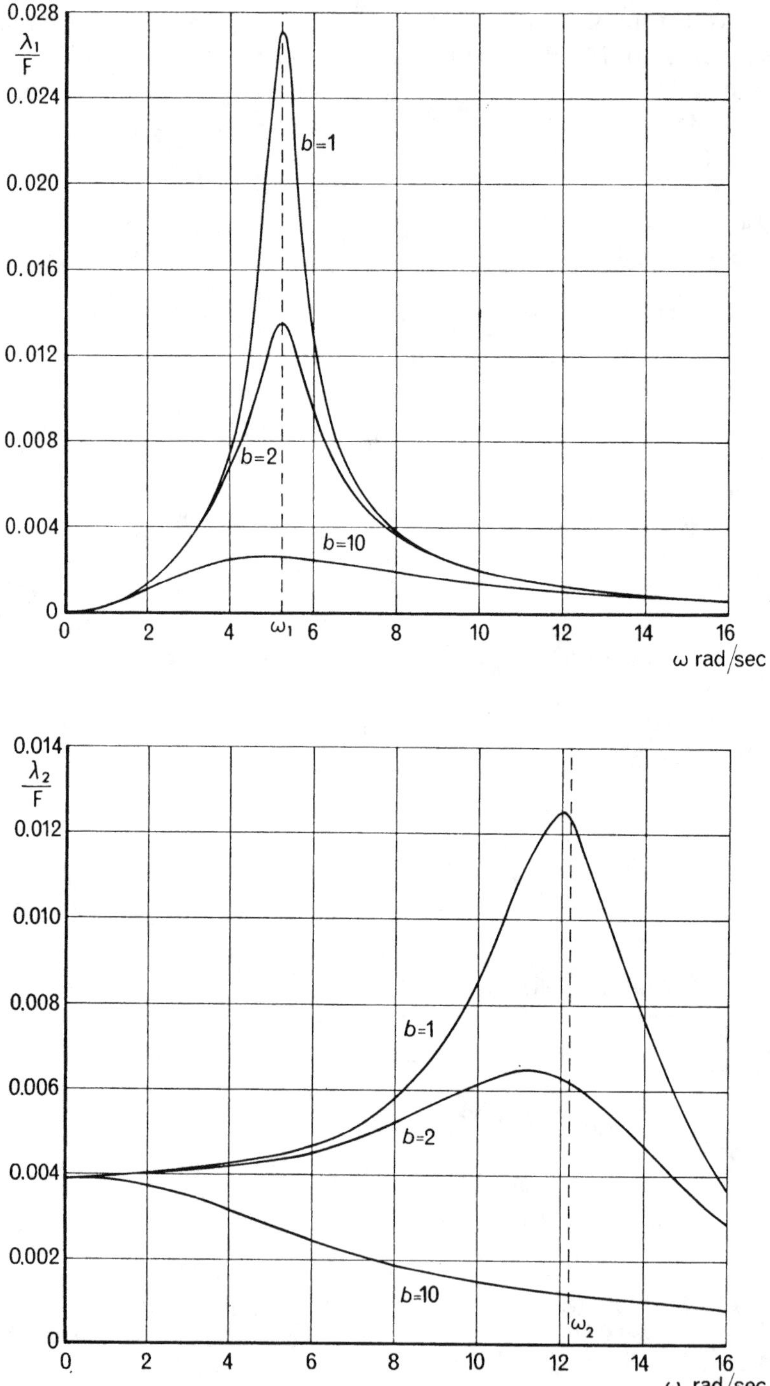
0.028
$\frac{\lambda_1}{F}$
0.024
0.020
0.016
0.012
0.008
0.004
0
b=1
b=2
b=10
0
2
4
ω_1
6
8
10
12
14
16
ω rad/sec
0.014
$\frac{\lambda_2}{F}$
0.012
0.010
0.008
0.006
0.004
0.002
0
b=1
b=2
b=10
ω_2
0
2
4
6
8
10
12
14
16
ω rad/sec

Fig. 5.4.1

In a multi-degree-of-freedom system the shape of each forced mode varies with frequency and for a complete description of the motion the shapes of the various modes, given by the column matrices $\boldsymbol{\varkappa}^{(r)}$, have to be specified also. If the rth forced mode of the system is scaled so that one component (the tth, $\varkappa_t^{(r)}$, say) is chosen to be unity, then λ_r will give the amplitude of the coordinate q_t, but only of that one, in the rth mode.

EXAMPLES 5.4

1. Verify that the forced modes $\boldsymbol{\varkappa}$ and forced modes of excitation $\boldsymbol{\Sigma}$ of Ex. 5.3.4 satisfy the orthogonality conditions in the forms (5.4.4) and (5.4.5).

2. The forces $\mathbf{Q} = \{1 \quad -1\} F \sin \omega t$ are applied to the system of Ex. 5.3.4. Express the response $\mathbf{q}$ in terms of the forced modes $\boldsymbol{\varkappa}$, taking $\omega = \sqrt{(k/M)}$ and $b = \sqrt{(kM)}$ as before.

3. An harmonic generalised force $\Phi_j \sin \omega t$ is applied at the generalised coordinate q_j of a system having viscous damping. Find an expression for the response at some coordinate q_i of the system. Would the response at q_j be the same as that found originally at q_i, if the same force were applied at q_i instead of q_j?

5.5 Complex receptances

The introduction of damping into the system has brought with it the necessity of discussing phase differences. A convenient way of doing this is through the use of complex numbers, and this section will be devoted to a discussion of this alternative method. This discussion will be couched in algebraic terms; the matter may be thought of as possessing a geometric interpretation also.†

Much of the method may be introduced by referring to the system of fig. 5.2.1 for which the equation of motion is

$$M\ddot{y} + b\dot{y} + ky = F \sin \omega t. \tag{5.5.1}$$

The first step is to introduce a complex quantity

$$u = x + iy, \tag{5.5.2}$$

whose imaginary part is y, and then to form an equation for u which has equation (5.5.1) as its imaginary part. The required equation is

$$M\ddot{u} + b\dot{u} + ku = F e^{i\omega t} \tag{5.5.3}$$

since the real and imaginary parts of this equation are respectively

$$M\ddot{x} + b\dot{x} + kx = F \cos \omega t \tag{5.5.4}$$

and equation (5.5.1). If we solve equation (5.5.3) for u then the solution y of equation (5.5.1) will be given by

$$y = \operatorname{Im}(u). \tag{5.5.5}$$

The trial solution

$$u = U e^{i\omega t} \tag{5.5.6}$$

† Bishop and Johnson, *The Mechanics of Vibration* (1960), §1.3.

satisfies equation (5.5.3) provided that

$$\{-\omega^2 M + i\omega b + k\}\, U = F, \tag{5.5.7}$$

or

$$U = \frac{F}{(k-\omega^2 M) + i\omega b}. \tag{5.5.8}$$

This gives

$$u = \frac{F\, e^{i\omega t}}{(k-\omega^2 M) + i\omega b}, \tag{5.5.9}$$

and therefore

$$y = \mathrm{Im}\left\{\frac{F\, e^{i\omega t}}{(k-\omega^2 M) + i\omega b}\right\}. \tag{5.5.10}$$

With the notation

$$\tan\zeta = \frac{\omega b}{k-\omega^2 M} \tag{5.5.11}$$

equation (5.5.10) becomes

$$y = \frac{F \sin(\omega t - \zeta)}{\{(k-\omega^2 M)^2 + \omega^2 b^2\}^{\frac{1}{2}}}, \tag{5.5.12}$$

which is exactly the same as the result obtained in § 5.2.

It should be emphasised that for the problem that has been discussed so far there is absolutely no benefit to be had from the introduction of complex quantities—in fact the real equations are easier to solve. Moreover, the real equations (5.5.11) and (5.5.12) give explicit expressions for the phase-difference and amplitude, whereas equation (5.5.10) only gives them implicitly. However, as we shall now show, the complex method of solution lends itself to extension far more easily than the real method.

First, let us agree to call the quantity u a 'complex displacement', and $F e^{i\omega t}$ a 'complex force'.† Equations (5.5.1) and (5.5.4) show that the real and imaginary parts of u are the (real) displacements due to the real and imaginary parts of the complex force. Having introduced these complex quantities we now write

$$\alpha = \frac{1}{(k-\omega^2 M) + i\omega b}, \tag{5.5.13}$$

so that equation (5.5.9) becomes,

$$u = \alpha F e^{i\omega t}, \tag{5.5.14}$$

which states that

$$(\text{complex displacement}) = \alpha \times (\text{complex force}).$$

This shows that α relates the complex displacement and force in the same way as a receptance relates (real) displacements and forces in undamped vibration. For this reason α will be called the 'complex receptance' of the system. It will be noticed that, when the system is undamped, α becomes the ordinary receptance $1/(k-\omega^2 M)$.

We shall now discuss some extensions of this simple problem, and as we do this we shall begin to see some of the advantages of the complex method. The advantages of the method arise from the fact, which may easily be verified, that the solution of

† The physical significance of these complex quantities is simply that they are combinations of two physical quantities in quadrature.

equation (5.5.3) for u remains the same, namely equation (5.5.9), when F becomes complex, and even when both F and ω are complex. Thus, even when F and ω are complex, the complex displacement and force are linked according to equation (5.5.14) by the complex receptance given in equation (5.5.13).

Suppose it is required to find the response of the system of fig. 5.2.1 to the force $F_0 \sin(\omega t+\phi)$. The equation of motion is then

$$M\ddot{y}+b\dot{y}+ky = F_0 \sin(\omega t+\phi). \tag{5.5.15}$$

Since $$(F_0 e^{i\phi})\, e^{i\omega t} = F_0 e^{i(\omega t+\phi)} = F_0 \cos(\omega t+\phi)+iF_0 \sin(\omega t+\phi), \tag{5.5.16}$$

equation (5.5.15) is the imaginary part of equation (5.5.3), where now F is the complex quantity

$$F = F_0 e^{i\phi}. \tag{5.5.17}$$

But the solution of equation (5.5.3) is always given by equation (5.5.14). Therefore

$$\begin{aligned} y &= \operatorname{Im}(\alpha F e^{i\omega t}) = \operatorname{Im}(\alpha F_0 e^{i(\omega t+\phi)}) \\ &= \operatorname{Im}\left\{\frac{F_0 e^{i(\omega t+\phi)}}{(k-\omega^2 M)+i\omega b}\right\} \\ &= \frac{F_0 \sin(\omega t+\phi-\zeta)}{\{(k-\omega^2 M)^2+\omega^2 b^2\}^{\frac{1}{2}}}. \end{aligned} \tag{5.5.18}$$

Since this expression is just what we would expect in view of equation (5.5.12), the fact that we can allow F to be complex in equation (5.5.3) is not particularly important. However, we shall find that the analogue of this procedure for the general system having n degrees of freedom is important. It will allow us to find the response of the system to a set of forces which are not all in phase.

Finally, suppose that it is required to find the response of the system of fig. 5.2.1 to the force $F_0 e^{-\tau t}\sin(\sigma t+\phi)$—a sinusoidal force whose amplitude varies exponentially.† Such a force might have to be taken into account in a discussion of the stability of the free vibration of a system of which the present system forms a part.‡ The equation of motion is

$$M\ddot{y}+b\dot{y}+ky = F_0 e^{-\tau t}\sin(\sigma t+\phi). \tag{5.5.19}$$

Since $$F_0 e^{i\phi} e^{i(\sigma+i\tau)t} = F_0 e^{-\tau t} e^{i(\sigma t+\phi)} = F_0 e^{-\tau t}\{\cos(\sigma t+\phi)+i\sin(\sigma t+\phi)\}, \tag{5.5.20}$$

equation (5.5.19) is the imaginary part of equation (5.5.3) where now

$$F = F_0 e^{i\phi}; \quad \omega = \sigma+i\tau. \tag{5.5.21}$$

Equation (5.5.3) still has the solution (5.5.14) so that

$$y = \operatorname{Im}(\alpha F e^{i\omega t}) = \operatorname{Im}\left\{\frac{F_0 e^{-\tau t} e^{i(\sigma t+\phi)}}{(k-\omega^2 M)+i\omega b}\right\}. \tag{5.5.22}$$

† Strictly, the terms 'amplitude', $(F_0 e^{-\tau t})$, 'phase' (ϕ) and 'frequency' (σ) should not be applied to any but sinusoidally varying quantities. We shall do so, however, since there is little likelihood of ambiguity.

‡ G. M. L. Gladwell and R. E. D. Bishop, 'The receptances of uniform and non-uniform rotating shafts', *J. Mech. Engng Sci.* **1**, no. 1 (June 1959), pp. 78–91, and Gladwell and Bishop 'The vibration of rotating shafts supported in flexible bearings', *J. Mech. Engng Sci.* **1**, no. 3 (Dec. 1959), pp. 195–206.

If the notation
$$\tan\xi = \frac{\sigma(b-2\tau M)}{k-(\sigma^2-\tau^2)\,M-\tau b} \tag{5.5.23}$$
is used then it is found that
$$y = \frac{F_0 e^{-\tau t}\sin(\sigma t+\phi-\xi)}{\{[k-(\sigma^2-\tau^2)\,M-\tau b]^2+\sigma^2(b-2\tau M)^2\}^{\frac{1}{2}}}. \tag{5.5.24}$$

It may be pointed out that in a discussion of stability it would probably be best to work entirely in terms of the complex quantities u, F and ω, rather than in terms of the real quantities y, F_0, ϕ, σ and τ. Equation (5.5.9) is far less cumbersome than equation (5.5.24). However, if an expression for y is required then probably the easiest way to obtain it is to proceed, as we have done, via the expression for u.

We shall now consider a system having a number of degrees of freedom and we shall find that the results obtained for the system having one degree of freedom may be extended quite naturally. The forced motion of the general system is governed by the Lagrange equations (5.3.4), and when the forces are all in phase and of frequency ω this becomes equation (5.3.5), namely,
$$\mathbf{A}\ddot{\mathbf{q}}+\mathbf{B}\dot{\mathbf{q}}+\mathbf{C}\mathbf{q} = \mathbf{\Phi}\sin\omega t. \tag{5.5.25}$$
With the system having but one degree of freedom, a complex quantity u was introduced whose imaginary part was y; now a complex column matrix $\mathbf{u}$ will be introduced such that
$$\mathbf{q} = \operatorname{Im}(\mathbf{u}). \tag{5.5.26}$$
The analogue of equation (5.5.3) is
$$\mathbf{A}\ddot{\mathbf{u}}+\mathbf{B}\dot{\mathbf{u}}+\mathbf{C}\mathbf{u} = \mathbf{\Phi}\,e^{i\omega t}, \tag{5.5.27}$$
which, since the matrices $\mathbf{A}$, $\mathbf{B}$ and $\mathbf{C}$ are real, has equation (5.5.25) as its imaginary part.

Equation (5.5.27) may be solved by means of the trial solution
$$\mathbf{u} = \mathbf{\Psi}\,e^{i\omega t}, \tag{5.5.28}$$
where the column matrix $\mathbf{\Psi}$ is allowed to be complex—that is, to contain complex elements. This leads to the equation
$$[(\mathbf{C}-\omega^2\mathbf{A})+i\omega\mathbf{B}]\,\mathbf{\Psi} = \mathbf{\Phi}, \tag{5.5.29}$$
in which the quantity in square brackets is a square matrix whose elements are complex. We denote this square matrix by $\mathbf{Z}$. Provided that $\mathbf{Z}$ is non-singular it will have an inverse $\mathbf{Z}^{-1}$, and the solution of (5.5.29) will be
$$\mathbf{\Psi} = \mathbf{Z}^{-1}\mathbf{\Phi}, \tag{5.5.30}$$
so that
$$\mathbf{u} = \mathbf{\Psi}\,e^{i\omega t} = \mathbf{Z}^{-1}\mathbf{\Phi}\,e^{i\omega t}. \tag{5.5.31}$$

This equation is the analogue of equation (5.5.14), and by comparing the two equations we see that the analogue of the complex number α is the complex matrix $\mathbf{Z}^{-1}$. We may therefore call $\mathbf{Z}^{-1}$ the 'complex receptance matrix' of the system and write
$$\mathbf{Z}^{-1} = \boldsymbol{\alpha}. \tag{5.5.32}$$

With this notation equation (5.5.31) becomes

$$\mathbf{u} = \boldsymbol{\alpha}\boldsymbol{\Phi}\, e^{i\omega t}, \tag{5.5.33}$$

or

(complex displacement matrix) = (complex matrix $\boldsymbol{\alpha}$) × (complex force matrix).

Equations (5.5.26) and (5.5.33) may now be combined to give the column matrix $\mathbf{q}$. Since the elements of $\boldsymbol{\alpha}$ are complex the displacements q_r will not be in phase. That is, they will have the form

$$q_1 = S_1 \sin(\omega t + \psi_1), \quad q_2 = S_2 \sin(\omega t + \psi_2), \quad \ldots, \quad q_n = S_n \sin(\omega t + \psi_n). \tag{5.5.34}$$

This is of course to be expected as the equations (5.5.34) are to be equivalent to equation (5.4.19).

All the theory for systems having n degrees of freedom has so far related to an arbitrary set of forces, all having the same frequency ω, and all being in phase. The use of complex quantities allows us to relax the latter restriction.† Any set of forces of frequency ω may be represented by quantities of the form

$$Q_r = R_r \sin(\omega t + \phi_r). \tag{5.5.35}$$

The equation governing the motion of the system due to these forces is

$$\mathbf{A}\ddot{\mathbf{q}} + \mathbf{B}\dot{\mathbf{q}} + \mathbf{C}\mathbf{q} = \mathbf{Q}, \tag{5.5.36}$$

where

$$\mathbf{Q} = \begin{bmatrix} Q_1 \\ Q_2 \\ \vdots \\ Q_n \end{bmatrix}. \tag{5.5.37}$$

This equation is the imaginary part of the equation

$$\mathbf{A}\ddot{\mathbf{u}} + \mathbf{B}\dot{\mathbf{u}} + \mathbf{C}\mathbf{u} = \boldsymbol{\Phi}\, e^{i\omega t} \tag{5.5.38}$$

for the particular choice

$$\Phi_r = R_r e^{i\phi_r}. \tag{5.5.39}$$

Equation (5.5.38) is the same as equation (5.5.27). Just as the solution of equation (5.5.3) is always equation (5.5.14) whether F is real or complex, so the solution of equation (5.5.27) is always equation (5.5.33), whether $\boldsymbol{\Phi}$ is real or complex. Therefore the solution of equation (5.5.36) is

$$\mathbf{q} = \operatorname{Im}(\boldsymbol{\alpha}\boldsymbol{\Phi}\, e^{i\omega t}), \tag{5.5.40}$$

where the elements of $\boldsymbol{\Phi}$ are given by equation (5.5.39).

We note in passing that the method outlined above probably provides the simplest way of finding the displacements due to the set of forces (5.5.35) even when

† If the system is acted upon by forces of different frequencies then the resultant displacement matrix will be the sum of the displacement matrices due to the forces of each frequency taken one frequency at a time.

there is no damping. Equation (5.5.40) is still valid in this case, but now the matrix $\boldsymbol{\alpha}$ is the real matrix

$$\boldsymbol{\alpha} = (\mathbf{C}-\omega^2\mathbf{A})^{-1} \tag{5.5.41}$$

and

$$\mathbf{q} = \begin{bmatrix} q_1 \\ q_2 \\ \vdots \\ q_n \end{bmatrix} = \boldsymbol{\alpha}\,\mathrm{Im}\,(\boldsymbol{\Phi}\,e^{i\omega t}) = \boldsymbol{\alpha}\begin{bmatrix} R_1 \sin(\omega t+\phi_1) \\ R_2 \sin(\omega t+\phi_2) \\ \vdots \\ R_n \sin(\omega t+\phi_n) \end{bmatrix}. \tag{5.5.42}$$

Finally, if it is required to find the response of the system to a set of forces of the type

$$Q_r = R_r e^{-\tau t}\sin(\sigma t+\phi_r), \tag{5.5.43}$$

then the corresponding equation for $\mathbf{u}$ is still equation (5.5.27) with

$$\Phi_r = R_r e^{i\phi_r}, \quad \omega = \sigma + i\tau. \tag{5.5.44}$$

The solution is again given by equation (5.5.33) so that

$$\mathbf{q} = \mathrm{Im}\,(\boldsymbol{\alpha\Phi}\,e^{i\omega t}) = e^{-\tau t}\,\mathrm{Im}\,(\boldsymbol{\alpha\Phi}\,e^{i\sigma t}). \tag{5.5.45}$$

This result is of importance in the study of systems which may become unstable.

We conclude this section by giving a series expression for the complex receptance matrix $\boldsymbol{\alpha}$. This will enable us to find an expression for the response due to an arbitrary set of forces of frequency ω which will reduce to equation (5.4.19) when the forces are all in phase. In § 5.3 it was shown that for each value of ω there are n column matrices $\boldsymbol{\varkappa}^{(r)}$, n angles θ_r and n column matrices $\boldsymbol{\Sigma}^{(r)}$ which are related by the equation

$$(\mathbf{C}-\omega^2\mathbf{A})\,\boldsymbol{\varkappa}^{(r)}.\sin(\omega t-\theta_r)+\omega.\mathbf{B}\boldsymbol{\varkappa}^{(r)}.\cos(\omega t-\theta_r) = \boldsymbol{\Sigma}^{(r)}.\sin\omega t. \tag{5.5.46}$$

Since all the quantities occurring in this equation are real, it is the imaginary part of the equation

$$(\mathbf{C}-\omega^2\mathbf{A})\,\boldsymbol{\varkappa}^{(r)}.e^{i(\omega t-\theta_r)}+i\omega.\mathbf{B}\boldsymbol{\varkappa}^{(r)}.e^{i(\omega t-\theta_r)} = \boldsymbol{\Sigma}^{(r)}.e^{i\omega t}. \tag{5.5.47}$$

This equation shows that

$$\mathbf{u} = \boldsymbol{\varkappa}^{(r)}.e^{i(\omega t-\theta_r)} \tag{5.5.48}$$

is the solution of equation (5.5.27) corresponding to

$$\boldsymbol{\Phi} = \boldsymbol{\Sigma}^{(r)}. \tag{5.5.49}$$

If therefore it is required to solve equation (5.5.27) for arbitrary $\boldsymbol{\Phi}$—real or complex—then the solution may be obtained by expressing $\boldsymbol{\Phi}$ in the form

$$\boldsymbol{\Phi} = \lambda_1\boldsymbol{\Sigma}^{(1)}+\lambda_2\boldsymbol{\Sigma}^{(2)}+\ldots+\lambda_n\boldsymbol{\Sigma}^{(n)}, \tag{5.5.50}$$

and the solution will be

$$\mathbf{u} = \sum_{r=1}^{n}\lambda_r.\boldsymbol{\varkappa}^{(r)}.e^{i(\omega t-\theta_r)}. \tag{5.5.51}$$

If $\boldsymbol{\Phi}$ is complex then the constants λ_r will also be complex, but they will still be given by equation (5.4.17), that is,

$$\lambda_r = \frac{\boldsymbol{\varkappa}'^{(r)}\boldsymbol{\Phi}}{\boldsymbol{\varkappa}'^{(r)}\boldsymbol{\Sigma}^{(r)}}. \tag{5.5.52}$$

By substituting for λ_r in equation (5.5.51) we obtain

$$\mathbf{u} = \left\{\frac{\boldsymbol{\kappa}^{(1)}\boldsymbol{\kappa}'^{(1)}.\,e^{-i\theta_r}}{\boldsymbol{\kappa}'^{(1)}\boldsymbol{\Sigma}^{(1)}} + \frac{\boldsymbol{\kappa}^{(2)}\boldsymbol{\kappa}'^{(2)}.\,e^{-i\theta_2}}{\boldsymbol{\kappa}'^{(2)}\boldsymbol{\Sigma}^{(2)}} + \ldots + \frac{\boldsymbol{\kappa}^{(n)}\boldsymbol{\kappa}'^{(n)}.\,e^{-i\theta_n}}{\boldsymbol{\kappa}'^{(n)}\boldsymbol{\Sigma}^{(n)}}\right\}\boldsymbol{\Phi}\,e^{i\omega t}. \quad (5.5.53)$$

This equation links the complex displacement matrix **u** with the complex force matrix $\boldsymbol{\Phi}\,e^{i\omega t}$. The expression in curly brackets, which is an $n \times n$ matrix with complex elements, is therefore the complex receptance matrix $\boldsymbol{\alpha}$. The result may be compared with equation (2.8.21). If the individual receptances α_{ij} are required they may be found by employing the previous notation

$$\boldsymbol{\kappa}^{(r)} = \begin{bmatrix} \kappa_1^{(r)} \\ \kappa_2^{(r)} \\ \vdots \\ \kappa_n^{(r)} \end{bmatrix}. \quad (5.5.54)$$

They are found to be given by

$$\alpha_{ij} = \frac{\kappa_i^{(1)}\kappa_j^{(1)}\,e^{-i\theta_1}}{\kappa'^{(1)}\Sigma^{(1)}} + \frac{\kappa_i^{(2)}\kappa_j^{(2)}\,e^{-i\theta_2}}{\kappa'^{(2)}\Sigma^{(2)}} + \ldots + \frac{\kappa_i^{(n)}\kappa_j^{(n)}\,e^{-i\theta_n}}{\kappa'^{(n)}\Sigma^{(n)}}, \quad (5.5.55)$$

which may be compared with equation (2.8.22).

In this section we have obtained two expressions for the complex receptance matrix of a damped system. The first of these gives $\boldsymbol{\alpha}$ as the inverse of the complex matrix **Z**. Although this result was obtained quite easily it should be pointed out that the inversion of a complex matrix is a very tedious business. The second expression for $\boldsymbol{\alpha}$ is that given in equation (5.5.53). This equation is the end product of a long chain of reasoning involving the modes of the system and the orthogonality principle. To obtain the series expansion it is necessary to find the n characteristic angles θ_r and the n forced modes $\boldsymbol{\kappa}^{(r)}$ and forced modes of excitation $\boldsymbol{\Sigma}^{(r)}$, and this is just about as difficult as inverting the matrix **Z**. In spite of this, the series form has several distinct advantages. First of all it permits a simpler discussion of the *nature* of damped forced vibration. Secondly, on account of the range of values taken by ω, it is sometimes possible to use just one or two of the terms in the series. If this is so then the receptance matrix may be calculated relatively easily as the individual terms in the series can be found independently of each other.

EXAMPLES 5.5

1. Solve Ex. 5.4.2 by direct inversion in the manner of equation (5.5.31) and so check the result found previously.

2.† A matrix **K** is defined as having the vectors $\boldsymbol{\kappa}^{(r)}$ as its columns. It may be used to specify a change of coordinates such that

$$\mathbf{q} = \mathbf{K}\boldsymbol{\nu},$$

where the column matrix $\boldsymbol{\nu}$ is that of 'damped principal coordinates'.

(*a*) Taking $\boldsymbol{\nu} = \operatorname{Im}(\mathbf{w})$, find the equation governing the complex vector **w**.

(*b*) Solve the equation, giving an expression for w_r, the rth element of **w**.

† Ex. 5.5.2 is intended for the reader who has read §4.2 of this book. For a full treatment of 'damped principal coordinates' (as they relate to hysteretic damping) see Bishop and Gladwell, *Phil. Trans.* A, **255** (1963), pp. 241–80.

CHAPTER 6

CONTINUOUS SYSTEMS

...To understand it, cling passionately to one another and think of faint lilies. *Patience*

The theory of vibration which is presented in the previous chapters relates to an idealised vibrating system having a finite number of degrees of freedom. This chapter is concerned mainly with continuous systems, that is, systems with an infinity of degrees of freedom. First it will be shown how receptances may be defined for continuous systems, then it will be shown how composite systems, whether continuous or discrete, may be analysed by the use of matrices. Finally, it will be shown how the theory developed for systems with a finite number of degrees of freedom may be applied to the vibration of continuous systems.

Throughout this chapter only undamped vibration will be discussed. But the reader will observe that much of the analysis may be generalised to cover damped vibrating systems. Throughout this chapter, therefore, all displacements and forces will be harmonic, and will have the same frequency and phase.

6.1 The receptances of continuous systems

The receptances of a discrete system—that is, a system with a finite number of degrees of freedom—were defined in § 2.2. Although they were first introduced as elements of the receptance matrix, their fundamental meaning is such that the receptance α_{xy} gives the displacement at the coordinate x due to a harmonic force at the coordinate y. Thus if a force $F \sin \omega t$ is applied at the coordinate y then the displacement at x is

$$q = \alpha_{xy} F \sin \omega t. \tag{6.1.1}$$

For a continuous system the concept of a receptance matrix cannot be used because, by definition, the system has an infinity of degrees of freedom. But a continuous system does have individual receptances which may be defined by equation (6.1.1). This may best be explained by means of an example.

The uniform shaft shown in fig. 6.1.1 is clamped at its left-hand end and free at the right-hand end. Let the torsional stiffness (moment per radian) per unit length be C, the moment of inertia of unit length be $\mathcal{J}$ and denote the angular displacement at time t of the element of the shaft distance x from O by $\theta(x, t)$. If the right-hand end of the shaft is excited by a torque $F \sin \omega t$ then it may be shown† that the equation governing $\theta(x, t)$ is

$$C \frac{\partial^2 \theta}{\partial x^2} = \mathcal{J} \frac{\partial^2 \theta}{\partial t^2}, \tag{6.1.2}$$

† Bishop and Johnson, *The Mechanics of Vibration*, p. 245. The equation is derived in § 6.5 also.

and that the conditions at the ends of the shaft are

$$\left.\begin{aligned} \theta = 0 \quad \text{at} \quad x = 0, \\ C\frac{\partial \theta}{\partial x} = F \sin \omega t \quad \text{at} \quad x = l. \end{aligned}\right\} \tag{6.1.3}$$

If the displacement is to be harmonic and have the same frequency and phase as the torque then it must be expressible in the form

$$\theta = X(x) \sin \omega t, \tag{6.1.4}$$

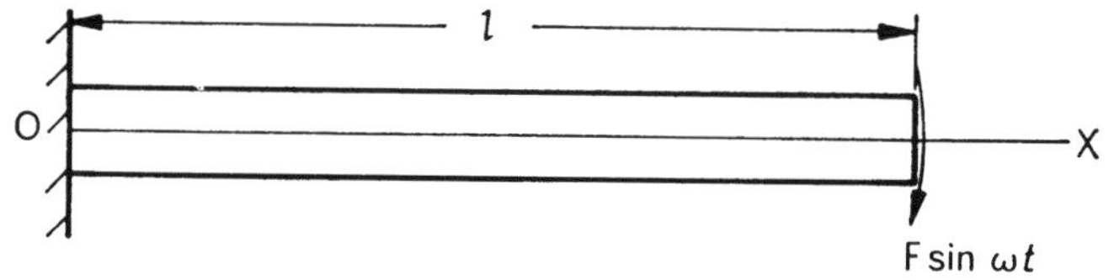

Fig. 6.1.1

where $X(x)$ is a function of x only. On substituting this expression into equation (6.1.2) we obtain

$$C\frac{d^2X}{dx^2} + \mathcal{J}\omega^2 X = 0, \tag{6.1.5}$$

whose general solution is

$$X(x) = A \cos \lambda x + B \sin \lambda x, \tag{6.1.6}$$

where A and B are arbitrary constants and the dimensionless quantity λl is given by

$$\lambda l = \omega \sqrt{\left(\frac{\mathcal{J} l^2}{C}\right)}. \tag{6.1.7}$$

This solution satisfies the end conditions provided that

$$A = 0, \quad B = \frac{F}{C\lambda \cos \lambda l}. \tag{6.1.8}$$

This means that the displacement of the shaft at the coordinate x is

$$\theta(x, t) = \frac{\sin \lambda x}{C\lambda \cos \lambda l} F \sin \omega t, \tag{6.1.9}$$

so that

$$\alpha_{xl} = \frac{\sin \lambda x}{C\lambda \cos \lambda l}. \tag{6.1.10}$$

If, instead of being fixed at the end $x = 0$, the shaft had been free, the end condition there would have been

$$\frac{\partial \theta}{\partial x} = 0 \quad \text{at} \quad x = 0, \tag{6.1.11}$$

and the constants A and B would have been

$$A = \frac{-F}{C\lambda \sin \lambda l}, \quad B = 0, \tag{6.1.12}$$

so that
$$\alpha_{xl} = \frac{-\cos \lambda x}{C\lambda \sin \lambda l}. \tag{6.1.13}$$

If the shaft is free at the end $x = l$ and acted on by a torque at $x = 0$ then the reader may verify that the relevant receptance, α_{x0}, is given by
$$\alpha_{x0} = \frac{-\cos \lambda(l-x)}{C\lambda \sin \lambda l}. \tag{6.1.14}$$

In § 6.4 we shall need to know the natural frequencies and principal modes of the shaft shown in fig. 6.1.1, and we shall therefore obtain them now.

By definition, the natural frequencies of the system are the frequencies at which it may vibrate freely, without the application of any torque. They are therefore the frequencies at which the shaft may vibrate, subject to the 'homogeneous' end conditions obtained by putting $F = 0$ in the conditions (6.1.3); that is to say
$$\left.\begin{aligned} \theta = 0 \quad &\text{at} \quad x = 0, \\ \frac{\partial \theta}{\partial x} = 0 \quad &\text{at} \quad x = l. \end{aligned}\right\} \tag{6.1.15}$$

The solution (6.1.6) satisfies these conditions provided that
$$A = 0, \quad \lambda B \cos \lambda l = 0. \tag{6.1.16}$$

If $B = 0$ there is no vibration at all, and therefore the natural frequencies are given by
$$\cos \lambda l = 0, \quad \text{or} \quad \lambda l = \frac{(2r-1)\,\pi}{2} \quad (r = 1, 2, \ldots), \tag{6.1.17}$$

so that
$$\omega_r = \sqrt{\left(\frac{C}{\mathcal{J} l^2}\right)}\,\frac{(2r-1)\,\pi}{2} \quad (r = 1, 2, \ldots). \tag{6.1.18}$$

The reader will notice that, just as it was found for discrete systems, the natural frequencies are the frequencies for which the receptances of the system become infinite.

The rth principal mode is, by definition, the shape which the shaft takes up when it vibrates freely with frequency ω_r. Equations (6.1.6), (6.1.16) and (6.1.17) show that this shape is given by the rth 'characteristic function'
$$\phi_r(x) = \sin \frac{(2r-1)\,\pi x}{2l}. \tag{6.1.19}$$

The receptance α_{xl} shown in equation (6.1.10) bears little resemblance to the receptance of a discrete system. It will now be shown that the receptance α_{xl}—and indeed any receptance α_{xy} of the shaft—may be expressed in a series form which is a direct generalisation of the expression which was found for receptances of discrete systems in § 2.8.

To obtain this expression we note that if the shaft is subjected to a distributed torque $f(x) \sin \omega t$ per unit length the equation of motion of the element of length δx at the section x is
$$\mathcal{J} \frac{\partial^2 \theta}{\partial t^2} \delta x = C \frac{\partial^2 \theta}{\partial x^2} \delta x + f(x)\, \delta x \sin \omega t. \tag{6.1.20}$$

The displacement of the shaft will still be of the form (6.1.4), so that X will now satisfy the equation

$$-J\omega^2 X - C\frac{d^2X}{dx^2} = f(x). \tag{6.1.21}$$

Suppose now that $f(x)$ may be expressed as an infinite series in the characteristic functions $\phi_r(x)$ given in equation (6.1.19), so that

$$f(x) = \sum_{r=1}^{\infty} P_r \phi_r(x). \tag{6.1.22}$$

Then equation (6.1.21) may be solved by assuming the trial solution

$$X = \sum_{r=1}^{\infty} p_r \phi_r(x). \tag{6.1.23}$$

Since $\phi_r(x)$ satisfies equation (6.1.5) with $\omega = \omega_r$, that is

$$C\frac{d^2\phi_r}{dx^2} + J\omega_r^2 \phi_r = 0, \tag{6.1.24}$$

we have

$$-C\frac{d^2X}{dx^2} = J\sum_{r=1}^{\infty} \omega_r^2 p_r \phi_r(x). \tag{6.1.25}$$

On substituting these expressions for X, d^2X/dx^2 and $f(x)$ into equation (6.1.21) and equating coefficients of $\phi_r(x)$ we obtain

$$J(\omega_r^2 - \omega^2)\, p_r = P_r. \tag{6.1.26}$$

This equation shows that the displacement due to any distribution of torque $f(x)$ may be found provided that the coefficients P_r are known.

The conditions which have to be satisfied by $f(x)$ in order that it may be expanded in the form (6.1.22), and the justification of the term-by-term differentiation employed in equation (6.1.25) are the subject-matter of Fourier analysis and will not be discussed here. It is sufficient for our purposes that predictions based on the use of the series expressions are found to be reliable.

If then it is assumed that the use of the series is legitimate then the coefficients may be found quite easily. For if equation (6.1.22) is multiplied by $\phi_s(x)$ and integrated from 0 to l, then

$$\int_0^l f(x)\,\phi_s(x)\,dx = \sum_{r=1}^{\infty} P_r \int_0^l \phi_r(x)\,\phi_s(x)\,dx. \tag{6.1.27}$$

But if $r \neq s$ then

$$\int_0^l \phi_r(x)\,\phi_s(x)\,dx = \int_0^l \sin\frac{(2r-1)\,\pi x}{2l}\sin\frac{(2s-1)\,\pi x}{2l}\,dx$$

$$= \frac{1}{2}\int_0^l \left\{\cos\frac{(r-s)\,\pi x}{l} - \cos\frac{(r+s-l)\,\pi x}{l}\right\} dx = 0, \tag{6.1.28}$$

and if $r = s$
$$\int_0^l \phi_r(x)\,\phi_s(x)\,dx = \int_0^l \sin^2\frac{(2s-1)\,\pi x}{2l}\,dx$$
$$= \frac{1}{2}\int_0^l \left\{1-\cos\frac{(2s-1)\,\pi x}{l}\right\}dx = \frac{l}{2}. \tag{6.1.29}$$

Substituting these results into equation (6.1.27), we obtain

$$P_s = \frac{2}{l}\int_0^l f(x)\,\phi_s(x)\,dx. \tag{6.1.30}$$

The last equation gives the coefficients P_s for a general distribution of torque. To find the receptance α_{xy} we need to find the coefficients corresponding to a concentrated torque $F\sin\omega t$ at the section y. Such a torque may be considered to be the limit of the distributed torque

$$f(x) = \begin{cases} F/h & y-h \leqslant x \leqslant y, \\ 0 & \text{otherwise,} \end{cases} \tag{6.1.31}$$

and therefore
$$P_s = \frac{2}{l}\lim_{h\to 0}\int_{y-h}^{y}\frac{F}{h}\phi_s(x)\,dx = \frac{2}{l}F\,\phi_s(y). \tag{6.1.32}$$

If we now substitute this result in equation (6.1.26) then equation (6.1.23) gives

$$X = \frac{2}{Jl}\sum_{r=1}^{\infty}\frac{\phi_r(x)\,\phi_r(y)}{\omega_r^2-\omega^2}F \tag{6.1.33}$$

or, in other words,
$$\alpha_{xy} = \frac{2}{Jl}\sum_{r=1}^{\infty}\frac{\phi_r(x)\,\phi_r(y)}{\omega_r^2-\omega^2}. \tag{6.1.34}\dagger$$

This expression may be compared with that given in equation (2.8.22).

Equation (6.1.34) shows that the receptance α_{xy} is symmetrical, that is

$$\alpha_{xy} = \alpha_{yx}. \tag{6.1.35}$$

It may be proved that, provided there are no gyroscopic or aerodynamic forces acting on them, the receptances of all vibrating systems, whether discrete or continuous, satisfy this reciprocal relation.

It has been shown that the modes of the uniform clamped-free shaft in torsion satisfy the orthogonality condition (6.1.28). In order to see what relation this condition bears to the orthogonality conditions which were found to hold for discrete systems we consider a slightly more general problem. Consider a shaft whose moment of inertia per unit length, J, and stiffness per unit length, C, are variable functions of x. The equation satisfied by the characteristic function $\phi_r(x)$ will be

$$\frac{d}{dx}\left(C(x)\frac{d\phi_r}{dx}\right) + J(x)\,\omega_r^2\,\phi_r(x) = 0, \tag{6.1.36}\ddagger$$

† It is found that the series for $f(x)$ corresponding to a point load, namely

$$(2F/l)\sum_{r=1}^{\infty}\phi_r(x)\,\phi_r(y),$$

does not converge, but that the series for α_{xy} does, and agrees with the expression for α_{xy} in closed form. See also Ex. 6.3.2.

‡ Strictly speaking, this equation is an approximate one.

and the conditions to be satisfied will be

$$\left.\begin{aligned} \phi_r &= 0 \quad \text{at} \quad x = 0, \\ \frac{d\phi_r}{dx} &= 0 \quad \text{at} \quad x = l. \end{aligned}\right\} \tag{6.1.37}$$

Therefore

$$(\omega_r^2 - \omega_s^2) \int_0^l \mathcal{J}(x)\, \phi_r(x)\, \phi s(x)\, dx = \int_0^l \left\{ \phi_r(x) \frac{d}{dx}\left(C(x) \frac{d\phi_s}{dx} \right) - \phi_s(x) \frac{d}{dx}\left(C(x) \frac{d\phi_r}{dx} \right) \right\} dx, \tag{6.1.38}$$

which when integrated by parts becomes

$$\left[C(x) \left\{ (\phi_r x) \frac{d\phi_s}{dx} - \phi_s(x) \frac{d\phi_r}{dx} \right\} \right]_0^l - \int_0^l C(x) \left\{ \frac{d\phi_r}{dx} \frac{d\phi_s}{dx} - \frac{d\phi_s}{dx} \frac{d\phi_r}{dx} \right\} dx, \tag{6.1.39}$$

and this is zero because both $\phi_r(x)$ and $\phi_s(x)$ satisfy the end conditions (6.1.37). Therefore, provided $\omega_r \neq \omega_s$, and this can be shown to hold provided $r \neq s$, we have

$$\int_0^l \mathcal{J}(x)\, \phi_r(x)\, \phi_s(x)\, dx = 0 \quad (r \neq s). \tag{6.1.40}$$

Moreover,

$$\begin{aligned} \int_0^l C(x) \frac{d\phi_r}{dx} \frac{d\phi_s}{dx} dx &= \left[C(x) \frac{d\phi_r}{dx} \phi_s(x) \right]_0^l - \int_0^l \frac{d}{dx}\left(C(x) \frac{d\phi_r}{dx} \right) \phi_s(x)\, dx \\ &= \omega_r^2 \int_0^l \mathcal{J}(x)\, \phi_r(x)\, \phi_s(x)\, dx \end{aligned} \tag{6.1.41}$$

by equation (6.1.36), so that

$$\int_0^l C(x) \frac{d\phi_r}{dx} \frac{d\phi_s}{dx} dx = 0 \quad (r \neq s), \tag{6.1.42}$$

and

$$\int_0^l C(x) \left(\frac{d\phi_r}{dx} \right)^2 dx = \omega_r^2 \int_0^l \mathcal{J}(x)\, [\phi_r(x)]^2\, dx. \tag{6.1.43}$$

Equations (6.1.40) and (6.1.42) are the counterparts for the shaft of the orthogonality relations (2.5.7) and (2.5.8) respectively.

The reader will notice that equations (6.1.40) and (6.1.42) hold for any shaft in torsion—not just the clamped-free shaft—provided that the end conditions at each end are

$$\text{either} \quad \theta = 0 \quad \text{or} \quad \partial\theta/\partial x = 0. \tag{6.1.44}$$

It may be shown that for any such shaft the receptance α_{xy} is given by

$$\alpha_{xy} = \sum_{r=1}^{\infty} \frac{\phi_r(x)\, \phi_r(y)}{a_r(\omega_r^2 - \omega^2)}, \tag{6.1.45}$$

where the $\phi_r(x)$ are the characteristic functions for the shaft and the constants a_r are given by

$$a_r = \int_0^l \mathcal{J}(x)\, [\phi_r(x)]^2\, dx. \tag{6.1.46}$$

Equations (6.1.40) and (6.1.42) are typical examples of orthogonal relations, and equation (6.1.45) is a typical example of a receptance for a continuous vibrating system.

We note in passing that the following two problems have mathematical formulations which are identical to that of the shaft undergoing torsional vibration (problem (i)), provided that the symbols which occur in the problems are suitably interpreted:

(ii) the transverse vibration of a uniform string in tension;

(iii) the longitudinal vibration of a uniform elastic bar.

Table 6.1.1 shows the scheme of re-interpretation of the symbols which occur in the problems. The condition at either end of the body is either 'end fixed' (θ, v or u zero) or 'end free' (zero torque, lateral force or longitudinal force).

TABLE 6.1.1

	Torsional vibrations of uniform shaft	Transverse vibrations of uniform string in tension	Longitudinal vibrations of uniform bar
Corresponding basic quantities	Angular displacement, θ	Lateral displacement, v	Longitudinal displacement, u
	Moment of inertia of unit length, J	Mass of unit length, μ	Mass of unit length, $A\rho$, with A = area of cross-section ρ = mass density
	Torsional stiffness (moment per radian) of unit length, C	Tension in string, S	Force per unit extensional strain, AE, with A = area of cross-section E = Young's modulus
Dimensionless frequency parameter	$\lambda l = \sqrt{\left(\frac{Jl^2}{C}\right)}\,\omega$	$\lambda l = \sqrt{\left(\frac{\mu l^2}{S}\right)}\,\omega$	$\lambda l = \sqrt{\left(\frac{\rho l^2}{E}\right)}\,\omega$

As a further example of a continuous vibrating system, consider a beam undergoing flexural vibration. If the beam vibrates in one plane, the horizontal plane, say, then the motion of any section of the beam is specified by two coordinates, not just one as in the problems which have been considered above. The two coordinates are the deflection $v(x, t)$ of the section and the slope $\partial v/\partial x$ of the beam at that section. In addition, the beam may be excited in two ways, by a harmonic transverse force or a harmonic bending couple. This means that for the vibration of beams the concept of a receptance takes on a wider meaning; it may link a deflection with a force or couple or may link a slope with a force or couple. This state of affairs is, of course, covered by our original conception of a receptance as a quantity linking *generalised* displacements and forces.

Consider the uniform cantilever beam shown in fig. 6.1.2. If A is the area of the cross-section, ρ the mass-density, E Young's modulus, and I the second moment of area of the cross-section of the beam about the neutral axis through its centroid, then it may be shown that the differential equation governing v is

$$\frac{\partial^2 v}{\partial t^2}+\frac{EI}{A\rho}\frac{\partial^4 v}{\partial x^4}=0. \qquad (6.1.47)\dagger$$

† Bishop and Johnson, *The Mechanics of Vibration* (1960), p. 283. See also Ex. 6.5.1.

If the beam is subject to a harmonic force $F \sin \omega t$ at the end $x = l$ then it may be shown that the end conditions are

$$\left.\begin{array}{c} v = 0 = \dfrac{\partial v}{\partial x} \quad \text{at} \quad x = 0, \\ \dfrac{\partial^2 v}{\partial x^2} = 0, \quad EI\dfrac{\partial^3 v}{\partial x^3} = -F \sin \omega t \quad \text{at} \quad x = l. \end{array}\right\} \tag{6.1.48}$$

If the deflection is to be harmonic and in phase with the force then

$$v(x, t) = X(x) \sin \omega t, \tag{6.1.49}$$

and substituting this into equation (6.1.47), we obtain

$$\frac{d^4 X}{dx^4} - \frac{A\rho\omega^2}{EI} X = 0. \tag{6.1.50}$$

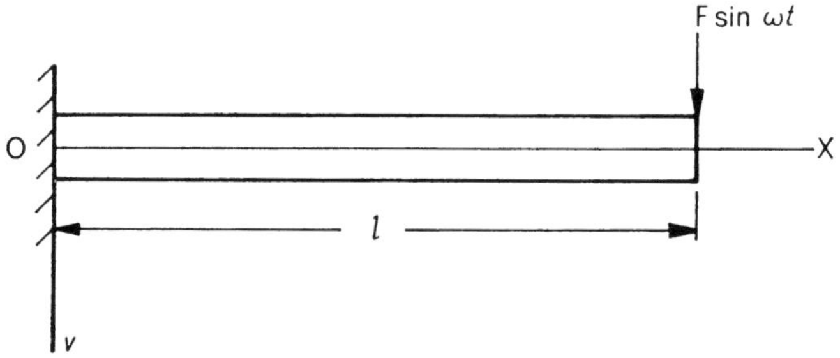

Fig. 6.1.2

The general form of X is therefore

$$X = A \cos \lambda x + B \sin \lambda x + C \cosh \lambda x + D \sinh \lambda x, \tag{6.1.51}$$

where the dimensionless frequency parameter λl is given by

$$\lambda l = \left[\frac{A\rho l^4 \omega^2}{EI}\right]^{\frac{1}{4}}, \tag{6.1.52}$$

and A, B, C and D are arbitrary constants. The solution satisfies the end conditions provided that

$$\left.\begin{array}{c} A = -C = \dfrac{-F(\sin \lambda l + \sinh \lambda l)}{2EI\lambda^3(1 + \cos \lambda l \cosh \lambda l)}, \\ B = -D = \dfrac{F(\cos \lambda l + \cosh \lambda l)}{2EI\lambda^3(1 + \cos \lambda l \cosh \lambda l)}. \end{array}\right\} \tag{6.1.53}$$

These values may be substituted into equation (6.1.51) to give

$$v(x, t) = \frac{-\{(\sin \lambda l + \sinh \lambda l)(\cos \lambda x - \cosh \lambda x) - (\cos \lambda l + \cosh \lambda l)(\sin \lambda x - \sinh \lambda x)\} F \sin \omega t}{2EI\lambda^3(1 + \cos \lambda l \cosh \lambda l)}. \tag{6.1.54}$$

If the receptance linking v and F is denoted by α_{xl}, and that linking $\partial v/\partial x$ with F by $\alpha_{x'l}$, then equation (6.1.54) shows that

$$\left.\begin{aligned}\alpha_{xl} &= \frac{-\{(\sin\lambda l+\sinh\lambda l)(\cos\lambda x-\cosh\lambda x)-(\cos\lambda l+\cosh\lambda l)(\sin\lambda x-\sinh\lambda x)\}}{2EI\lambda^3(1+\cos\lambda l\cosh\lambda l)},\\ \alpha_{x'l} &= \frac{(\sin\lambda l+\sinh\lambda l)(\sin\lambda x+\sinh\lambda x)+(\cos\lambda l+\cosh\lambda l)(\cos\lambda x-\cosh\lambda x)}{2EI\lambda^2(1+\cos\lambda l\cosh\lambda l)}.\end{aligned}\right\} \tag{6.1.55}$$

By considering the vibration due to a harmonic bending couple $H\sin\omega t$ at $x = l$ one may find in a similar way the receptances, which may be denoted by $\alpha_{xl'}$ and $\alpha_{x'l'}$, respectively, linking the deflection and slope at section x with a couple at $x = l$.

Our purpose in this section has been twofold—to show that the concept of a receptance occurs naturally in the vibration of continuous systems, and to compare the receptances of continuous systems with those of discrete. It is not our purpose to discuss any further how the receptances of continuous systems may be found. Bishop and Johnson† have derived and tabulated the receptances (in both closed and series forms) corresponding to all the various end conditions of all the systems which have been discussed in this section.

EXAMPLES 6.1

1. Find the natural frequencies and principal modes of torsional oscillation of a uniform shaft free at both ends. Normalise the modes so that equation (6.1.29) is satisfied, even for the rigid body mode, and express the receptance α_{xy} as a series of the form (6.1.45).‡

2. A clamped-free uniform shaft is subjected to the triangular distribution of torque

$$f(\xi) = \begin{cases} F(\xi-y+h)/h^2 & y-h \leqslant \xi \leqslant y,\\ F(-\xi+y+h)/h^2 & y \leqslant \xi \leqslant y+h,\\ 0 & \text{otherwise,}\end{cases}$$

Find the total torque applied to the shaft and the coefficients P_s using equation (6.1.30) and show that the series for $f(\xi)$ converges. Show further that, as $h\to 0$, the expression for the deflection at the section x tends to that given by equation (6.1.34). [This latter result justifies the use of the series expression for α_{xy}.]

3. Find the frequency equation for transverse vibration of a uniform beam free at both ends, and find an approximate value for the lowest (non-zero) natural frequency. Do the same for a clamped-clamped beam.

4. Find the principal modes of the uniform free-free beam and normalise them so that

$$\int_0^l \phi_r(x)\,\phi_s(x)\,dx = \begin{cases} 0 & r \neq s,\\ l & r = s.\end{cases}$$

† *The Mechanics of Vibration*, chs. 4, 6, 7. The notation used there for torsional vibration is a little different; our J and C are replaced by $J\rho$ and JG respectively.

‡ The series expressions for the receptances of certain continuous free-free systems given by Bishop and Johnson in *The Mechanics of Vibration* are incomplete in that the rigid-body terms have been omitted. This remark applies to a taut string, a shaft executing torsional and longitudinal vibrations, and a beam in transverse vibration.

Show that, in addition, the beam has two independent rigid body modes and choose two so that the above equation is satisfied by *any* two modes of the system.

5. Using the results of the preceding example, express the receptance α_{xy} of a uniform free-free beam as an infinite series.†

Find the end receptances α_{00}, α_{0l} and α_{ll} of a uniform free-free beam in closed form. Check that for small values of ω the result agrees with that given by the series in Ex. 6.1.5.

6.2 Composite systems

In this section it will be shown how a composite system consisting of a number of continuous or discrete systems linked together may be analysed when the receptances of its component parts are known. This analysis is important because it means that provided the receptances of certain simple systems—such as those mentioned in the previous section—are known (and many have in fact been tabulated) a great number of more complicated systems can be analysed.

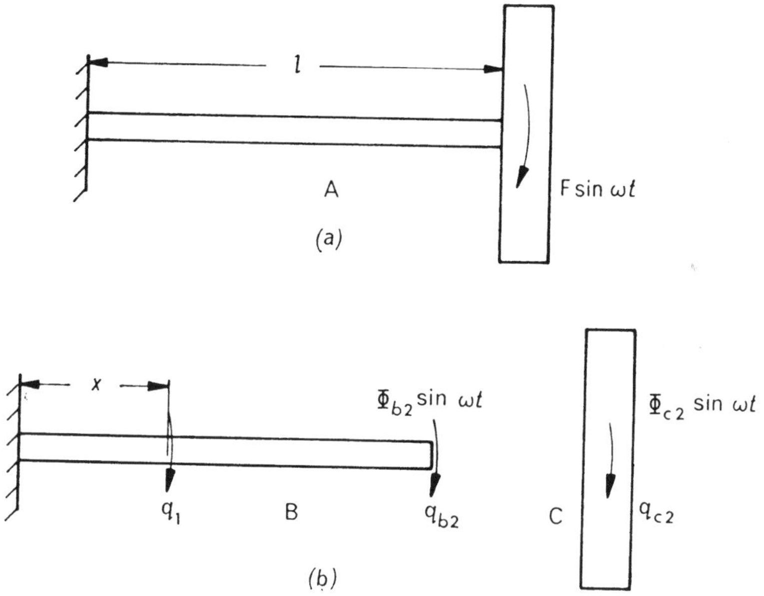

Fig. 6.2.1

Consider the vibrating system shown in fig. 6.2.1 (*a*). It consists of the uniform shaft discussed in section 6.1 with a flywheel of moment of inertia I attached at the free end. We shall now show how the motion of the system—and in particular the displacement at the section x due to a torque $F \sin \omega t$ applied at the end—can be deduced from the receptances of the shaft which were found in the last section.

If a harmonic torque $F \sin \omega t$ is applied at the flywheel the whole system will execute harmonic vibration in phase with the torque. The flywheel and the shaft will be acted on by certain torques whose magnitudes may be found by using the

† See footnote ‡ on p. 184 opposite.

fact that the total torque on the system is $F \sin \omega t$. This condition may be expressed by saying that the torques will be in equilibrium with a torque $-F \sin \omega t$ and is therefore called the 'equilibrium condition'. The motion of the end of the shaft may be found by using the receptance α_{ll} given in equation (6.1.10)—with $x = l$. The motion of the flywheel may be found by writing down the equation of motion for the flywheel. The vibration of the system may then be expressed in terms of F by using the fact that the motion of the end of the shaft is the same as that of the flywheel; this is the 'compatibility condition' for the system.

In carrying out the analysis which is described above we shall use a notation which is applicable to the analysis of any composite system. Denote the 'subsystems' consisting of the shaft and the flywheel by B and C respectively (fig. 6.2.1 (b)) and denote the composite system of the shaft-plus-flywheel by A. Let the angular displacements of the shaft at the section x and at the end be q_1 and q_{b2} respectively, and let the displacement of the flywheel be q_{c2}. Suppose that the amplitudes of the forces applied to the shaft and the flywheel are Φ_{b2} and Φ_{c2} respectively. Then the compatibility condition is

$$q_{b2} = q_{c2} = q_2, \tag{6.2.1}$$

and the equilibrium condition is

$$\Phi_{b2} + \Phi_{c2} = F. \tag{6.2.2}$$

The receptances α_{xl} and α_{ll} give the displacements q_1 and q_{b2} respectively due to excitation at the coordinate q_{b2}; call them β_{12} and β_{22}. Let the single receptance of the flywheel be γ_{22}. The convention is that subsystems $B, C, \ldots$, with receptances $\beta, \gamma, \ldots$, combine to form a system A whose receptances are denoted by α. The receptances are given by equation (6.1.10), so that

$$\beta_{12} = \frac{\sin \lambda x}{C\lambda \cos \lambda l}, \quad \beta_{22} = \frac{\tan \lambda l}{C\lambda}. \tag{6.2.3}$$

The equation of motion of the flywheel is

$$I\ddot{q}_{c2} = \Phi_{c2} \sin \omega t, \tag{6.2.4}$$

which gives, for steady oscillation with the driving frequency,

$$q_{c2} = -\frac{1}{I\omega^2} \Phi_{c2} \sin \omega t, \tag{6.2.5}$$

and therefore

$$\gamma_{22} = -\frac{1}{I\omega^2}. \tag{6.2.6}$$

These are the only receptances that are required because

$$q_1 = \beta_{12} \Phi_{b2} \sin \omega t, \tag{6.2.7}$$

$$q_{b2} = \beta_{22} \Phi_{b2} \sin \omega t, \tag{6.2.8}$$

$$q_{c2} = \gamma_{22} \Phi_{c2} \sin \omega t, \tag{6.2.9}$$

and when used in conjunction with equations (6.2.1) and (6.2.2) these equations give

$$\Phi_{b2} = \frac{\gamma_{22}}{\beta_{22} + \gamma_{22}} F \tag{6.2.10}$$

and
$$q_1 = \frac{\beta_{12}\gamma_{22}}{\beta_{22}+\gamma_{22}} F\sin\omega t; \quad q_2 = \frac{\beta_{22}\gamma_{22}}{\beta_{22}+\gamma_{22}} F\sin\omega t. \tag{6.2.11}$$

These equations express the displacements q_1 and q_2 in terms of the torque $F\sin\omega t$, and the multiplying factors are therefore receptances. They are the receptances α_{12} and α_{22} of the composite system. That is

$$\alpha_{12} = \frac{\beta_{12}\gamma_{22}}{\beta_{22}+\gamma_{22}}, \quad \alpha_{22} = \frac{\beta_{22}\gamma_{22}}{\beta_{22}+\gamma_{22}}, \tag{6.2.12}$$

and these equations express the receptances of the composite system in terms of those of the shaft and flywheel separately.

The frequency equation of the composite system may be obtained by noticing that if the system executes free vibration (with $F = 0$) then equations (6.2.8) and (6.2.9) combine with the conditions of compatibility and equilibrium to give

$$(\beta_{22}+\gamma_{22})\,\Phi_{b2}\sin\omega t = 0. \tag{6.2.13}$$

This equation shows that unless
$$\beta_{22}+\gamma_{22} = 0, \tag{6.2.14}$$
the force amplitude Φ_{b2}, and hence Φ_{c2}, q_1 and q_2, will all be zero. The last equation is therefore the frequency equation. Equation (6.2.12) shows that it may also be obtained as the equation giving the values of λ (and hence ω) for which the receptances become infinite. Substituting for β_{22} and γ_{22} their values given in equations (6.2.3) and (6.2.6), we obtain the frequency equation

$$\frac{\tan\lambda l}{C\lambda} - \frac{1}{I\omega^2} = 0, \tag{6.2.15}$$

and if ω is expressed in terms of λ by using equation (6.1.7), this becomes

$$\cot\lambda l = \frac{I}{\mathcal{J}l}\lambda l. \tag{6.2.16}$$

Composite systems may be classified according to (*a*) the number of component subsystems they possess, and (*b*) the manner in which these subsystems are linked together. Two quite different systems may in fact be split up into the same number of subsystems linked together in the same way. If two such systems are analysed in the manner described above for the shaft and flywheel, then the results will be identical in form in the two cases. This may be illustrated by means of the following simple example.

The vibrating system of fig. 6.2.2 (*a*) consists of a light spring of stiffness k to which a mass M is attached. The problem is to find the displacements of the mid-point of the spring and of the mass, due to a force $F\sin\omega t$ applied to the mass. The composite system may be divided up as shown in fig. 6.2.2 (*b*); it consists of two parts linked together by a single coordinate just as the shaft and flywheel were. The reader may verify that if the proper meanings are given to the symbols q_1, q_{b2}, Φ_{b2}, etc., the analysis is just the same as before—the compatibility equation is equation (6.2.1) and the equilibrium condition is equation (6.2.2), and so on.

Therefore, as before, the receptances α_{12} and α_{22} of the composite system are given by equations (6.2.12), and the frequency equation is (6.2.14). It is in this sense that the problem is comparable with that of the shaft and flywheel. The reader may verify that, for the problem of fig. 6.2.2,

$$\beta_{12} = \frac{1}{2k}, \quad \beta_{22} = \frac{1}{k}, \quad \gamma_{22} = \frac{-1}{M\omega^2}, \tag{6.2.17}$$

which give
$$\alpha_{12} = \frac{1}{2(k-M\omega^2)}, \quad \alpha_{22} = \frac{1}{k-M\omega^2}, \tag{6.2.18}$$

and the frequency equation
$$k - M\omega^2 = 0. \tag{6.2.19}$$

These results could, of course, have been obtained by the methods of Chapter 2 which would here have been simpler.

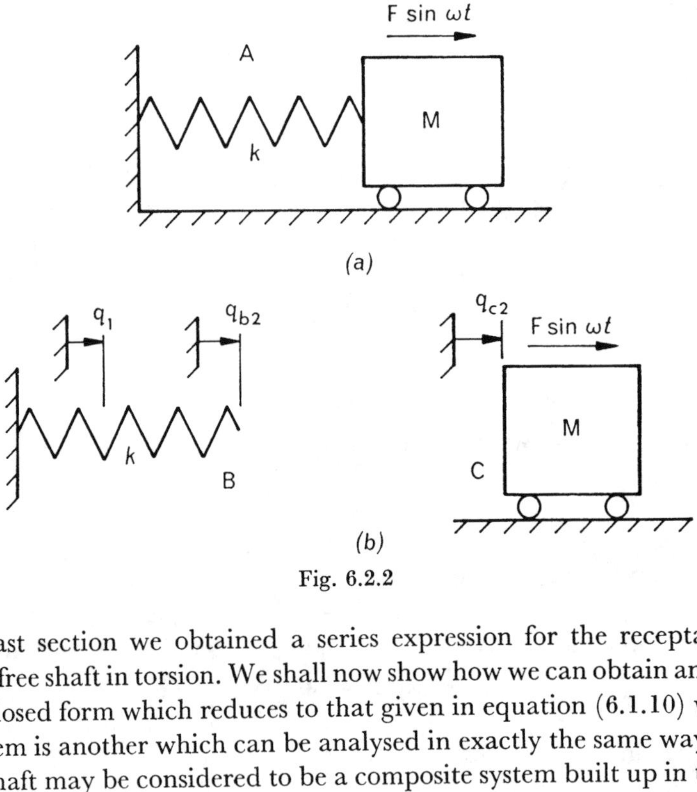

Fig. 6.2.2

In the last section we obtained a series expression for the receptance α_{xy} of a clamped-free shaft in torsion. We shall now show how we can obtain an expression for α_{xy} in closed form which reduces to that given in equation (6.1.10) when $y = l$. This problem is another which can be analysed in exactly the same way as the last two. The shaft may be considered to be a composite system built up in the manner indicated in fig. 6.2.3 (*b*). The subsystems *B* and *C* are now a clamped-free shaft of length y and a free-free shaft of length $l-y$. If $x < y$ the receptance α_{xy} of the whole shaft may be obtained by finding the receptance α_{12} linking a coordinate q_1 of *B* to excitation at q_2. If $x > y$ the receptance may be obtained by finding the receptance α_{32} where q_3 denotes the deflection at a section x $(x > y)$ of *C*.

For this problem the receptances β may be obtained by putting $l = y$ in equation (6.1.10) so that

$$\beta_{12} = \frac{\sin \lambda x}{C\lambda \cos \lambda y}, \quad \beta_{22} = \frac{\sin \lambda y}{C\lambda \cos \lambda y}, \tag{6.2.20}$$

and the γ receptances may be obtained by putting $l-y$ instead of l and $x-y$ instead of x in equation (6.1.14), so that

$$\left.\begin{aligned}\gamma_{22} &= \frac{-\cos\lambda(l-y)}{C\lambda\sin\lambda(l-y)},\\ \gamma_{32} &= \frac{-\cos\lambda[(l-y)-(x-y)]}{C\lambda\sin\lambda(l-y)} = \frac{-\cos\lambda(l-x)}{C\lambda\sin(l-y)}.\end{aligned}\right\} \tag{6.2.21}$$

If we substitute these values into equations (6.2.12) we obtain

$$\alpha_{12} = \frac{\sin\lambda x\cos\lambda(l-y)}{C\lambda\cos\lambda l}, \quad \alpha_{22} = \frac{\sin\lambda y\cos\lambda(l-y)}{C\lambda\cos\lambda l}. \tag{6.2.22}$$

To obtain α_{32} we notice that, by analogy with equation (6.2.10),

$$q_3 = \gamma_{32}\Phi_{c2}\sin\omega t = \frac{\gamma_{32}\beta_{22}}{\beta_{22}+\gamma_{22}}F\sin\omega t, \tag{6.2.23}$$

so that

$$\alpha_{32} = \frac{\gamma_{32}\beta_{22}}{\beta_{22}+\gamma_{22}}. \tag{6.2.24}$$

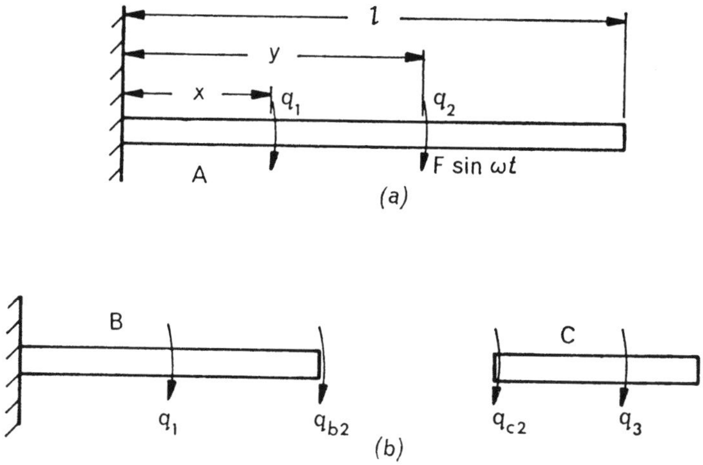

Fig. 6.2.3

When the values of β_{22}, γ_{22} and γ_{32} are substituted into this, it becomes

$$\alpha_{32} = \frac{\sin\lambda y\cos\lambda(l-x)}{C\lambda\cos\lambda l}. \tag{6.2.25}$$

Equations (6.2.22) and (6.2.25) show therefore that

$$\alpha_{xy} = \left\{\begin{aligned}&\frac{\sin\lambda x\cos\lambda(l-y)}{C\lambda\cos\lambda l} \quad \text{if} \quad 0 \leqslant x \leqslant y,\\ &\frac{\sin\lambda y\cos\lambda(l-x)}{C\lambda\cos\lambda l} \quad \text{if} \quad y \leqslant x \leqslant l.\end{aligned}\right\} \tag{6.2.26}$$

Similar expressions may be found in tables for the receptances of a shaft in torsion under other end conditions, and for various other types of system.†

In this section we have analysed three composite systems which superficially are very different but which are all equivalent for the purposes of the analysis which has been described. The three systems may be considered as examples of the abstract 'box' type of composite system shown in fig. 6.2.4.

For the three systems that have been discussed and for any others which are examples of this abstract composite system, the receptances α_{12}, α_{22} and α_{32} are given by equations (6.2.12) and (6.2.24), and the frequency equation is equation (6.2.14).

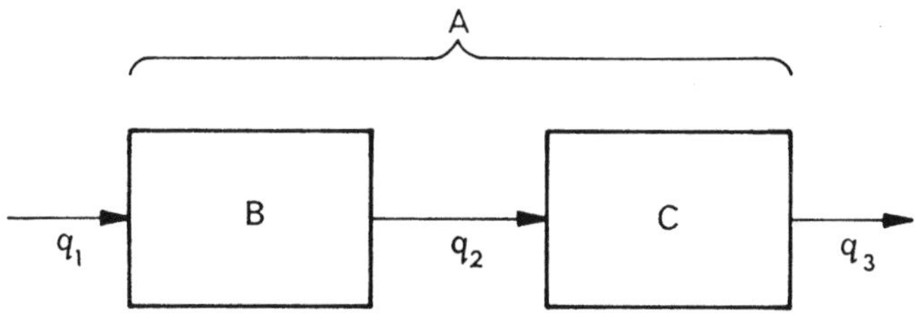

Fig. 6.2.4

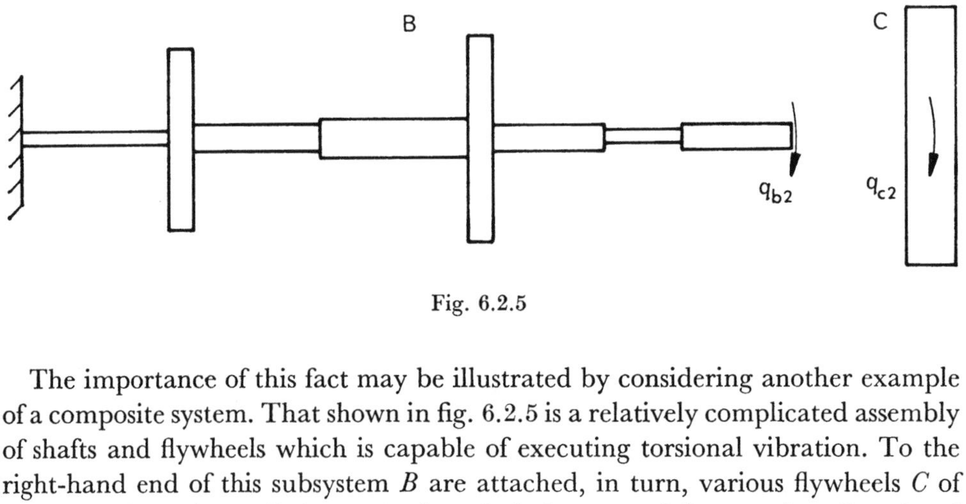

Fig. 6.2.5

The importance of this fact may be illustrated by considering another example of a composite system. That shown in fig. 6.2.5 is a relatively complicated assembly of shafts and flywheels which is capable of executing torsional vibration. To the right-hand end of this subsystem B are attached, in turn, various flywheels C of differing dimensions. It is required to find the natural frequencies of torsional vibration of the whole system when each of the flywheels is attached. A complete analysis of each system in turn would be an extremely long process, and the analysis presented in this section makes this unnecessary. Each of the composite systems formed by attaching one of the flywheels to the subsystem B is an example of the box type of composite system shown in fig. 6.2.4. If the receptance β_{22} is calculated for a range of values of ω, then the natural frequencies of each system can be found

† Bishop and Johnson, *The Mechanics of Vibration*. The receptances α_{xy}, $\alpha_{x'y'}$, etc., of beams in flexural vibration are given by Gladwell and Bishop, 'Interior receptances of beams', *J. Mech. Engng Sci.* **2**, no. 1 (1960), pp. 1–15.

by plotting β_{22} against $-\gamma_{22}$ as shown in fig. 6.2.6.† A convenient way of carrying out the calculations involved in finding the end receptance β_{22} of a system of the type shown in fig. 6.2.5 will be described in the following section.

Before leaving the system shown in fig. 6.2.4, let us obtain the receptances α_{11} and α_{31}. To obtain these we notice that if a force $F_1 \sin \omega t$ is applied at q_1 and no force is applied at q_2 then

$$\left.\begin{aligned} q_1 &= (\beta_{11}F_1 + \beta_{12}\Phi_{b2}) \sin \omega t, \\ q_{b2} &= (\beta_{21}F_1 + \beta_{22}\Phi_{b2}) \sin \omega t, \\ q_{c2} &= \gamma_{22}\Phi_{c2} \sin \omega t, \\ q_3 &= \gamma_{32}\Phi_{c2} \sin \omega t. \end{aligned}\right\} \tag{6.2.27}$$

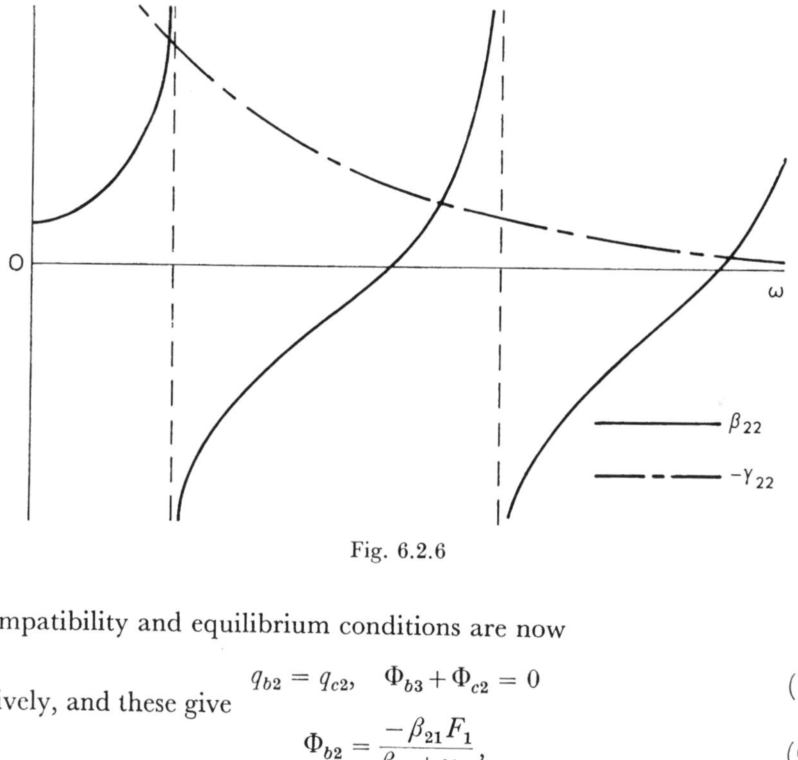

Fig. 6.2.6

The compatibility and equilibrium conditions are now

$$q_{b2} = q_{c2}, \quad \Phi_{b3} + \Phi_{c2} = 0 \tag{6.2.28}$$

respectively, and these give

$$\Phi_{b2} = \frac{-\beta_{21}F_1}{\beta_{22}+\gamma_{22}}, \tag{6.2.29}$$

$$\left.\begin{aligned} q_1 &= \left(\beta_{11} - \frac{\beta_{12}^2}{\beta_{22}+\gamma_{22}}\right) F_1 \sin \omega t, \\ q_3 &= \frac{\gamma_{32}\beta_{21}}{\beta_{22}+\gamma_{22}} F_1 \sin \omega t, \end{aligned}\right\} \tag{6.2.30}$$

so that

$$\alpha_{11} = \beta_{11} - \frac{\beta_{12}^2}{\beta_{22}+\gamma_{22}}, \quad \alpha_{31} = \frac{\gamma_{32}\beta_{21}}{\beta_{22}+\gamma_{22}}. \tag{6.2.31}$$

† The convention used in fig. 6.2.6 differs from that used in Bishop and Johnson, *The Mechanics of Vibration*, in that the axes are interchanged. That the receptance curves must have the general shapes shown is demonstrated in §3.3 of that book.

In arriving at these relations it has been assumed that $\beta_{12} = \beta_{21}$ which, as we showed in § 2.2, must be the case for the passive systems to which all our discussions have been limited.

The reader will observe that two systems may have different numbers of degrees of freedom and yet may be equivalent for the analysis which has been described. In fact, of the systems that have been examined in detail, those of figs. 6.2.1 and 6.2.3 have infinite freedom while that of fig. 6.2.2 has only one degree of freedom. The reader will observe also that a system can often be divided into subsystems in a number of ways. This is evident from our treatment of the system shown in fig. 6.2.3.

Until now, all the systems which have been considered in this section have consisted of a number of parts linked at single coordinates. A number of systems having two components linked by two coordinates are shown in fig. 6.2.7. For the stepped cantilever in transverse vibration shown in fig. 6.2.7 (*a*) the coordinates linking the two uniform portions are the deflection and the slope at the join. Fig. 6.2.7 (*b*) shows a beam (which may or may not be flexible) performing small transverse vibrations while supported on two light springs; in this case the system could be split into two subsystems B and C as shown. The continuous beam in fig. 6.2.7 (*c*) can be split, for the purposes of analysis, into a pinned-pinned beam B and two internal supports C. These three composite systems all conform to the general 'box' arrangement of fig. 6.2.7 (*d*).

Suppose that the frequency equations of the systems shown in fig. 6.2.7 are required. We may proceed in the following way. The equations of motion of B due to the excitation at q_1 and q_2 are

$$\left.\begin{aligned} q_{b1} &= (\beta_{11}\Phi_{b1}+\beta_{12}\Phi_{b2})\sin\omega t,\\ q_{b2} &= (\beta_{21}\Phi_{b1}+\beta_{22}\Phi_{b2})\sin\omega t.\end{aligned}\right\} \tag{6.2.32}$$

Similarly, the equations of motion of C are

$$\left.\begin{aligned} q_{c1} &= (\gamma_{11}\Phi_{c1}+\gamma_{12}\Phi_{c2})\sin\omega t,\\ q_{c2} &= (\gamma_{21}\Phi_{c1}+\gamma_{22}\Phi_{c2})\sin\omega t.\end{aligned}\right\} \tag{6.2.33}$$

The compatibility and equilibrium conditions are

$$q_{b1} = q_{c1}, \quad q_{b2} = q_{c2} \tag{6.2.34}$$

and
$$\Phi_{b1}+\Phi_{c1} = 0 = \Phi_{b2}+\Phi_{c2} \tag{6.2.35}$$

respectively. When combined with equations (6.2.32) and (6.2.33) these give

$$\left.\begin{aligned} (\beta_{11}+\gamma_{11})\,\Phi_{b1}+(\beta_{12}+\gamma_{12})\,\Phi_{b2} &= 0,\\ (\beta_{21}+\gamma_{21})\,\Phi_{b1}+(\beta_{22}+\gamma_{22})\,\Phi_{b2} &= 0.\end{aligned}\right\} \tag{6.2.36}$$

These equations show that unless

$$(\beta_{11}+\gamma_{11})(\beta_{22}+\gamma_{22})-(\beta_{12}+\gamma_{12})^2 = 0, \tag{6.2.37}$$

the amplitudes Φ_{b1} and Φ_{b2} will both be zero, and if this occurs the system will not vibrate at all. The last equation is therefore the frequency equation for the system.

The system of fig. 6.2.7 (c) is such that the conditions at the supports are

$$q_1 = 0 = q_2. \tag{6.2.38}$$

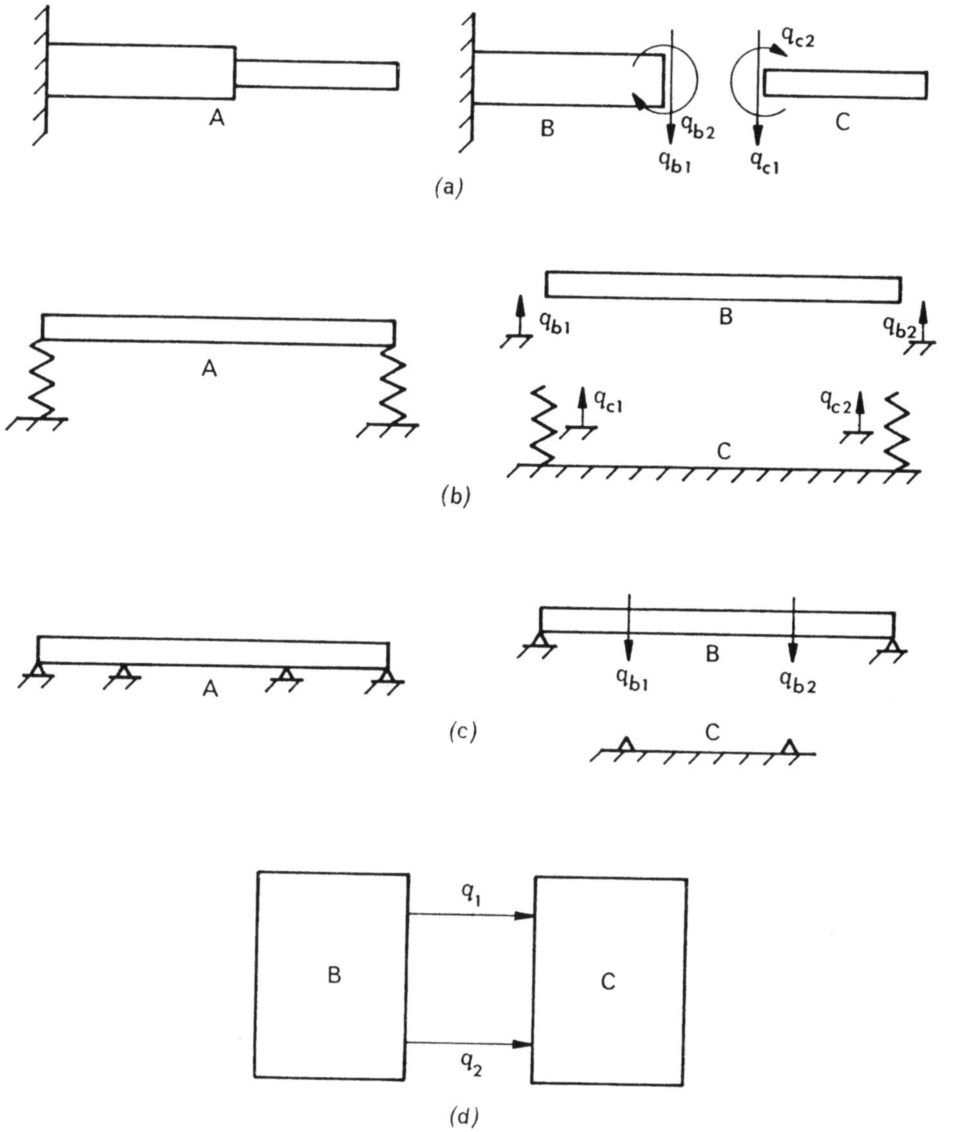

Fig. 6.2.7

For this system, then, we have

$$\gamma_{11} = 0 = \gamma_{12} = \gamma_{22}, \tag{6.2.39}$$

and the frequency equation reduces to

$$\beta_{11}\beta_{22} - \beta_{12}^2 = 0. \tag{6.2.40}$$

The examples which have been given are sufficient to show how simple composite systems can be analysed. The receptances and frequency equations of many 'box' type composite systems are tabulated elsewhere.†

EXAMPLES 6.2

1. Find the frequency equation and the receptances α_{13}, α_{23} and α_{33} for the system shown in the diagram.

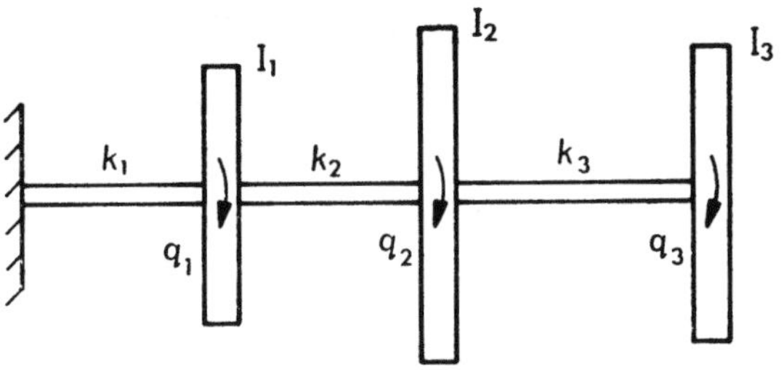

Ex. 6.2.1

2. A torsional system consists of a shaft of stiffness k to which is attached a flywheel whose moment of inertia is I. A flywheel $2I$ is to be attached to the system as shown by means of a shaft k'. Show that, if $x = I\omega^2/k$ and x_1, x_2 are the values of x corresponding to the natural frequencies then $x_1 < \frac{1}{3}$, $x_2 > 1$ for all values of k', and find the value of k'/k if $x_1 = 0{\cdot}3$ and $x_2 = 7$.

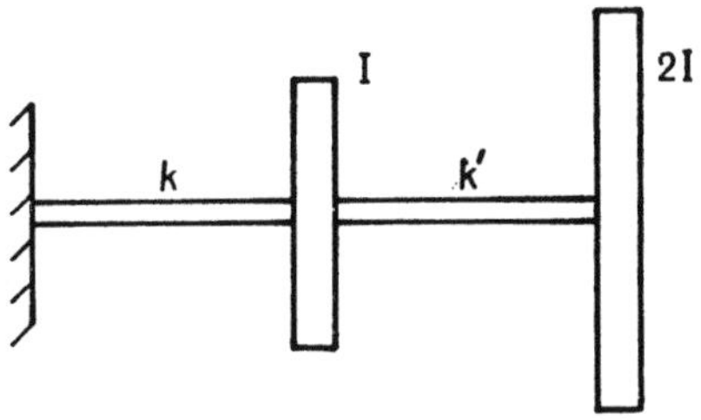

Ex. 6.2.2

3. Use the receptances obtained in Ex. 6.1.6 to write down the frequency equation of the uniform flexible beam of fig. 6.2.7 (b) when the springs are identical and of stiffness k. Identify the two parts into which the equation can be factorised and describe another way in which they may be obtained.

4. Find the first natural frequency of the system shown in fig. 6.2.1 where $I = \mathcal{J}l/2$ by solving equation (6.2.16). Use Newton's iterative method (see equations (9.5.3) and (9.5.4)).

6.3 Further discussion of composite systems

For certain composite systems there are ways of simplifying the analysis involved in the determination of the natural frequencies, principal modes and receptances. The simplest such system is the torsional system consisting of a set of flywheels on

† Bishop and Johnson, *The Mechanics of Vibration*, ch. 1.

a light shaft, the analysis for which is due essentially to Holzer.† One of the quickest ways of finding approximate values for the natural frequencies of a torsional system composed of massive shafts with or without attached flywheels, is to approximate the system by a set of flywheels on a light shaft and then use Holzer's method. Holzer's analysis gives the *exact* natural frequencies of the approximate system of flywheels on a light shaft, but these are not, of course, the same as the natural frequencies of the original system. The errors which are introduced when a continuous system is 'idealised' in this way are discussed in § 6.4.

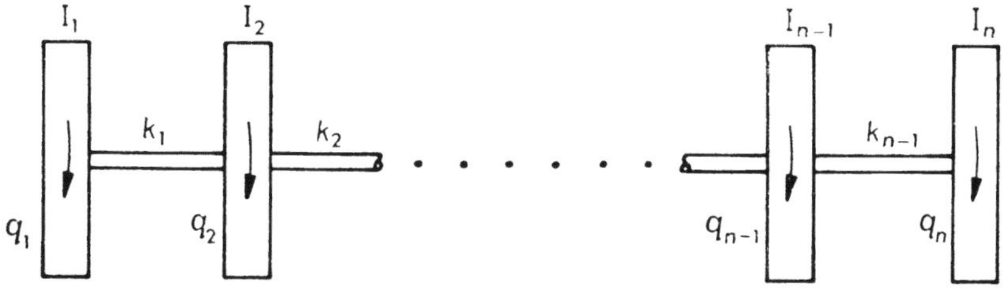

Fig. 6.3.1

Fig. 6.3.1 shows a typical torsional system. Let the displacements of the flywheels be denoted by $q_1, q_2, \ldots, q_n$ so that the kinetic and potential energies are

$$\left.\begin{aligned} T &= \tfrac{1}{2}\{I_1\dot{q}_1^2 + I_2\dot{q}_2^2 + \ldots + I_n\dot{q}_n^2\}, \\ V &= \tfrac{1}{2}\{k_1(q_2-q_1)^2 + k_2(q_3-q_2)^2 + \ldots + k_{n-1}(q_n-q_{n-1})^2\}. \end{aligned}\right\} \tag{6.3.1}$$

If just the natural frequencies and principal modes of the system are required these may be obtained by the method discussed in Chapter 2. Thus the frequency equation for the system is

$$\begin{vmatrix} k_1 - I_1\omega^2 & -k_1 & 0 & \ldots & \ldots & 0 \\ -k_1 & k_1+k_2-I_2\omega^2 & -k_2 & \ldots & \ldots & 0 \\ 0 & -k_2 & k_2+k_3-I_3\omega^2 & \ldots & \ldots & 0 \\ \cdots & \cdots & \cdots & \cdots & \cdots & \cdots \\ 0 & 0 & 0 & \ldots & k_{n-2}+k_{n-1}-I_{n-1}\omega^2 & -k_{n-1} \\ 0 & 0 & 0 & \ldots & -k_{n-1} & k_{n-1}-I_n\omega^2 \end{vmatrix} = 0. \tag{6.3.2}$$

When expanded this is an equation of degree n in ω^2 in which the coefficients may be found by an elegant method due to Crossley and Germen.‡ When a natural frequency has been found the corresponding principal mode may be found very easily in a manner shown below.

More information may be required, however, than can be obtained easily from the analysis of Chapter 2, and this is where the Holzer technique (sometimes in

† H. Holzer, *Die Berechnung der Drehschwingungen* (Berlin, 1921).

‡ F. R. E. Crossley and U. Germen, 'A method of evaluation of a large determinant', *J. Appl. Mech.* **27** (1960), pp. 350–1; see also the Discussion in **28** (1961), pp. 157–8.

a modified form) is useful. For instance, it might be required to find not only the natural frequencies and principal modes of the system, but also the changes produced in them by changes in the system—for example a change in the moment of inertia of one of the flywheels, or the stiffness of one of the connecting shafts.

Holzer's technique may be explained as follows. Suppose that a torque

$$Q_n = \Phi_n \sin \omega t \tag{6.3.3}$$

is applied at the end flywheel whose rotation is q_n. The Lagrangian equations of the system are

$$\left.\begin{aligned} k_1(\Psi_1-\Psi_2) &= I_1\omega^2\Psi_1, \\ k_1(\Psi_2-\Psi_1)+k_2(\Psi_2-\Psi_3) &= I_2\omega^2\Psi_2, \\ \cdots\cdots\cdots\cdots\cdots\cdots\cdots & \cdots\cdots\cdots \\ k_{n-2}(\Psi_{n-1}-\Psi_{n-2})+k_{n-1}(\Psi_{n-1}-\Psi_n) &= I_{n-1}\omega^2\Psi_{n-1}, \\ k_{n-1}(\Psi_n-\Psi_{n-1}) &= I_n\omega^2\Psi_n+\Phi_n. \end{aligned}\right\} \tag{6.3.4}$$

By adding the first m of these equations together we obtain

$$k_m(\Psi_m-\Psi_{m+1}) = \sum_{r=1}^{m} I_r\omega^2\Psi_r \quad (m = 1, 2, \ldots, n-1), \tag{6.3.5}$$

while by adding all the equations we obtain

$$\sum_{r=1}^{n} I_r\omega^2\Psi_r = -\Phi_n. \tag{6.3.6}$$

Now ω will be a natural frequency of the system if the system can vibrate freely, that is with $\Phi_n = 0$, at that frequency. In other words, the natural frequencies are given by the condition

$$\sum_{r=1}^{n} I_r\omega^2\Psi_r = 0. \tag{6.3.7}$$

Later on we shall refer to the quantity

$$R \equiv \sum_{r=1}^{n} I_r\omega^2\Psi_r. \tag{6.3.8}$$

This is the 'residual torque' $-\Phi_n$, and it is the amplitude of the torque which would have to be applied at q_n to keep the system in equilibrium.

The way in which the condition (6.3.7) is used is to construct a 'Holzer Table' for the system. For the system of fig. 6.3.2 this would be as shown in Table 6.3.1, though, in any practical computation the values of the moments of inertia and torsional stiffnesses would be expressed numerically and not through the symbols I and k. The table has seven columns. Columns (1) and (6) contain the values of I and k. When a trial value of ω^2 has been chosen column (2) is filled in. The computation then proceeds along each row from column (3) to column (7). The first element of Ψ is taken arbitrarily to be unity. The succeeding elements are obtained in sequence by subtracting the entry in column (7) of row (m) from Ψ_m to give Ψ_{m+1} according to the equation

$$\Psi_{m+1} = \Psi_m - \frac{1}{k_m}\sum_{r=1}^{m} I_r\omega^2\Psi_r. \tag{6.3.9}$$

The entry in column (5) at any stage gives the sum of all the elements in column (4) up to that point. The final entry in column (5) is the 'residual' torque R defined in equation (6.3.8). The aim is to choose ω^2 to make R vanish. When ω has been chosen in this way column (3) gives the corresponding mode. Thus the Holzer table may be used to find the modes even if the natural frequencies have been found in some other way.

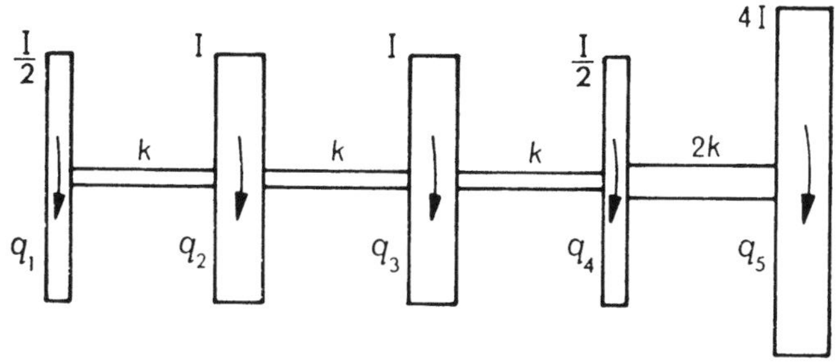

Fig. 6.3.2

TABLE 6.3.1

$\omega^2 = 0{\cdot}3k/I$

No. ...	(1)	(2)	(3)	(4)	(5)	(6)	(7)
m, r	I_r	$I_r\omega^2$	Ψ_r	$I_r\omega^2\Psi_r$	$\sum_{r=1}^{m} I_r\omega^2\Psi_r$	k_m	$\frac{1}{k_m}\sum_{r=1}^{m} I_r\omega^2\Psi_r$
1	$\frac{1}{2}I$	$0{\cdot}15k$	1	$0{\cdot}15k$	$0{\cdot}15k$	k	0·15
2	I	$0{\cdot}3k$	0·85	$0{\cdot}255k$	$0{\cdot}405k$	k	0·405
3	I	$0{\cdot}3k$	0·445	$0{\cdot}133k$	$0{\cdot}538k$	k	0·538
4	$\frac{1}{2}I$	$0{\cdot}15k$	−0·093	$-0{\cdot}014k$	$0{\cdot}524k$	$2k$	0·262
5	$4I$	$1{\cdot}2k$	−0·355	$-0{\cdot}426k$	$0{\cdot}098k$	—	—

In this presentation of the Holzer technique, we first gave the algebraic theory upon which the method is based and we then showed how the necessary computation could be made in a systematic fashion. An alternative approach is to dispense with the algebraic treatment of equations (6.3.4) to (6.3.8), to draw up the table (working from left to right across columns (1) to (7) in turn) and to use purely physical arguments in making the various entries.† The reader who wishes to adopt this alternative approach and who is new to this subject may find the reasoning simpler if he lowers the entries in columns (6) and (7) (which relate to shafts) so that they fall between the entries in column (5); the entries in the table will then form rows corresponding to a flywheel, a shaft, a flywheel, and so on.

The ways in which the various trial values of ω^2 are chosen vary according to whether the computation is being done by a high-speed computer or a desk calculator. For a high-speed computer the usual way is to programme the calculations so that the machine automatically calculates the residual torque R for regular intervals of ω^2, and every time a change of sign occurs it reverses and proceeds with smaller intervals of ω^2 until a change of sign occurs again; and this can be continued

† For example, the numbers in column (7) represent amplitudes of angles of twist.

until the residual torque is smaller than some predetermined value. For desk calculations this method will clearly be long and wasteful, and several methods have been suggested for finding successive approximations. A typical example is the so-called 'Rayleigh–Kohn' method.†

The procedure is simply to take some trial value of ω^2, say ω_1^2, to find the set of ratios $\Psi_1 : \Psi_2 : \ldots : \Psi_n$ from the Holzer table and then to use this mode to calculate a new value, ω_c^2 by computing Rayleigh's quotient, that is by equating the potential and kinetic energies for this mode (see § 2.10).

Rayleigh's quotient may be calculated very simply using the data already calculated in the Holzer table.‡ If ω is a *true* natural frequency and $\Psi_1, \Psi_2, \ldots, \Psi_n$ are the *true* amplitudes of vibration for that mode then the kinetic and potential energies for this mode will be equal and therefore

$$\tfrac{1}{2}\omega^2 \sum_{m=1}^{n} I_m \Psi_m^2 = \frac{1}{2} \sum_{m=1}^{n-1} k_m(\Psi_m - \Psi_{m+1})^2. \tag{6.3.10}$$

Rayleigh's approximation is obtained by equating the kinetic and potential energies for an approximate mode; and for the approximate mode and frequency obtained from the Holzer table this gives

$$\frac{\omega_c^2}{\omega_1^2} = \frac{\sum_{m=1}^{n-1} k_m(\Psi_m - \Psi_{m+1})^2}{\sum_{m=1}^{n} I_m \omega_1^2 \Psi_m^2}. \tag{6.3.11}$$

But

$$\sum_{m=1}^{n} I_m \omega_1^2 \Psi_m^2 = \Sigma\,(\text{col. }3), (\text{col. }4) \tag{6.3.12}$$

and, because of equations (6.3.5) and (6.3.8), we have

$$\begin{aligned}\sum_{m=1}^{n-1} k_m(\Psi_m - \Psi_{m+1})^2 &= \omega_1^2 \sum_{m=1}^{n-1}\left[(\Psi_m - \Psi_{m+1}) \sum_{r=1}^{m} I_r \Psi_r\right] \\ &= \omega_1^2\left[\sum_{m=1}^{n} I_m \Psi_m^2 - \Psi_n \sum_{m=1}^{n} I_m \Psi_m\right] \\ &= \sum_{m=1}^{n} I_m \omega_1^2 \Psi_m^2 - R\Psi_n \\ &= \Sigma\,(\text{col. }3)\,(\text{col. }4) - R\Psi_n,\end{aligned} \tag{6.3.13}$$

so that

$$\frac{\omega_c^2}{\omega_1^2} = \frac{\Sigma\,(\text{col. }3)\,(\text{col. }4) - R\Psi_n}{\Sigma\,(\text{col. }3)\,(\text{col. }4)}. \tag{6.3.14}$$

This may be written in an alternative form in view of the fact that

$$\begin{aligned}\sum_{m=1}^{n-1} k_m(\Psi_m - \Psi_{m+1})^2 &= \sum_{m=1}^{n-1} \frac{1}{k_m}\left(\sum_{r=1}^{m} I_r \omega_1^2 \Psi_r\right)^2 \\ &= \Sigma\,(\text{col. }5)\,(\text{col. }7).\end{aligned} \tag{6.3.15}$$

† S. H. Crandall and W. G. Strang, 'An improvement of the Holzer table based on a suggestion of Rayleigh's', *J. Appl. Mech.* **24** (1957), pp. 228–30; for a discussion with further references see *J. Appl. Mech.* **25** (1958), pp. 160–1.

‡ S. Mahalingam, 'An improvement of the Holzer method', *J. Appl. Mech.* **25** (1958), pp. 618–20.

Thus, by the equation preceding (6.3.13),

$$\frac{\omega_c^2}{\omega_1^2} = \frac{\Sigma\,(\text{col. }5)\,(\text{col. }7)}{\Sigma\,(\text{col. }5)\,(\text{col. }7) + R\Psi_n}. \tag{6.3.16}$$

It may be verified that, for Table 6.3.1,

$$\omega_c^2 = 0{\cdot}3\frac{k}{I}\frac{0{\cdot}612}{0{\cdot}577} = 0{\cdot}318\frac{k}{I}, \tag{6.3.17}$$

and that for this value of ω the residual torque is $0{\cdot}013k$.

For a discussion of the conditions under which the successive approximations converge to a natural frequency, and ways of increasing the speed of convergence, the reader is referred to the papers of Crandall,† Crandall and Strang‡ and Mahalingam§ and also to the book by Crandall.|| A few remarks are made on the subject later on in connection with the 'split' Holzer table.

The Holzer analysis can be used to find the effect of *changes* in the system on the natural frequencies.¶ Let ω be an exact natural frequency of the system and $\Psi_1, \Psi_2, \ldots, \Psi_n$ the amplitudes of vibration in the corresponding mode. Our purpose is now to find the effect on ω of a small change δI_s in the moment of inertia of the sth flywheel. The change in ω may be found by differentiating equation (6.3.10). This gives

$$\delta(\omega^2)\sum_{m=1}^{n} I_m \Psi_m^2 + \omega^2\left[\sum_{m=1}^{n} 2I_m\Psi_m\frac{\partial\Psi_m}{\partial I_s}\delta I_s + \delta I_s\Psi_s^2\right]$$
$$-\sum_{m=1}^{n-1} 2k_m(\Psi_m - \Psi_{m+1})\frac{\partial}{\partial I_s}(\Psi_m - \Psi_{m+1})\,\delta I_s = 0. \tag{6.3.18}$$

On substituting for $k_m(\Psi_m - \Psi_{m+1})$ from equation (6.3.5) we find that

$$\sum_{m=1}^{n-1} k_m(\Psi_m - \Psi_{m+1})\frac{\partial}{\partial I_s}(\Psi_m - \Psi_{m+1}) = \omega^2\sum_{m=1}^{n-1}\sum_{r=1}^{m} I_r\Psi_r\frac{\partial}{\partial I_s}(\Psi_m - \Psi_{m+1})$$
$$= \omega^2\sum_{m=1}^{n-1} I_m\Psi_m\frac{\partial\Psi_m}{\partial I_s} + \omega^2\sum_{m=1}^{n-1}\sum_{r=1}^{m-1} I_r\Psi_r\frac{\partial\Psi_m}{\partial I_s} - \omega^2\sum_{m=1}^{n-1}\sum_{r=1}^{m} I_r\Psi_r\frac{\partial\Psi_{m+1}}{\partial I_s}. \tag{6.3.19}$$

The last two sums differ only in one term—as may be seen by expanding them—and, because the mode is an exact one so that equation (6.3.7) is satisfied, their difference is

$$-\omega^2\sum_{r=1}^{n-1} I_r\Psi_r\frac{\partial\Psi_n}{\partial I_s} = \omega^2 I_n\Psi_n\frac{\partial\Psi_n}{\partial I_s}. \tag{6.3.20}$$

It follows that

$$\sum_{m=1}^{n-1} k_m(\Psi_m - \Psi_{m+1})\frac{\partial}{\partial I_s}(\Psi_m - \Psi_{m+1}) = \omega^2\sum_{m=1}^{n} I_m\Psi_m\frac{\partial\Psi_m}{\partial I_s}, \tag{6.3.21}$$

† S. H. Crandall, 'Iterative procedures related to relaxation methods for eigenvalue problems', *Proc. Roy. Soc.* A, **207** (1951), pp. 416–23.

‡ Crandall and Strang, *J. Appl. Mech.* **24**. § S. Mahalingam, *J. Appl. Mech.* **25**.

|| Crandall, *Engineering Analysis* (McGraw-Hill, 1956), p. 112.

¶ W. A. Tuplin, 'The effect of changes in a torsionally vibrating system on the natural frequencies of the system', *Phil. Mag.* ser. 7, **21** (1936), pp. 1097–111. This analysis has been presented more simply by Mahalingam, *J. Appl. Mech.* **25**.

which when substituted into equation (6.3.18) give Tuplin's formula

$$\delta(\omega^2) = -\omega^2 \frac{\delta I_s \Psi_s'^2}{\sum_{m=1}^{n} I_m \Psi_m'^2}. \tag{6.3.22}$$

If the frequency ω has been calculated from a Holzer table the denominator can be calculated from it (see equation (6.3.12)).

This result may be used for finding another approximation from a Holzer table.† Suppose that a table is constructed for a certain frequency ω_1 and the residual torque is found to be

$$R = \sum_{m=1}^{n} I_m \omega_1^2 \Psi'_m. \tag{6.3.23}$$

This residual can be nullified by replacing I_n by a suitable moment of inertia I'_n. Then ω_1 becomes the exact natural frequency of a modified system B for which

$$I'_m = I_m \quad (m = 1, 2, \ldots, n-1)$$

and

$$I'_n = \frac{-\sum_{m=1}^{n-1} I_m \omega_1^2 \Psi'_m}{\omega_1^2 \Psi'_n}. \tag{6.3.24}$$

The change in the moment of inertia is therefore

$$\delta I_n = I_n - I'_n = \frac{\sum_{m=1}^{n} I_m \omega_1^2 \Psi'_m}{\omega_1^2 \Psi'_n} = \frac{R}{\omega_1^2 \Psi'_n}. \tag{6.3.25}$$

The system B can be converted to the original system A by adding δI_n to the last flywheel. The corresponding change in the frequency is given by Tuplin's formula. We thus obtain the second approximation

$$\omega_2^2 = \omega_1^2 - \omega_1^2 \frac{\delta I_n \Psi_n'^2}{\sum_{m=1}^{n} I'_m \Psi_m'^2}. \tag{6.3.26}$$

In terms of the data obtained in the Holzer table for ω_1 this last result may be written as

$$\frac{\omega_2^2}{\omega_1^2} = \frac{\Sigma\,(\text{col. }5)\,(\text{col. }7) - R\Psi'_n}{\Sigma\,(\text{col. }5)\,(\text{col. }7)}. \tag{6.3.27}$$

For Table 6.3.1 this gives

$$\omega_2^2 = \frac{0{\cdot}3k}{I}\,\frac{0{\cdot}647}{0{\cdot}612} = \frac{0{\cdot}317k}{I}, \tag{6.3.28}$$

which has a residual $0{\cdot}019k$.

Tuplin's formula (6.3.22) gives the change in natural frequency resulting from a small change in the moment of inertia of one of the flywheels. If other changes are made—for example if another system is attached to the original system—then a more detailed analysis is needed. The relevant receptances of the system will then

† S. Mahalingam, *ibid.*

have to be calculated. Certain receptances of the system can be obtained from the Holzer table directly. For, since the only excitation is $Q_n = \Phi_n \sin \omega t$ at q_n we have

$$\alpha_{1n} = \frac{\Psi_1}{\Phi_n}, \quad \alpha_{2n} = \frac{\Psi_2}{\Phi_n}, \quad \ldots, \quad \alpha_{nn} = \frac{\Psi_n}{\Phi_n}. \tag{6.3.29}$$

Suppose that various systems are separately to be attached to the original system at the single coordinate q_n. If the end receptance of the system which is added is γ the frequency equation of the whole will be

$$\alpha_{nn} = -\gamma. \qquad (6.3.30)\dagger$$

Thus if α_{nn} is calculated for a number of values of ω the frequency equations of all the systems may be obtained by plotting α_{nn} against $-\gamma$. If a number of such composite systems have to be analysed this method of analysis will be considerably simpler than setting up the frequency equation of each system separately.

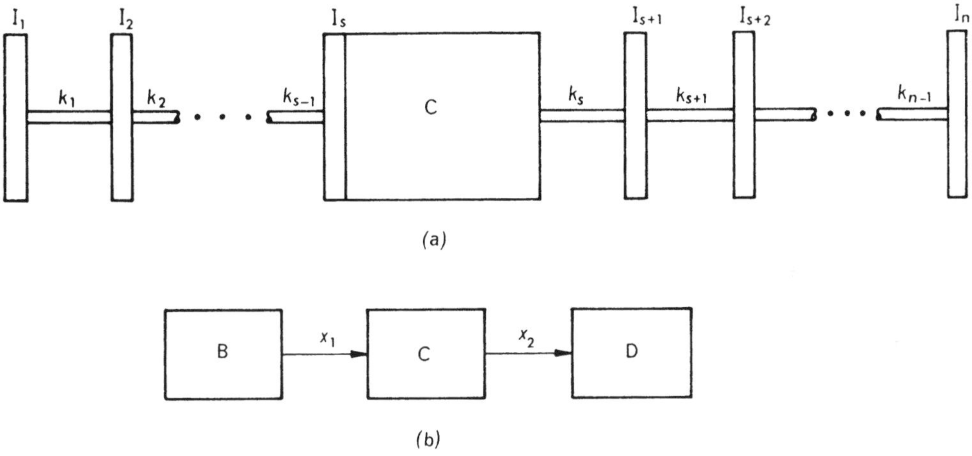

Fig. 6.3.3

Suppose now that some subsystem C is inserted in the original system at some interior point—at the sth flywheel say. Then if the two parts into which the original system is thereby broken are denoted by B and D, the composite system may be represented diagrammatically by fig. 6.3.3 (b) and the frequency equation will be

$$(\beta_{11} + \gamma_{11})(\gamma_{22} + \delta_{22}) - \gamma_{12}^2 = 0. \tag{6.3.31}$$

The receptances of C we shall suppose to be known so that it is necessary to calculate only β_{11} and δ_{22}. These are end receptances of the subsystems B and D and may therefore be obtained from Holzer tables compiled for these systems. For the subsystem B we suppose that harmonic excitation is applied at the flywheel I_s, while for D we suppose that it is applied at the end of the shaft k_s. For D we work

† The original system is now properly to be regarded as a subsystem. Nevertheless, the symbol α will be retained for its receptances.

the Holzer table backwards from an assumed value of q_n to find $q_{n-1}, q_{n-2}, \ldots, q_s$ and finally the residual torque at q_s. For D the equations will be

$$\left.\begin{aligned} k_{n-1}(\Psi_n - \Psi_{n-1}) &= \omega^2 I_n \Psi_n, \\ k_{n-2}(\Psi_{n-1} - \Psi_{n-2}) + k_{n-1}(\Psi_{n-1} - \Psi_n) &= \omega^2 I_{n-1} \Psi_{n-1}, \\ \cdots\cdots\cdots\cdots\cdots\cdots\cdots\cdots\cdots\cdots & \\ k_s(\Psi_{s+1} - \Psi_s) + k_{s+1}(\Psi_{s+1} - \Psi_{s+2}) &= \omega^2 I_{s+1} \Psi_{s+1}, \\ k_s(\Psi_s - \Psi_{s+1}) &= \Phi_s, \end{aligned}\right\} \tag{6.3.32}$$

which give

$$\left.\begin{aligned} \Psi_{m-1} &= \Psi_m - \frac{1}{k_{m-1}} \sum_{r=m}^{n} I_r \omega^2 \Psi_r \quad (m = s+1, \ldots, n), \\ \Phi_{sd} &= - \sum_{r=s+1}^{n} I_r \omega^2 \Psi_r. \end{aligned}\right\} \tag{6.3.33}$$

Table 6.3.2 shows a split Holzer table for the system of fig. 6.3.2 when the break is made at q_3. The table shows that

$$\beta_{11} = \frac{0{\cdot}445}{-0{\cdot}538k} = \frac{-0{\cdot}827}{k}; \quad \delta_{22} = \frac{-0{\cdot}86x}{-1{\cdot}26kx} = \frac{0{\cdot}683}{k}, \tag{6.3.34}$$

where the subscripts of the receptances relate to the linking coordinates x of fig. 6.3.3 (*b*) and not to the coordinates q_r denoting the rotations of the flywheels. Thus the composite system of fig. 6.3.3 (*a*) may be analysed by evaluating the receptances β_{11} and δ_{22} for suitable values of ω and finding the roots of the frequency equation (6.3.31).

TABLE 6.3.2

$\omega^2 = 0{\cdot}3k/I$

No. m, r	I_r	$I_r\omega^2$	Ψ_r	$I_r\omega^2\Psi_r$	$\sum_{r=1}^{m} I_r\omega^2\Psi_r$	k_m	$\frac{1}{k_m}\sum_{r=1}^{m} I_r\omega^2\Psi_r$
1	$\frac{1}{2}I$	$0{\cdot}15k$	1	$0{\cdot}15k$	$0{\cdot}15k$	k	$0{\cdot}15$
2	I	$0{\cdot}3k$	$0{\cdot}85$	$0{\cdot}255k$	$0{\cdot}405k$	k	$0{\cdot}405$
3	I	$0{\cdot}3k$	$0{\cdot}445$	$0{\cdot}133k$	$0{\cdot}538k$	—	—
3	—	—	$-0{\cdot}86x$	—	—	—	—
4	$\frac{1}{2}I$	$0{\cdot}15k$	$0{\cdot}4x$	$0{\cdot}06kx$	$1{\cdot}26kx$	k	$1{\cdot}26x$
5	$4I$	$1{\cdot}2k$	x	$1{\cdot}2kx$	$1{\cdot}2kx$	$2k$	$0{\cdot}6x$
m, r No.	I_r	$I_r\omega^2$	Ψ_r	$I_r\omega^2\Psi_r$	$\sum_{r=m}^{n} I_r\omega^2\Psi_r$	k_{m-1}	$\frac{1}{k_{m-1}}\sum_{r=m}^{n} I_r\omega^2\Psi_r$

The reason why the starting value of Ψ_n in Table 6.3.2 has been taken as x and not 1 is that by so doing we may use this split Holzer table to analyse the original (unbroken) system. For the original system, without the subsystem C, the condition which has to be satisfied at q_s if the system is split there is

$$q_{sb} = q_{sd}. \tag{6.3.35}$$

Thus Table 6.3.2 will be a split Holzer table for the original system if x is chosen so that

$$q_{sb} = 0{\cdot}445 = q_{sd} = -0{\cdot}86x, \tag{6.3.36}$$

that is if
$$x = -0{\cdot}517. \tag{6.3.37}$$

The residual torque at q_s will then be

$$R = -\Phi_s = -(\Phi_{sb}+\Phi_{sd}) = -[(-0{\cdot}538k)+(-1{\cdot}26)(-0{\cdot}517)k] = -0{\cdot}113k. \tag{6.3.38}$$

The natural frequencies of the system are still those frequencies which make this residual torque vanish. The split Holzer table may be used to find the receptances of the system due to excitation at q_s, for

$$\alpha_{ms} = \frac{\Psi_m}{\Phi_s} \quad (m = 1, 2, \ldots, n). \tag{6.3.39}$$

Sometimes it is an advantage to use a split Holzer table for the analysis of the original (complete) system. This is because the speed of convergence of successive approximations obtained by the Rayleigh–Kohn method is increased when the residual torque is placed at that flywheel which has the largest amplitude in the corresponding principal mode.† This might be expected on intuitive grounds, since the system will be most sensitive to a torque if it is placed at the point with the largest amplitude.

Holzer's method has been formulated in an alternative way,‡ and it has been extended to damped vibration.§

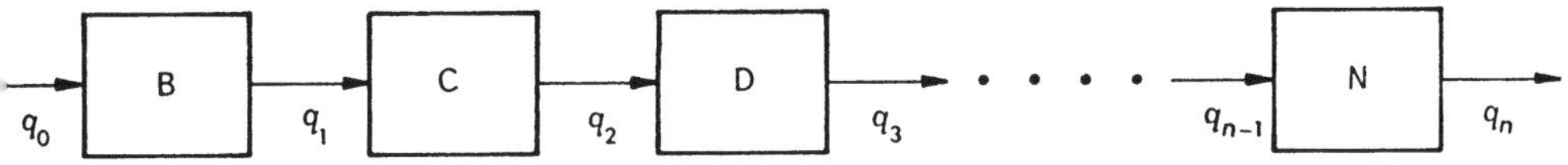

Fig. 6.3.4

The Holzer analysis may also be extended to apply to any chain of systems linked by single coordinates—such as a chain of *massive* shafts and flywheels.|| The system A shown in fig. 6.3.4 is composed of subsystems $B, C, D, \ldots, N$ linked together by coordinates $q_1, q_2, \ldots, q_{n-1}$. It is required to calculate the receptances $\alpha_{0n}, \alpha_{1n}, \ldots, \alpha_{nn}$ of the system in terms of the receptances $\beta, \gamma, \delta, \ldots, \nu$ of the subsystems $B, C, D, \ldots, N$; and the calculation is to be carried out for some particular (numerical) driving frequency ω. For the sake of definiteness we suppose that the system is free at the coordinate q_0, although the analysis for a system fixed at q_0 is very similar. The only external force acting on the system is the force

$$Q_n = \Phi_n \sin \omega t \tag{6.3.40}$$

† Crandall and Strang, *J. Appl. Mech.* **24**.

‡ E. Saibel, 'Free and forced vibration of composite systems', *Proc. Symp. on Spectral Theory and Differential Problems* (Oklahoma A. and M., 1951), pp. 333–43.

§ J. P. den Hartog and J. P. Li, 'Forced torsional vibrations with damping: an extension of Holzer's method', *J. Appl. Mech.* **13** (1946), pp. 276–80; see also K. E. Bisshopp, 'Forced torsional vibration of systems with distributed mass and internal and external damping', *J. Appl. Mech.* **26** (1958), pp. 8–12.

|| Bishop and Johnson, *The Mechanics of Vibration* (1960), § 6.6.

at q_n. Using the notation introduced in § 6.2 we have the following equations for the motion of B

$$\Psi_0 = \beta_{01}\Phi_{b1}, \quad \Psi_1 = \beta_{11}\Phi_{b1}. \tag{6.3.41}$$

Here Φ_{b1} is the amplitude of the force

$$Q_{b1} = \Phi_{b1}\sin\omega t \tag{6.3.42}$$

acting on the subsystem B at q_{b1}, Ψ_1 is the amplitude of the displacement

$$q_1 = q_{b1} = q_{c1} = \Psi_1\sin\omega t, \tag{6.3.43}$$

and β_{01}, β_{11} are the receptances of the unattached subsystem B. If some arbitrary value is assigned to Ψ_0 the first of equations (6.3.41) gives the value of Φ_{b1} and the second then gives the value of Ψ_1. Analysing the motion of $C, D, \ldots, N$ in the same way we obtain the following set of equations

$$\left.\begin{array}{ll} \Psi_0 & \Phi_{b1} = \Psi_0/\beta_{01}, \\ \Psi_1 = \beta_{11}\Phi_{b1} & \Phi_{c1} = -\Phi_{b1}, \\ (\Psi_1 = \gamma_{11}\Phi_{c1} + \gamma_{12}\Phi_{c2}) & \Phi_{c2} = (\Psi_1 - \gamma_{11}\Phi_{c1})/\gamma_{12}, \\ \Psi_2 = \gamma_{21}\Phi_{c1} + \gamma_{22}\Phi_{c2} & \Phi_{d2} = -\Phi_{c2}, \\ \ldots\ldots\ldots\ldots & \ldots\ldots\ldots\ldots \end{array}\right\} \tag{6.3.44}$$

which terminate with

$$\left.\begin{array}{ll} \ldots\ldots\ldots\ldots & \ldots\ldots\ldots\ldots \\ (\Psi_{n-1} = \nu_{n-1,n-1}\Phi_{N,n-1} + \nu_{n-1,n}\Phi_n) & \Phi_n = (\Psi_{n-1} - \nu_{n-1,n-1}\Phi_{N,n-1})/\nu_{n-1,n}, \\ \Psi_n = \nu_{n,n-1}\Phi_{N,n-1} + \nu_{nn}\Phi_n. & \end{array}\right\} \tag{6.3.45}$$

These equations give all the amplitudes $\Psi_1, \Psi_2, \ldots, \Psi_n$ and Φ_n in the form

$$(\text{a numerical factor}) \times \Psi_0.$$

The receptances $\alpha_{0n}, \alpha_{1n}, \ldots, \alpha_{nn}$ are given by

$$\alpha_{0n} = \frac{\Psi_0}{\Phi_n}, \quad \alpha_{1n} = \frac{\Psi_1}{\Phi_n}, \quad \ldots, \quad \alpha_{nn} = \frac{\Psi_n}{\Phi_n}, \tag{6.3.46}$$

and these expressions do not depend on the value given to Ψ_0.

This analysis may be used in the same way as the Holzer technique. Thus the frequency equation of the system is

$$\Phi_n = 0, \tag{6.3.47}$$

and for a system composed of uniform shafts and flywheels, for which the end receptances, such as β_{11} and β_{12}, are simple trigonometrical functions, this probably provides the best way of finding the natural frequencies *exactly*. The technique is to take a number of values of ω and calculate $\Phi_{b1}, \Psi_1, \Phi_{c1}, \Phi_{c2}, \Psi_2, \ldots, \Phi_n, \Psi_n$ in turn; the natural frequencies are given by the zeros of Φ_n and the principal modes are given by the corresponding sets of ratios

$$\Psi_0 : \Psi_1 : \ldots : \Psi_n. \tag{6.3.48}$$

The displacement at any interior point x of any one of the subsystems (say C, which might be a shaft) may be found if the receptances γ_{x1}, γ_{x2} are known, since

$$\Psi_x = \gamma_{x1}\Phi_{c1} + \gamma_{x2}\Phi_{c2}. \tag{6.3.49}$$

In this way, provided that the receptances of the subsystems are known, the displacement of every point of the system may be found.

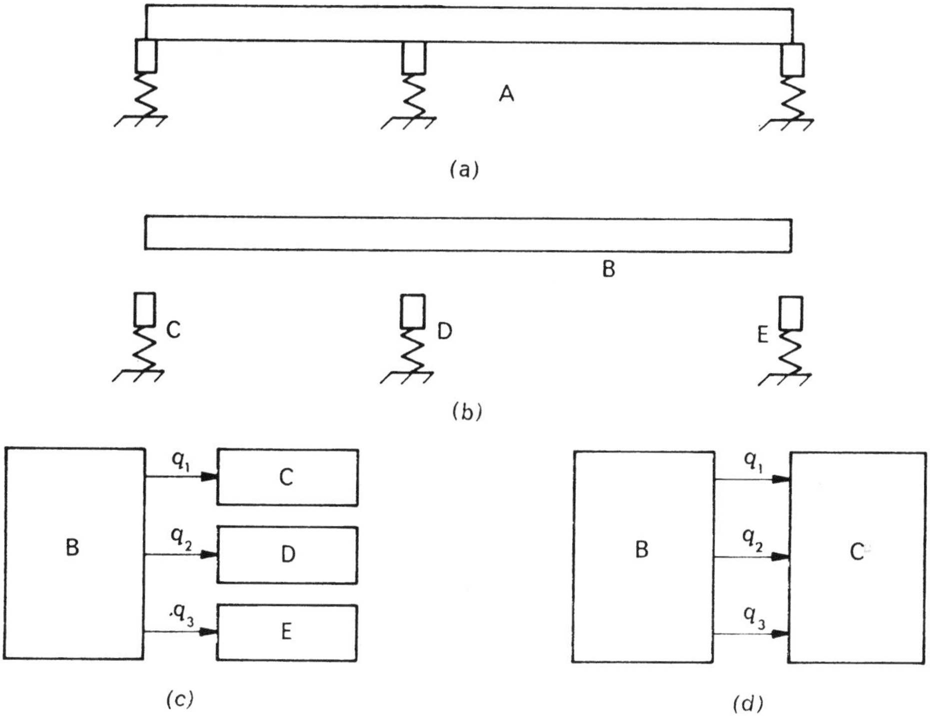

Fig. 6.3.5

As in the Holzer analysis, the effect of changes in the system may be determined by calculating the appropriate receptances. Thus if a new system P is attached to the system at the right-hand end the new frequency equation will be

$$\alpha_{nn} + \pi_{nn} = 0. \tag{6.3.50}$$

If one of the intermediate systems $C, D, \ldots$ is modified the effect on the natural frequency may be determined as before by finding the end receptances of the parts of the system on either side of the co-ordinate at which the change is made.

A composite system consisting of two parts linked by a number, say n, coordinates, is another for which there is a particularly appropriate way of presenting the analysis. This will be discussed by reference to an example.

The system shown in fig. 6.3.5 consists of a uniform beam resting on flexible supports. For convenience, just three supports have been considered; but the reader will notice that the analysis which follows may easily be generalised to apply to a beam resting on any number of supports. The only reason why the beam has been taken to be uniform is that the receptances of uniform beams have been tabulated, but the method of analysis would be exactly the same for a non-uniform beam resting on flexible supports. It will be assumed that the supports are attached to the beam at single points so that they only exert transverse forces, and not couples as well, on the beam. If this is done then the system may be analysed as in fig. 6.3.5 (*b*). The analysis may be represented schematically as in fig. 6.3.5 (*c*) or as in fig. 6.3.5 (*d*). The latter will be adopted because it is simpler. In this representation the three parts making up the subsystem *C* are unconnected, which means that

$$\gamma_{12} = 0 = \gamma_{23} = \gamma_{31}. \tag{6.3.51}$$

Suppose it is required to find the natural frequencies of the system shown in fig. 6.3.5. The responses at q_1, q_2 and q_3 for *B* due to the harmonic excitations Q_{b1}, Q_{b2} and Q_{b3} at these coordinates are given by the matrix equation

$$\begin{bmatrix} q_{b1} \\ q_{b2} \\ q_{b3} \end{bmatrix} = \begin{bmatrix} \beta_{11} & \beta_{12} & \beta_{13} \\ \beta_{21} & \beta_{22} & \beta_{23} \\ \beta_{31} & \beta_{32} & \beta_{33} \end{bmatrix} \begin{bmatrix} Q_{b1} \\ Q_{b2} \\ Q_{b3} \end{bmatrix}. \tag{6.3.52}$$

The receptances shown are the only ones which are relevant to the analysis. Bearing in mind the changed meaning of the term, we shall refer to the square matrix as a 'receptance matrix' $\boldsymbol{\beta}$ and write equation (6.3.52) as

$$\mathbf{q}_b = \boldsymbol{\beta}\mathbf{Q}_b. \tag{6.3.53}$$

Equally, for the subsystem *C* we have

$$\begin{bmatrix} q_{c1} \\ q_{c2} \\ q_{c3} \end{bmatrix} = \begin{bmatrix} \gamma_{11} & 0 & 0 \\ 0 & \gamma_{22} & 0 \\ 0 & 0 & \gamma_{33} \end{bmatrix} \begin{bmatrix} Q_{c1} \\ Q_{c2} \\ Q_{c3} \end{bmatrix}, \tag{6.3.54}$$

which may be written

$$\mathbf{q}_c = \boldsymbol{\gamma}\mathbf{Q}_c. \tag{6.3.55}$$

There are now compatibility and equilibrium conditions at each support, and if the system is vibrating freely they may be expressed in the two matrix equations

$$\mathbf{q}_b = \mathbf{q}_c = \mathbf{q}; \quad \mathbf{Q}_b + \mathbf{Q}_c = \mathbf{0} \tag{6.3.56}$$

respectively. Equations (6.3.53), (6.3.55) and (6.3.56) may now be combined to give

$$(\boldsymbol{\beta} + \boldsymbol{\gamma})\,\mathbf{Q}_b = \mathbf{0}, \tag{6.3.57}$$

which shows that the frequency equation is

$$|\boldsymbol{\beta} + \boldsymbol{\gamma}| = 0. \tag{6.3.58}$$

For the specific system under consideration this equation is

$$\begin{vmatrix} \beta_{11}+\gamma_{11} & \beta_{12} & \beta_{13} \\ \beta_{21} & \beta_{22}+\gamma_{22} & \beta_{23} \\ \beta_{31} & \beta_{32} & \beta_{33}+\gamma_{33} \end{vmatrix} = 0. \tag{6.3.59}$$

It is clear that if A is any system which may be analysed in the form shown in fig. 6.3.5 (d), in which there are n coordinates linking B and C, and if $\boldsymbol{\beta}$ and $\boldsymbol{\gamma}$ are the receptance matrices with elements β_{ij} and γ_{ij} $(i,j = 1, 2, ..., n)$, then the frequency equation of A will be given by equation (6.3.58). Thus if the supports are supposed to exert not only forces, but also couples on the beams then the frequency equation will still have the form given in equation (6.3.58), but now there will be six coordinates—three deflections and three slopes—linking the beam with the supports.

Suppose it is required to find the receptances, or the receptance matrix, of the composite system of beam and supports. To find these we suppose that harmonic forces Q_1, Q_2, Q_3 (making up a force matrix $\mathbf{Q}$) are applied at q_1, q_2, q_3. The equilibrium conditions are now

$$\mathbf{Q} = \mathbf{Q}_b + \mathbf{Q}_c \tag{6.3.60}$$

which, when combined with equations (6.3.53), (6.3.55) and the first of equations (6.3.56), give

$$\mathbf{Q} = \boldsymbol{\beta}^{-1}\mathbf{q}_b + \boldsymbol{\gamma}^{-1}\mathbf{q}_c = (\boldsymbol{\beta}^{-1} + \boldsymbol{\gamma}^{-1})\,\mathbf{q}. \tag{6.3.61}$$

But for the composite system

$$\mathbf{q} = \boldsymbol{\alpha}\mathbf{Q}, \tag{6.3.62}$$

and therefore

$$\boldsymbol{\alpha}^{-1} = \boldsymbol{\beta}^{-1} + \boldsymbol{\gamma}^{-1}, \tag{6.3.63}$$

which is equivalent to

$$\boldsymbol{\alpha} = (\boldsymbol{\beta}^{-1} + \boldsymbol{\gamma}^{-1})^{-1}. \tag{6.3.64}$$

Generally, just one or two of the receptances α_{ij} would be needed, but if the whole matrix $\boldsymbol{\alpha}$ were required then three operations of matrix inversion would have to be carried out.

As a final example we consider a composite system consisting of a chain of doubly-linked systems, such as a set of massive beams undergoing transverse vibration. The end receptances of such a system may be calculated by an extension of the method described for a singly-linked chain.

Myklestad† has presented a method for solving beam problems in which the first step is to represent the segments of the beam approximately by a set of masses on a *light* beam segment. The analysis of the approximate system is a special case of the analysis which will now be described.

The system A of fig. 6.3.6 consists of subsystems $B, C, ..., N$ linked by coordinates $q_1, q_1'; q_2, q_2'; ...; q_{n-1}, q_{n-1}'$. We suppose the system to be free at the left-hand end and acted on by generalised forces

$$\left.\begin{aligned} Q_n &= \Phi_n \sin \omega t, \\ Q_n' &= \Phi_n' \sin \omega t, \end{aligned}\right\} \tag{6.3.65}$$

at the right-hand end. Our object is to present a method by which the numerical values of the amplitudes of the displacements q_r, q_r', due to these forces, may be

† N. O. Myklestad, 'A new method for calculating natural modes of uncoupled bending vibrations of airplane wings and other types of beams', *J. Aero. Sci.* **11** (1944), pp. 153–62; see also Myklestad, 'Numerical analysis of forced vibrations of beams', *J. Appl. Mech.* **20** (1953), pp. 53–6. Further discussion of this method is given by R. E. D. Bishop, 'Myklestad's method for non-uniform vibrating beams', *The Engineer* (14 and 21 Dec. 1956); see also S. Mahalingam, 'An improvement of the Myklestad method for flexural vibration problems', *J. Aero/Space Sci.* **26** (1959), pp. 46–50.

found for a given (numerical) value of ω the driving frequency. The method is applicable, in particular, to a set of uniform massive beams because the receptances of uniform beams are tabulated functions. For beams the displacements q_r and q'_r will be the deflection and slope respectively, at the severed cross-section, and the forces Q_r and Q'_r will be the shearing force and bending moment at this rth junction.

We start by assigning arbitrary values Ψ_0, Ψ'_0 (not numerical values) to the amplitudes of q_0 and q'_0. Then, since the only forces acting on B are $\Phi_{b1}\sin\omega t$ and $\Phi'_{b1}\sin\omega t$ we have

$$\left.\begin{aligned}\Psi_0 &= \beta_{01}\Phi_{b1}+\beta_{01'}\Phi'_{b1},\\ \Psi'_0 &= \beta_{0'1}\Phi_{b1}+\beta_{0'1'}\Phi'_{b1}.\end{aligned}\right\} \tag{6.3.66}$$

These two equations may be written as the single matrix equation

$$\begin{bmatrix}\Psi_0\\ \Psi'_0\end{bmatrix} = \begin{bmatrix}\beta_{01} & \beta_{01'}\\ \beta_{0'1} & \beta_{0'1'}\end{bmatrix}\begin{bmatrix}\Phi_{b1}\\ \Phi'_{b1}\end{bmatrix}, \tag{6.3.67}$$

which may be abbreviated to $\quad \boldsymbol{\Psi}_0 = \boldsymbol{\beta}_{01}\boldsymbol{\Phi}_{b1},$ (6.3.68)

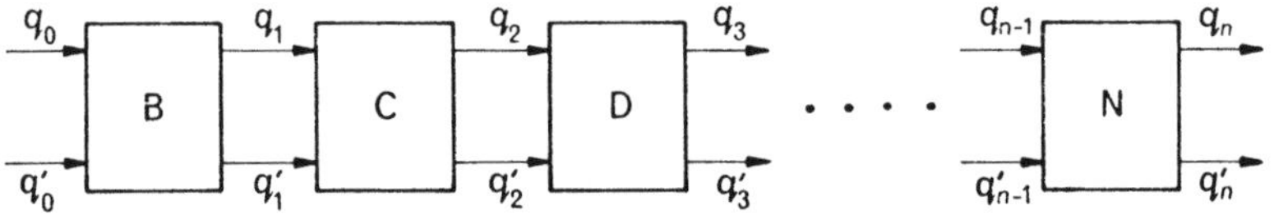

Fig. 6.3.6

where now $\boldsymbol{\beta}_{01}$ is the 2×2 receptance matrix shown, and $\boldsymbol{\Psi}_0$ and $\boldsymbol{\Phi}_{b1}$ are column matrices. This equation may be solved to give

$$\boldsymbol{\Phi}_{b1} = \boldsymbol{\beta}_{01}^{-1}\boldsymbol{\Psi}_0. \tag{6.3.69}$$

Turning now to the 'inboard' side of B, we have

$$\boldsymbol{\Psi}_1 = \begin{bmatrix}\Psi_1\\ \Psi'_1\end{bmatrix} = \begin{bmatrix}\beta_{11} & \beta_{11'}\\ \beta_{1'1} & \beta_{1'1'}\end{bmatrix}\begin{bmatrix}\Phi_{b1}\\ \Phi'_{b1}\end{bmatrix} = \boldsymbol{\beta}_{11}\boldsymbol{\Phi}_{b1}. \tag{6.3.70}$$

By comparing these equations with those obtained for the singly-linked chain we see that provided $\boldsymbol{\beta}_{01}$, $\boldsymbol{\beta}_{11}$ are interpreted as 2×2 matrices, and $\boldsymbol{\Psi}_0$, $\boldsymbol{\Psi}_1$, $\boldsymbol{\Phi}_{b1}$ as column matrices, the analyses are exactly the same. In fact, provided the reinterpretation is carried right through, all the previous equations ((6.3.42)–(6.3.45)) apply. Thus for the system C we have

$$\boldsymbol{\Phi}_{c1} = -\boldsymbol{\Phi}_{b1}; \quad \boldsymbol{\Phi}_{c2} = \boldsymbol{\gamma}_{12}^{-1}(\boldsymbol{\Psi}_1-\boldsymbol{\gamma}_{11}\boldsymbol{\Phi}_{c1}), \tag{6.3.71}$$

and the equations terminate with the analogues of equations (6.3.45). In this way we obtain all the amplitude matrices $\boldsymbol{\Psi}_1, \boldsymbol{\Psi}_2, \ldots, \boldsymbol{\Psi}_n, \boldsymbol{\Phi}_n$ in the form

$$\text{(a numerical } 2\times 2 \text{ matrix)}\times\boldsymbol{\Psi}_0.$$

The final equations will therefore have the form

$$\left.\begin{aligned}\boldsymbol{\Psi}_r &= \mathbf{A}_r\boldsymbol{\Psi}_0 \quad (r = 1, 2, \ldots, n),\\ \boldsymbol{\Phi}_n &= \mathbf{B}\boldsymbol{\Psi}_0.\end{aligned}\right\} \tag{6.3.72}$$

The second equation may be solved to give

$$\boldsymbol{\Psi}_0 = \mathbf{B}^{-1}\boldsymbol{\Phi}_n, \tag{6.3.73}$$

and then the first becomes

$$\boldsymbol{\Psi}_r = \mathbf{A}_r\mathbf{B}^{-1}\boldsymbol{\Phi}_n. \tag{6.3.74}$$

This equation gives all the receptances corresponding to excitation at q_n and q_n' for, when expanded, it is

$$\boldsymbol{\Psi}_r = \begin{bmatrix}\Psi_r \\ \Psi_r'\end{bmatrix} = \begin{bmatrix}\alpha_{rn} & \alpha_{rn'} \\ \alpha_{r'n} & \alpha_{r'n'}\end{bmatrix}\begin{bmatrix}\Phi_n \\ \Phi_n'\end{bmatrix}. \tag{6.3.75}$$

The required receptances are thus the elements of the matrix $\mathbf{A}_r\mathbf{B}^{-1}$.

Theoretically the method may be used to find the natural frequencies of the system. These will be the frequencies for which the system may vibrate with

$$\Phi_n = 0 = \Phi_n', \tag{6.3.76}$$

so that the frequency equation is

$$|\mathbf{B}| = 0. \tag{6.3.77}$$

Plotting Φ_n and Φ_n' against ω may not be easy, but it must be remembered that this is a fundamentally difficult problem.

We note that the method may be generalised to apply to a chain of systems each linked by n coordinates; the only modification is that the symbols $\boldsymbol{\beta}_{01}$, $\boldsymbol{\beta}_{11}$, etc., now represent $n \times n$ matrices.

An alternative method of solution for beam systems, and composite systems generally, has been developed by Falk, Fuhrke, Pestel and others.† This method, which is based on the use of 'transfer' or 'transmission' matrices, may be considered as an extension and generalisation of Myklestad's method. Its chief merits are that it may easily be couched in matrix terms, that the matrices involved do not increase in size with the number of subsystems in the system,‡ and that it may be applied to a large variety of systems.

We shall now give a very brief introduction to the method.

The essence of the method may be described by referring to fig. 6.3.6, and assuming that it represents a set of beams joined end to end. In the receptance method, the displacement and slope at each joining coordinate are expressed in terms of the forces and moments acting. In the method of transmission matrices the four quantities—displacement, slope, force and moment—appear on an equal footing, and their values at one joining coordinate are expressed in terms of those at the previous one. The matrix linking the two sets of four quantities is a transmission matrix.

Before finding the transmission matrix for the simple case of a uniform beam we note that the receptance and transmission matrix methods use different sign conventions for forces and moments. The two are compared in figs. 6.3.7 (*a*) and (*b*),

† For references to the writings of these and other workers on the subject the reader is referred to H. Marguerre, *Matrices of Transmission in Beam Problems*. This is ch. 2 of I. N. Sneddon and R. Hill (eds.), *Progress in Solid Mechanics*, vol. I (North-Holland, 1962); an elementary introduction is given by J. M. Prentis and F. A. Leckie, *Mechanical Vibrations: An Introduction to Matrix Methods*, ch. 5 (Longmans, 1963).

‡ It thus resembles the method last described: in the method of transmission matrices the matrices are usually 4×4 or 6×6.

which show a force $F\sin\omega t$ applied at the join of two subsystems B and C. Fig. 6.3.7 (*a*) shows the sign convention used in the Receptance Method; the forces and moments are given by the equations:

$$\left.\begin{aligned} \Phi_{1b}\sin\omega t &= -(EI)_b\frac{\partial^3 v}{\partial x^3}, & \Phi_{1c}\sin\omega t &= (EI)_c\frac{\partial^3 v}{\partial x^3},\\ \Phi_{2b}\sin\omega t &= (EI)_b\frac{\partial^2 v}{\partial x^2}, & \Phi_{2c}\sin\omega t &= -(EI)_c\frac{\partial^2 v}{\partial x^2},\\ \Phi_{1b}+\Phi_{1c} &= F, & \Phi_{2b}+\Phi_{2c} &= 0. \end{aligned}\right\} \tag{6.3.78}$$

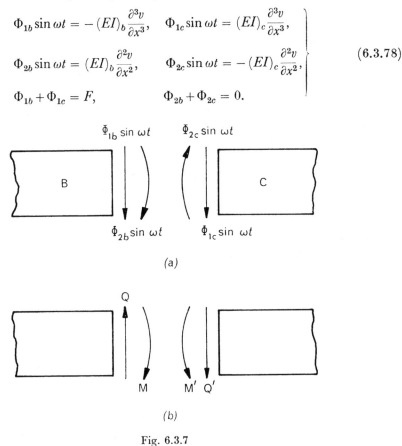

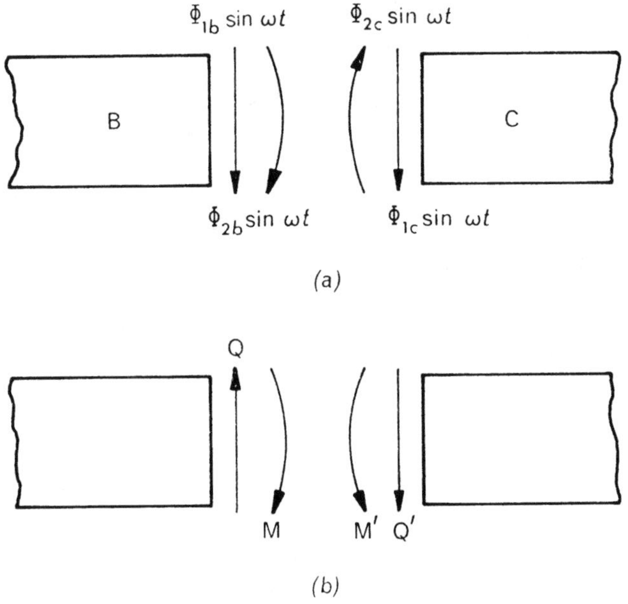

Fig. 6.3.7

Fig. 6.3.7 (*b*) shows the sign convention used in the Transmission Matrix Method; the forces and moments are given by the equations:

$$\left.\begin{aligned} Q\sin\omega t &= (EI)_b\frac{\partial^3 v}{\partial x^3}, & Q'\sin\omega t &= (EI)_c\frac{\partial^3 v}{\partial x^3},\\ M\sin\omega t &= (EI)_b\frac{\partial^2 v}{\partial x^2}, & M'\sin\omega t &= (EI)_c\frac{\partial^2 v}{\partial x^2},\\ Q'-Q &= F, & M'-M &= 0. \end{aligned}\right\} \tag{6.3.79}$$

It will be seen that the sign convention in fig. 6.3.7 (*a*) is based on the idea of generalised forces; thus a positive moment applied to a subsystem is associated with a positive slope. The convention of fig. 6.3.7 (*b*) is based, instead, on the ideas of shear forces and bending moments, as used in beam theory.

Returning now to fig. 6.3.6 and using this new sign convention, let us find expressions for the displacement, slope, bending moment and shear force at the cross-section '1' of the uniform beam B. These expressions are required in terms of the like quantities at the section '0'. To do this, we return to the general solution (6.1.51) and apply the conditions

$$v = v_0, \quad \partial v/\partial x = \theta_0, \quad M = M_0, \quad Q = Q_0 \quad \text{when} \quad x = 0. \tag{6.3.80}$$

In this way, we find that

$$X = v(x) = v_0 \tfrac{1}{2}(\cosh \lambda x + \cos \lambda x) + \theta_0 l \frac{1}{2\lambda l}(\sinh \lambda x + \sin \lambda x)$$
$$+ \frac{M_0 l^2}{EI} \frac{1}{2(\lambda l)^2}(\cosh \lambda x - \cos \lambda x) + \frac{Q_0 l^3}{EI} \frac{1}{2(\lambda l)^3}(\sinh \lambda x - \sin \lambda x) \tag{6.3.81}$$

for any section of subsystem B, distant x from the end '0'.

This last equation may now be used to express the four quantities v_1, $\theta_1 l$, $M_1 l^2/EI$ and $Q_1 l^3/EI$ (which all have the same dimension) at the end $x = l$ in terms of the corresponding quantities v_0, $\theta_0 l$, $M_0 l^2/EI$ and $Q_0 l^3/EI$ at $x = 0_0$. For instance, the expression for $\theta_1 l$ is

$$\theta_1 l = (\lambda l)\, \tfrac{1}{2}(\sinh \lambda l - \sin \lambda l)\, v_0 + \tfrac{1}{2}(\cosh \lambda l + \cos \lambda l)\, \theta_0 l$$
$$+ \frac{1}{2(\lambda l)}(\sinh \lambda l + \sin \lambda l) \frac{M_0 l^2}{EI} + \frac{1}{2(\lambda l)^2}(\cosh \lambda l - \cos \lambda l) \frac{Q_0 l^3}{EI}. \tag{6.3.82}$$

The four equations may be assembled into the single equation

$$\begin{bmatrix} v \\ \theta l \\ \dfrac{Ml^2}{EI} \\ \dfrac{Ql^3}{EI} \end{bmatrix}_1 = \begin{bmatrix} C_0 & S_1 & C_2 & S_3 \\ (\lambda l)^4 S_3 & C_0 & S_1 & C_2 \\ (\lambda l)^4 C_2 & (\lambda l)^4 S_3 & C_0 & S_1 \\ (\lambda l)^4 S_1 & (\lambda l)^4 C_2 & (\lambda l)^4 S_3 & C_0 \end{bmatrix} \begin{bmatrix} v \\ \theta l \\ \dfrac{Ml^2}{EI} \\ \dfrac{Ql^3}{EI} \end{bmatrix}_0 \tag{6.3.83}$$

in which the square matrix is the transmission matrix for the beam. The quantities C_0, S_1, etc., are given by

$$\left.\begin{aligned} C_0 &= \tfrac{1}{2}(\cosh \lambda l + \cos \lambda l) = \tfrac{1}{2}F_9; \qquad S_1 = \frac{1}{2(\lambda l)}(\sinh \lambda l + \sin \lambda l) = \frac{1}{2}\frac{F_7}{\lambda l}; \\ C_2 &= \frac{1}{2(\lambda l)^2}(\cosh \lambda l - \cos \lambda l) = -\frac{1}{2}\frac{F_{10}}{(\lambda l)^2}; \quad S_3 = \frac{1}{2(\lambda l)^3}(\sinh \lambda l - \sin \lambda l) \\ &\qquad = -\frac{1}{2}\frac{F_8}{(\lambda l)^3}. \end{aligned}\right\} \tag{6.3.84}$$

The advantage of using C_0, S_1, etc., is that when ω (or λl) is small, each is approximately unity; on the other hand the F's are tabulated functions.†

It is not proposed to discuss the Transmission Matrix Method further, as a very comprehensive survey of the method and some of its many applications has recently been undertaken by Marguerre,‡ and the reader is referred to this (but see Ex. 6.3.6).

† *The Mechanics of Vibration* (1960), table 7.1 (*d*).

‡ H. Marguerre, *ibid.*; see also J. M. Prentis and F. A. Leckie, *ibid.*; E. C. Pestel and F. A. Leckie, *Matrix Methods in Elastomechanics* (McGraw-Hill, 1963).

EXAMPLES 6.3

1. Use the general expression given in § 2.8 for the receptance α_{xx} to prove that for a Holzer table with excitation at q_s the amplitude Ψ_s and the residual torque R will have opposite signs just below a natural frequency and the same signs just above.

2. Prove that the second approximation ω_c given by equation (6.3.16) is always greater than ω_2 given by equation (6.3.27).

3. Show that when the excitation is applied at q_s the formulae ((6.3.16) and (6.3.27)) giving ω_c and ω_2 are the same as before except that Ψ_s replaces Ψ_n.

4. Table 6.3.1 suggests that the greatest amplitude in the first mode of the system of fig. 6.3.2 will occur at q_1. Using the Rayleigh–Kohn method with excitation at q_1 find the next two approximations to the first mode of the system starting from the value $\omega^2 = 0{\cdot}318k/I$ obtained in equation (6.3.17).

5. The true first natural frequency of the system of fig. 6.3.2 is given by $\omega_1^2 = 0{\cdot}32044k/I$ and the corresponding mode is $1:0{\cdot}83978:0{\cdot}41046:-0{\cdot}15038:-0{\cdot}41876$. It is desired to increase the first natural frequency so that $\omega_1^2 = 0{\cdot}33k/I$ by altering the moment of inertia of one of the flywheels. Use Tuplin's formula to find how this can be done with the least amount of alteration. Investigate the accuracy of the formula by finding the true first natural frequency of the system (when it is altered in conformity with Tuplin's formula) and the exact alteration which should be made.

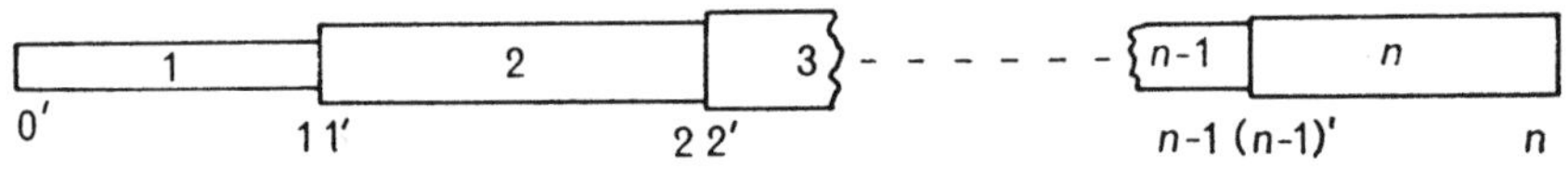

Ex. 6.3.6

6. The figure shows a set of uniform beams, joined end-to-end. Individual members of the set have different values of l, EI and $A\rho$. The quantities l_s and EI_s are a standard length and a standard flexural rigidity respectively. Let $\{\mathbf{v}\}_i$ and $\{\mathbf{v}\}_{i'}$ represent the column vectors

$$\{v \quad \theta l_s \quad Ml_s^2/EI_s \quad Ql_s^3/EI_s\}$$

evaluated on either side of the ith junction.

(a) Write down the equation relating $\{\mathbf{v}\}_i$ and $\{\mathbf{v}\}_{i'}$.

(b) Find the transmission matrix $\mathbf{T}_i$ linking $\{\mathbf{v}\}_{(i-1)'}$ and $\{\mathbf{v}\}_i$ for the ith component beam.

(c) Hence find the transmission matrix $\mathbf{T}$ (expressed in terms of the $\mathbf{T}_i$) linking $\{\mathbf{v}\}_n$ and $\{\mathbf{v}\}_{0'}$.

(d) Assuming that the composite beam has 4 component uniform beams, find its frequency equation when it is (i) clamped-free, (ii) pinned-clamped; giving the equation in terms of the elements of $\mathbf{T}$.

6.4 The approximation of continuous systems by lumped-mass systems

It is well known that the behaviour of a continuous system may be approximated in some sense by the behaviour of a lumped-mass system. Our aim in this section will be to formulate some more precise statements concerning the degree to which a given continuous system can be approximated.

We shall begin our discussion by considering a simple system, a uniform clamped-free shaft in torsion. An exact analysis of this problem was given in § 6.1 and it was shown in particular that the natural frequencies of the system are given by

$$\omega_r = \sqrt{\left(\frac{C}{\mathcal{J}l^2}\right)}\frac{(2r-1)\pi}{2} \quad (r = 1, 2, \ldots), \tag{6.4.1}$$

and the characteristic functions are

$$\phi_r(x) = \sin\frac{(2r-1)\pi x}{2l}. \tag{6.4.2}$$

We now ask 'How closely can the behaviour of the system be approximated by a system consisting of n massive flywheels on a light shaft?' and 'What size should the flywheels be and where should they be placed to give the best approximation?' The answer to the second question depends on the exact meaning that is given to the words 'the best approximation'. We shall merely consider certain straightforward ways of distributing the flywheels and discuss the degree of approximation obtained by using them.

There are a number of ways of idealising a segment of a shaft (fig. 6.4.1 (a)) in this manner and three of the simpler ways are indicated in fig. 6.4.1 (b), (c) and (d). In all these models the light shaft on which the flywheel (or flywheels) is placed is assumed to have a uniform stiffness per unit length; and, in this section, it will be represented by a line as shown. In the first model (fig. 6.4.1 (b)), used by Rayleigh,† the mass of each of the n segments into which the shaft is supposed to be divided is concentrated into two equal flywheels, one at each end. In the second (fig. 6.4.1 (c)), which is apparently due to Lagrange and is investigated by Duncan,‡ it is concentrated into a flywheel at the centre. In the third (fig. 6.4.1 (d)) the inertia is placed at an end of the segment. If the notation of § 6.1 is used then, in each case

$$k = \frac{Cn}{l}, \quad I = \frac{\mathcal{J}l}{n}. \tag{6.4.3}$$

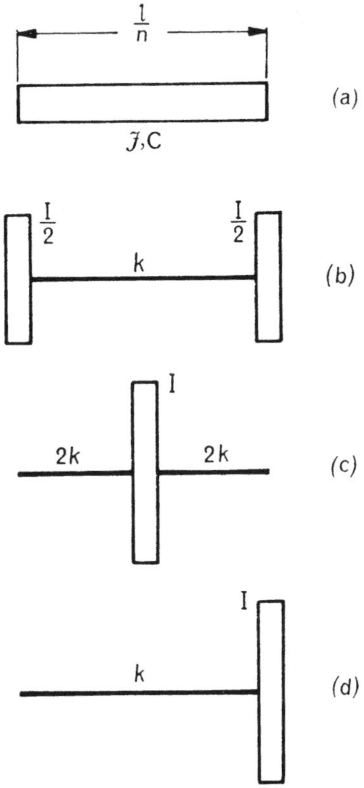

Fig. 6.4.1

For the uniform clamped-free shaft, the three representations lead to the systems shown in fig. 6.4.2.

† Lord Rayleigh, *The Theory of Sound*, art. 120.
‡ W. J. Duncan, 'A critical examination of the representation of massive and elastic bodies by rigid masses elastically connected', *Quart. J. Mech. Appl. Math.* **15** (1952), pp. 97–108.

The natural frequencies and principal modes of each of the three approximate systems may be found exactly and simply for any value of n. The analysis is very similar in the three cases and we shall give the analysis in full only for the first. For this system the kinetic and potential energies are

$$\left.\begin{aligned} T &= \tfrac{1}{2}I\{\dot{q}_1^2+\dot{q}_2^2+\ldots+\dot{q}_{n-1}^2+\tfrac{1}{2}\dot{q}_n^2\}, \\ V &= \tfrac{1}{2}k\{q_1^2+(q_2-q_1)^2+\ldots+(q_n-q_{n-1})^2\}, \end{aligned}\right\} \tag{6.4.4}$$

so that

$$\mathbf{A} = I\begin{bmatrix} 1 & 0 & 0 & \ldots & 0 & 0 \\ 0 & 1 & 0 & \ldots & 0 & 0 \\ 0 & 0 & 1 & \ldots & 0 & 0 \\ \cdots & \cdots & \cdots & \cdots & \cdots & \cdots \\ 0 & 0 & 0 & \ldots & 1 & 0 \\ 0 & 0 & 0 & \ldots & 0 & \tfrac{1}{2} \end{bmatrix}, \quad \mathbf{C} = k\begin{bmatrix} 2 & -1 & 0 & \ldots & 0 & 0 \\ -1 & 2 & -1 & \ldots & 0 & 0 \\ 0 & -1 & 2 & \ldots & 0 & 0 \\ \cdots & \cdots & \cdots & \cdots & \cdots & \cdots \\ 0 & 0 & 0 & \ldots & 2 & -1 \\ 0 & 0 & 0 & \ldots & -1 & 1 \end{bmatrix} \tag{6.4.5}$$

and the frequency equation is

$$\begin{vmatrix} 2-I\omega^2/k & -1 & 0 & \ldots & 0 & 0 \\ -1 & 2-I\omega^2/k & -1 & \ldots & 0 & 0 \\ 0 & -1 & 2-I\omega^2/k & \ldots & 0 & 0 \\ \cdots & \cdots & \cdots & \cdots & \cdots & \cdots \\ 0 & 0 & 0 & \ldots & 2-I\omega^2/k & -1 \\ 0 & 0 & 0 & \ldots & -1 & 1-I\omega^2/2k \end{vmatrix} = 0. \tag{6.4.6}$$

It is convenient to write

$$2-I\omega^2/k = \mu \tag{6.4.7}$$

and the frequency equation is then

$$\Delta_n \equiv \begin{vmatrix} \mu & -1 & 0 & \ldots & 0 & 0 \\ -1 & \mu & -1 & \ldots & 0 & 0 \\ 0 & -1 & \mu & \ldots & 0 & 0 \\ \cdots & \cdots & \cdots & \cdots & \cdots & \cdots \\ 0 & 0 & 0 & \ldots & \mu & -1 \\ 0 & 0 & 0 & \ldots & -1 & \tfrac{1}{2}\mu \end{vmatrix} = 0. \tag{6.4.8}$$

It may easily be verified that, if

$$\mu = 2\cos\theta, \tag{6.4.9}$$

then

$$\Delta_n = \cos n\theta. \tag{6.4.10†}$$

This means that the values of θ giving the natural frequencies are

$$\theta_{rn} = \frac{(2r-1)\pi}{2n} \quad (r = 1, 2, \ldots, n), \tag{6.4.11}$$

† See Ex. 6.4.1.

and therefore the natural frequencies ω_{rn} of the system are given by

$$\omega_{rn}^2 = \frac{2k}{I}\left\{1-\cos\frac{(2r-1)\pi}{2n}\right\} \quad (r = 1, 2, \ldots, n), \tag{6.4.12}$$

and when k, I are expressed in terms of C and $\mathcal{J}$ this equation may be written

$$\omega_{rn} = \sqrt{\left(\frac{C}{\mathcal{J}l^2}\right)}\, 2n \sin\frac{(2r-1)\pi}{4n} \quad (r = 1, 2, \ldots, n). \tag{6.4.13}$$

Therefore

$$\frac{\omega_{rn}}{\omega_r} = \frac{2n \sin (2r-1)\pi/4n}{(2r-1)\pi/2}, \tag{6.4.14}$$

so that (the right-hand member being of the form $(\sin x)/x$)

$$\frac{\omega_r - \omega_{rn}}{\omega_r} = \frac{1}{6}\left[\frac{(2r-1)\pi}{4n}\right]^2 + (\text{higher inverse powers of } n), \tag{6.4.15}$$

and in the limit

$$n^2\left\{\frac{\omega_r - \omega_{rn}}{\omega_r}\right\} = \frac{(2r-1)^2\pi^2}{96}. \tag{6.4.16}$$

This result shows that the proportional error in the frequency of any given mode ultimately varies inversely as the square of the number of segments, and for a given value of n increases rapidly as r increases. Tables 6.4.1 and 6.4.2 show the percentage error

$$\epsilon_{rn} = 100\left\{\frac{\omega_r - \omega_{rn}}{\omega_r}\right\} \tag{6.4.17}$$

and the convergence of $n^2\epsilon_{rn}$ for the first two modes.

TABLE 6.4.1

ERROR FOR THE FIRST MODE OF A UNIFORM CLAMPED-FREE SHAFT

n	$\sqrt{\left(\frac{\mathcal{J}l^2}{C}\right)}\,\omega_{1n}$	ϵ_{1n}	$n^2\epsilon_{1n}$
1	1·4142	9·9684	9·968
2	1·5307	2·5548	10·202
3	1·5529	1·1384	10·246
4	1·5607	0·6413	10·261
5	1·5643	0·4107	10·267
10	1·5692	0·1028	10·277
∞	1·5708	0	10·281

To find the principal mode of the approximate system (fig. 6.4.2(*a*)) corresponding to the frequency ω_{rn} we have to solve the equations

$$\left.\begin{aligned} \mu\Psi_1 - \Psi_2 &= 0, \\ -\Psi_{s-1} + \mu\Psi_s - \Psi_{s+1} &= 0 \quad (s = 2, 3, \ldots, n-1), \\ -\Psi_{n-1} + \tfrac{1}{2}\mu\Psi_n &= 0, \end{aligned}\right\} \tag{6.4.18}$$

for which the determinant of coefficients is Δ_n. The trial solution

$$\Psi_s = A \sin s\theta \quad (s = 1, 2, \ldots, n) \tag{6.4.19}$$

satisfies the equations provided that

$$\mu = 2\cos\theta, \tag{6.4.20}$$

and θ has one of the values given by equation (6.4.11). Thus the rth principal mode is given by $\mathbf{\Psi}^{(r)}$ where

$$\Psi_s^{(r)} = A_r \sin \frac{(2r-1)\,\pi s}{2n} \quad (s = 1, 2, \ldots, n). \tag{6.4.21}$$

By putting $x = sl/n$ in equation (6.4.2) we see that equation (6.4.21) (with $A_r = 1$) gives the displacement of the continuous shaft at the sth point of subdivision when it vibrates in its rth mode. In other words, the approximate mode agrees with the exact mode at the points of subdivision.

TABLE 6.4.2

ERROR FOR THE SECOND MODE OF A UNIFORM CLAMPED-FREE SHAFT

n	$\sqrt{\left(\frac{Jl^2}{C}\right)}\,\omega_{2n}$	ϵ_{2n}	$n^2\epsilon_{2n}$
1	—	—	—
2	3·6955	21·5786	86·315
3	4·2426	9·9684	89·715
4	4·4446	5·6835	90·936
5	4·5399	3·6602	91·505
6	4·5922	2·5505	91·817
10	4·6689	0·9227	92·270
30	4·7075	0·1028	92·494
∞	4·7124	0	92·528

If the system of fig. 6.4.2 (c) is analysed in a similar way it is found that it has the same natural frequencies as the first, namely those given by equation (6.4.13), and that its rth principal mode $\mathbf{\Psi}^{(r)}$ is given by

$$\Psi_s^{(r)} = \sin \frac{(2s-1)\,(2r-1)\,\pi}{4n} \quad (s = 1, 2, \ldots, n), \tag{6.4.22}$$

so that there is again agreement with the exact modes at the flywheels.

For the system shown in fig. 6.4.2 (d) it is found that the natural frequencies are

$$\omega_{rn} = \sqrt{\left(\frac{C}{Jl^2}\right)}\, 2n \sin \frac{(2r-1)\,\pi}{4n+2}. \tag{6.4.23}$$

The principal mode $\mathbf{\Psi}^{(r)}$ is now given by

$$\Psi_s^{(r)} = \sin \frac{s(2r-1)\,\pi}{2n+1} \quad (s = 1, 2, \ldots, n), \tag{6.4.24}$$

and this does not agree with the exact mode at the flywheels. From equation (6.4.23) it may be shown that, for this system, the proportional error is

$$\frac{\omega_r - \omega_{rn}}{\omega_r} = \frac{1}{2n+1} + \frac{n(2r-1)^2\pi^2}{12(2n+1)^3}, \tag{6.4.25}$$

so that in the limit

$$n\left\{\frac{\omega_r - \omega_{rn}}{\omega_r}\right\} = \tfrac{1}{2}. \tag{6.4.26}$$

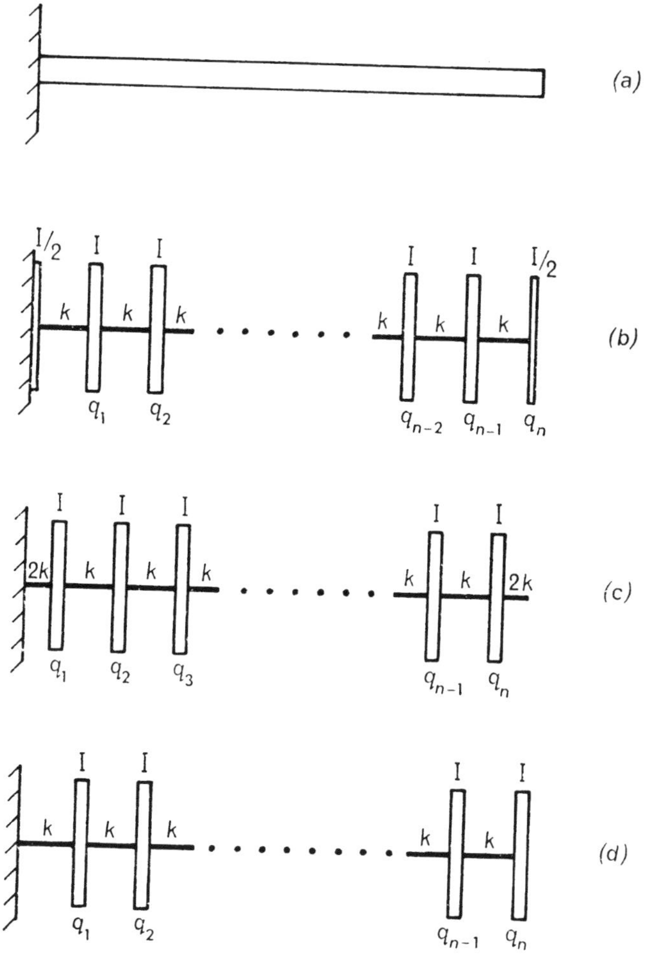

Fig. 6.4.2

Thus the proportional error in the frequency ultimately varies as the number of segments. Tables 6.4.3 and 6.4.4 show the percentage error ϵ_{rn} and the convergence of $n\epsilon_{rn}$ for the first two modes. The proportional error in ω_2 is seen to be now only a little greater than that in ω_1, but both are much greater than the errors incurred by using the systems of figs. 6.4.2 (*b*) and (*c*).

The reader may verify that for a uniform shaft built in at both ends, or free at both ends, the three ways of replacing the continuous shaft by flywheels on a light shaft corresponding to the representations shown in figs. 6.4.1 (*b*), (*c*) and (*d*), give results which are similar to those obtained for the clamped-free shaft. Thus for the first two representations the error in the natural frequencies ultimately varies inversely as n^2, and for the third varies inversely as n.

TABLE 6.4.3

ERROR FOR THE FIRST MODE OF A UNIFORM CLAMPED-FREE SHAFT INCURRED BY USING THE APPROXIMATION SHOWN IN FIG. 6.4.2(d)

n	$\sqrt{\left(\frac{Jl^2}{C}\right)}\omega_{1n}$	ϵ_{1n}	$n\epsilon_{1n}$
1	1·0	36·338	36·338
2	1·2361	21·309	42·619
3	1·3351	15·004	45·013
4	1·3892	11·562	46·249
5	1·4231	9·401	47·006
10	1·4939	4·898	48·983
∞	1·5708	0	50·000

TABLE 6.4.4

ERROR FOR THE SECOND MODE WHEN THE SAME APPROXIMATION IS USED

n	$\sqrt{\left(\frac{Jl^2}{C}\right)}\omega_{2n}$	ϵ_{2n}	$n\epsilon_{2n}$
1	—	—	—
2	3·2361	31·329	62·657
3	3·7409	20·615	61·845
4	4·0000	15·117	60·468
5	4·1541	11·847	59·239
10	4·4504	5·560	55·602
∞	4·7124	0	50·000

Until now we have discussed only simple torsional systems for which the exact solutions may be found both for the system and its approximate representations. (The corresponding results for a stretched string and a bar executing longitudinal vibration may be found by using Table 6.1.1.) In the remainder of this section we shall discuss, in general terms, the approximate analysis of other vibrating systems.

Duncan† considered the approximation of a uniform cantilever beam executing transverse vibration by the model shown in fig. 6.4.3(c) (the 'Duncan' model). In this model the mass of the beam is concentrated into n equal masses $m = A\rho l/n$ situated at the mid-points of n equal segments of length l/n of a massless beam. He showed (empirically) that the errors incurred by calculating the natural frequencies from the model are ultimately proportional to $1/n^2$. Livesley‡ has shown that if

† Duncan, *Quart. J. Mech. Appl. Math.* **5**.

‡ R. K. Livesley, 'The equivalence of continuous and discrete mass distributions in certain vibration problems', *Quart. J. Mech. Appl. Math.* **8** (1955), pp. 353–60.

a simply-supported (i.e. pinned-pinned) beam is approximated by using the model shown in fig. 6.4.3 (*b*) (the 'Rayleigh' model) then the errors in the natural frequencies are ultimately proportional to $1/n^4$, and that

$$\lim_{n\to\infty} n^4\left(\frac{\omega_r-\omega_{rn}}{\omega_r}\right) = \frac{(\lambda_r l)^4}{1440}, \tag{6.4.27}$$

where $\lambda_r l$ is the value of the dimensionless parameter λl (see equation (6.1.52)) in the rth mode of the (original) beam: for a pinned-pinned beam $\lambda_r l = r\pi$. It has since been shown† that an exact analytical solution may be found for any model of the types shown in fig. 6.4.3 (*b*) and (*c*), under any end-conditions (clamped, pinned, sliding or free). In addition, if the ends are clamped, pinned or sliding (but not free) the errors in the natural frequencies are always given by

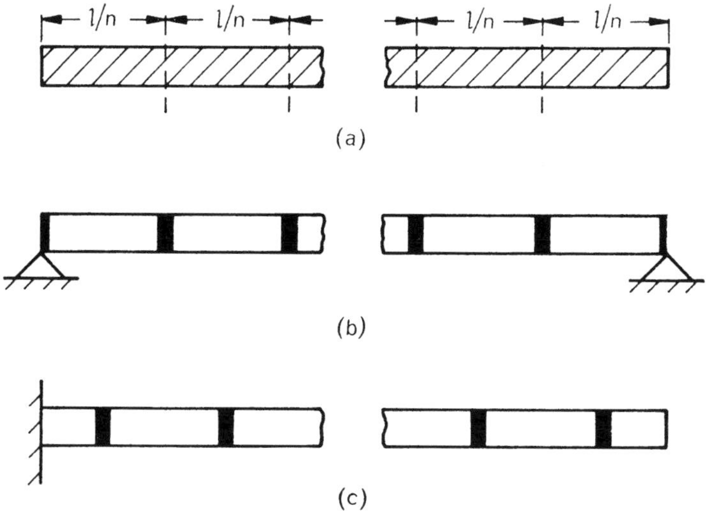

Fig. 6.4.3

equation (6.4.27), where $\lambda_r l$ has the value appropriate to the mode and end-conditions. If the beam has a free end the errors incurred by using the models are in opposite directions and

$$\epsilon_{rn}\,(\text{Duncan}) = -\tfrac{1}{2}\epsilon_{rn}\,(\text{Rayleigh}). \tag{6.4.28}$$

For the Rayleigh model of any beam with one free end the proportional error is given by

$$\frac{\omega_r-\omega_{rn}}{\omega_r} \approx \frac{1}{3}\frac{\lambda_r l}{n^2}, \tag{6.4.29}$$

and for a free-free beam the error is twice this.

Duncan‡ also discussed the approximate analysis of general torsional systems. He considered a shaft whose moment of inertia $\mathcal{J}(x)$ and stiffness $C(x)$ varied along

† G. M. L. Gladwell, 'The approximation of uniform beams in transverse vibration by sets of masses elastically connected', *Proc. 4th Nat. Congr. Appl. Mech.* (Amer. Soc. Mech. Eng., 1962), p. 169.

‡ Duncan, *Quart. J. Mech. Appl. Math.* **5**.

its length, as in fig. 6.4.4. He supposed the shaft to be divided into n segments of equal length and the whole moment of inertia of each segment condensed into a rigid flywheel at the mid-point of the segment—according to the generalisation of the model shown in fig. 6.4.2 (c). The moment of inertia of the flywheel replacing the sth segment $((s-1)\,l/n \leqslant x \leqslant sl/n)$ will be

$$I_s = \int_{(s-1)\,l/n}^{sl/n} \mathcal{J}(x)\,dx. \tag{6.4.30}$$

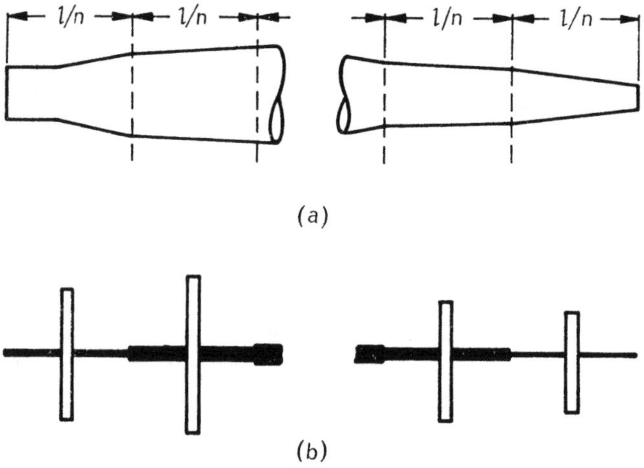

Fig. 6.4.4

Also, he supposed 'the elastic specification of the segmented body to be the same as that of the actual body' but he did not assign a precise meaning to this phrase. Presumably he meant that the stiffness coefficients between certain points should be the same for both bodies. If these points are chosen to be the points of subdivision $x = sl/n$ $(s = 0, 1, 2, \ldots, n)$, then it means that the sth segment must be replaced by a shaft of stiffness k_s, where

$$k_s = \int_{(s-1)\,l/n}^{sl/n} \frac{dx}{C(x)}. \tag{6.4.31}$$

Having approximated the body, Duncan found the discrepancy between the kinetic and potential energies of the actual and approximate systems, then, using Rayleigh's Principle, he showed that the errors in the natural frequencies would ultimately vary as $1/n^2$. Duncan's argument was rather vague, and a lot more analysis would be needed before it could be formulated rigorously. However, his conclusion can be shown to be correct by using a method developed by Fox† and considering the set of Lagrangian equations for the discrete system as a 'finite-difference' approximation to the differential equation of the non-uniform shaft.

† L. Fox, *The Numerical Solution of Two-Point Boundary Problems in Ordinary Differential Equations* (Oxford University Press, 1957), ch. 10.

The kind of torsional system more often encountered in practice is one consisting of a number of uniform shafts (of different diameters) together with a number of flywheels as in fig. 6.4.5 (*a*). Such a system may be approximated by replacing each uniform shaft by a number of flywheels, as described earlier in this section (see fig. 6.4.2 (*b*), (*c*)). To find the errors incurred by calculating the natural frequencies from this model (fig. 6.4.5 (*b*)) we may use the analysis of § 6.2. It was shown there that the frequency equation of any composite system may be set up in terms of the receptances (or certain receptances) of its component parts. Thus the frequency equation of the (original) shaft system may be set up in terms of the

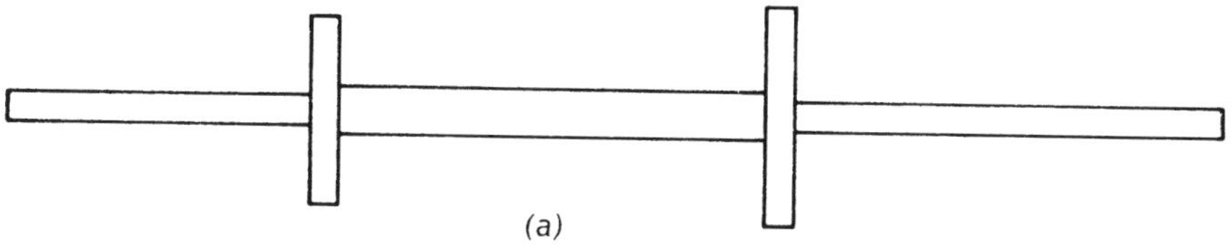

(a)

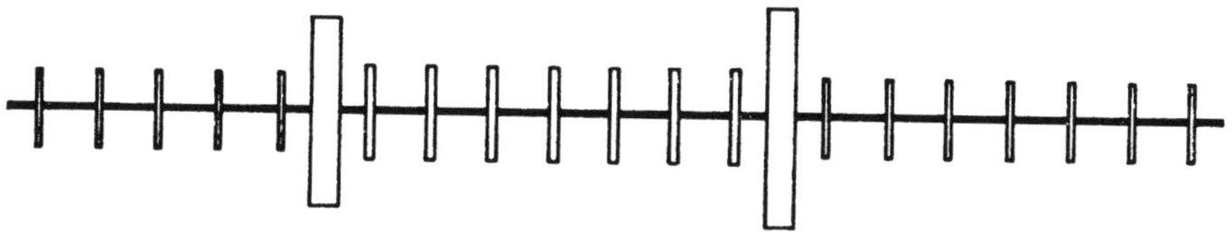

(b)

Fig. 6.4.5

(end) receptances of the individual shafts and flywheels. We now make use of the following fundamental fact. If each set of flywheels on a light shaft which replaces any massive shaft is considered as a single subsystem, then the frequency equation for the complete approximate system will have exactly the same *form* as that of the original system, and the only difference will be that the original receptances will be replaced by the receptances of the approximate subsystems. The frequency equations, and hence the natural frequencies, of the original and approximate systems may therefore be compared if the discrepancies between the receptances of a uniform shaft and its approximation are known. Of course, in practice, the natural frequencies of a system of flywheels on a light shaft (the approximate system) will not be calculated by the receptance methods described in § 6.2, but by methods similar to those described in § 6.3—the methods of Holzer or Crossley and Germen. However in the discussion of the errors involved, it is convenient to suppose that the receptance method is used. It may be proved that whichever representation is used (fig. 6.4.1 (*b*) or (*c*)) the discrepancy between an end-receptance of a uniform shaft and the corresponding receptance of the model is ultimately proportional to $1/n^2$. From this it may be deduced that if the shaft system of fig. 6.4.5 (*a*) is approximated by a model of the type shown in fig. 6.4.5 (*b*) then, provided that a fixed proportion of the total number, N, of flywheels is used on each shaft, the errors incurred by calculating the natural frequencies from the approximate system will ultimately be proportional to $1/N^2$. The argument has a wide range of application and may be used, for example, to deduce corresponding results for beam systems in transverse vibration.

The results which have been discussed in this section have both theoretical and practical value. They furnish the assurance that if (a large class of) continuous vibrating systems are approximated by discrete models of the type described then, provided that a sufficient number of degrees of freedom is used, the natural frequencies and receptances may theoretically be determined to any desired accuracy. They thus provide a theoretical justification for the methods of Holzer, Myklestad, etc., as applied to continuous vibrating systems (see § 6.3). The practical value of the results has been pointed out by Duncan.† Suppose, for example, that a torsional system of the type shown in fig. 6.4.5 (*a*) is approximated first by dividing it into m segments and then into n segments, and suppose that the approximations to the true rth natural frequency ω_r of the system obtained in this way are ω_{rm} and ω_{rn} respectively. The general result states that provided m and n are large it is approximately true that

$$m^2(\omega_r - \omega_{rm}) = A_r = n^2(\omega_r - \omega_{rn}), \tag{6.4.32}$$

where the term A_r will in general not be known. If this equation is treated as an *exact* equation then ω_r has the value

$$\omega_r = \frac{n^2\omega_{rn} - m^2\omega_{rm}}{n^2 - m^2}. \tag{6.4.33}$$

This formula provides a method for combining two approximations. Let us apply it to the data given in Table 6.4.2. Taking $m = 4$, $n = 5$ we find

$$\sqrt{\left(\frac{Jl^2}{C}\right)}\,\omega_2 = \frac{25(4{\cdot}5399) - 16(4{\cdot}4446)}{9} = 4{\cdot}7093, \tag{6.4.34}$$

which is better than the value obtained from $n = 30$, and corresponds to $n = 38$ in the table. This technique is called 'the deferred approach to the limit'.‡ Provided of course that the necessary modifications are made, it may be used when the error is proportional to some other inverse power of n, the fourth, for example.

A problem of some practical importance is to know how many flywheels are needed in the approximation of a continuous shaft system if a certain number of its natural frequencies are to be found with a certain percentage accuracy. Let us consider the problem in relation to the approximation of the system of fig. 6.4.2 (*a*) by that of fig. 6.4.2 (*b*). Table 6.4.1 shows that to obtain the first natural frequency to within about a 1 per cent error one has to take $n = 3+$. Table 6.4.2 shows that $n = 3$ will give a 10 per cent error in the second natural frequency and that to obtain an error of only 1 per cent one has to take $n = 10$.

These facts are particular examples of certain results which are found to hold quite generally for a large class of systems. First, if a continuous vibrating system is approximated by a discrete system having n degrees of freedom the percentage

† Duncan, *Quart. J. Mech. Appl. Math.* **5**.

‡ L. F. Richardson and J. A. Gaunt, 'The deferred approach to the limit', *Phil. Trans.* A, **226** (1927), pp. 299–361.

errors ϵ_{rn} given by equation (6.4.17) will increase with r. Thus if a certain natural frequency is obtained with some percentage error ϵ the errors in the lower natural frequencies will be less than ϵ. Secondly, to obtain a certain number, r say, of the natural frequencies of a continuous system to some desired accuracy one has generally to use many more than r degrees of freedom in the discrete model. Thus there are a number of degrees of freedom which are merely 'ballast'; they are there to ensure that the other r degrees of freedom accurately represents the first r degrees of freedom of the continuous system. Exactly how many more degrees of freedom must be used depends of course on the system and on the accuracy required. For the system of fig. 6.4.2 (*b*) an estimate may be obtained from equation (6.4.15). Suppose that the first r natural frequencies are required to within an error of ϵ per cent. Then n must be chosen so that

$$\frac{1}{6}\left\{\frac{(2r-1)\,\pi}{4n}\right\}^2 100 < \epsilon. \tag{6.4.35}$$

That is

$$n > \frac{10}{4\sqrt{6}}\frac{(2r-1)\,\pi}{\sqrt{\epsilon}} \doteqdot \frac{(2r-1)\,\pi}{\sqrt{\epsilon}}. \tag{6.4.36}$$

Thus to obtain an error of 1 per cent one must take about $6r$ degrees of freedom, and to obtain an error of 2 per cent one must take a little more than $4r$. The results already quoted from Tables 6.4.1 and 6.4.2 show that if r is small a 1 per cent error may be obtained with less than $6r$ degrees of freedom; $6r$ gives the number required for large values of r. Reasonable accuracy has been found to be obtainable from the use of $4r$ degrees of freedom, but this result is based on experience (of aircraft vibration) not on theory.

Duncan gives another rule. 'To find the frequency in some particular mode of oscillation correct to 1 per cent one should use about 13 segments per *complete* wavelength.' For the first mode of the system of fig. 6.4.1 (*a*) the complete wavelength is $4l$ so that 13 segments corresponds to $n = 3+$, and this agrees with the accuracy 1·1 per cent obtained for $n = 3$. For the second mode the complete wavelength is $4l/3$ so that 13 segments corresponds to $n = 3 \times 13/4 \doteqdot 10$, which again is correct.

EXAMPLES 6.4

1. Demonstrate the truth of the result quoted in equations (6.4.9) and (6.4.10).

2. Approximate a free-free shaft using Rayleigh's representation (see fig. 6.4.1 (*b*)) with $n+1$ flywheels and find the natural frequencies and principal modes. Show that the errors in the natural frequencies are ultimately proportional to $1/n^2$. How many flywheels are needed to obtain an error of at most 1 per cent in the first natural frequency?

3. Using equation (6.4.32) is equivalent to assuming that

$$\omega_{rn} = \omega_r + A_r/n^2.$$

A better approximation is to assume that

$$\omega_{rn} = \omega_r + A_r/n^2 + B_r/n^4.$$

Using this formula obtain the analogue of equation (6.4.32) giving ω_r in terms of three approximations ω_{rm}, ω_{rn}, ω_{rp}. Apply the formula to Table 6.4.2 with $m = 3$, $n = 4$, $p = 5$.

4. Horvay and Ormondroyd† have shown that, in a certain sense, the 'best' way of replacing a uniform shaft with a flywheel attached (see the figure) by a light shaft with a flywheel at each end is to use Rayleigh's representation for the shaft with

$$I = \mathcal{J}l, \quad k = C/l + \tfrac{1}{6}\mathcal{J}l\omega^2.$$

Compare the approximation for the first natural frequency obtained by using this model for a clamped-free shaft (with $I' = 0$) with that obtained by using the ordinary Rayleigh representation.

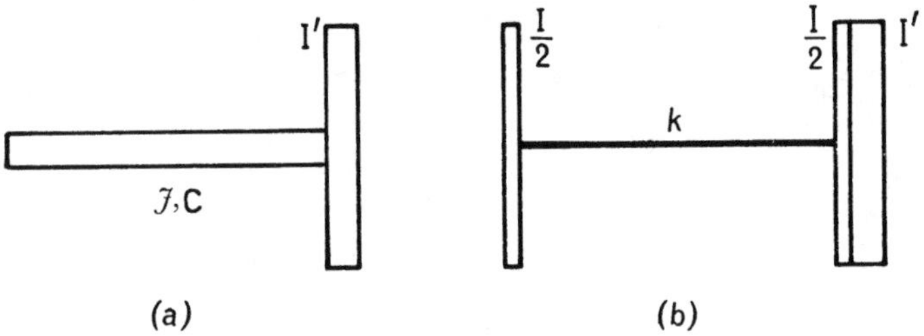

Ex. 6.4.4

5. Show that Rayleigh's representation with

$$I = \frac{\mathcal{J}l}{n}, \quad k = \frac{nC}{l} + \frac{1}{6}\frac{\mathcal{J}l}{n}\omega^2$$

(in conformity with the Horvay–Ormondroyd result) leads to natural frequencies which are greater than the true natural frequencies (whether for a free-free, a clamped-free or a clamped-clamped shaft). Show also that the errors are still proportional to $1/n^2$. What number should replace $\frac{1}{6}$ if the errors are to be proportional to $1/n^4$?

6.5 The Principle of Virtual Work, the Rayleigh–Ritz and other methods

It was shown in § 2.0 that the equations of motion of a continuous vibrating system can be expressed by means of the Principle of Virtual Work, namely (see equation (2.0.4))

$$\delta W_e + \delta W_{in} = \delta V. \tag{6.5.1}$$

We shall now show how this principle can be used to obtain the exact partial differential equation governing the system, using a clamped-free shaft subject to some distributed torque $f(x, t)$ per unit length, as an example.

Suppose that $\theta(x, t)$ is the actual angular displacement of the section of the shaft at position x and

$$\delta\theta = \epsilon\eta(x, t) \tag{6.5.2}$$

is any virtual displacement. The function $\eta(x, t)$ must satisfy the kinematic boundary condition, which in this case is

$$\eta = 0 \quad \text{at} \quad x = 0, \tag{6.5.3}$$

† G. Horvay and J. Ormondroyd, 'Static and dynamic spring constants', *J. Appl. Mech.* **10** (1943), pp. A213–19; see also 'The appropriate lumped constants of vibrating shaft systems', *J. Appl. Mech.* **10** (1943), pp. A220–4, and the Discussion, **12** (1945), p. A56.

but is otherwise arbitrary. Using a dot and dash to denote differentiation with respect to t and x respectively, we obtain

$$\delta W_e = \int_0^l f(x,t)\,\epsilon\eta(x,t)\,dx, \tag{6.5.4}$$

$$\delta W_{in} = -\int_0^l \mathcal{J}(x)\,\ddot{\theta}(x,t)\,\epsilon\eta(x,t)\,dx, \tag{6.5.5}$$

$$\delta V = \frac{1}{2}\int_0^l C(x)\,\{\theta'(x,t)+\epsilon\eta'(x,t)\}^2\,dx - \frac{1}{2}\int_0^l C(x)\,\{\theta'(x,t)\}^2\,dx$$

$$= \int_0^l C(x)\,\theta'(x,t)\,\epsilon\eta'(x,t)\,dx, \tag{6.5.6}$$

neglecting terms involving ϵ^2. The last integral may be integrated by parts and written

$$\delta V = \epsilon[C(x)\,\theta'(x,t)\,\eta(x,t)]_0^l - \epsilon\int_0^l \{C(x)\,\theta'(x,t)\}'\,\eta(x,t)\,dx. \tag{6.5.7}$$

On substituting these various expressions in equation (6.5.1) and making use of the boundary condition satisfied by η we obtain

$$\int_0^l [\{C(x)\,\theta'(x,t)\}' - \mathcal{J}(x)\,\ddot{\theta}(x,t) + f(x,t)]\,\eta(x,t)\,dx - C(l)\,\theta'(l,t)\,\eta(l,t) = 0. \tag{6.5.8}$$

We now argue as follows:

(i) Equation (6.5.8) holds for arbitrary $\eta(x,t)$ satisfying the end condition (6.5.3).

(ii) It therefore holds for all $\eta(x,t)$ satisfying not only this condition but also the additional condition

$$\eta = 0 \quad \text{at} \quad x = l. \tag{6.5.9}$$

(iii) For all such $\eta(x,t)$ the last term in equation (6.5.8) is zero, and if the integral is to be zero for arbitrary such $\eta(x,t)$ the integrand must be zero everywhere. Therefore

$$\frac{\partial}{\partial x}\left\{C(x)\frac{\partial\theta}{\partial x}\right\} - \mathcal{J}(x)\frac{\partial^2\theta}{\partial t^2} + f(x,t) = 0. \tag{6.5.10}$$

This is the differential equation satisfied by θ.

(iv) On substituting this result in equation (6.5.8) we obtain

$$C(l)\,\theta'(l,t)\,\eta(l,t) = 0. \tag{6.5.11}$$

This must be zero for all $n(x,t)$ *satisfying the single condition* (6.5.3), and therefore

$$C(l)\,\theta'(l,t) = 0. \tag{6.5.12}$$

It is a general feature of the Principle that it yields not only the governing differential equation, but also the remaining, dynamic, end condition.

In the discussion, in § 2.0, of the use of the 'Rayleigh–Ritz' or 'assumed-modes' method for reducing a continuous system to discrete form, no mention was made of the way in which the modes should be chosen. This matter will now be discussed, mainly by providing a review of the literature on the subject.

There are two essential criteria governing the choice of modes:

(i) They must satisfy the kinematic boundary conditions (see § 2.0).

(ii) They must impose as little constraint as possible on the motion of the system.

Consider the problem in relation to a clamped-free shaft in torsion. The problem is to find a set of modes $\psi(x)$ satisfying first the kinematic condition

$$\psi_r(0) = 0. \tag{6.5.13}$$

Secondly, these modes must be such that any angular displacement θ, or at least any that is likely to be encountered in the (low) frequency range which is being considered, can be expressed approximately in the form

$$\theta(x) = \sum_{r=1}^{n} q_r \psi_r(x). \tag{6.5.14}$$

The simplest set of functions satisfying the boundary condition is obtained by taking

$$\psi_r(x) = (x/l)^r \quad (r = 1, 2, \ldots, n). \tag{6.5.15}$$

These functions also satisfy the second criterion because it may be proved† that any function which satisfies the kinematic boundary condition and which is well-behaved in the interval $0 \leqslant x \leqslant l$ can be approximated to any specified accuracy by a series of the type (6.5.14) merely by taking n large enough. On the other hand, it may easily be verified that inaccurate results would be obtained if the term in $\psi_1(x)$ were omitted; this would be equivalent to imposing an appreciable constraint on the system, and the second criterion would thus be violated.

More accurate results can be obtained with a given number of assumed modes by demanding that they satisfy not only the kinematic boundary conditions, but also the dynamic boundary conditions. For a clamped-free shaft the functions would have to be chosen to satisfy the condition (6.5.12), that is

$$C(l)\,\psi_r'(l) = 0. \tag{6.5.16}$$

This condition and the previous one may be satisfied by putting

$$\psi_r(x) = (r+1)\,(x/l)^r - r(x/l)^{r+1} \quad (r = 1, 2, \ldots, n). \tag{6.5.17}$$

Duncan‡ has determined functions of this type for shafts in torsional vibration and beams in transverse vibration, for all the various end conditions, clamped-free, pinned-pinned, etc.

Even greater accuracy may be obtained by choosing the modes to be certain linear combinations of Duncan's functions. Thus one may choose the first function to be identical with the first Duncan function; for the second, one may combine the first two Duncan functions to give a node in roughly the expected place; and for the third, one may combine the first three Duncan functions to give two nodes in roughly the correct places.

† This may easily be deduced from Weierstrass's theorem; see, for example, I. P. Natanson, *Theory of Functions of a Real Variable* (Ungar, 1955), p. 109.

‡ W. J. Duncan, 'The representation of aircraft wings, tails and fuselages by semi-rigid structures in dynamic and static problems', *Aero Res. Counc. Rep. and Memo.* 1904 (1943).

If the principal modes of vibration are known for one system they can be used to advantage in the analysis of similar systems. Thus the modes of a uniform beam or shaft can be used in the analysis of non-uniform beams.† Other sets of functions for beam and shaft problems have been devised by Williams‡ and Rauscher.§

Provided that the modes are chosen properly, it is possible to obtain more accurate natural frequencies and principal modes of a system, by using a given number of modes than by using the same number of lumped masses. As an example, consider the use of the modes (6.5.17) with a uniform clamped-free shaft. The coefficients a_{rs}, c_{rs} are in this case given by

$$\left.\begin{aligned} a_{rs} &= J\int_0^l \psi_r(x)\,\psi_s(x)\,dx, \\ c_{rs} &= C\int_0^l \psi_r'(x)\,\psi_s'(x)\,dx. \end{aligned}\right\} \tag{6.5.18}$$

Introducing the dimensionless parameter $\xi = x/l$ we find

$$\left.\begin{aligned} a_{rs} &= Jl\int_0^1 \psi_r(\xi)\,\psi_s(\xi)\,d\xi, \\ c_{rs} &= (C/l)\int_0^1 \psi_r'(\xi)\,\psi_s'(\xi)\,d\xi, \end{aligned}\right\} \tag{6.5.19}$$

where in this last expression dashes denote differentiation with respect to ξ. Thus

$$\left.\begin{aligned} a_{11} &= Jl\int_0^1 (2\xi-\xi^2)^2\,d\xi = Jl(8/15), \\ c_{11} &= (C/l)\int_0^1 (2-2\xi)^2\,d\xi = (C/l)\,(4/3). \end{aligned}\right\} \tag{6.5.20}$$

If only this first mode is used the approximation obtained for the first natural frequency is

$$\omega_1 = \sqrt{(c_{11}/a_{11})} = 1{\cdot}5811\sqrt{\frac{C}{Jl^2}}. \tag{6.5.21}$$

This special case of the Rayleigh–Ritz method in which first one mode is used is usually called Rayleigh's method. The multiplier in equation (6.5.21) should be 1·5708, so that the approximation is extremely good, having a percentage error of 0·645. An inspection of Table 6.4.1 shows that this error is slightly greater than that incurred by taking four masses in a lumped mass model.

If the first two modes are used it is found that

$$A = Jl\begin{bmatrix} 8/15 & 13/30 \\ 13/30 & 13/35 \end{bmatrix}, \quad C = C/l\begin{bmatrix} 4/3 & 1 \\ 1 & 6/5 \end{bmatrix}, \tag{6.5.22}$$

which leads to the approximations

$$\omega_1 = 1{\cdot}5710\sqrt{\frac{C}{Jl^2}}, \quad \omega_2 = 4{\cdot}854\sqrt{\frac{C}{Jl^2}}. \tag{6.5.23}$$

† For example, Bishop and Johnson, *The Mechanics of Vibration*, chs. 6 and 7.

‡ D. Williams, 'The use of the principle of minimum potential energy in problems of static equilibrium', *Aero Res. Counc. Rep. and Memo.* 1827 (1937).

§ M. Rauscher, 'Station functions and air density variations in flutter analysis', *J. Aero. Sci.* **16** (1945), pp. 345–53.

The first frequency is now given almost exactly, and the second is in error by about 3·008 %. Table 6.4.2 shows that at least six masses would have to be used in a lumped-mass model to obtain an error as small as this.

It will be noticed that the approximate values for the natural frequencies which are shown in equations (6.5.21) and (6.5.23) are in each case too great. Since the fundamental assumption of the Rayleigh–Ritz method, that the system can vibrate only in certain specified modes, is equivalent to the imposition of a set of constraints, this is only to be expected. The proof of the result for discrete systems is given in Theorem 4.5.1. This feature provides a contrast with the lumped mass method in which the approximate values may sometimes be too great and sometimes too small.

If the Rayleigh–Ritz equations have been completely solved using a certain number of assumed modes, the effect of adding a further mode may sometimes be ascertained by using the 'escalator method',† the principle of which is as follows. Suppose that the equation

$$(\mathbf{C}-\omega^2\mathbf{A})\,\boldsymbol{\Psi} = \mathbf{0} \tag{6.5.24}$$

has been solved completely and that the matrix $\mathbf{X}$ of principal modes has been found. This matrix will satisfy equations (2.5.10) and (2.5.13), namely

$$\mathbf{X}'\mathbf{A}\mathbf{X} = \mathbf{L}, \quad \mathbf{X}'\mathbf{C}\mathbf{X} = \boldsymbol{\Omega}\mathbf{L}, \tag{6.5.25}$$

where $\mathbf{L}$ and $\boldsymbol{\Omega}$ are the diagonal matrices

$$\mathbf{L} = \begin{bmatrix} a_1 & 0 & \dots & 0 \\ 0 & a_2 & \dots & 0 \\ \dots & \dots & \dots & \dots \\ 0 & 0 & \dots & a_n \end{bmatrix}, \quad \boldsymbol{\Omega} = \begin{bmatrix} \omega_1^2 & 0 & \dots & 0 \\ 0 & \omega_2^2 & \dots & 0 \\ \dots & \dots & \dots & \dots \\ 0 & 0 & \dots & \omega_n^2 \end{bmatrix}. \tag{6.5.26}$$

The addition of a further mode will lead to the set of equations

$$\begin{bmatrix} \mathbf{C} & \mathbf{D} \\ \mathbf{D}' & c \end{bmatrix}\begin{bmatrix} \boldsymbol{\Psi} \\ x \end{bmatrix} - \omega^2 \begin{bmatrix} \mathbf{A} & \mathbf{B} \\ \mathbf{B}' & a \end{bmatrix}\begin{bmatrix} \boldsymbol{\Psi} \\ x \end{bmatrix} = \mathbf{0}, \tag{6.5.27}$$

where $\mathbf{B}$, $\mathbf{D}$ are column vectors of order n and a, c, x are scalars. The equations may be written

$$\left.\begin{aligned} (\mathbf{C}-\omega^2\mathbf{A})\,\boldsymbol{\Psi} + (\mathbf{D}-\omega^2\mathbf{B})\,x &= \mathbf{0}, \\ (\mathbf{D}'-\omega^2\mathbf{B}')\,\boldsymbol{\Psi} + (c-\omega^2 a)\,x &= 0, \end{aligned}\right\} \tag{6.5.28}$$

of which the second is simply a scalar equation. Now write

$$\boldsymbol{\Psi} = \mathbf{X}\boldsymbol{\Pi} \tag{6.5.29}$$

in line with equation (2.4.9), premultiply the first equation by $\mathbf{X}'$ and use equations (6.5.25). The result is

$$\left.\begin{aligned} \mathbf{L}(\boldsymbol{\Omega}-\omega^2\mathbf{I})\,\boldsymbol{\Pi} + \mathbf{X}'(\mathbf{D}-\omega^2\mathbf{B})\,x &= \mathbf{0}, \\ (\mathbf{D}'-\omega^2\mathbf{B}')\mathbf{X}\boldsymbol{\Pi} + (c-\omega^2 a)\,x &= 0. \end{aligned}\right\} \tag{6.5.30}$$

Now put

$$\mathbf{X}'(\mathbf{D}-\omega^2\mathbf{B}) = \mathbf{Z} = \{z_1, z_2, \dots, z_n\}, \tag{6.5.31}$$

† J. Morris, *The Escalator Method* (Chapman and Hall, 1947), ch. 10; J. Morris and J. W. Head, 'The escalator process for the solution of Lagrangian frequency equations', *Phil. Mag.*, ser. 7, **35** (1944), p. 735.

and the equations become

$$\left.\begin{aligned}\mathbf{L}(\boldsymbol{\Omega}-\omega^2\mathbf{I})\,\boldsymbol{\Pi}+\mathbf{Z}x &= \mathbf{0},\\ \mathbf{Z}'\boldsymbol{\Pi}+(c-\omega^2 a)\,x &= 0.\end{aligned}\right\} \tag{6.5.32}$$

The first equation gives $$\boldsymbol{\Pi} = -\mathbf{L}^{-1}(\boldsymbol{\Omega}-\omega^2\mathbf{I})^{-1}\mathbf{Z}x, \tag{6.5.33}$$

which when substituted into the second yields

$$\mathbf{Z}'\mathbf{L}^{-1}(\boldsymbol{\Omega}-\omega^2\mathbf{I})^{-1}\mathbf{Z}x = (c-\omega^2 a)\,x, \tag{6.5.34}$$

which may be written

$$\sum_{r=1}^{n}\frac{z_r^2}{a_r(\omega_r^2-\omega^2)} = c-\omega^2 a. \tag{6.5.35}$$

This equation can conveniently be solved by using Newton's method (see equations (9.5.3), (9.5.4)). Further discussion of the method, with details of its many applications and a guide to the relevant computation procedure, is to be found in the work of Morris.†

The principle of the Rayleigh–Ritz method can be applied to discrete systems as well as to continuous ones. We begin our discussion of this matter by a consideration of Rayleigh's quotient theorem (see § 2.10). This states that the quotient

$$\omega_c^2 = \frac{\boldsymbol{\Psi}'\mathbf{C}\boldsymbol{\Psi}}{\boldsymbol{\Psi}'\mathbf{A}\boldsymbol{\Psi}} \tag{6.5.36}$$

is stationary for the principal modes of a system, and that the stationary value corresponding to the rth mode $\boldsymbol{\Psi}^{(r)}$ is the square of the rth natural frequency ω_r.

The chief usefulness of Rayleigh's theorem lies in the fact that it provides an easy means of calculating an approximation to the fundamental frequency of a system. If *any* mode shape $\boldsymbol{\Psi}$ is substituted in the expression (6.5.36) the result will be an upper bound for ω_1^2; if it is a rough approximation to the true first mode $\boldsymbol{\Psi}^{(1)}$, the result will be a good approximation to ω_1^2.

The chief drawbacks of Rayleigh's theorem as an alternative to the *direct* calculation of the modes and frequencies from the equations

$$|\mathbf{C}-\omega^2\mathbf{A}| = 0, \quad (\mathbf{C}-\omega^2\mathbf{A})\,\boldsymbol{\Psi} = \mathbf{0} \tag{6.5.37}$$

are as follows:

(1) It can be expected to give *accurate* results only for the fundamental frequency of simple systems for which it is possible to guess the approximate shape of the mode.

(2) It is difficult to apply to higher frequencies of a system.

(3) It contains no 'built-in' means of calculating an improved estimate of the mode shape from the calculated value of the frequency.

(4) Its systematic use involves the characteristic value problem given by equations (6.5.37); these are the equations which will be obtained when one seeks the stationary values of the quotient.

If the Rayleigh–Ritz approach is applied to the problem these drawbacks are largely overcome. The basic step is, as usual, that one chooses a family of modes

† Morris, *The Escalator Method.*

$\mathbf{\Theta}^{(1)}, \mathbf{\Theta}^{(2)}, \ldots, \mathbf{\Theta}^{(m)}$ and considers only those vibrations of the system which may be expressed as a sum of vibrations in these modes. That is, one assumes that

$$\mathbf{\Psi} = g_1 \mathbf{\Theta}^{(1)} + g_2 \mathbf{\Theta}^{(2)} + \ldots + g_m \mathbf{\Theta}^{(m)}, \tag{6.5.38}$$

where $g_1, g_2, \ldots, g_m$ are a set of coordinates. This equation may be rewritten as

$$\mathbf{\Psi} = \mathbf{\Theta\Gamma}, \tag{6.5.39}$$

where $\mathbf{\Theta}$ denotes the matrix of order $n \times m$ having the $\mathbf{\Theta}^{(r)}$ as columns and

$$\mathbf{\Gamma} = \{g_1, g_2, \ldots, g_m\}. \tag{6.5.40}$$

The next step is that one forms the quotient (6.5.36) and finds the values of the coordinates $g_1, g_2, \ldots, g_m$ which make its value stationary. The quotient is

$$\frac{\mathbf{\Psi}'\mathbf{C}\mathbf{\Psi}}{\mathbf{\Psi}'\mathbf{A}\mathbf{\Psi}} = \frac{\mathbf{\Gamma}'(\mathbf{\Theta}'\mathbf{C}\mathbf{\Theta})\mathbf{\Gamma}}{\mathbf{\Gamma}'(\mathbf{\Theta}'\mathbf{A}\mathbf{\Theta})\mathbf{\Gamma}} = \frac{\mathbf{\Gamma}'\mathbf{C}^*\mathbf{\Gamma}}{\mathbf{\Gamma}'\mathbf{A}^*\mathbf{\Gamma}}, \tag{6.5.41}$$

where $\mathbf{C}^*$, $\mathbf{A}^*$ are square matrices of order m. The equations obtained by demanding that the quotient be stationary are simply

$$|\mathbf{C}^* - \omega^2 \mathbf{A}^*| = 0 \tag{6.5.42}$$

and

$$(\mathbf{C}^* - \omega^2 \mathbf{A}^*)\,\mathbf{\Gamma} = \mathbf{0}, \tag{6.5.43}$$

where

$$\mathbf{A}^* = \mathbf{\Theta}'\mathbf{A}\mathbf{\Theta}, \quad \mathbf{C}^* = \mathbf{\Theta}'\mathbf{C}\mathbf{\Theta}. \tag{6.5.44}$$

The basic assumption of the Rayleigh–Ritz method, equation (6.5.38), is that one considers the system to be constrained so that it can vibrate only in combinations of a certain set of m modes. If in fact $m = n$, and the modes $\mathbf{\Theta}^{(r)}$ are linearly independent, then equation (6.5.38) will not put any constraint at all upon the system; this is equivalent to choosing a new set of coordinates and it may be a simplification to do this in some cases, but the resulting equation (6.5.42) will still be of order n. If $m < n$ the method leads to an equation of order m, the first few of whose characteristic values will be approximations to the first few of the natural frequencies of the original system. The accuracy of the approximations will depend on the choice of the modes $\mathbf{\Theta}^{(r)}$ and the number, m, of modes which are considered. The important criterion for the choice of the modes is that the constraint introduced by the assumption (6.5.38) must not have too much effect on the vibration in the principal modes which are sought.

It will be noticed that: (i) the method requires far less guess-work than the straightforward Rayleigh method, (ii) it provides a means of calculating the corresponding mode (equation (6.5.43) gives the values of the coordinates g to be used in equation (6.5.38)) and (iii) it may be applied to frequencies other than the fundamental.

There are numerous other methods for solving vibration problems related to continuous systems. These include Stodola's method,† the complementary energy

† J. P. den Hartog, *Mechanical Vibrations* (McGraw-Hill, 4th edn., 1956), p. 156.

method,† and methods employing influence coefficients and an integral equation formulation. An excellent review of these and other methods, and a comparison with 'lumped-mass' and 'assumed-mode' methods, may be found in a paper by Minhinnick,‡ who also gives an exhaustive bibliography of literature covering the period up to 1953. Other general discussions may be found in textbooks on Aeroelasticity.§

At the time when Minhinnick's review was written (1952–53), digital computers were not generally available. Their advent has shifted the centre of difficulty in problems of vibration analysis, particularly those relating to large and complicated structures. Previously the greatest difficulty was encountered in the solution of the matrix equations, and therefore much time and energy were expended in devising methods which did not involve the solution of high-order matrix equations. Minhinnick summarises the situation by saying 'the reduction of the order of the matrix is the most important consideration'. The Rayleigh–Ritz or complementary energy method using specially constructed functions and numerical integration, and the solution of the integral equation, again using special functions and numerical integration or collocation were some of the preferred methods. In spite of the fact that they involved much more calculation in the initial stages, they were preferred to lumped-mass methods because they could be used with far fewer degrees of freedom.

Now that computers are in widespread use, matrix problems offer very little difficulty. Most computers have ready-made programmes for the inversion and characteristic value problems of large matrices. *Now* a method is judged mainly on the amount of work which is involved in the preparation of the equations for the computer, i.e. the determination of the matrices from the given data. The result is that, at least in preliminary calculations, lumped-mass models are used wherever possible because they involve so much less work in the pre-computer stage. An additional argument in favour of the simpler lumped-mass methods is that, at present, the greatest errors incurred in the analytical determination of natural modes and frequencies arise from errors in the calculation of the mass, and (more particularly) the stiffness, data. Until greater accuracy can be introduced in this stage of the process it is quite reasonable to use lumped-mass methods.

The more sophisticated methods still have their place where calculation must be performed using a desk machine, or where a more precise analysis must be made. For some of these methods the preliminary determination of the matrix coefficients can be achieved by an auxiliary matrix programme.||

† P. A. Libby and R. C. Sauer, 'Comparison of the Rayleigh–Ritz and complementary energy methods in vibration analysis', *J. Aero. Sci.* **16**, no. 11 (Nov. 1949), p. 700; A. I. van de Vooren and J. H. Greidanus, 'Complementary energy method in vibration analysis', *J. Aero. Sci.* **17**, no. 7 (July 1950), p. 454. Other references are given in these papers.

‡ I. T. Minhinnick, 'The theoretical determination of normal modes and frequencies of vibration', *AGARD Rep.* no. 36 (1956).

§ Bisplinghoff, Ashley and Halfman, *Aeroelasticity*. Y. C. Fung, *The Theory of Aeroelasticity* (Wiley, 1955); R. H. Scanlan and R. Rosenbaum, *Aircraft Vibration and Flutter* (Macmillan, 1951).

|| See, for example, S. U. Benscoter and M. L. Gossard, 'Matrix methods for calculating cantilever beam deflections', *N.A.C.A. Tech. Note* 1827 (1949); J. S. Archer, 'A stiffness matrix method of natural mode analysis', *Proc. Nat. Specialists Meeting* (*Inst. Aero. Sci.*), *Fort Worth, Texas*, pp. 88–97.

EXAMPLES 6.5

1. Use the Principle of Virtual Work, that is equation (6.5.1), to obtain the equations of motion and the dynamic end conditions for a non-uniform cantilever undergoing transverse vibration. It may be assumed that

$$V = \frac{1}{2}\int_0^l EI\left(\frac{\partial^2 v}{\partial x^2}\right)^2 dx.$$

Particular attention should be paid to the argument used to obtain the dynamic end conditions.

2. Find a set of 'Duncan functions' for a non-uniform cantilever in transverse vibration. Assume that the free end is not 'pointed'.

Would the functions be suitable if a concentrated load and/or moment were applied at the free end? If not, construct suitable sets of functions for use in such situations.

3. Find the lowest natural frequency of the system shown in fig. 6.2.1 when $I = \mathcal{J}l/2$ by the Rayleigh–Ritz method with

(i) one principal mode of the clamped-free shaft;
(ii) two principal modes of the clamped-free shaft;
(iii) one of the functions (6.5.17);
(iv) two of these functions.

4. Find approximations to the first natural frequency of the system specified by the matrices (6.4.5) with $n = 7$ by using first one, and then both the modes

$$\boldsymbol{\Theta}^{(1)} = \{1,\ 2,\ 3,\ 4,\ 5,\ 6,\ 7\},$$
$$\boldsymbol{\Theta}^{(2)} = \{1,\ 4,\ 9,\ 16,\ 25,\ 36,\ 49\},$$

in a Rayleigh–Ritz analysis. Compare with the results shown in Table 6.4.1 by putting $I = \mathcal{J}l/7$, $k = 7C/l$.

6.6 Branch mode analysis

In this section we shall describe a further method which has been found to be particularly useful for the analysis of composite structures. The method, which was initially devised for the determination of the free-free natural frequencies and normal modes of an aircraft in still air, is designed for use with an electronic computer. It is a hybrid of the 'lumped-mass' and 'assumed modes' methods which combines the simplicity of the former with some of the advantages of the latter (see § 6.5). In this section we shall give a general discussion of the method, leaving the details of its application to a particular problem to § 6.7.

The essence of the procedure is to divide the problem into a number of parts. In the simplest form of the method the procedure would be as follows:

(1) Replace the structure by a lumped-mass system.
(2) Impose a sequence of sets of constraints on the system in turn. These sets of constraints are usually chosen so that, when one set is imposed, only one component of the system can vibrate with distortion, the others either remaining fixed or vibrating as rigid bodies (but in either case remaining attached to the distorting component).

(3) Determine a certain number of the lower natural frequencies and principal modes of these constrained systems.

(4) Use these modes, which are called 'branch modes' (together with the appropriate rigid-body modes, if the system is free) in a Rayleigh–Ritz analysis of the whole system.

The method has several advantages over a pure 'lumped-mass' or a pure 'assumed-modes' approach. The former technique raises the difficulty that accurate results can be obtained only by using a very large number of degrees of freedom. Even when computers are available this situation is to be avoided whenever possible as the computing time for one characteristic value of a matrix of order n is proportional to n^2 at least. The main difficulty in an 'assumed-modes' approach is that, for a complicated system, it is difficult to determine a suitable set of modes—that is, a set of modes which bear some resemblance to the possible vibrations of the system. The Branch Mode Method overcomes both these difficulties and has several other advantages in addition. It avoids excessively large matrices by substituting a number of characteristic value problems relating to small matrices for one characteristic value problem relating to one large matrix. The saving in time may be illustrated by a simple example. Suppose that the system has four components, each of which is approximated by a lumped-mass system having 20 masses, and suppose that the first five natural frequencies of the system are required. For a lumped-mass solution applied to the whole system the computing time would be *very roughly* $T_1 = 5\times 80^2$. If the branch mode method is used, and if, for the sake of argument, five modes of each constrained system (each 'branch') are used in the final calculations, the time taken would be

$$T_2 = 4\times 5\times 20^2 + 5\times 20^2 = (5/16)\,T_1. \qquad (6.6.1)$$

It avoids the difficulty of determining a suitable set of modes by including the preliminary calculations on the branches; these are specifically designed to furnish a set of modes which are related to the system.† Provided that the branch modes are properly chosen it is found that the Rayleigh–Ritz analysis of the lumped-mass system (stage 4) produces errors which are negligible compared to those incurred by replacing the original system by the lumped-mass system (stage 1). This, combined with the fact that it is so much quicker, is the chief reason why the method is used.‡

Another advantage of the method is that, even with modern computers, there is a limit to the number of degrees of freedom that may be taken into account in a characteristic value problem (i.e. a limit on n). The branch mode method enables one to obtain greater accuracy while still remaining within this limit. Also, the

† Suitable modes for a composite system may also be built up by piecing together admissible modes for each component by means of the equilibrium and compatibility relations which hold at the joins between subsystems. This is described by W. C. Hurty, 'Vibrations of structural systems by component mode synthesis', *Proc. Amer. Soc. Civil Eng.* **86** (1960), EM4, pp. 51–69.

‡ B. A. Hunn, 'A method of calculating space free resonant modes of an aircraft', *J. Roy. Aero. Soc.* **57** (1953), pp. 420–2; B. A. Hunn, 'A method of calculating the normal modes of an aircraft', *Quart. J. Mech. Appl. Math.* **8** (1955), pp. 38–58; A. D. N. Smith, 'The use of linear algebra programmes in the field of aircraft flutter', *R.A.E. Structures Report*, D4/14827/ADNS.

effect on the whole structure of a modification in one of the components is much more easily determined. Finally, because smaller matrices are employed, the calculations are less liable to error. Also if an error occurs, the effect is less time-wasting, since only one, or at most, two calculations are affected.

We shall now discuss the various stages of the method which were outlined at the beginning of the section. The first stage, the way in which a continuous system should be approximated by a lumped-mass model, was considered, at least for simple systems, in § 6.4. For these systems some estimates were given for the number of degrees of freedom which are needed to obtain a certain number of natural frequencies with a specified accuracy. The approximation of more complicated systems by lumped-mass models is not our concern here (this is a part of stage 3 of the complete procedure of vibration analysis as described in § 2.0).

The second stage in the branch mode method is perhaps the most important. We shall now consider some simple systems and show how the branches are chosen.

6.6.1 Two-component free-free systems

(*a*) *Torsional systems.* In § 6.4, methods are described for approximating a massive shaft by a set of flywheels on a light shaft. Suppose that a system consisting of two such massive shafts joined end to end is approximated in this way using $n+1$ flywheels for the first shaft, B (it is not assumed that the flywheels have equal moments of inertia or that the stiffnesses of the connecting shafts are equal) and $n'+1$ for the other, C, as in fig. 6.6.1.†

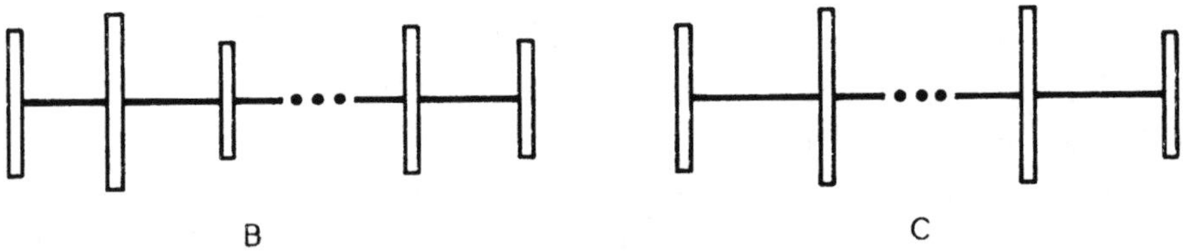

Fig. 6.6.1

The composite system will have only $(n+1)+(n'+1)-1 = n+n'+1$ degrees of freedom because the middle flywheels have coalesced. To find a suitable set of branches we argue as follows. For an *exact* analysis of the system, its motion must be described in terms of $n+n'+1$ linearly independent modes, one of which must be the rigid-body mode. We must therefore determine the branches so that the aggregate of branch modes contains just this number of linearly independent modes. Then for the final stage (4) an appropriate selection may be made from this set. For the system shown in fig. 6.6.1 there are two simple ways of choosing the branches:

(α) by demanding that one of the components vibrates with distortion while the other vibrates as a rigid body rigidly attached to the distorting component;

† The detailed application of the method to a simple system of this type is discussed in § 6.7.

(β) by demanding that one of the components vibrates with distortion while the other remains fixed.

These are indicated in fig. 6.6.2.

It will be noticed that in case α the rigid-body mode of the composite system is included in both sets of branch modes, but that the remaining modes are linearly independent. In case β, on the other hand, the rigid body mode does not occur in either set and so has to be added in separately. Rigid body modes can be considered as branch modes of the system obtained by constraining the whole system to be rigid. It will be noticed that the compatibility condition $q_{b2} = q_{c2}$ at the junction of the subsystem is satisfied in both α and β.

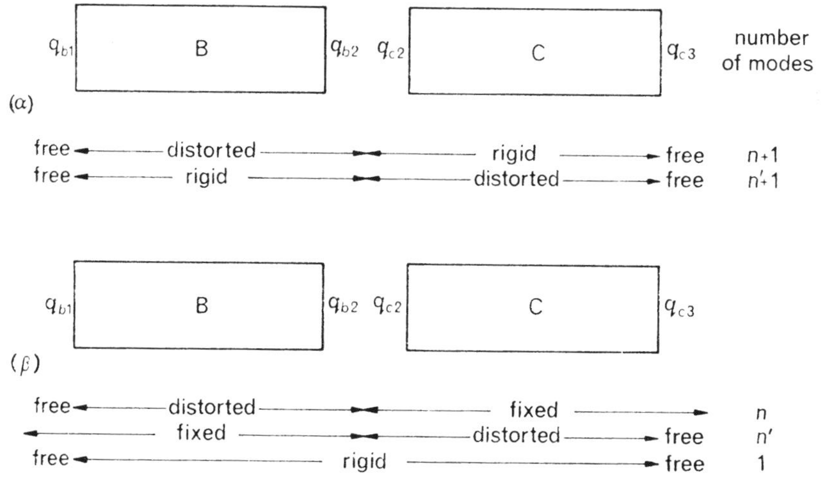

Fig. 6.6.2

(*b*) *Beams in transverse vibration.* Such systems may be analysed in ways which are similar to those just described. If *B* and *C* represent lumped-mass models of beams, e.g. sets of concentrated masses on massless elastic beams, then there will be *two* linking coordinates between the subsystems, the deflection and slope at the join. In forming the branches the analogues of either α or β may be used. The only differences will be that the 'fixed' condition will be replaced by 'clamped' (to ensure the satisfaction of both the compatibility conditions) and there will be two rigid body modes instead of one.

It will be understood that the number of linking coordinates may be greater than two. Thus the bending-and-twisting vibration of a beam of channel section, or a wing-fuselage system, could be handled by schemes such as α or β. These schemes can in fact be used for any two-component free-free systems. It can be verified also that they can be used for any two-component system which has at least one rigid-body mode, such as a sliding-free beam system.

If one component, *B*, is much larger than the other, it would be advisable to take account of this and use a division into branches such as that shown in fig. 6.6.3. This scheme has the advantage that it provides just the right number of modes.

6.6.2 Two-component supported-free systems

If the system shown in fig. 6.6.1 is fixed at the left-hand end, the composite system will have $n+(n'+1)-1$, that is $n+n'$, degrees of freedom. A set of $n+n'$ linearly independent branch modes may be found as shown in fig. 6.6.4.

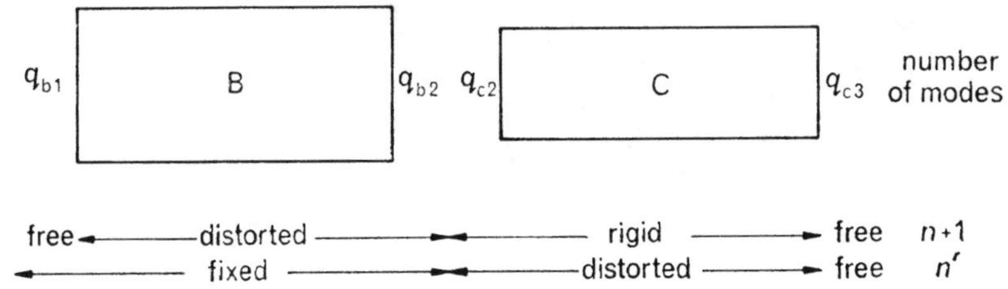

Fig. 6.6.3

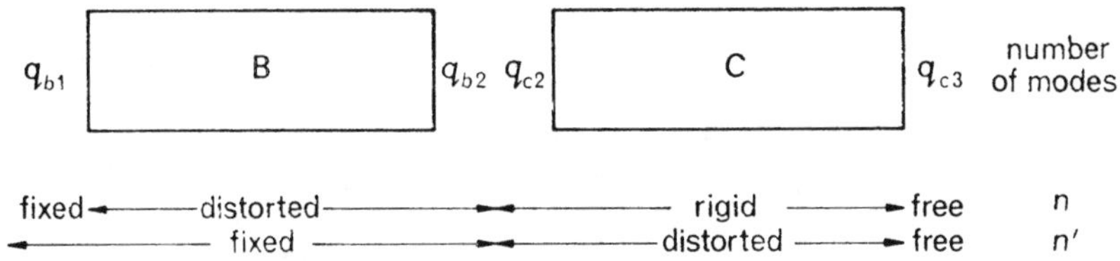

Fig. 6.6.4

It may be verified that this method of division may be used for any two-component, supported-free system, such as a clamped-free beam. Again the number of linking coordinates may be greater than two if necessary.

6.6.3 Two-component supported-supported system

No simple set of branch modes may be obtained for a system supported at both ends. Thus for the torsional system fixed at both ends, the analysis of fig. 6.6.5 is

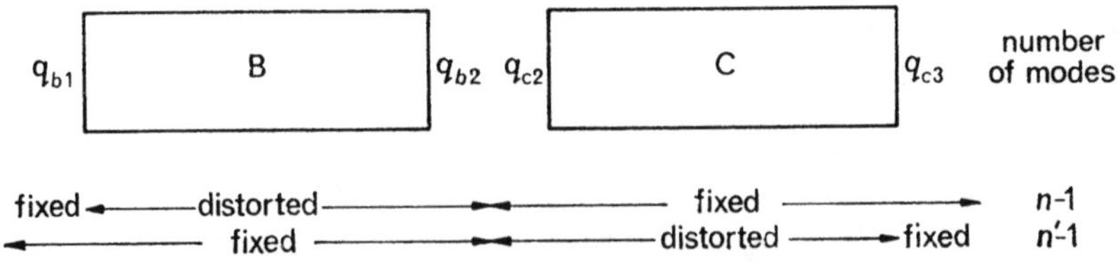

Fig. 6.6.5

insufficient because it gives only $n+n'-2$ modes, whereas the composite system has $n+n'-1$ degrees of freedom. The inadequacy may also be seen from the fact that all the branch modes give zero deflection at the joining coordinate, though clearly the deflection of the composite system will not always be zero at this point. A set of branch modes may be obtained by dividing the system into three branches. This is illustrated for a torsional system having 7 degrees of freedom in fig. 6.6.6.

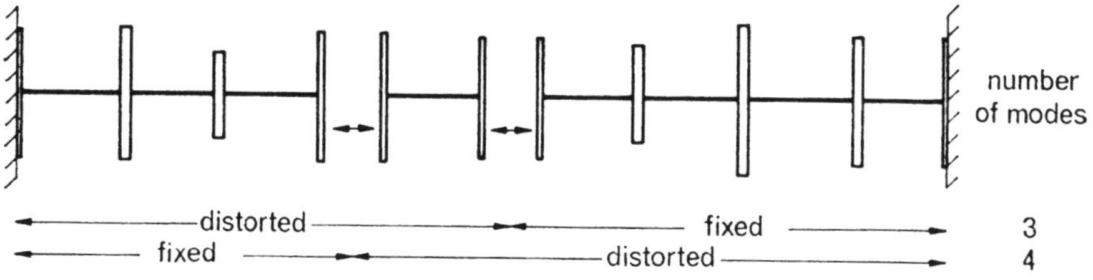

Fig. 6.6.6

6.6.4 Systems with many components

Systems having more than two components may be analysed in various ways similar to those shown above. In particular multi-component systems which have at least one rigid-body mode may always be analysed by the analogue of scheme β in fig. 6.6.2—that is, by assuming that just one component vibrates with distortion while the others remain rigid. A method of analysis for a 3-component supported-free system is shown in fig. 6.6.7.

To conclude the discussion of stage 2 we take note of two matters. First, a general criterion governing the formation of the branches is that the aggregate of the sets of branch modes must be 'complete'. That is, it must be possible to express any

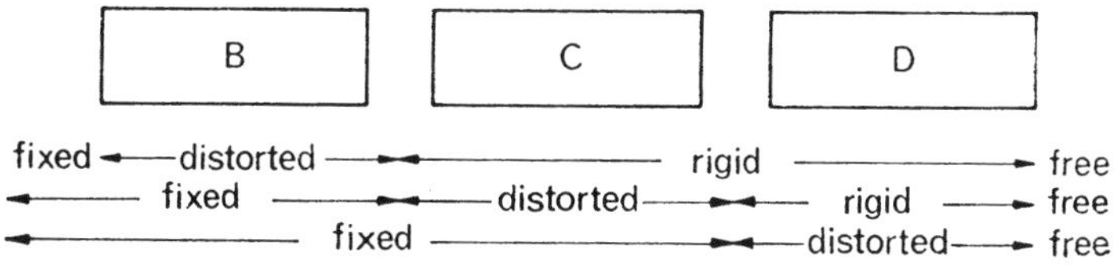

Fig. 6.6.7

motion of the system as a sum of motions in the various branch modes. This is the criterion that was applied in subsection (i) (*a*) above. Secondly, it is not essential that the modes that are used in the final stage of the analysis should be chosen in the manner described above. Rough approximations to the principal modes of the branches could be used instead of exact principal modes, without much loss in accuracy. The important point is always that the modes should bear some resemblance to the possible vibrations of the system.

The only matter connected with stage 3 which has to be discussed is the criterion for deciding *how many* modes have to be taken from each set of branch modes for use in the final stage of the analysis. This is largely a matter which has to be decided on the basis of experience. One rule that has been suggested for aircraft flutter analysis, in calculating the lowest modes, is to take those branch modes whose natural frequencies are less than twice the highest natural frequency required.

We shall now make some general remarks about the final stage in the analysis, the Rayleigh–Ritz analysis of the system using a suitable selection of the branch modes. Suppose that the lumped mass model of the system has n degrees of freedom, and that it is analysed into branches $B, C, \ldots, S$. (A branch can be labelled by the same letter that is used to denote the subsystem which is allowed to vibrate with distortion, in the branch.) Suppose further that the numbers of modes taken from the branches are $m_1, m_2, \ldots, m_r$ respectively, and that the system has m_0 rigid body modes. The sets of branch modes may be represented as the columns of matrices $\mathbf{X}_b, \mathbf{X}_c, \ldots, \mathbf{X}_s$, having orders $n \times m_1, n \times m_2, \ldots, n \times m_r$ respectively, and the rigid body modes as the columns of a matrix $\mathbf{X}_0$ of order $n \times m_0$.

The basic assumption of the Rayleigh–Ritz analysis lies in equation (6.5.39), namely

$$\mathbf{\Psi} = \mathbf{\Theta\Gamma}. \tag{6.6.2}$$

In the present problem the matrix $\mathbf{\Theta}$ is

$$\mathbf{\Theta} = [\mathbf{X}_0, \mathbf{X}_b, \mathbf{X}_c, \ldots, \mathbf{X}_s], \tag{6.6.3}$$

which may be conveniently partitioned into rigid body modes and other branch modes by writing it in the form

$$\mathbf{\Theta} = [\mathbf{X}_0, \mathbf{X}_a], \tag{6.6.4}$$

where

$$\mathbf{X}_a = [\mathbf{X}_b, \mathbf{X}_c, \ldots, \mathbf{X}_s]. \tag{6.6.5}$$

This last matrix is of order $n \times m$, where

$$m = \sum_{t=1}^{r} m_t. \tag{6.6.6}$$

The total number of modes, $m_0 + m$, will in general be much less than n. If $\mathbf{\Gamma}$ is partitioned in the same way, into coordinates relating to rigid body modes and coordinates relating to other branch modes, that is

$$\mathbf{\Gamma} = \begin{bmatrix} \mathbf{\Gamma}_0 \\ \mathbf{\Gamma}_a \end{bmatrix}, \tag{6.6.7}$$

where $\mathbf{\Gamma}_0$, $\mathbf{\Gamma}_a$ are column matrices of orders m_0 and m respectively, then equation (6.6.2) becomes

$$\mathbf{\Psi} = [\mathbf{X}_0 \quad \mathbf{X}_a] \begin{bmatrix} \mathbf{\Gamma}_0 \\ \mathbf{\Gamma}_a \end{bmatrix}. \tag{6.6.8}$$

The inertia matrix used in the Rayleigh–Ritz analysis will be (see equation (6.5.44))

$$\mathbf{A}^* = \mathbf{\Theta}'\mathbf{A\Theta}. \tag{6.6.9}$$

This may be partitioned in the form

$$\begin{bmatrix} \mathbf{X}_0' \\ \mathbf{X}_b' \\ \vdots \\ \mathbf{X}_s' \end{bmatrix} \mathbf{A}[\mathbf{X}_0 \quad \mathbf{X}_b \quad \ldots \quad \mathbf{X}_s] = \begin{bmatrix} \mathbf{A}_{00} & \mathbf{A}_{0b} & \ldots & \mathbf{A}_{0s} \\ \mathbf{A}_{b0} & \mathbf{A}_{bb} & \ldots & \mathbf{A}_{bs} \\ \ldots & \ldots & \ldots & \ldots \\ \mathbf{A}_{s0} & \mathbf{A}_{sb} & \ldots & \mathbf{A}_{ss} \end{bmatrix}, \tag{6.6.10}$$

where $$\mathbf{A}_{00} = \mathbf{X}_0'\mathbf{A}\mathbf{X}_0, \quad \mathbf{A}_{bb} = \mathbf{X}_b'\mathbf{A}\mathbf{X}_b, \quad \mathbf{A}_{0b} = \mathbf{X}_0'\mathbf{A}\mathbf{X}_b = \mathbf{A}_{b0}', \text{ etc.} \tag{6.6.11}$$

are matrices of orders $m_0 \times m_0$, $m_1 \times m_1$, $m_0 \times m_1$, etc. The complete matrix $\mathbf{A}^*$ is a square matrix of order $m_0 + m$.

The matrix $\mathbf{A}^*$ may also be partitioned in the form

$$\begin{bmatrix} \mathbf{A}_{00} & \vdots & \mathbf{A}_{0b} & \cdots & \mathbf{A}_{0s} \\ \cdots & & \cdots & & \cdots \\ \mathbf{A}_{b0} & \vdots & \mathbf{A}_{bb} & \cdots & \mathbf{A}_{bs} \\ \cdots & & \cdots & & \cdots \\ \mathbf{A}_{s0} & \vdots & \mathbf{A}_{sb} & \cdots & \mathbf{A}_{ss} \end{bmatrix} = \begin{bmatrix} \mathbf{A}_{00} & \mathbf{A}_{0a} \\ \mathbf{A}_{a0} & \mathbf{A}_{aa} \end{bmatrix}, \tag{6.6.12}$$

where $$\mathbf{A}_{aa} = \mathbf{X}_a'\mathbf{A}\mathbf{X}_a, \quad \mathbf{A}_{0a} = \mathbf{X}_0'\mathbf{A}\mathbf{X}_a = \mathbf{A}_{a0}' \tag{6.6.13}$$

are matrices of orders $m \times m$ and $m_0 \times m$ respectively.

The stability matrix
$$\mathbf{C}^* = \mathbf{\Theta}'\mathbf{C}\mathbf{\Theta} \tag{6.6.14}$$

may be partitioned in similar ways, namely

$$\begin{bmatrix} \mathbf{X}_0' \\ \mathbf{X}_b' \\ \vdots \\ \mathbf{X}_s' \end{bmatrix} \mathbf{C}[\mathbf{X}_0 \quad \mathbf{X}_b \quad \cdots \quad \mathbf{X}_s] = \begin{bmatrix} \mathbf{C}_{00} & \vdots & \mathbf{C}_{0b} & \cdots & \mathbf{C}_{0s} \\ \cdots & & \cdots & & \cdots \\ \mathbf{C}_{b0} & \vdots & \mathbf{C}_{bb} & \cdots & \mathbf{C}_{bs} \\ \cdots & & \cdots & & \cdots \\ \mathbf{C}_{s0} & \vdots & \mathbf{C}_{sb} & \cdots & \mathbf{C}_{ss} \end{bmatrix} = \begin{bmatrix} \mathbf{C}_{00} & \mathbf{C}_{0a} \\ \mathbf{C}_{a0} & \mathbf{C}_{aa} \end{bmatrix}, \tag{6.6.15}$$

but now $\mathbf{C}_{00}$, $\mathbf{C}_{0a}$ and $\mathbf{C}_{a0}$ will all be zero. This follows from the fact that $\mathbf{X}_0$, being made up of columns which are rigid-body modes of the system, satisfies the equations

$$\mathbf{C}\mathbf{X}_0 = \mathbf{0} = \mathbf{X}_0'\mathbf{C}. \tag{6.6.16}$$

Thus $$\mathbf{C}_{00} = \mathbf{X}_0'\mathbf{C}\mathbf{X}_0 = \mathbf{0}; \quad \mathbf{C}_{0a} = \mathbf{X}_0'\mathbf{C}\mathbf{X}_a = \mathbf{0} = \mathbf{C}_{a0}'. \tag{6.6.17}$$

The Rayleigh–Ritz equation
$$(\mathbf{C}^* - \omega^2\mathbf{A}^*)\mathbf{\Gamma} = \mathbf{0} \tag{6.6.18}$$
now becomes

$$\begin{bmatrix} \mathbf{0} & \mathbf{0} \\ \mathbf{0} & \mathbf{C}_{aa} \end{bmatrix} \begin{bmatrix} \mathbf{\Gamma}_0 \\ \mathbf{\Gamma}_a \end{bmatrix} - \omega^2 \begin{bmatrix} \mathbf{A}_{00} & \mathbf{A}_{0a} \\ \mathbf{A}_{a0} & \mathbf{A}_{aa} \end{bmatrix} \begin{bmatrix} \mathbf{\Gamma}_0 \\ \mathbf{\Gamma}_a \end{bmatrix} = \mathbf{0}. \tag{6.6.19}$$

The first matrix in this last equation is singular, and for computational purposes it is convenient to work with an equation containing non-singular matrices. Such an equation may be obtained as follows. Multiply out: this gives

$$\left.\begin{aligned} \mathbf{A}_{00}\mathbf{\Gamma}_0 + \mathbf{A}_{0a}\mathbf{\Gamma}_a &= \mathbf{0}, \\ \mathbf{C}_{aa}\mathbf{\Gamma}_a - \omega^2(\mathbf{A}_{a0}\mathbf{\Gamma}_0 + \mathbf{A}_{aa}\mathbf{\Gamma}_a) &= \mathbf{0}. \end{aligned}\right\} \tag{6.6.20}$$

The first equation yields $$\mathbf{\Gamma}_0 = -\mathbf{A}_{00}^{-1}\mathbf{A}_{0a}\mathbf{\Gamma}_a, \tag{6.6.21}$$

which when substituted in the second gives

$$[\mathbf{C}_{aa} - \omega^2(\mathbf{A}_{aa} - \mathbf{A}_{a0}\mathbf{A}_{00}^{-1}\mathbf{A}_{0a})]\,\mathbf{\Gamma}_a = \mathbf{0}. \tag{6.6.22}$$

The frequency equation for the system is thus

$$|\mathbf{C}_{aa}-\omega^2(\mathbf{A}_{aa}-\mathbf{A}_{a0}\mathbf{A}_{00}^{-1}\mathbf{A}_{0a})| = 0, \tag{6.6.23}$$

which is in the standard form

$$|\mathbf{C}-\omega^2\mathbf{A}| = 0. \tag{6.6.24}$$

It will be noticed that the matrix $\mathbf{A}_{aa}-\mathbf{A}_{a0}\mathbf{A}_{00}^{-1}\mathbf{A}_{0a}$ is symmetric because $\mathbf{A}_{aa}$ is symmetric and

$$\mathbf{A}_{0a} = \mathbf{A}'_{a0}. \tag{6.6.25}$$

Numerical methods for solving equations of the form (6.6.24) are discussed in Chapters 8 and 9.

If the branch modes are chosen to be exact principal modes of the (lumped-mass) system when it is subjected to a certain set of constraints, some simplification is introduced into the analysis. The following result holds.

Theorem 6.6.1. Two branch modes belonging to the same branch system are **A**-orthogonal and **C**-orthogonal, where **A** and **C** are the inertia and stability matrices of the *whole* system.

Proof. Suppose that the branch is obtained from the original system by applying $n-n_b$ constraints. For the branch system there will be $n-n_b$ independent relations between the coordinates $q_1, q_2, \ldots, q_n$, so that they may all be expressed in terms of n_b of them, and without loss of generality we may assume these to be the first n_b coordinates. In matrix notation this means that

$$\mathbf{q} = \mathbf{B}_b\mathbf{q}, \tag{6.6.26}$$

where $\mathbf{B}_b$ is a singular matrix having zeros in the last $n-n_b$ columns. For example, if $n = 5$ and the constraints are

$$q_3 = q_4 = q_5, \tag{6.6.27}$$

then

$$\begin{bmatrix} q_1 \\ q_2 \\ q_3 \\ q_4 \\ q_5 \end{bmatrix} = \begin{bmatrix} 1 & 0 & 0 & 0 & 0 \\ 0 & 1 & 0 & 0 & 0 \\ 0 & 0 & 1 & 0 & 0 \\ 0 & 0 & 1 & 0 & 0 \\ 0 & 0 & 1 & 0 & 0 \end{bmatrix} \begin{bmatrix} q_1 \\ q_2 \\ q_3 \\ q_4 \\ q_5 \end{bmatrix}. \tag{6.6.28}$$

The kinetic and potential energies of the branch system will be given by

$$\left.\begin{aligned} 2T &= \dot{\mathbf{q}}'\mathbf{A}\dot{\mathbf{q}} = \dot{\mathbf{q}}'\mathbf{B}_b\mathbf{A}\mathbf{B}_b\dot{\mathbf{q}}, \\ 2V &= \mathbf{q}'\mathbf{C}\mathbf{q} = \mathbf{q}'\mathbf{B}'_b\mathbf{C}\mathbf{B}_b\mathbf{q}. \end{aligned}\right\} \tag{6.6.29}$$

The matrices $\mathbf{B}'_b\mathbf{A}\mathbf{B}_b$ and $\mathbf{B}'_b\mathbf{C}\mathbf{B}_b$ will have zeros in the last $n-n_b$ rows and columns, and may thus be written

$$\mathbf{B}'_b\mathbf{A}\mathbf{B}_b = \begin{bmatrix} \mathbf{A}_b & \mathbf{0} \\ \mathbf{0} & \mathbf{0} \end{bmatrix}, \quad \mathbf{B}_b\mathbf{C}\mathbf{B}_b = \begin{bmatrix} \mathbf{C}_b & \mathbf{0} \\ \mathbf{0} & \mathbf{0} \end{bmatrix}. \tag{6.6.30}$$

The modes of the branch may be defined by column vectors $\mathbf{\Psi}_b^{*(r)}$ of order n_b. These modes will satisfy the usual orthogonality relations

$$\mathbf{\Psi}_b^{*'(r)}\mathbf{A}_b\mathbf{\Psi}_b^{*(s)} = 0 = \mathbf{\Psi}_b^{*'(r)}\mathbf{C}_b\mathbf{\Psi}_b^{*(s)}, \tag{6.6.31}$$

with respect to $\mathbf{A}_b$ and $\mathbf{C}_b$. The mode vector $\mathbf{\Psi}_b^{*(r)}$ gives the ratios which exist between the n_b independent coordinates in the rth mode of the branch system. The ratios which exist between all the n coordinates will be given by the vector

$$\mathbf{\Psi}_b^{(r)} = \mathbf{B}_b \begin{bmatrix} \mathbf{\Psi}_b^{*(r)} \\ 0 \end{bmatrix}. \tag{6.6.32}$$

The orthogonality of the modes $\mathbf{\Psi}_b^{(r)}$ now follows from the fact that

$$\begin{aligned} \mathbf{\Psi}_b^{\prime(r)}\mathbf{A}\mathbf{\Psi}_b^{(s)} &= [\mathbf{\Psi}_b^{*\prime(r)} \quad \mathbf{0}]\,\mathbf{B}_b'\mathbf{A}\mathbf{B}_b \begin{bmatrix} \mathbf{\Psi}_b^{*(s)} \\ \mathbf{0} \end{bmatrix} \\ &= [\mathbf{\Psi}_b^{*\prime(r)} \quad \mathbf{0}] \begin{bmatrix} \mathbf{A}_b & \mathbf{0} \\ \mathbf{0} & \mathbf{0} \end{bmatrix} \begin{bmatrix} \mathbf{\Psi}_b^{*(s)} \\ \mathbf{0} \end{bmatrix} = 0. \end{aligned} \tag{6.6.33}$$

An exactly similar argument leads to the $\mathbf{C}$-orthogonality of the modes.

Since the matrix $\mathbf{X}_b$ is made up from the modes $\mathbf{\Psi}_b^{(r)}$, in fact from the columns $\mathbf{\Psi}_b^{(1)}, \mathbf{\Psi}_b^{(2)}, \ldots, \mathbf{\Psi}_b^{(m_1)}$, the $\mathbf{A}$-orthogonality and $\mathbf{C}$-orthogonality of these modes implies that the matrices

$$\mathbf{A}_{bb} = \mathbf{X}_b'\mathbf{A}\mathbf{X}_b, \quad \mathbf{C}_{bb} = \mathbf{X}_b'\mathbf{C}\mathbf{X}_b \tag{6.6.34}$$

will be diagonal (see equation (2.5.10)). Similarly, the matrices $\mathbf{A}_{cc}, \ldots, \mathbf{A}_{ss}$; $\mathbf{C}_{cc}, \ldots, \mathbf{C}_{ss}$ may all be shown to be diagonal. In general, two branch modes belonging to *different* branches will be neither $\mathbf{A}$-orthogonal nor $\mathbf{C}$-orthogonal, so that the matrices $\mathbf{A}_{bc}$, $\mathbf{C}_{bc}$, etc., will not be diagonal.

In the general case, as indicated in equation (6.6.19), the rigid-body modes of the system (if it has any), must be taken into account. But it will now be shown that, if the branches are chosen in the manner described under 'Systems with Many Components' (so that just one component vibrates with distortion while the remaining vibrate as rigid bodies rigidly attached to the distorting component), then the rigid body modes can be neglected. The reason for this is that if the branches are chosen in this way, all the rigid modes of the body will belong to *each* set of branch modes and will therefore be $\mathbf{A}$-orthogonal and $\mathbf{C}$-orthogonal to all the other branch modes. Therefore

$$\mathbf{A}_{0a} = \mathbf{0} = \mathbf{A}_{a0}, \tag{6.6.35}$$

and this means that the equation (6.6.19) separates into the two equations

$$\omega^2\mathbf{A}_{00}\mathbf{\Gamma}_0 = \mathbf{0}, \tag{6.6.36}$$

$$(\mathbf{C}_{aa} - \omega^2\mathbf{A}_{aa})\,\mathbf{\Gamma}_a = \mathbf{0}, \tag{6.6.37}$$

of which the first concerns just the rigid-body modes and the second the principal modes which involve distortion.

If the branches are chosen in this way, and if, in addition, they are chosen so that they do not *overlap*, there is another simplification in that *any* two of the branch modes are $\mathbf{C}$-orthogonal, so that $\mathbf{C}_{aa}$ is a diagonal matrix. The requirement that the branches do not overlap means that it must be possible to divide the whole system into a number of parts so that each part vibrates with distortion when it is considered as part of just one branch and vibrates without distortion in all the others. The **C**-orthogonality of the modes then follows from the fact that an off-diagonal term in **C**, say c_{rs}, is a measure of the work done by the stresses, σ_r, associated with a branch mode r when they are moved through the displacements, d_s, of a branch mode s, that is

$$c_{rs} \simeq \sum_{\text{whole system}} \sigma_r d_s. \tag{6.6.38}$$

If r and s belong to different branches then wherever σ_r is non-zero (due to distorsion in mode r) d_s will be zero since there will be no distortion in mode s there, so that c_{rs} will be zero. When $\mathbf{C}_{aa}$ is diagonal, equation (6.6.37) is a straightforward characteristic value equation of the form

$$(\mathbf{B} - \lambda \mathbf{I})\,\mathbf{\Psi} = \mathbf{0}. \tag{6.6.39}$$

We conclude this discussion with the remark that the characteristic equation of branch mode analysis should be considered to be the equation (6.6.19), and not the simpler equations (6.6.36) and (6.6.37). In addition it should be remembered that the matrices $\mathbf{C}_{aa}$ and $\mathbf{A}_{aa}$ will have the special forms discussed above (e.g. $\mathbf{A}_{bb}$, $\mathbf{A}_{cc}$, etc., diagonal) only if the branch modes are chosen in the manner described.

6.7 A simple application of branch mode analysis

In this section we shall apply the branch mode method to the simple two-component free-free torsional system shown in fig. 6.7.1.

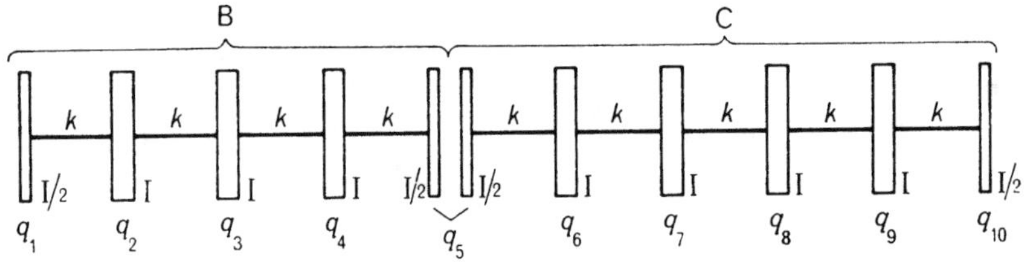

Fig. 6.7.1

We have chosen this system for two reasons. First, all the branch modes can be found exactly using the analysis of §6.4 (equations (6.4.4)–(6.4.21)), and the principal modes of the whole system can be found using a similar analysis (see Ex. 6.4.2). Secondly, when k and I have the values

$$k = \frac{9C}{l}, \quad I = \frac{\mathcal{J}l}{9}, \tag{6.7.1}$$

the system is a lumped-mass model of a uniform free-free shaft of stiffness C and moment of inertia $\mathcal{J}$ per unit length (see equation (6.4.3)). Since the errors incurred by replacing such a shaft by the lumped-mass model have been calculated (in Ex. 6.4.2) it will be possible to compare these errors with those incurred by using the branch mode method to analyse the lumped-mass model.

The inertia and stability matrices for the system are

$$\mathbf{A} = I\begin{bmatrix} \frac{1}{2} & 0 & 0 & \dots & 0 & 0 \\ 0 & 1 & 0 & \dots & 0 & 0 \\ 0 & 0 & 1 & \dots & 0 & 0 \\ \dots & \dots & \dots & \dots & \dots & \dots \\ 0 & 0 & 0 & \dots & 1 & 0 \\ 0 & 0 & 0 & \dots & 0 & \frac{1}{2} \end{bmatrix}, \quad \mathbf{C} = k\begin{bmatrix} 1 & -1 & 0 & \dots & 0 & 0 \\ -1 & 2 & -1 & \dots & 0 & 0 \\ 0 & -1 & 2 & \dots & 0 & 0 \\ \dots & \dots & \dots & \dots & \dots & \dots \\ 0 & 0 & 0 & \dots & 2 & -1 \\ 0 & 0 & 0 & \dots & -1 & 1 \end{bmatrix}, \tag{6.7.2}$$

both being of order 10.

The branches will be chosen using method β, that is, by assuming that one component distorts while the other remains fixed. It would actually be easier to use method α, but to do so would not be so instructive. The branch B is obtained by imposing the constraints

$$q_5 = 0 = q_6 = q_7 = q_8 = q_9 = q_{10}, \tag{6.7.3}$$

so that the matrix $\mathbf{B}_b$ to be used in the equation

$$\mathbf{q} = \mathbf{B}_b\mathbf{q} \tag{6.7.4}$$

is

$$\mathbf{B}_b = \mathbf{diag}\,(1 \quad 1 \quad 1 \quad 1 \quad 0 \quad 0 \quad 0 \quad 0 \quad 0 \quad 0). \tag{6.7.5}$$

That is to say $\mathbf{B}_b$ is a diagonal matrix (of order 10), the elements in whose principal diagonal are those given.† The inertia and stability matrices for the branch system are

$$\mathbf{B}_b'\mathbf{A}\mathbf{B}_b = \begin{bmatrix} A_b & 0 \\ 0 & 0 \end{bmatrix}, \quad \mathbf{B}_b'\mathbf{C}\mathbf{B}_b = \begin{bmatrix} C_b & 0 \\ 0 & 0 \end{bmatrix}. \tag{6.7.6}$$

If the multiplication is carried out it is found that

$$\mathbf{A}_b = I\begin{bmatrix} \frac{1}{2} & 0 & 0 & 0 \\ 0 & 1 & 0 & 0 \\ 0 & 0 & 1 & 0 \\ 0 & 0 & 0 & 1 \end{bmatrix}, \quad \mathbf{C}_b = k\begin{bmatrix} 1 & -1 & 0 & 0 \\ -1 & 2 & -1 & 0 \\ 0 & -1 & 2 & -1 \\ 0 & 0 & -1 & 2 \end{bmatrix}. \tag{6.7.7}$$

These matrices have been determined using matrix multiplication because that is the way they would have to be determined in a more complicated problem, and because matrix multiplication is convenient for computer programming. In the present problem the matrices could have been taken straight from equation (6.4.5) by putting $n = 4$ and reversing the order of the coordinates.

Usually the modes $\mathbf{\Psi}_b^{*(r)}$ would have to be calculated too: in the present problem they can be obtained from equation (6.4.21), again by putting $n = 4$ and reversing

† In the same notation, we could express $\mathbf{A}$ as $I\,\mathbf{diag}\,(\frac{1}{2}, 1, 1, 1, 1, 1, 1, 1, 1, \frac{1}{2})$.

the order of the coordinates. We shall use just two of the branch modes. The first two modes $\boldsymbol{\Psi}_b^{*(r)}$ are

$$\boldsymbol{\Psi}_b^{*(1)} = a_{1b}\begin{bmatrix}\sin\pi/2\\ \sin 3\pi/8\\ \sin\pi/4\\ \sin\pi/8\end{bmatrix}, \quad \boldsymbol{\Psi}_b^{*(2)} = a_{2b}\begin{bmatrix}\sin 3\pi/2\\ \sin 9\pi/8\\ \sin 3\pi/4\\ \sin 3\pi/8\end{bmatrix}, \tag{6.7.8}$$

where a_{1b} and a_{2b} are multiplying constants whose values we shall select later. The corresponding branch modes are given by

$$\boldsymbol{\Psi}_b^{(r)} = \mathbf{B}_b\begin{bmatrix}\boldsymbol{\Psi}_b^{*(r)}\\ \mathbf{0}\end{bmatrix}, \tag{6.7.9}$$

and in the present problem, with $\mathbf{B}_b$ given by equation (6.7.5), they will be obtained merely by adding six zeros on to the appropriate $\boldsymbol{\Psi}_b^{*(r)}$; thus, for example,

$$\boldsymbol{\Psi}_b^{(1)} = a_{1b}\{\sin\pi/2, \sin 3\pi/8, \sin\pi/4, \sin\pi/8, 0, 0, 0, 0, 0, 0\}. \tag{6.7.10}$$

The arbitrary constants a_{1b}, a_{2b} will be chosen later so that the modes are suitably normalised.

The branch modes belonging to the branch system C may be determined in a similar way. The matrix $\mathbf{B}_c$ is

$$\mathbf{B}_c = \mathbf{diag}\,(0 \quad 0 \quad 0 \quad 0 \quad 0 \quad 1 \quad 1 \quad 1 \quad 1 \quad 1), \tag{6.7.11}$$

and the inertia and stability matrices are

$$\mathbf{B}_c'\mathbf{A}\mathbf{B}_c = \begin{bmatrix}\mathbf{0} & \mathbf{0}\\ \mathbf{0} & \mathbf{A}_c\end{bmatrix}, \quad \mathbf{B}_c'\mathbf{C}\mathbf{B}_c = \begin{bmatrix}\mathbf{0} & \mathbf{0}\\ \mathbf{0} & \mathbf{C}_c\end{bmatrix}, \tag{6.7.12}$$

where

$$\mathbf{A}_c = I\begin{bmatrix}1 & 0 & 0 & 0 & 0\\ 0 & 1 & 0 & 0 & 0\\ 0 & 0 & 1 & 0 & 0\\ 0 & 0 & 0 & 1 & 0\\ 0 & 0 & 0 & 0 & \frac{1}{2}\end{bmatrix}, \quad \mathbf{C}_c = k\begin{bmatrix}2 & -1 & 0 & 0 & 0\\ -1 & 2 & -1 & 0 & 0\\ 0 & -1 & 2 & -1 & 0\\ 0 & 0 & -1 & 2 & -1\\ 0 & 0 & 0 & -1 & 1\end{bmatrix}. \tag{6.7.13}$$

The first two branch modes, which may be found by putting $n = 5$ in equation (6.4.21), are

$$\left.\begin{aligned}\boldsymbol{\Psi}_c^{(1)} &= a_{1c}\{0, 0, 0, 0, 0, \sin\pi/10, \sin\pi/5, \sin 3\pi/10, \sin 2\pi/5, \sin\pi/2\},\\ \boldsymbol{\Psi}_c^{(2)} &= a_{2c}\{0, 0, 0, 0, 0, \sin 3\pi/10, \sin 3\pi/5, \sin 9\pi/10, \sin 6\pi/5, \sin 3\pi/2\}.\end{aligned}\right\} \tag{6.7.14}$$

We shall now carry out the Rayleigh–Ritz analysis of the system using the four modes $\boldsymbol{\Psi}_b^{(1)}$, $\boldsymbol{\Psi}_b^{(2)}$, $\boldsymbol{\Psi}_c^{(1)}$, $\boldsymbol{\Psi}_c^{(2)}$ and the rigid body mode $\mathbf{X}_0$ which in this case is simply

$$\mathbf{X}_0 \equiv \boldsymbol{\Psi}_0^{(1)} = \{1, 1, 1, 1, 1, 1, 1, 1, 1, 1\}. \tag{6.7.15}$$

The first step is to calculate the inertia and stability matrices

$$\begin{bmatrix} \mathbf{A}_{00} & \mathbf{A}_{0a} \\ \mathbf{A}_{a0} & \mathbf{A}_{aa} \end{bmatrix} \quad \text{and} \quad \begin{bmatrix} \mathbf{0} & \mathbf{0} \\ \mathbf{0} & \mathbf{C}_{aa} \end{bmatrix}. \tag{6.7.16}$$

The general theory given in § 6.6 states that $\mathbf{C}_{bb}$ and $\mathbf{C}_{cc}$ are diagonal matrices. Also, since the branches do not overlap, $\mathbf{C}_{bc}$ is zero. This means that $\mathbf{C}_{aa}$, which is given by

$$\mathbf{C}_{aa} = \begin{bmatrix} \mathbf{C}_{bb} & \mathbf{C}_{bc} \\ \mathbf{C}_{cb} & \mathbf{C}_{cc} \end{bmatrix}, \tag{6.7.17}$$

is a diagonal matrix. It is convenient to choose the arbitrary multipliers so that $\mathbf{C}_{aa}$ is the unit matrix, or rather $k \times$ (unit matrix). The matrix $\mathbf{C}_{aa}$ is

$$\mathbf{C}_{aa} = \begin{bmatrix} \mathbf{X}'_b \\ \mathbf{X}'_c \end{bmatrix} \mathbf{C}[\mathbf{X}_b \quad \mathbf{X}_c] = \begin{bmatrix} \mathbf{\Psi}'^{(1)}_b \\ \mathbf{\Psi}'^{(2)}_b \\ \mathbf{\Psi}'^{(1)}_c \\ \mathbf{\Psi}'^{(2)}_c \end{bmatrix} \mathbf{C}[\mathbf{\Psi}^{(1)}_b \quad \mathbf{\Psi}^{(2)}_b \quad \mathbf{\Psi}^{(1)}_c \quad \mathbf{\Psi}^{(2)}_c]$$

$$= \mathbf{diag}\,(\mathbf{\Psi}'^{(1)}_b\mathbf{C}\mathbf{\Psi}^{(1)}_b,\ \mathbf{\Psi}'^{(2)}_b\mathbf{C}\mathbf{\Psi}^{(2)}_b,\ \mathbf{\Psi}'^{(1)}_c\mathbf{C}\mathbf{\Psi}^{(1)}_c,\ \mathbf{\Psi}'^{(2)}_c\mathbf{C}\mathbf{\Psi}^{(2)}_c). \tag{6.7.18}$$

Working to seven decimal places, we find

$$\left.\begin{aligned} a_{1b} &= 1{\cdot}8122549, & a_{2b} &= 0{\cdot}6363793; \\ a_{1c} &= 2{\cdot}0214712, & a_{2c} &= 0{\cdot}6965515. \end{aligned}\right\} \tag{6.7.19}$$

The matrix $\mathbf{A}_{aa}$ may be calculated in a similar way. It is

$$\mathbf{A}_{aa} = \begin{bmatrix} \mathbf{A}_{bb} & \mathbf{A}_{bc} \\ \mathbf{A}_{cb} & \mathbf{A}_{cc} \end{bmatrix}. \tag{6.7.20}$$

The general theory states that $\mathbf{A}_{bb}$ and $\mathbf{A}_{cc}$ are diagonal. In the present problem it is found that $\mathbf{A}_{bc}$ (and therefore $\mathbf{A}_{cb}$) is zero. This follows from the fact that the branches B and C are quite independent of each other, being linked by a flywheel which is always fixed in all the branch modes of B and C; it may easily be verified by calculating $\mathbf{A}_{bc}$ from the equation

$$\mathbf{A}_{bc} = \mathbf{X}'_b\mathbf{A}\mathbf{X}_c. \tag{6.7.21}$$

If the constants $a_{1b}, a_{2b}, a_{1c}, a_{2c}$ are given the values shown in equation (6.7.19) it is found that

$$\mathbf{A}_{aa} = I \times \mathbf{diag}\,(6{\cdot}585536,\ 0{\cdot}809957,\ 10{\cdot}215864,\ 1{\cdot}212960). \tag{6.7.22}$$

Finally, it is found that

$$\mathbf{A}_{00} = \mathbf{X}'_0\mathbf{A}\mathbf{X}_0 = 9I \tag{6.7.23}$$

and

$$\mathbf{A}_{0a} = \mathbf{X}'_0\mathbf{A}[\mathbf{X}_b \quad \mathbf{X}_c]$$

$$= I[4{\cdot}555411 \quad 0{\cdot}476204 \quad 6{\cdot}381532 \quad 0{\cdot}683529], \tag{6.7.24}$$

so that the matrix $\mathbf{A}_{a0}\mathbf{A}_{00}^{-1}\mathbf{A}_{0a}$ has the value

$$\frac{I}{9}\begin{bmatrix} 4\cdot555\,411 \\ 0\cdot476\,204 \\ 6\cdot381\,532 \\ 0\cdot683\,529 \end{bmatrix} [4\cdot555\,411 \quad 0\cdot476\,204 \quad 6\cdot381\,532 \quad 0\cdot683\,529]$$

or

$$I\begin{bmatrix} 2\cdot305\,753 & 0\cdot241\,034 & 3\cdot230\,056 & 0\cdot345\,973 \\ 0\cdot241\,034 & 0\cdot025\,197 & 0\cdot337\,657 & 0\cdot036\,167 \\ 3\cdot230\,056 & 0\cdot337\,657 & 4\cdot524\,883 & 0\cdot484\,662 \\ 0\cdot345\,973 & 0\cdot036\,167 & 0\cdot484\,662 & 0\cdot051\,912 \end{bmatrix}. \tag{6.7.25}$$

The frequency equation

$$|\mathbf{C}_{aa} - \omega^2(\mathbf{A}_{aa} - \mathbf{A}_{a0}\mathbf{A}_{00}^{-1}\mathbf{A}_{0a})| = 0 \tag{6.7.26}$$

may thus be expressed in the form

$$|\mathbf{B} - \lambda\mathbf{I}_4| = 0, \tag{6.7.27}$$

where $\mathbf{I}_4$ denotes the unit matrix of order four,

$$\lambda = k/I\omega^2 \tag{6.7.28}$$

and

$$\mathbf{B} = (\mathbf{A}_{aa} - \mathbf{A}_{a0}\mathbf{A}_{00}^{-1}\mathbf{A}_{0a})/I$$

$$= \begin{bmatrix} 4\cdot262\,783 & -0\cdot241\,034 & -3\cdot230\,056 & -0\cdot345\,973 \\ -0\cdot241\,034 & 0\cdot784\,760 & -0\cdot337\,657 & -0\cdot036\,167 \\ -3\cdot230\,056 & -0\cdot337\,657 & 5\cdot690\,981 & -0\cdot484\,662 \\ -0\cdot345\,973 & -0\cdot036\,167 & -0\cdot484\,662 & 1\cdot161\,048 \end{bmatrix}. \tag{6.7.29}$$

The greatest characteristic value of equation (6.7.27) is found to be

$$\lambda_1 = 8\cdot290\,290, \tag{6.7.30}$$

and the corresponding characteristic vector $\mathbf{\Gamma}_a^{(1)}$ is

$$\mathbf{\Gamma}_a^{(1)} = \{0\cdot798\,345, \quad 0\cdot019\,209, \quad -1, \quad 0\cdot029\,142\}. \qquad (6.7.31)\dagger$$

Thus the value obtained for the first natural frequency of the lumped-mass system is

$$\omega_{1(\mathrm{B.M.})} = \frac{1}{\sqrt{8\cdot290\,290}}\sqrt{\frac{k}{I}} = 0\cdot347\,308\sqrt{\frac{k}{I}}. \tag{6.7.32}$$

When k and I are given the values shown in equation (6.7.1), the value of ω_1 obtained from the branch mode analysis ($\omega_{1(\mathrm{B.M.})}$), the exact ω_1 for the lumped-mass model ($\omega_{1(\mathrm{L.M.})}$) and the exact ω_1 for the uniform shaft (ω_1) are respectively

$$\omega_{1(\mathrm{B.M.})} = 3\cdot125\,77\sqrt{\frac{C}{\mathcal{J}l^2}}, \quad \omega_{1(\mathrm{L.M.})} = 3\cdot125\,67\sqrt{\frac{C}{\mathcal{J}l^2}}, \quad \omega_1 = 3\cdot141\,59\sqrt{\frac{C}{\mathcal{J}l^2}}. \tag{6.7.33}$$

† These values may easily be found by applying the iterative technique described in §8.5.

It is seen that the difference between $\omega_{1(\text{B.M.})}$ and $\omega_{1(\text{L.M.})}$ is very much less than that between $\omega_{1(\text{L.M.})}$ and ω_1.

We must now find the approximation to the first principal mode of the lumped-mass system. To do this we must find $\mathbf{\Gamma}_0^{(1)}$, for the proportions in which the modes making up $\mathbf{X}_0$ and $\mathbf{X}_a$ occur in the first principal mode are given by the elements of $\mathbf{\Gamma}_0^{(1)}$ and $\mathbf{\Gamma}_a^{(1)}$ respectively. This is the meaning of the equation (see equation (6.6.8))

$$\mathbf{\Psi}^{(1)} = [\mathbf{X}_0 \quad \mathbf{X}_a] \begin{bmatrix} \mathbf{\Gamma}_0^{(1)} \\ \mathbf{\Gamma}_a^{(1)} \end{bmatrix} = \mathbf{X}_0\mathbf{\Gamma}_0^{(1)} + \mathbf{X}_a\mathbf{\Gamma}_a^{(1)}. \tag{6.7.34}$$

To find $\mathbf{\Gamma}_0^{(1)}$ we use equation (6.6.21), which gives

$$\begin{aligned}\mathbf{\Gamma}_0^{(1)} &= -\mathbf{A}_{00}^{-1}\mathbf{A}_{0a}\mathbf{\Gamma}_a^{(1)} \\ &= \frac{-1}{9l}\, l[4{\cdot}555\,411,\ 0{\cdot}476\,204,\ 6{\cdot}381\,532,\ 0{\cdot}683\,529] \begin{bmatrix} 0{\cdot}798\,345 \\ 0{\cdot}019\,209 \\ -1 \\ 0{\cdot}029\,142 \end{bmatrix} \\ &= 0{\cdot}301\,742. \end{aligned} \tag{6.7.35}$$

Using this value of $\mathbf{\Gamma}_0^{(1)}$ and the value of $\mathbf{\Gamma}_a^{(1)}$ given in equation (6.7.31) we find

$$\mathbf{\Psi}_{(\text{B.M.})}^{(1)} = [\mathbf{\Psi}_0^{(1)}, \mathbf{\Psi}_b^{(1)}, \mathbf{\Psi}_b^{(2)}, \mathbf{\Psi}_c^{(1)}, \mathbf{\Psi}_c^{(2)}] \begin{bmatrix} 0{\cdot}301\,742 \\ 0{\cdot}798\,345 \\ 0{\cdot}019\,209 \\ -1 \\ 0{\cdot}029\,142 \end{bmatrix}. \tag{6.7.36}$$

This mode is shown in line (i) of Table 6.7.1 (*a*). Line (ii) shows the true first mode of the lumped-mass model (which, so far as the points of subdivision are concerned, is also the true mode of the continuous system, as was shown in equation (6.4.21)), while line (iii) shows the approximate mode scaled so that it agrees with the true mode at the joining coordinate q_5. The greatest error occurs at q_6, where it is about 1·6 per cent.

The values obtained for the higher frequencies of the lumped-mass model are shown in Table 6.7.2, in which are shown also the true natural frequencies of the lumped-mass and continuous systems, together with the various percentage errors. It will be seen that, for the first three frequencies, the errors incurred in the branch mode analysis (col. IV) are small compared to those incurred by using the lumped-mass model instead of the continuous system (col. V).

The approximation obtained for the second mode is shown in Table 6.7.1 (*b*), while the true and approximate modes are shown in fig. 6.7.2. It will be seen from the curves that the accuracy of the approximate modes falls off, being excellent for the first and poor for the fourth.

The approximate modes and frequencies of the lumped-mass model have been obtained on the assumption that the only branch modes which are excited are

TABLE 6.7.1 (*a*)

	q_1	q_2	q_3	q_4	q_5	q_6	q_7	q_8	q_9	q_{10}
(i)	1·736323	1·633738	1·333432	0·866703	0·301742	−0·306505	−0·867143	−1·327389	−1·632721	−1·740028
(ii)	1	0·93969	0·76604	0·5	0·17365	−0·17365	−0·5	−0·76604	−0·93969	−1
(iii)	0·99924	0·94020	0·76738	0·49878	0·17365	−0·17639	−0·49903	−0·76390	−0·93962	−1·00137

TABLE 6.7.1 (*b*)

(i)	1·06163	0·78383	0·14409	−0·49296	−0·97717	−0·91249	−0·55168	0·12304	0·80559	1·09389
(ii)	1	0·76604	0·17365	−0·5	−0·93969	−0·93969	−0·5	0·17365	0·76604	1·0
(iii)	1·0209	0·75376	0·13856	−0·47405	−0·93969	−0·87749	−0·53052	0·11832	0·77468	1·0519

(i) = Unscaled approximate mode. (ii) = True mode. (iii) = Approximate mode scaled to give agreement with the true mode at q_5.

TABLE 6.7.2

r	I $\sqrt{\left(\frac{C}{Jl^2}\right)}\omega_{r(\text{B.M.})}$	II $\sqrt{\left(\frac{C}{Jl^2}\right)}\omega_{r(\text{L.M.})}$	III $\sqrt{\left(\frac{C}{Jl^2}\right)}\omega_r$	IV $\frac{\text{I}-\text{II}}{\text{II}}\times 100$	V $\frac{\text{III}-\text{II}}{\text{III}}\times 100$	VI $\frac{\text{III}-\text{I}}{\text{III}}\times 100$
1	3·12577	3·12567	3·14159	0·003	0·51	0·50
2	6·18546	6·15636	6·28319	0·47	2·0	1·5
3	9·0375	9·0000	9·4248	0·42	4·5	4·1
4	12·722	11·570	12·566	10	7·9	−1·2

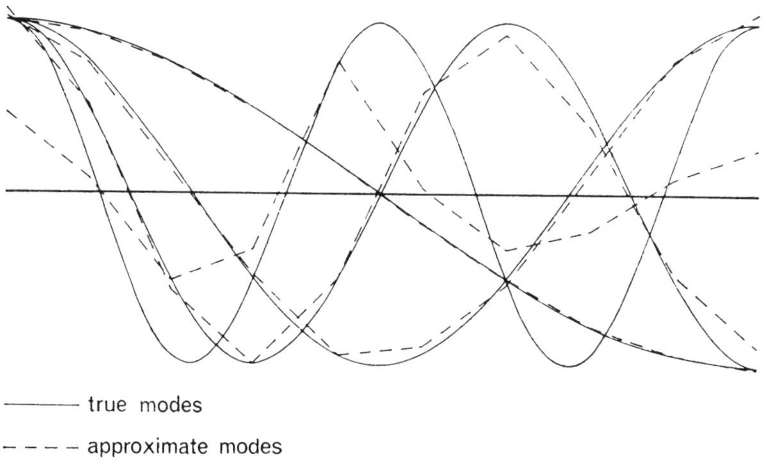

Fig. 6.7.2

$\mathbf{\Psi}_0^{(1)}$, $\mathbf{\Psi}_b^{(1)}$, $\mathbf{\Psi}_b^{(2)}$, $\mathbf{\Psi}_c^{(1)}$, $\mathbf{\Psi}_c^{(2)}$. We shall now investigate how far this assumption is justified for the present problem. The true modes of the system are known (Ex. 6.4.2, $n = 9$). Therefore the true proportions in which the various branch modes of B and C occur in any mode may be found by writing

$$\mathbf{q} = p_0\mathbf{\Psi}_0^{(1)} + p_{b1}\mathbf{\Psi}_b^{(1)} + \ldots + p_{b4}\mathbf{\Psi}_b^{(4)} + p_{c1}\mathbf{\Psi}_c^{(1)} + \ldots + p_{c5}\mathbf{\Psi}_c^{(5)}, \tag{6.7.37}$$

where $\mathbf{q}$ denotes any one of the (known) true modes of the system. The coefficients $p_0, p_{b1}, \ldots, p_{c5}$ may be found as follows. First, since all the branch modes have zero in the fifth place, we have

$$q_5 = p_0. \tag{6.7.38}$$

Secondly, multiplying through by $\mathbf{\Psi}_b'^{(r)}\mathbf{C}$ and $\mathbf{\Psi}_c'^{(r)}\mathbf{C}$ and then using the orthogonality of the modes, we find

$$\mathbf{\Psi}_b'^{(r)}\mathbf{C}\mathbf{q} = p_{br}(\mathbf{\Psi}_b'^{(r)}\mathbf{C}\mathbf{\Psi}_b^{(r)}), \quad \mathbf{\Psi}_c'^{(r)}\mathbf{C}\mathbf{q} = p_{cr}(\mathbf{\Psi}_c'^{(r)}\mathbf{C}\mathbf{\Psi}_c^{(r)}). \tag{6.7.39}$$

If the modes are normalised so that

$$\mathbf{\Psi}_b'^{(r)}\mathbf{C}\mathbf{\Psi}_b^{(r)} = k = \mathbf{\Psi}_c'^{(r)}\mathbf{C}\mathbf{\Psi}_c^{(r)} \tag{6.7.40}$$

then

$$p_{br} = (1/k)\,(\mathbf{\Psi}_b'^{(r)}\mathbf{C}\mathbf{q}), \quad p_{cr} = (1/k)\,(\mathbf{\Psi}_c'^{(r)}\mathbf{C}\mathbf{q}). \tag{6.7.41}$$

TABLE 6.7.3 (*a*). MODE PROPORTIONS: (i) TRUE; (ii) APPROXIMATE

		p_0	p_{b1}	p_{b2}	p_{b3}	p_{b4}	p_{c1}	p_{c2}	p_{c3}	p_{c4}	p_{c5}
1	(i)	0·301652	0·797848	0·019202	0·005406	0·001348	−1	0·029132	0·008658	0·003420	0·000947
	(ii)	0·301742	0·798345	0·019209	0	0	−1	0·029142	0	0	0
2	(i)	−0·972772	1	−0·349034	−0·077830	−0·018578	0·768412	−0·719454	−0·132864	−0·048270	−0·013109
	(ii)	−0·977174	1	−0·355993	0	0	0·769977	−0·738742	0	0	0
3	(i)	0·15578	−0·25488	0·78070	0·06919	0·01509	−0·21574	−1	0·13934	0·04112	0·01061
	(ii)	0·32369	−0·26441	0·84808	0	0	−0·22394	−1	0	0	0
4	(i)	−0·43025	0·32866	1	−0·25106	−0·04469	0·28568	0·48379	−0·91566	−0·13407	−0·03123
	(ii)	−0·6499	0·4879	1	0	0	0·4269	0·6235	0	0	0

TABLE 6.7.3 (*b*). TRUE MODE PROPORTIONS

5	0·33084	−0·24536	−0·41035	0·72972	0·07140	−0·21566	−0·28736	−1	0·26921	0·04946
6	−0·19909	0·14545	0·19891	1	−0·09719	0·12856	0·15470	0·26710	−0·97677	−0·06553
7	0·47429	−0·29046	−0·06233	−0·31002	1	−0·36342	−0·62432	−0·48900	−0·80818	0·81862
8	−0·05910	0·04266	0·05097	0·08084	1	0·03787	0·04229	0·05455	0·09353	−0·99805
9	0·24733	−0·17832	−0·21034	−0·31479	−0·89646	−0·15838	−0·17556	−0·22122	−0·34456	−1

Table 6.7.3 (*a*) shows the true proportions in which the branch modes occur in the first four principal modes. It also shows the approximate proportions which are used in the branch mode analysis. Table 6.7.3 (*b*) shows the proportions in which they occur in the remaining five modes. The table shows the proportions in which the branch modes occur when they are normalised according to the rule (6.7.40). This normalisation has the effect of reducing the absolute magnitudes of the mode elements in the higher modes, so that if, in a certain motion, p_{b1} and p_{b4} are of the same order of magnitude, the contribution of $\mathbf{\Psi}_b^{(4)}$ to the motion will be much less than that of $\mathbf{\Psi}_b^{(1)}$.

Table 6.7.3 (*a*) shows that to obtain the third mode of the system accurately one would have to take into account $\mathbf{\Psi}_c^{(3)}$, and possibly $\mathbf{\Psi}_b^{(3)}$ also, and to calculate the fourth one would have to include, in addition, the effect of $\mathbf{\Psi}_c^{(4)}$. Thus to calculate the fourth mode of this simple system it might be easier to work with the original coordinates, rather than to calculate the branch modes. Table 6.7.3 (*b*) shows that calculation of the higher modes requires that all the branch modes be taken into account.

CHAPTER 7

THE SOLUTION OF LINEAR EQUATIONS AND THE INVERSION OF MATRICES

In a contemplative fashion,
  And a tranquil frame of mind,
Free from every kind of passion,
  Some solution let us find. *The Gondoliers*

The methods described in this chapter for the solution of linear equations spring from that taught to school children, elimination of the variables until a single equation in a single unknown is reached. In their hands, the method often leads to the conclusion that $0 = 0$. To avoid this conclusion being drawn when it does not reflect any special property of the equations, the process will be systematised. The number of arithmetic operations (essentially multiplications) required to solve a set of equations of order n varies as n^3. For quite moderate values of n, say 20, this provides plenty of opportunity for mistakes. A running check technique will be described to detect faults before much work based on wrong values has been carried out.

The large number of operations involved has a further consequence which is still of importance when automatic digital computers are used to carry out the arithmetic and checking techniques. This is that the number of round-off errors that are made can affect the result of a calculation to a very great extent. When work is carried out by hand, a skilled computer will see what is happening, and take avoiding action if it becomes necessary. The machine is not endowed with any native intelligence, and if the programmer does not provide for any necessary avoiding action, the machine will continue to run. It may well produce plausible results, which are nevertheless quite wrong. It is, therefore, necessary to consider in some detail the possibility that the arithmetical process, as it is carried out, may show sensitivity of various types which are not reflections of properties of the original physical system, but due entirely to an unfortunate choice of the way in which the equations are treated.

There still remains the possibility that the equations to be solved have an inherent sensitivity to small changes in the coefficients of the equations or in the elements of the right-hand side. Such sensitivity is known as ill-conditioning (see § 7.5) and is usually due to the nature of the physical system, although formulation of the problem in a different way may show that the ill-condition is not relevant to the problem that has really to be solved. If the equations that have to be solved are really ill-conditioned a solution cannot be found without a great deal of labour, and its sensitivity to the initial data is such that its value is dubious, except in the case that the ill-conditioning is a mathematical device introduced to exploit this very sensitivity (see § 9.6).

7.1 Basic technique

Suppose that there are n equations in n unknowns. We shall write them

$$\left.\begin{aligned} a_{11}x_1+a_{12}x_2+\ldots+a_{1n}x_n &= y_1,\\ a_{21}x_1+a_{22}x_2+\ldots+a_{2n}x_n &= y_2,\\ \ldots\ldots\ldots\ldots\ldots\ldots\ldots\ldots\ldots & \\ a_{n1}x_1+a_{n2}x_2+\ldots+a_{nn}x_n &= y_n.\end{aligned}\right\} \tag{7.1.1}$$

From the first equation the first variable, x_1, can be found in terms of the others. This value can be substituted in the other equations to eliminate x_1 from them. The modified second equation can then be used to find x_2 in terms of $x_3, x_4, \ldots, x_n$, and substitution of this value into modified equations three, four, and so on, eliminates x_2 from them. The modified third equation can then be used to eliminate x_3 from the other doubly modified equations, and so on. This ends when the nth equation has been modified $n-1$ times and it then involves only x_n. From this, x_n can be found. The value of x_n can then be substituted in the final version of equation $n-1$, which involves only x_n and x_{n-1}, to give a value for x_{n-1}. This process of back-substitution can be continued until all the variables have been evaluated in turn.

During the elimination process, the variable being eliminated at any stage is the current 'pivotal variable'. The equation from which it is found in terms of the later variables is called the current 'pivotal equation' and the coefficient of the pivotal variable in the pivotal equation is called the 'pivot' or the 'pivotal coefficient'. The set of equations used in the back-substitution consists of the pivotal equations. The elimination process is closely connected with the triangulation process of § 1.5.

Consider the equations

$$2x_1+3x_2+x_3 = 13, \tag{7.1.2}$$

$$3x_1-x_2-2x_3 = 5, \tag{7.1.3}$$

$$-4x_1+6x_2+x_3 = 1. \tag{7.1.4}$$

From equation (7.1.2) we find

$$x_1 = 6{\cdot}5-1{\cdot}5x_2-0{\cdot}5x_3.$$

When this is substituted in equations (7.1.3) and (7.1.4) it gives

$$(-1-4{\cdot}5)\,x_2+(-2-1{\cdot}5)\,x_3 = -14{\cdot}5, \tag{7.1.5}$$

$$(6+6)\,x_2+(1+2)\,x_3 = 27, \tag{7.1.6}$$

and now, on substituting for x_2 from (7.1.5) into (7.1.6) we find

$$\left\{3-\frac{(12)\,(3{\cdot}5)}{5{\cdot}5}\right\}x_3 = 27-\frac{(12)\,(4{\cdot}5)}{5{\cdot}5}. \tag{7.1.7}$$

The final set of equations is the triangular set consisting of the pivotal equations

$$2x_1+3x_2+x_3 = 13, \qquad (7.1.2)$$

$$-5{\cdot}5x_2-3{\cdot}5x_3 = -14{\cdot}5, \qquad (7.1.5)$$

$$\frac{-25{\cdot}5}{5{\cdot}5}x_3 = \frac{-25{\cdot}5}{5{\cdot}5}. \qquad (7.1.7)$$

Back substitution then yields the values in turn

$$x_3 = 1$$

from equation (7.1.7),
$$x_2 = -\{-14{\cdot}5+(3{\cdot}5)\,(1)\}/5{\cdot}5$$
$$= 2$$

from (7.1.5), and
$$x_1 = \{13-(3)\,(2)-1\}/2$$
$$= 3$$

from (7.1.2).

EXAMPLES 7.1

1. The system of Ex. 2.1.2 is such that $m = 1 = l$. If excitation is of the type prescribed, with $P = \sin 4t$ pdl, find the response vector $\{\theta, \phi, \psi\}$. (Take $g = 32$ ft/sec^2.)

7.2 Reduction in the amount of recording

In the method described above, the equations are written down may times in modified versions. During the process of elimination, about $n^3/6$ numbers have to be recorded for a set of n equations in n unknowns. An examination of the way in which the pivotal equations—the final equations—are derived from the original equations will show how this number may be reduced considerably.

In the example, we obtained the reduced equations (7.1.5) and (7.1.6) by substituting for x_1 from (7.1.2) into (7.1.3) and (7.1.4). The first point to notice is that equations (7.1.5) and (7.1.6) can also be derived by forming linear combinations of the original equations. We have

$$\left.\begin{aligned}\textbf{equation } (7.1.2) &= \textbf{equation } (7.1.2),\\ \textbf{equation } (7.1.5) &= \textbf{equation } (7.1.3)-1{\cdot}5\textbf{ equation } (7.1.2),\\ \textbf{equation } (7.1.6) &= \textbf{equation } (7.1.4)+2\textbf{ equation } (7.1.2).\end{aligned}\right\} \qquad (7.2.1)$$

In addition we may express the final pivotal equation, (7.1.7), in terms of equations (7.1.5) and (7.1.6); in fact

$$\textbf{equation } (7.1.7) = \textbf{equation } (7.1.6)+\frac{12}{5{\cdot}5}\textbf{ equation } (7.1.5). \qquad (7.2.2)$$

Combining these results we have the following scheme

$$
\begin{aligned}
&\textbf{equation } (7.1.2) = \textbf{equation } (7.1.2),\\
&\textbf{equation } (7.1.5) = \textbf{equation } (7.1.3) - 1{\cdot}5\ \textbf{equation } (7.1.2),\\
&\textbf{equation } (7.1.7) = \textbf{equation } (7.1.4) + 2\ \textbf{equation } (7.1.2)\\
&\qquad\qquad + \frac{12}{5{\cdot}5}\ \textbf{equation } (7.1.5).
\end{aligned} \tag{7.2.3}
$$

This scheme may be generalised to apply to a set of n equations. We find that:

(i) The first pivotal equation is the first of the original equations.

(ii) The second pivotal equation is the second original equation minus a multiple of the first pivotal equation.

(iii) The third pivotal equation is the third original equation minus a multiple of the first pivotal equation minus a multiple of the second pivotal equation.

(iv) In general, the rth pivotal equation is formed from the rth original equation by subtracting multiples of the previous pivotal equations.

These results must now be expressed in matrix form. Denote the matrix of coefficients of the original equations by $\mathbf{A}$, and the matrix of coefficients of the pivotal equations by $\mathbf{U}$. The matrix $\mathbf{U}$ has non-zero coefficients only above the diagonal, and is therefore called 'upper triangular'. In the example

$$
\mathbf{U} = \begin{bmatrix} 2 & 3 & 1 \\ 0 & -5{\cdot}5 & -3{\cdot}5 \\ 0 & 0 & -4{\cdot}6363 \end{bmatrix}. \tag{7.2.4}
$$

The rows of $\mathbf{U}$ are related to those of $\mathbf{A}$ in the same way as the pivotal equations are related to the original equations. Therefore, denoting the rth rows of $\mathbf{A}$ and $\mathbf{U}$ by $\boldsymbol{\rho}_r(\mathbf{A})$ and $\boldsymbol{\rho}_r(\mathbf{U})$ respectively, we have

$$
\left.
\begin{aligned}
\boldsymbol{\rho}_1(\mathbf{U}) &= \boldsymbol{\rho}_1(\mathbf{A}),\\
\boldsymbol{\rho}_2(\mathbf{U}) &= \boldsymbol{\rho}_2(\mathbf{A}) - l_{21}\boldsymbol{\rho}_1(\mathbf{U}),\\
\boldsymbol{\rho}_3(\mathbf{U}) &= \boldsymbol{\rho}_3(\mathbf{A}) - l_{31}\boldsymbol{\rho}_1(\mathbf{U}) - l_{32}\boldsymbol{\rho}_2(\mathbf{U}),\\
&\cdots\cdots\cdots\cdots\cdots\cdots\cdots\cdots\cdots\cdots\\
\boldsymbol{\rho}_r(\mathbf{U}) &= \boldsymbol{\rho}_r(\mathbf{A}) - l_{r1}\boldsymbol{\rho}_1(\mathbf{U}) - l_{r2}\boldsymbol{\rho}_2(\mathbf{U}) - \ldots - l_{r,r-1}\boldsymbol{\rho}_{r-1}(\mathbf{U}),\\
&\qquad (r = 3, 4, \ldots, n).
\end{aligned}
\right\} \tag{7.2.5}
$$

Now we rearrange these equations to give the rows of $\mathbf{A}$ in terms of the rows of $\mathbf{U}$, so that

$$
\boldsymbol{\rho}_r(\mathbf{A}) = l_{r1}\boldsymbol{\rho}_1(\mathbf{U}) + l_{r2}\boldsymbol{\rho}_2(\mathbf{U}) + \ldots + l_{r,r-1}\boldsymbol{\rho}_{r-1}(\mathbf{U}) + \boldsymbol{\rho}_r(\mathbf{U}) \quad (r = 1, 2, \ldots, n), \tag{7.2.6}
$$

and if we introduce the lower triangular matrix $\mathbf{L}$ formed from the rows

$$
\boldsymbol{\rho}_r(\mathbf{L}) = [l_{r1}, l_{r2}, \ldots, l_{r,r-1}, 1, 0, 0, \ldots, 0], \tag{7.2.7}
$$

we may write

$$
\boldsymbol{\rho}_r(\mathbf{A}) = \boldsymbol{\rho}_r(\mathbf{L})\,\mathbf{U} \quad (r = 1, 2, \ldots, n). \tag{7.2.8}
$$

These n equations may be combined to give the single equation

$$\mathbf{A} = \mathbf{LU} \tag{7.2.9}$$

which, when expanded, is

$$\begin{bmatrix} a_{11} & a_{12} & \dots & a_{1n} \\ a_{21} & a_{22} & \dots & a_{2n} \\ \dots & \dots & \dots & \dots \\ \dots & \dots & \dots & \dots \\ a_{n1} & a_{n2} & \dots & a_{nn} \end{bmatrix} = \begin{bmatrix} 1 & 0 & 0 & \dots & 0 \\ l_{21} & 1 & 0 & \dots & 0 \\ l_{31} & l_{32} & 1 & \dots & 0 \\ \dots & \dots & \dots & \dots & \dots \\ l_{n1} & l_{n2} & l_{n3} & \dots & 1 \end{bmatrix} \begin{bmatrix} u_{11} & u_{12} & u_{13} & \dots & u_{1n} \\ 0 & u_{22} & u_{23} & \dots & u_{2n} \\ 0 & 0 & u_{33} & \dots & u_{3n} \\ \dots & \dots & \dots & \dots & \dots \\ 0 & 0 & 0 & \dots & u_{nn} \end{bmatrix}. \tag{7.2.10}$$

The elimination process can thus be seen to be equivalent to factorising **A** into the product of two triangular factors. In this form the process is known as a method of triangular resolution.

We must now show how this factorisation may be used in the solution of the equations. The original set of equations is

$$\mathbf{Ax} = \mathbf{y}, \tag{7.2.11}$$

and when **A** is factorised, this becomes

$$\mathbf{LUx} = \mathbf{y}. \tag{7.2.12}$$

Denote the right-hand side of the final set of pivotal equations by **z** so that

$$\mathbf{Ux} = \mathbf{z} \tag{7.2.13}$$

and therefore

$$\mathbf{Lz} = \mathbf{y}. \tag{7.2.14}$$

The solution thus divides into three stages. First find the elements of **L** and **U**. Secondly, solve equation (7.2.14) for **z**, and finally solve equation (7.2.13) for **x**.

The elements of **L** and **U** are most easily found in the order $\boldsymbol{\rho}_1(\mathbf{U})$, $\mathbf{col}_1(\mathbf{L})$, $\boldsymbol{\rho}_2(\mathbf{U})$, $\mathbf{col}_2(\mathbf{L})$ and so on, and the calculation scheme for $n = 4$ is as follows:

$$\begin{array}{llll} u_{11} = a_{11} & u_{12} = a_{12} & u_{13} = a_{13} & u_{14} = a_{14} \\ l_{21} = \dfrac{a_{21}}{u_{11}} & u_{22} = a_{22} - l_{21}u_{12} & u_{23} = a_{23} - l_{21}u_{13} & u_{24} = a_{24} - l_{21}u_{14} \\ l_{31} = \dfrac{a_{31}}{u_{11}} & l_{32} = \dfrac{a_{32} - l_{31}u_{12}}{u_{22}} & u_{33} = a_{33} - l_{31}u_{13} - l_{32}u_{23} & u_{34} = a_{34} - l_{31}u_{14} - l_{32}u_{24} \\ l_{41} = \dfrac{a_{41}}{u_{11}} & l_{42} = \dfrac{a_{42} - l_{41}u_{12}}{u_{22}} & l_{43} = \dfrac{a_{43} - l_{41}u_{13} - l_{42}u_{23}}{u_{33}} & u_{44} = a_{44} - l_{41}u_{14} - l_{42}u_{24} - l_{43}u_{34}. \end{array} \tag{7.2.15}$$

The diagonal elements of **L** have been omitted because they are all ones. The reader will notice that the products which occur in the expression for any element of **L** or **U** involve elements of **L** coming only from the particular row concerned and elements of **U** only from the column concerned. For example, the numerator of l_{43} involves l_{41} and l_{42}, from the same row as l_{43}, and u_{13} and u_{23}, from the same column. In addition—and this is more important—all the elements of **L** and **U** are formed from elements occurring *earlier* in the same row or column. This makes it possible to find the l's and u's systematically without having to solve simultaneous equations.

The next step is to form the column vector **z**. By comparing equation (7.2.14) with the sth column of equation (7.2.9), namely

$$\mathrm{col}_s(\mathbf{A}) = \mathbf{L}\,\mathrm{col}_s(\mathbf{U}), \tag{7.2.16}$$

we see that **z** is obtained from **y** just as a column of **U** is obtained from the corresponding column of **A**. Thus for $n = 4$ we would have

$$\left.\begin{aligned} z_1 &= y_1, \\ z_2 &= y_2 - l_{21}z_1, \\ z_3 &= y_3 - l_{31}z_1 - l_{32}z_2, \\ z_4 &= y_4 - l_{41}z_1 - l_{42}z_2 - l_{43}z_3, \end{aligned}\right\} \tag{7.2.17}$$

and this set of calculations would normally appear to the right of the scheme above, forming a fifth column in the set of equations (7.2.15).

The calculations for the example of § 7.1 would appear as follows:

2	3	1	13
3	−1	−2	5
−4	6	1	1

Original quantities

2	3	1	13
$\frac{3}{2}$ $= 1{\cdot}5$	$-1-(1{\cdot}5)\,(3)$ $= -5{\cdot}5$	$-2-(1{\cdot}5)\,(1)$ $= -3{\cdot}5$	$5-(1{\cdot}5)\,(13)$ $= -14{\cdot}5$
$-\frac{4}{2}$ $= -2$	$\frac{6-(-2)\,(3)}{-5{\cdot}5}$ $= -2{\cdot}1818$	$1-(-2)\,(1)-(-2{\cdot}1818)\,(-3{\cdot}5)$ $= -4{\cdot}6363$	$1-(-2)\,(13)-(-2{\cdot}1818)\,(-14{\cdot}5)$ $= -4{\cdot}6361$

Derived quantities

In carrying out the arithmetic of any step, there is no need to write down the separate products occurring in any scalar product, since desk calculating machines can be used to accumulate sums or differences of products as they are formed by multiplication. There is thus no need to truncate the partial products, and each of the quantities calculated introduces but one rounding-off, that made when the value of the element of **L**, **U** or **z** is recorded. The number of elements to be recorded in this way of doing the elimination is reduced to about n^2.

The calculations occurring in the back-substitution, that is the solution of equation (7.2.13), are similar to those already described. The formulae for the elements of **x** are

$$\left.\begin{aligned} x_n &= z_n/u_{nn}, \\ x_{n-1} &= (z_{n-1} - u_{n-1,n}x_n)/u_{n-1,n-1}, \\ x_{n-2} &= (z_{n-2} - u_{n-2,n}x_n - u_{n-2,n-1}x_{n-1})/u_{n-2,n-2}, \\ &\ldots\ldots\ldots\ldots\ldots\ldots\ldots\ldots \\ \text{and, in general,} \\ x_r &= (z_r - u_{rn}x_n - u_{r,n-1}x_{n-1} - \ldots - u_{r,r+1}x_{r+1})/u_{rr}. \end{aligned}\right\} \tag{7.2.18}$$

None of the partial products need be recorded, the scalar products being accumulated on the desk calculator as the multiplications are performed.

Using equations (7.2.18) and the results tabulated in the last example we find that

$$\begin{aligned} x_3 &= -4{\cdot}6361/(-4{\cdot}6363) \\ &= 1{\cdot}0000, \\ x_2 &= \{-14{\cdot}5 - (-3{\cdot}5)\,(1{\cdot}0000)\}/(-5{\cdot}5) \\ &= 2{\cdot}0000, \\ x_1 &= \{13 - (1)\,(1{\cdot}0000) - (3)\,(2{\cdot}0000)\}/2 \\ &= 3{\cdot}0000. \end{aligned}$$

The organisation of the elimination in this way is due to Doolittle† although many American authors prefer to ascribe it to Crout.‡ It appears to have been rediscovered several times, by Cholesky,§ Aitken|| and Banachiewicz,¶ among others.

In the procedure which has been described, the diagonal elements of **L** have been chosen to be unity. There does not seem to be any particular advantage in this and the factorisation could be as readily carried out with unit values for the diagonal elements of **U**. In fact, both triangles can be given unit diagonal elements by introducing a third factor. We may define a modified upper triangular matrix **V** by the equation

$$v_{ij} = u_{ij}/u_{ii}, \tag{7.2.19}$$

and introduce a diagonal matrix **D**, with diagonal elements given by

$$d_{ii} = u_{ii}. \tag{7.2.20}$$

Then

$$\mathbf{U} = \mathbf{DV}, \tag{7.2.21}$$

† M. H. Doolittle, 'Method employed in the solution of normal equations and the adjustment of a triangulation', *U.S. Coast Geodetic Survey Report* (1878), pp. 115–20.

‡ P. D. Crout, 'A short method of evaluating determinants and solving systems of linear equations with real or complex coefficients', *Trans. A.I.E.E.* **60** (1941), pp. 1235–40.

§ About 1916. Cholesky's procedure is described by Benoit, 'Note sur une méthode, etc.' (Procédé du Commandant Cholesky), *Bull. géodésique* (*Toulouse*) (1924), pp. 67–77.

|| A. C. Aitken, 'On the evaluation of determinants, the formation of their adjugates and the practical solution of simultaneous linear equations', *Proc. Edinb. Math. Soc.* ser. 2, **3** (1932), pp. 207–19.

¶ T. Banachiewicz, 'Méthode de résolution numérique des équations linéaires, du calcul des déterminants et des inverses et de réduction des forme quadratiques', *Bull. Acad. Polon. Sci.*, ser. A (1938), pp. 393–404.

and the original matrix is factorised into the triple product

$$\mathbf{A} = \mathbf{LDV}. \tag{7.2.22}$$

For the matrix **U** of equation (7.2.4), **D** and **V** are found to be

$$\mathbf{D} = \begin{bmatrix} 2 & 0 & 0 \\ 0 & -5{\cdot}5 & 0 \\ 0 & 0 & -4{\cdot}6363 \end{bmatrix}, \quad \mathbf{V} = \begin{bmatrix} 1 & 1{\cdot}5 & 0{\cdot}5 \\ 0 & 1 & 0{\cdot}6364 \\ 0 & 0 & 1 \end{bmatrix}.$$

The way in which the factorisation has been changed by an arbitrary act of choice shows the non-uniqueness of matrix factorisation, and even within the form of equation (7.2.22) there is much latitude. If no restrictions are placed on **L**, **D** and **V** except that they shall be respectively lower triangular, diagonal and upper triangular, then the total number of elements which are not necessarily zero and which have to be determined is n^2+2n. On the other hand, only n^2 equations are obtained between the elements when the three matrices are multiplied out. The n^2 equations may be arranged in the form

$$\left.\begin{aligned} l_{11}d_{11}v_{11} &= a_{11}, \\ l_{r1}d_{11}v_{11} &= a_{r1} \quad (r = 2, 3, \ldots, n), \\ l_{11}d_{11}v_{1s} &= a_{1s} \quad (s = 2, 3, \ldots, n); \end{aligned}\right\} \tag{7.2.23}$$

$$\left.\begin{aligned} l_{22}d_{22}v_{22} &= a_{22} - l_{21}d_{11}v_{12}, \\ l_{r2}d_{22}v_{22} &= a_{r2} - l_{r1}d_{11}v_{12} \quad (r = 3, 4, \ldots, n), \\ l_{22}d_{22}v_{2s} &= a_{2s} - l_{21}d_{11}v_{1s} \quad (s = 3, 4, \ldots, n); \end{aligned}\right\} \tag{7.2.24}$$

and so on. Clearly the first equation needs supplementing before l_{11}, d_{11} and v_{11} can be determined. In the last example l_{11} and v_{11} were chosen to be unity, but other choices are possible. One convenient factorisation stems from the choice

$$\left.\begin{aligned} l_{rr} = v_{rr} &= \sqrt{|a_{rr} - l_{r1}d_{11}v_{1r} - l_{r2}d_{22}v_{2r} - \ldots - l_{r,r-1}d_{r-1,r-1}v_{r-1,r}|}, \\ d_{rr} &= \operatorname{sign}\,(a_{rr} - l_{r1}d_{11}v_{1r} - l_{r2}d_{22}v_{2r} - \ldots - l_{r,r-1}d_{r-1,r-1}v_{r-1,r}). \end{aligned}\right\} \tag{7.2.25}$$

The elements of **D** will then be ± 1, or possibly zero. Further elements of **L** and **V** are then determined from the equations

$$\left.\begin{aligned} l_{rs} &= (a_{rs} - l_{r1}d_{11}v_{1s} - l_{r2}d_{22}v_{2s} - \ldots - l_{r,s-1}d_{s-1,s-1}v_{s-1,s})/d_{ss}v_{ss}, \\ v_{rs} &= (a_{rs} - l_{r1}d_{11}v_{1s} - l_{r2}d_{22}v_{2s} - \ldots - l_{r,r-1}d_{r-1,r-1}v_{r-1,s})/d_{rr}v_{rr}, \end{aligned}\right\} \tag{7.2.26}$$

where $r > s$ in the first equation and $r < s$ in the second. In the calculation the elements are found in the following order d_{11}, l_{11}, v_{11}; $\boldsymbol{\rho}_1(\mathbf{V})$, $\mathbf{col}_1(\mathbf{L})$; d_{22}, l_{22}, v_{22}; $\boldsymbol{\rho}_2(\mathbf{V})$, $\mathbf{col}_2(\mathbf{L})$; and so on.

For the matrix **A** of the example given in § 7.1 the calculation scheme would be as follows

$+2$ $l_{11} = v_{11} = \sqrt{2}$ $d_{11} = +1$	$v_{12} = \dfrac{3}{1{\cdot}4142}$ $v_{12} = 2{\cdot}1213$	$v_{13} = \dfrac{1}{1{\cdot}4142}$ $v_{13} = 0{\cdot}7071$
$l_{21} = \dfrac{3}{1{\cdot}4142}$ $l_{21} = 2{\cdot}1213$	$-1-(2{\cdot}1213)\ (1)\ (2{\cdot}1213)$ $= -5{\cdot}4999$ $l_{22} = v_{22} = \sqrt{(5{\cdot}4999)}$ $= 2{\cdot}3452$ $d_{22} = -1$	$v_{23} = \dfrac{-2-(2{\cdot}1213)\ (1)\ (0{\cdot}7071)}{(-1)\ (2{\cdot}3452)}$ $v_{23} = 1{\cdot}4924$
$l_{31} = \dfrac{-4}{1{\cdot}4142}$ $= -2{\cdot}8285$	$l_{32} = \dfrac{6-(-2{\cdot}8285)\ (1)\ (2{\cdot}1213)}{(-1)\ (2{\cdot}3452)}$ $= -5{\cdot}1169$	$1-(-2{\cdot}8285)\ (1)\ (0{\cdot}7071)$ $-(-5{\cdot}1169)\ (-1)\ (1{\cdot}4924)$ $= -4{\cdot}6363$ $l_{33} = v_{33} = \sqrt{(4{\cdot}6363)} = 2{\cdot}1532$ $d_{33} = -1$

EXAMPLES 7.2

1. (*a*) Using a step-by-step process of 'triangulation' similar to that described in § 1.5 (*j*), find the matrices **U** and **L** corresponding to

$$\mathbf{A} = \begin{bmatrix} 4 & -6 & -3 \\ -3 & 0 & -3 \\ -1 & -2 & -1 \end{bmatrix}.$$

(*b*) Repeat this factorisation, setting the calculations out as in equation (7.2.15).

2. Solve the equation $\mathbf{Ax} = \mathbf{y}$ where **A** is the matrix given in Ex. 7.2.1 and $\mathbf{y} = \{1, 1, 1\}$.

3. Factorise the matrix **A** of Ex. 7.2.1 into matrices **L**, **D** and **V** in the manner of equations (7.2.25) and (7.2.26).

7.3 Checks

A running check can be provided which will detect many mistakes at the end of the calculation row or column in which they are made. Errors in the original matrix will not be detected by this and it is possible (and will therefore sometimes happen now that extensive calculations are being undertaken) that compensating errors will occur. None the less it is a useful precaution to use the check, which is based on the associativity of matrix multiplication.

Consider the set of equations (7.1.1), namely

$$\mathbf{Ax} = \mathbf{y}. \tag{7.3.1}$$

The 'augmented' matrix of the set is the matrix consisting of **A** and **y**. Let **r** denote the column vector whose elements are the sums of the respective rows of the augmented matrix. Then

$$\mathbf{r} = \mathbf{A}\boldsymbol{l}+\mathbf{y}, \tag{7.3.2}$$

where $\boldsymbol{l}$ denotes the column of ones. Similarly, the sums of the rows of the modified augmented matrix $(\mathbf{U}, \mathbf{z})$ are given by

$$\mathbf{r}^0 = \mathbf{U}\boldsymbol{l}+\mathbf{z}. \tag{7.3.3}$$

But

$$\begin{aligned} \mathbf{A}\boldsymbol{l}+\mathbf{y} &= (\mathbf{LU})\,\boldsymbol{l}+\mathbf{y} \\ &= \mathbf{L}(\mathbf{U}\boldsymbol{l})+\mathbf{Lz} \\ &= \mathbf{L}(\mathbf{U}\boldsymbol{l}+\mathbf{z}), \end{aligned} \tag{7.3.4}$$

so that

$$\mathbf{r} = \mathbf{L}\mathbf{r}^0. \tag{7.3.5}$$

This equation gives a relation between the sums of the elements in the rows of the augmented matrix and the corresponding sums for $(\mathbf{U}, \mathbf{z})$. By comparing it with equations (7.2.14) and (7.2.16) we notice that the relation of $\mathbf{r}$ to $\mathbf{r}^0$ is the same as that of $\mathbf{y}$ to $\mathbf{z}$, or of any column of $\mathbf{A}$ to the corresponding column of $\mathbf{U}$. This makes the checking procedure for the row sums very simple. First calculate $\mathbf{r}$; then calculate the sth element of $\mathbf{r}^0$ along with the sth row of $(\mathbf{U}, \mathbf{z})$ and compare it with the actual sum of the sth row of $(\mathbf{U}, \mathbf{z})$.

There is a similar check for the column sums. Let $\mathbf{c}$ and $\mathbf{c}^0$ be the row vectors whose elements are the sums of the columns of $\mathbf{A}$ and $\mathbf{L}$. Then

$$\mathbf{c} = \boldsymbol{l}'\mathbf{A}, \quad \mathbf{c}^0 = \boldsymbol{l}'\mathbf{L}, \tag{7.3.6}$$

and therefore

$$\mathbf{c} = \mathbf{c}^0\mathbf{U}. \tag{7.3.7}$$

This equation relates the sums of the elements of the columns of $\mathbf{A}$ and $\mathbf{L}$. By comparing it with equation (7.2.8) we see that the relation of $\mathbf{c}$ to $\mathbf{c}^0$ is the same as that of any row of $\mathbf{A}$ to the corresponding row of $\mathbf{L}$. Therefore $\mathbf{c}^0$ can be calculated from $\mathbf{c}$ at the same time as $\mathbf{L}$ is calculated from $\mathbf{A}$.

For the equations that have been used in the previous examples we find

$$\mathbf{r} = \{19,\ 5,\ 4\}, \quad \mathbf{c} = [1,\ 8,\ 0].$$

The checking procedure now runs as follows. First we calculate the first row of $(\mathbf{U}, \mathbf{z})$. We then calculate r_1^0, and check that the actual row sum is equal to it. There should be exact equality, since nothing but copying has been done. Then the first column of $\mathbf{L}$ is found, and its check sum is found from equation (7.3.7) to be

$$c_1^0 = \tfrac{1}{2} = 0{\cdot}5.$$

The sum of the elements actually recorded in this column of $\mathbf{L}$ is $-0{\cdot}5$, but when this sum is supplemented with the unrecorded diagonal element 1 there is complete agreement. As a rule the numbers recorded will have been rounded-off, and the sum of the recorded elements may differ from the check-sum calculated from equation (7.3.7) up to $\tfrac{1}{2}n$ in the last decimal place recorded, if a fixed number of decimal places is written down after rounding-off. If a fixed number of significant figures is being recorded, then only the largest rounding-off errors in the recorded results can cause a discrepancy between the two versions of the column-sum, and this must be estimated as the calculation proceeds.

The second row of $(\mathbf{U}, \mathbf{z})$ is then found and the sum of the numbers in this row is compared with the check value of r_2^0, namely

$$r_2^0 = 5-(1\cdot5)\,(19) = -23\cdot5.$$

Once again, the actual sum is the same as the derived check-sum.

There will normally be a discrepancy between the calculated row sum, $r_i^{(0)}$, and the actual row sum,

$$u_{ii}+u_{i,\,i+1}+\ldots+u_{in}+z_i.$$

There are two distinct contributions to this, one from rounding of the numbers actually recorded, the other from ignoring the subdiagonal elements of $\mathbf{U}$. If the rounding-off is in the range $\pm e$ for each element recorded, the contribution is bounded in modulus by $(n-i+3)\,e$. (It must be remembered that z_i and $r_i^{(0)}$ contribute here.) The contribution to the discrepancy from the subdiagonal terms may be more serious than this. The exact subdiagonal elements that would be found in row i of $\mathbf{U}$ would in general be different from zero if they were calculated, because of the rounding of the elements of $\mathbf{L}$. The exact value of u_{ij} would be

$$u_{ij} = \frac{a_{ij}-l_{i1}u_{1j}-\ldots-l_{ij}u_{jj}}{l_{ii}} \quad (i>j). \tag{7.3.8}$$

If l_{ij} were recorded exactly, this would be zero. Its actual value is

$$u_{ij} = -\frac{\delta_{ij}u_{jj}}{l_{ii}}, \tag{7.3.9}$$

where

$$\text{recorded } l_{ij} = \text{exact } l_{ij}+\delta_{ij}, \tag{7.3.10}$$

so that

$$|\delta_{ij}| \leqslant e.$$

The total discrepancy in the row-sum check is thus bounded in modulus by

$$\left\{\frac{1}{|l_{ii}|}\sum_{j=1}^{i-1}|u_{jj}|+n-i+3\right\}e.$$

For the column-sum check, the possible discrepancy is

$$\left\{\frac{i}{|u_{jj}|}\sum_{i<j}|l_{ii}|+n-i+2\right\}e.$$

After the calculation of the second column of $\mathbf{L}$, the check sum c_2^0 is derived from equation (7.3.7), which yields

$$c_2^0 = \{8-(0\cdot5)\,(3)\}/(-5\cdot5) = -1\cdot1818.$$

This equals the sum of the recorded elements and the unrecorded diagonal one.

Lastly, the third row of $(\mathbf{U}, \mathbf{z})$ and the corresponding check sum r_3^0 are found; the latter is

$$r_3^0 = 4-(-2)\,(19)-(-2\cdot1818)\,(-23\cdot5) = -9\cdot2723.$$

The actual row sum is $-9\cdot2724$ which is sufficiently close to be within the range of variation due to rounding-off.

The checks which have been described are all concerned with the elimination, that is the calculation of **U**, **L** and **z**. To check the back substitution, the calculation of **x** from **z**, we use the true row-sums recorded during the elimination; for

$$\begin{aligned}\mathbf{r}^0 &= \mathbf{U}\boldsymbol{l}+\mathbf{z}\\ &= \mathbf{U}\boldsymbol{l}+\mathbf{U}\mathbf{x}\\ &= \mathbf{U}(\boldsymbol{l}+\mathbf{x}).\end{aligned} \tag{7.3.11}$$

If, therefore, the row-sums $\mathbf{r}^0$ are treated in the same way as **z** during the back-substitution, the derived values forming the column $\mathbf{x}^0$ should differ from the values of the elements of **x** by one in each element. The actual difference will include the effects of rounding-off, which should not be more than one rounding error, provided the true values of the elements of **x**, plus one, are used to derive the successive check values.

In the previous example $\mathbf{r}^0$ was found to be

$$\mathbf{r}^0 = \begin{bmatrix} 19 \\ -23{\cdot}5 \\ -9{\cdot}2724 \end{bmatrix}. \tag{7.3.12}$$

The last element of $\mathbf{r}^0$ is the value of the sum of the recorded elements in the third row of **U** and **z**, not the value derived from the check sum. When x_3 has been found a check number is found from the last equation of the set represented by (7.3.11), namely

$$\begin{aligned}x_3^0 &= r_3^0/u_{33}\\ &= -9{\cdot}2724/(-4{\cdot}6363)\\ &= 2{\cdot}0000.\end{aligned} \tag{7.3.13}$$

This is one different from x_3, and is recorded. Then x_2 is found, and the corresponding check number, x_2^0, from the penultimate equation of the set (7.3.11), namely

$$\begin{aligned}x_2^0 &= (r_2^0-u_{23}z^0)/u_{22}\\ &= \{-23{\cdot}5-(-3{\cdot}5)\,(2{\cdot}0000)\}/(-5{\cdot}5)\\ &= 3{\cdot}0000.\end{aligned} \tag{7.3.14}$$

This is exactly one different from x_2, and is recorded. Finally x_1 is found, and then x_1^0, from

$$\begin{aligned}x_1^0 &= (r_1^0-u_{13}z^0-u_{12}y^0)/u_{11}\\ &= \{19-(1)\,(2{\cdot}0000)-(3)\,(3{\cdot}0000)\}/2\\ &= 4{\cdot}0000.\end{aligned} \tag{7.3.15}$$

This again differs from the true solution, x_1, by one.

As a final check against blunders it is useful to substitute the solution into the original equations. This is not a very good check in all cases, because the values obtained after the substitution may be very insensitive to certain sorts of error, but a gross blunder in transcription may be shown up in this way. Moreover, the residual $(\mathbf{y}-\mathbf{A}\mathbf{x})$ may be used to bound the error in the computed result (see § 7.11), or to correct it.

The complete work sheet for the solution is shown on p. 264.

	A			y	r
	2	3	1	13	19
	3	−1	−2	5	5
	−4	6	1	1	4
c	1	8	0		

	U			z	r^0	x	x^0
	2	3	1	13	19	$\frac{13-(1)\,(1\cdot0000)-(3)\,(2\cdot0000)}{2}$ $=3\cdot0000$	$\frac{19-(1)\,(2\cdot0000)-(3)\,(3\cdot0000)}{2}$ $=4\cdot0000$
L	$\frac{3}{2}$ $=1\cdot5$	$-1-(1\cdot5)\,(3)$ $=-5\cdot5$	$-2-(1\cdot5)\,(1)$ $=-3\cdot5$	$5-(1\cdot5)\,(13)$ $=-14\cdot5$	$5-(1\cdot5)\,(19)$ $=-23\cdot5$	$\frac{-14\cdot5-(-3\cdot5)\,(1\cdot0000)}{-5\cdot5}$ $=2\cdot0000$	$\frac{-23\cdot5-(-3\cdot5)\,(2\cdot0000)}{-5\cdot5}$ $=3\cdot0000$
	$\frac{-4}{2}$ $=-2$	$\frac{6-(-2)\,(3)}{-5\cdot5}$ $=-2\cdot1818$	$1-(-2)\,(1)$ $-(-2\cdot1818)\,(-3\cdot5)$ $=-4\cdot6363$	$1-(-2)\,(13)$ $-(-2\cdot1818)\,(-14\cdot5)$ $=-4\cdot6361$	$4-(-2)\,(19)$ $-(-2\cdot1818)\,(-23\cdot5)$ $=-9\cdot2723$	$\frac{-4\cdot6361}{-4\cdot6363}$ $=1\cdot0000$	$\frac{-9\cdot2723}{-4\cdot6363}$ $=2\cdot0000$
c^0	$\frac{1}{2}$ $=0\cdot5$	$\frac{8-(0\cdot5)\,(3)}{-5\cdot5}$ $=-1\cdot1818$					

In practice, a work sheet would show only the results of the calculation, not the equations from which they are calculated.

EXAMPLES 7.3

1. Draw up a table like the lower of the two on the worksheet that is set out in this section. Place numbers in the various spaces to show the order in which the table is best filled. Place an asterisk in those spaces whose entries provide a check on other entries in the table.

2. Solve the equations

$$\begin{aligned} 4x_1 - 6x_2 - 3x_3 &= 1, \\ -3x_1 - 3x_3 &= 1, \\ -x_1 - 2x_2 - x_3 &= 1, \end{aligned}$$

by the method of this section, making all the various checks.

7.4 Symmetric factorisation

When the matrix $\mathbf{A}$ is symmetric, there is an advantage in taking the factors of the matrix to be

$$\mathbf{A} = \mathbf{LL}', \tag{7.4.1}$$

or

$$\mathbf{A} = \mathbf{LDL}', \tag{7.4.2}$$

where $\mathbf{L}$ is lower triangular and $\mathbf{D}$ is diagonal. In the second case various supplementary conditions may be imposed on $\mathbf{L}$ and $\mathbf{D}$; for example $\mathbf{L}$ may be taken to have unit diagonal elements. Only one triangle need be recorded, and the use of these factors in further matrix algebra, instead of the unsymmetric factors of equation (7.2.9), may preserve the symmetry of the resulting matrices (see § 8.2 on the reduction of $\mathbf{A} - \lambda\mathbf{C}$ to standard form).

The diagonal elements of $\mathbf{L}$ in equation (7.4.1) are found to involve square roots. For positive definite symmetric matrices it never becomes necessary to take the square root of a negative number, but for general symmetric matrices this may occur. If it does, one or more whole columns of $\mathbf{L}$ will be purely imaginary, as will the corresponding rows of $\mathbf{L}'$, but this will not lead to numbers which are not either real or purely imaginary occurring later. Imaginary elements may be avoided altogether by using the factorisation shown in equation (7.4.2).

The sequence of calculations occurring in the factorisation (7.4.1) for a (positive definite) symmetric matrix of order 4 may be obtained by considering the equation

$$\begin{bmatrix} l_{11} & & & \\ l_{21} & l_{22} & & \\ l_{31} & l_{32} & l_{33} & \\ l_{41} & l_{42} & l_{43} & l_{44} \end{bmatrix} \begin{bmatrix} l'_{11} & l'_{12} & l'_{13} & l'_{14} \\ & l'_{22} & l'_{23} & l'_{24} \\ & & l'_{33} & l'_{34} \\ & & & l'_{44} \end{bmatrix} = \begin{bmatrix} a_{11} & a_{12} & a_{13} & a_{14} \\ a_{21} & a_{22} & a_{23} & a_{24} \\ a_{31} & a_{32} & a_{33} & a_{34} \\ a_{41} & a_{42} & a_{43} & a_{44} \end{bmatrix}. \tag{7.4.3}$$

It is found to be more convenient to record $\mathbf{L}'$, rather than $\mathbf{L}$, and the sequence of calculations is

$$\left.\begin{aligned} &l'_{11} = \sqrt{(a_{11})}, \quad l'_{12} = \frac{a_{12}}{l'_{11}}, \quad l'_{13} = \frac{a_{13}}{l'_{11}}, \quad l'_{14} = \frac{a_{14}}{l'_{11}}; \\ &l'_{22} = \sqrt{(a_{22}-l'^2_{12})}, \quad l'_{23} = \frac{a_{23}-l'_{12}l'_{13}}{l'_{22}}, \quad l'_{24} = \frac{a_{24}-l'_{12}l'_{14}}{l'_{22}}; \\ &l'_{33} = \sqrt{(a_{33}-l'^2_{13}-l'^2_{23})}, \quad l'_{34} = \frac{a_{34}-l'_{13}l'_{14}-l'_{23}l'_{24}}{l'_{33}}; \\ &l'_{44} = \sqrt{(a_{44}-l'^2_{14}-l'^2_{24}-l'^2_{34})}, \end{aligned}\right\} \tag{7.4.4}$$

and may be carried out by rows or by columns. It does not matter which sign is taken for the square roots as the effects of sign vanish on multiplying $\mathbf{L}$ by $\mathbf{L}'$.

For the matrix

$$\mathbf{A} = \begin{bmatrix} 2 & 3 & 4 \\ 3 & 6 & 7 \\ 4 & 7 & 10 \end{bmatrix}$$

of order 3, the elements of the first row of $\mathbf{L}'$ are

$$l'_{11} = \sqrt{2} = 1{\cdot}414\,214,$$

$$l'_{12} = 3/\sqrt{2} = 2{\cdot}121\,320,$$

$$l'_{13} = 4/\sqrt{2} = 2{\cdot}828\,428.$$

The elements of the second row are

$$l'_{22} = \sqrt{(a_{22}-l'^2_{12})} = 1{\cdot}224\,745,$$

$$l'_{23} = \frac{a_{32}-l'_{12}l'_{13}}{l'_{22}} = 0{\cdot}816\,496,$$

and the remaining element is

$$l'_{33} = \sqrt{(a_{33}-l'^2_{13}-l'^2_{23})} = 1{\cdot}154\,699.$$

Thus

$$\mathbf{L}' = \begin{bmatrix} 1{\cdot}414\,214 & 2{\cdot}121\,320 & 2{\cdot}828\,428 \\ & 1{\cdot}224\,745 & 0{\cdot}816\,496 \\ & & 1{\cdot}154\,699 \end{bmatrix}.$$

If the equation

$$\mathbf{Ax} = \mathbf{y} \tag{7.4.5}$$

is solved, as in § 7.2, by dividing the problem into the two parts

$$\mathbf{Lz} = \mathbf{y}, \tag{7.4.6}$$

$$\mathbf{L}'\mathbf{x} = \mathbf{z}, \tag{7.4.7}$$

then the equations are

$$\left.\begin{aligned} z_1 &= y_1/l'_{11}, \\ z_2 &= (y_2 - l'_{12}z_1)/l'_{22}, \\ &\cdots\cdots\cdots\cdots\cdots\cdots \\ z_r &= (y_r - l'_{1r}z_1 - l'_{2r}z_2 - \ldots - l'_{r-1,r}z_{r-1})/l'_{rr}, \end{aligned}\right\} \qquad (7.4.8)$$

and

$$\left.\begin{aligned} x_n &= z_n/l'_{nn}, \\ x_{n-1} &= (z_{n-1} - l'_{n-1,n}x_n)/l'_{n-1,n-1}, \\ &\cdots\cdots\cdots\cdots\cdots\cdots \\ x_r &= (z_r - l'_{rn}x_n - l'_{r,n-1}x_{n-1} - \ldots - l'_{r,r+1}x_{r+1})/l'_{rr}. \end{aligned}\right\} \qquad (7.4.9)$$

The two calculations may be checked by using row sums, in which the right-hand sides of the equations may be included, since

$$\begin{aligned} \mathbf{A}\boldsymbol{l} + \mathbf{y} &= \mathbf{L}\mathbf{L}'\boldsymbol{l} + \mathbf{L}\mathbf{z} \\ &= \mathbf{L}(\mathbf{L}'\boldsymbol{l} + \mathbf{z}), \end{aligned} \qquad (7.4.10)$$

and

$$\begin{aligned} \mathbf{L}'\boldsymbol{l} + \mathbf{z} &= \mathbf{L}'\boldsymbol{l} + \mathbf{L}'\mathbf{x} \\ &= \mathbf{L}'(\boldsymbol{l} + \mathbf{x}). \end{aligned} \qquad (7.4.11)$$

Equation (7.4.10) shows that the row sum of $(\mathbf{L}', \mathbf{z})$ is related to the row sum of $(\mathbf{A}, \mathbf{y})$ in the same way as any column of $\mathbf{L}'$ is related to the corresponding column of $\mathbf{A}$. This is why it is more convenient to record $\mathbf{L}'$ than $\mathbf{L}$—so that the check sums can be found by adding along rows both for $(\mathbf{A}, \mathbf{y})$ and $(\mathbf{L}', \mathbf{z})$. Equation (7.4.11) shows that if the check sums for $(\mathbf{L}', \mathbf{z})$ are treated as an extra column of right-hand sides, and used in back-substitution, the resulting numbers will each be one more than the corresponding elements of $\mathbf{x}$.

The equations below may be solved by the square-root factorisation method

$$\begin{aligned} 2x_1 + 3x_2 + 4x_3 &= 16, \\ 3x_1 + 6x_2 + 7x_3 &= 28, \\ 4x_1 + 7x_2 + 10x_3 &= 36. \end{aligned}$$

The row sums are $\quad r_1 = 25, \quad r_2 = 44, \quad r_3 = 57.$

The first row of $(\mathbf{L}', \mathbf{z}, \mathbf{r}^0)$ is then found as follows:

$$\begin{aligned} l'_{11} &= \sqrt{a_{11}} &&= 1{\cdot}414\,214, \\ l'_{12} &= a_{12}/l'_{11} &&= 2{\cdot}121\,320, \\ l'_{13} &= a_{13}/l'_{11} &&= 2{\cdot}828\,426, \\ z_1 &= y_1/l'_{11} &&= 11{\cdot}313\,705, \\ r_1^0 &= r_1/l'_{11} &&= 17{\cdot}677\,664. \end{aligned}$$

The actual row sum of $(\mathbf{L}', \mathbf{z})$ is 17·677665, which agrees well with the derived value. The actual value is used in further calculation:

$$l'_{22} = \sqrt{(a_{22}-l'^2_{12})} \qquad = 1{\cdot}224\,745,$$

$$l'_{23} = (a_{23}-l'_{12}l'_{13})/l'_{22} \qquad = 0{\cdot}816\,499,$$

$$z_2 = (y_2-l'_{12}z_1)/l'_{22} \qquad = 3{\cdot}265\,995,$$

$$r^0_2 = (r_2-l'_{12}r^0_1)/l'_{22} \qquad = 5{\cdot}307\,240.$$

The actual row sum is 5·307239 which is used for later calculation.

$$l'_{33} = \sqrt{(a_{33}-l'^2_{13}-l'^2_{23})} \qquad = 1{\cdot}154\,702,$$

$$z_3 = (y_3-l'_{13}z_1-l'_{23}z_2)/l'_{33} \qquad = 1{\cdot}154\,706,$$

$$r^0_3 = (r_3-l'_{13}r^0_1-l'_{23}r^0_2)/l'_{33} \qquad = 2{\cdot}309\,407.$$

The actual row sum is 2·309408 which is the value used later. The back substitution is

$$x_3 = z_3/l'_{33} \qquad = 1{\cdot}000\,003,$$

$$x^0_3 = r^0_3/l'_{33} \qquad = 2{\cdot}000\,003,$$

$$x_2 = (z_2-l'_{23}x_3)/l'_{22} \qquad = 2{\cdot}000\,003,$$

$$x^0_2 = (r^0_2-l'_{23}x^0_3)/l'_{22} \qquad = 3{\cdot}000\,003,$$

$$x_1 = (z_1-l'_{13}x_3-l'_{12}x_2)/l'_{11} \qquad = 2{\cdot}999\,987,$$

$$x^0_1 = (r^0_1-l'_{13}x^0_3-l'_{12}x^0_2)/l'_{11} \qquad = 3{\cdot}999\,987.$$

The calculation may be displayed thus:

A			y	r	L′			z	r⁰
2	3	4	16	25	1·414214	2·121320	2·828426	11·313705	17·677665
3	6	7	28	44		1·224745	0·816499	3·265995	5·307239
4	7	10	36	57			1·154702	1·154706	2·309408

$x_1 = 2{\cdot}999987 \quad x_2 = 2{\cdot}000003 \quad x_3 = 1{\cdot}000003$

$x^0_1 = 3{\cdot}999987 \quad x^0_2 = 3{\cdot}000003 \quad x^0_3 = 2{\cdot}000003$

Factorisation into a triple product, as in equation (7.4.2), involves either more recording than the square root factorisation, or the calculation of sums of triple products. It is useful to be able to evaluate the three factors in order to simplify certain characteristic-value problems (see § 8.2), but they are not needed in solving equations or inverting a matrix.

If the diagonal elements of $\mathbf{L}$ are taken to be ones there are various ways of finding the three factors. The first method is to carry out an ordinary triangular factorisation, as in equation (7.2.9), and then to take the diagonal elements of $\mathbf{U}$ to make the diagonal factor $\mathbf{D}$. For it may easily be verified by an examination of equations (7.2.15) that if $\mathbf{A}$ is symmetric and

$$\mathbf{A} = \mathbf{LU}, \tag{7.4.12}$$

then
$$\mathbf{U} = \mathbf{D}\mathbf{L}', \tag{7.4.13}$$
where
$$\mathbf{D} = \begin{bmatrix} u_{11} & 0 & \dots & 0 \\ 0 & u_{22} & \dots & 0 \\ \dots & \dots & \dots & \dots \\ 0 & 0 & \dots & u_{nn} \end{bmatrix}. \tag{7.4.14}$$

This factorisation is particularly useful when $\mathbf{A}$ is singular. When this occurs, $|\mathbf{A}| = 0$, and therefore, since $|\mathbf{L}'| = 1$, $|\mathbf{U}| = 0$. But
$$|\mathbf{U}| = u_{11}u_{22}\dots u_{nn}, \tag{7.4.15}$$
and therefore one or more of the elements of $\mathbf{D}$ will be zero. Thus the singular matrix $\mathbf{A}$ is expressed in the form
$$\mathbf{A} = \mathbf{L}\mathbf{D}\mathbf{L}', \tag{7.4.16}$$
where $\mathbf{L}$ and $\mathbf{L}'$ are non-singular and $\mathbf{D}$ is a diagonal matrix with one or more zeros.

When $\mathbf{A}$ is a positive semi-definite matrix there is another way of factorising it which is sometimes convenient. The elements u_{rr} will be positive or zero (some will be zero because $\mathbf{A}$ will be singular). The matrix $\mathbf{D}$ may therefore be factorised into the three diagonal factors
$$\mathbf{D} = \mathbf{C}\mathbf{D}^0\mathbf{C}', \tag{7.4.17}$$
where
$$c_r = \begin{cases} d_r^{\frac{1}{2}} = u_{rr}^{\frac{1}{2}} & \text{if} \quad d_r \neq 0, \\ 1 & \text{if} \quad d_r = 0, \end{cases} \tag{7.4.18}$$
and
$$d_r^0 = \begin{cases} 1 & \text{if} \quad d_r \neq 0, \\ 0 & \text{if} \quad d_r = 0. \end{cases} \tag{7.4.19}$$
Then
$$\begin{aligned} \mathbf{A} &= (\mathbf{L}\mathbf{C})\,\mathbf{D}^0(\mathbf{C}'\mathbf{L}') \\ &= \mathbf{L}^0\mathbf{D}^0\mathbf{L}^{0\prime}, \end{aligned} \tag{7.4.20}$$
where $\mathbf{D}^0$ is a diagonal matrix composed of ones and noughts, and $\mathbf{L}^0$ and $\mathbf{L}^{0\prime}$ are non-singular.

When the diagonal elements of $\mathbf{L}$ are taken to be ones an alternative way of finding the three factors is to take the square root factors as in equation (7.4.1), divide each row of $\mathbf{L}'$ by its diagonal element, so that
$$\begin{bmatrix} l_{11} & l_{21} & \dots & l_{n1} \\ 0 & l_{22} & \dots & l_{n2} \\ \dots & \dots & \dots & \dots \\ 0 & 0 & \dots & l_{nn} \end{bmatrix} = \begin{bmatrix} l_{11} & 0 & \dots & 0 \\ 0 & l_{22} & \dots & 0 \\ \dots & \dots & \dots & \dots \\ 0 & 0 & \dots & l_{nn} \end{bmatrix} \begin{bmatrix} 1 & l_{21}/l_{11} & \dots & l_{n1}/l_{11} \\ 0 & 1 & \dots & l_{n2}/l_{22} \\ \dots & \dots & \dots & \dots \\ 0 & 0 & \dots & 1 \end{bmatrix}, \tag{7.4.21}$$
and then take
$$\mathbf{D} = \begin{bmatrix} l_{11}^2 & 0 & \dots & 0 \\ 0 & l_{22}^2 & \dots & 0 \\ \dots & \dots & \dots & \dots \\ 0 & 0 & \dots & l_{nn}^2 \end{bmatrix}. \tag{7.4.22}$$

This may be illustrated by means of the example following equation (7.4.4). From that triangle, we obtain

$$\mathbf{L}' = \begin{bmatrix} 1 & 1\cdot499\,999 & 1\cdot999\,999 \\ 0 & 1 & 0\cdot666\,669 \\ 0 & 0 & 1 \end{bmatrix},$$

$$\mathbf{D} = \begin{bmatrix} 2\cdot000\,001 & 0 & 0 \\ 0 & 1\cdot500\,000 & 0 \\ 0 & 0 & 1\cdot333\,337 \end{bmatrix}.$$

A fourth method is to proceed from the equations which are represented by the matrix equation (7.4.2). These are

$$\left.\begin{aligned} a_{11} &= l_{11} d_{11} l'_{11}, \\ a_{12} &= l_{11} d_{11} l'_{12}, \\ &\cdots\cdots\cdots\cdots \\ a_{1r} &= l_{11} d_{11} l'_{1r}; \end{aligned}\right\} \tag{7.4.23}$$

$$\left.\begin{aligned} a_{22} &= l_{21} d_{11} l'_{12} + l_{22} d_{22} l'_{22}, \\ a_{23} &= l_{21} d_{11} l'_{13} + l_{22} d_{22} l'_{23}, \\ &\cdots\cdots\cdots\cdots\cdots\cdots \\ a_{2r} &= l_{21} d_{11} l'_{1r} + l_{22} d_{22} l'_{2r}; \end{aligned}\right\} \tag{7.4.24}$$

and so on.

Remembering that $l_{rr} = l'_{rr} = 1$ we may solve these equations and find that

$$\left.\begin{aligned} d_{11} &= a_{11}, \\ l'_{12} &= a_{12}/d_{11}, \\ &\cdots\cdots\cdots\cdots \\ l'_{1r} &= a_{1r}/d_{11}; \end{aligned}\right\} \tag{7.4.25}$$

$$\left.\begin{aligned} d_{22} &= a_{22} - d_{11} l'^2_{12}, \\ l'_{23} &= (a_{23} - d_{11} l'_{12} l'_{13})/d_{22}, \\ &\cdots\cdots\cdots\cdots\cdots\cdots \\ l'_{2r} &= (a_{2r} - d_{11} l'_{12} l'_{1r})\, d_{22}; \end{aligned}\right\} \tag{7.4.26}$$

$$\left.\begin{aligned} d_{33} &= a_{33} - d_{11} l'^2_{13} - d_{22} l'^2_{23}, \\ l'_{34} &= (a_{34} - d_{11} l'_{13} l'_{14} - d_{22} l'_{23} l'_{24})/d_{33}, \\ &\cdots\cdots\cdots\cdots\cdots\cdots\cdots\cdots \\ l'_{3r} &= (a_{3r} - d_{11} l'_{13} l'_{1r} - d_{22} l'_{23} l'_{2r})/d_{rr}; \end{aligned}\right\} \tag{7.4.27}$$

and so on.

For the matrix of the example on p. 266 steps of the calculation are

$$d_{11} = 2,$$
$$l'_{12} = 3/2 = 1{\cdot}500\,000,$$
$$l'_{13} = 4/2 = 2{\cdot}000\,000,$$
$$d_{22} = 6-(2)\,(1{\cdot}5)^2 = 1{\cdot}500\,000,$$
$$l'_{23} = \{7-(2)\,(1{\cdot}5)\,(2)\}/1{\cdot}5 = 0{\cdot}666\,667,$$
$$d_{33} = 10-2(2)^2-(1{\cdot}5)\,(0{\cdot}666667)^2 = 1{\cdot}333\,333.$$

This process may be checked by the use of row sums, the check sums being formed from the column of row-sums of $\mathbf{A}$ by the same operations which yield the other columns of $\mathbf{L}'$, since they satisfy the equation

$$\mathbf{LDr}^0 = \mathbf{r}, \tag{7.4.28}$$

where $\mathbf{r}^0 = \mathbf{L}'\boldsymbol{l}$ and $\mathbf{r} = \mathbf{A}\boldsymbol{l}$. This equation is of the same form as that from which $\mathbf{L}'$ is found, namely equation (7.4.2). If a desk machine with two accumulators is available, this process presents no difficulty, but otherwise forming the triangular factors and then extracting the diagonal factor, as in equations (7.4.12)–(7.4.15), is easier.

EXAMPLES 7.4

1. Solve the equations

$$\begin{aligned} 4x_1-6x_2-3x_3 &= 1,\\ -6x_1-6x_3 &= 2,\\ -3x_1-6x_2-3x_3 &= 3, \end{aligned}$$

using the square root factorisation method of equation (7.4.1).

2. (*a*) Show that, if $\mathbf{A}$ is a positive definite symmetric matrix, the factors $\mathbf{L}$ and $\mathbf{L}'$ in equation (7.4.1) are real.

(*b*) Find $\mathbf{L}'$ for the matrix

$$\mathbf{A} = \begin{bmatrix} 4 & 6 & 3\\ 6 & 12 & 6\\ 3 & 6 & 9 \end{bmatrix}.$$

3. (*a*) Prove that if $\mathbf{A}$ is symmetric and equal to $\mathbf{LU}$, then $\mathbf{U} = \mathbf{DL}'$ where

$$\mathbf{D} = \begin{bmatrix} u_{11} & 0 & \dots & 0\\ 0 & u_{22} & \dots & 0\\ \dots & \dots & \dots & \dots\\ 0 & 0 & \dots & u_{nn} \end{bmatrix}.$$

(*b*) If

$$\begin{bmatrix} 2 & -1 & -3\\ -1 & 1 & 2\\ -3 & 2 & 5 \end{bmatrix} = \mathbf{LDL}',$$

find the factors $\mathbf{L}$ and $\mathbf{D}$. All the elements in the principal diagonal of $\mathbf{L}$ are to be ones.

4. (*a*) Show that if $\mathbf{A}$ is positive semi-definite, the elements u_{rr} will be positive or zero.
(*b*) Express the matrix of Ex. 7.4.3 as the product $\mathbf{L}^0\mathbf{D}^0\mathbf{L}^{0\prime}$ in the manner of equation (7.4.20).

5. Factorise the matrix

$$\mathbf{A} = \begin{bmatrix} 4 & -6 & -3 \\ -6 & 0 & -6 \\ -3 & -6 & -3 \end{bmatrix}$$

in the form $\mathbf{LDL}'$ using the method of equations (7.4.21) and (7.4.22).

7.5 Some difficulties

If the arithmetic involved could be carried out exactly, there would be few difficulties about solving sets of linear equations. Unfortunately, the number of figures required to represent the numbers produced in the course of the elimination grows very rapidly with the size of the system, and even with modern automatic digital computers it is not practicable to use numbers with very many figures, so that it becomes necessary to consider the effect of rounding-off numbers upon the precision of the results.

When the result of the calculation is very sensitive to rounding-off it will be sensitive to small changes in the initial data, the coefficients in the equations and the right-hand sides. Such equations are called 'ill-conditioned'. If enough figures could be used, it would be possible to obtain a close approximation to the solution of the equations, treating the given values as exact. However, the initial values are usually subject to uncertainties, such as errors of observation or rounding errors in replacing rational fractions by decimal fractions. Therefore within the range of variation of the given data there lies an infinity of equations, each of which has its solution. Often we cannot say which of the solutions of these neighbouring equations is the 'true' solution of the given set of equations, and if the equations are ill-conditioned, these possible solutions may spread over a considerable range. In such a case it is necessary to be very chary of drawing conclusions from a purported solution.

The equations

$$\left.\begin{aligned} x_1/2 + x_2/3 + x_3/4 &= 65, \\ x_1/3 + x_2/4 + x_3/5 &= 47, \\ x_1/4 + x_2/5 + x_3/6 &= 37, \end{aligned}\right\} \tag{7.5.1}$$

have the solution $x_1 = 60, \quad x_2 = 60, \quad x_3 = 60.$

When the coefficients in the equations are expressed as decimal fractions and rounded to two decimal places, the equations become

$$\left.\begin{aligned} 0{\cdot}50x_1 + 0{\cdot}33x_2 + 0{\cdot}25x_3 &= 65, \\ 0{\cdot}33x_1 + 0{\cdot}25x_2 + 0{\cdot}20x_3 &= 47, \\ 0{\cdot}25x_1 + 0{\cdot}20x_2 + 0{\cdot}17x_3 &= 37. \end{aligned}\right\} \tag{7.5.2}$$

The exact solution of this set of equations is

$$x_1 = 1325/28, \quad x_2 = 3375/28, \quad x_3 = 175/28,$$

and rounding these off to four decimal places, we obtain

$$x_1 = 47{\cdot}3214, \quad x_2 = 120{\cdot}5357, \quad x_3 = 6{\cdot}2500.$$

In no coefficient was the change more than $0{\cdot}00\dot{3}$ yet the solution is completely changed. These equations are very ill-conditioned. (They are a segment of the notoriously ill-conditioned equations associated with the Hilbert matrix $a_{ij} = (i+j^{-1})^{-1}$.)

The solution of the equations

$$\mathbf{Ax} = \mathbf{y} \tag{7.5.3}$$

may be written

$$\mathbf{x} = \mathbf{A}^{-1}\mathbf{y}. \tag{7.5.4}$$

If m is the numerically greatest element in $\mathbf{A}^{-1}$ a change of δy_r in y_r will affect at least one element of the solution by an amount equal to $m\delta y_r$. The inverse of the matrix $\mathbf{A}$ of the equations (7.5.1) is

$$\mathbf{A}^{-1} = \begin{bmatrix} 72 & -240 & 180 \\ -240 & 900 & -720 \\ 180 & -720 & 600 \end{bmatrix},$$

so that a change in any element of $\mathbf{y}$ will affect the solution by up to 900 times as much.

The effect of a change in the elements of $\mathbf{A}$ may be estimated as follows. Denote the change in $\mathbf{A}$ by $\boldsymbol{\delta}\mathbf{A}$, and the consequent change in $\mathbf{x}$ by $\boldsymbol{\delta}\mathbf{x}$, then

$$(\mathbf{A}+\boldsymbol{\delta}\mathbf{A})\,(\mathbf{x}+\boldsymbol{\delta}\mathbf{x}) = \mathbf{y}, \tag{7.5.5}$$

so that

$$\mathbf{A}\boldsymbol{\delta}\mathbf{x} = \mathbf{y}-\mathbf{Ax}-\boldsymbol{\delta}\mathbf{A}\,.\,\mathbf{x}-\boldsymbol{\delta}\mathbf{A}\,.\,\boldsymbol{\delta}\mathbf{x}. \tag{7.5.6}$$

If we ignore $\boldsymbol{\delta}\mathbf{A}\,.\,\boldsymbol{\delta}\mathbf{x}$ and use equation (7.4.1) we find

$$\mathbf{A}\,.\,\boldsymbol{\delta}\mathbf{x} = -\boldsymbol{\delta}\mathbf{A}\,.\,\mathbf{x}, \tag{7.5.7}$$

whence

$$\boldsymbol{\delta}\mathbf{x} = -\mathbf{A}^{-1}\,.\,\boldsymbol{\delta}\mathbf{A}\,.\,\mathbf{x}. \tag{7.5.8}$$

The change in x_k due to a change δa_{ij} in a_{ij} is

$$\delta x_k = -(\mathbf{A}^{-1})_{ki}x_j\delta a_{ij}. \tag{7.5.9}$$

In the example $x_1 = x_2 = x_3 = 60$, and the numerically greatest and least elements of $\mathbf{A}^{-1}$ are 900 and 72. Therefore δa_{ij} may be multiplied by numbers between 54 000 and 4320, which accords with the observed sensitivity of the solution to changes in $\mathbf{A}$.

How does this sensitivity become apparent during the elimination process of solving equations (7.5.1) or (7.5.2)? Large multipliers do not occur during the elimination, nor during the back-substitution. The sensitivity is shown by the occurrence of small divisors, the diagonal elements of **U**, as may be seen in the displayed calculation:

A			y	L	U			z	x
1/2	1/3	1/4	65	1	1/2	1/3	1/4	65	60
1/3	1/4	1/5	47	2/3	1	1/36	1/30	11/3	60
1/4	1/5	1/6	37	1/2	6/5	1	1/600	1/10	60

The value of x_3 is found as (1/10)/(1/600). Each of the small numbers in this fraction was found by subtracting much larger numbers, so that a change which is a small proportion of, say, a_{33} will cause an equal change in u_{33}, but this will be a larger proportion of u_{33}, and hence a larger proportion of x_3.

The sensitivity to small changes in **A** is due to the appearance of small diagonal elements in **U** as a result of cancellation of leading digits. Small elements in the off-diagonal positions of **L** or **U** do not cause the same trouble. From equation (7.2.9) it follows that

$$|\mathbf{A}| = |\mathbf{L}|\,.\,|\mathbf{U}| = |\mathbf{U}| \tag{7.5.10}$$

and small diagonal elements of **U** lead to a small value for $|\mathbf{U}|$ and $|\mathbf{A}|$. Hence, the rows and columns of **A** must be nearly linearly dependent. Geometrically, this means that the planes defined by equations (7.5.1) meet at a small solid angle, so that a small change in one plane shifts the point where all three meet a long way. The general state of affairs is similar, small diagonal elements in **U**, occurring through cancellation, are often a sign that the solution is sensitive to small changes in the original data, and the equations are ill-conditioned.

There is another difficulty which often arises, which, although similar to ill-conditioning, should be distinguished from it. Very often, when an attempt is made to solve a set of equations by the method of § 7.2, large errors accumulate (for example through the cancellation of leading digits) and the solution obtained is far from the true one. The difficulty is similar to ill-conditioning because, like ill-conditioning, it often shows itself in small diagonal elements of **U**. But it is not the same as ill-conditioning because the equations need not be sensitive to the original data. In fact, the difficulty can usually be overcome by re-arranging the equations and variables so that the small diagonal elements are replaced by larger elements with more significant figures. If it is possible to improve matters in this way the equations are not ill-conditioned, it is merely that the first few equations are not suitable for finding the first few variables in terms of the others.

It has been shown by Wilkinson† that this and the other difficulties described below do not arise with positive definite symmetric matrices, and the straightforward elimination processes described in § 7.2 or § 7.4 are then quite satisfactory.

The coefficients in the following equations are given to four significant figures:

$$0{\cdot}0001024x_1 + 0{\cdot}6873x_2 = 1, \tag{7.5.11}$$

$$0{\cdot}4096x_1 + 0{\cdot}1234x_2 = 2. \tag{7.5.12}$$

† J. H. Wilkinson, 'Error analysis of direct methods of matrix inversion,' *J. Assn Comput. Mach.* **8**, (1961), no. 3, pp. 281–330.

Using equation (7.5.11) as pivot to eliminate x_1 from (7.5.12), we derive

$$-2749x_2 = -3998, \tag{7.5.13}$$

whence

$$x_2 = 1{\cdot}454.$$

Substitution of this value in the pivotal equation (7.5.11) gives

$$x_1 = \{1-(0{\cdot}6873)(1{\cdot}454)\}/0{\cdot}0001024 = 6{\cdot}502. \tag{7.5.14}$$

In this last step, since a small denominator is used, the numerator must be small to yield a value of x_1 of order unity. The numerator becomes small by cancellation of leading digits, and its accuracy is low, about one significant digit, and so is the accuracy of x_1. Moreover, in the step leading to equation (7.5.13), the influence of the coefficient of x_2 in equation (7.5.12) has been almost entirely lost; any value for the coefficient between $-0{\cdot}7$ and $+0{\cdot}3$ would have given the same equation (7.5.13). In the back substitution to find x_1 this coefficient has no further influence, and hence has very little influence on the solution.

On the other hand, if equation (7.5.12) is used to eliminate x_1 from (7.5.11) the reduced equation is

$$0{\cdot}6873x_2 = 0{\cdot}9995, \tag{7.5.15}$$

in which the influence of the second coefficient in (7.5.12) has again been lost (in fact this coefficient does not influence the value of x_2 greatly with the other coefficients having the given values). The value of x_2 found is

$$x_2 = 1{\cdot}454$$

as before. However, substituting this in the pivotal equation leads to

$$x_1 = \{2-(0{\cdot}1234)(1{\cdot}454)\}/0{\cdot}4096 = 4{\cdot}445.$$

The second coefficient in equation (7.5.12) now has its proper influence. Moreover, since the denominator is not small, the numerator does not have to become small by cancellation and there is no loss of accuracy.

This example shows that the difficulty does not arise only from inaccuracy in the small pivot, but that even if this were to be found to many significant figures, trouble could be expected to occur during the back-substitution if the pivotal variable did not have a very large value.

A similar difficulty can arise if one variable has a large value. During the back substitution the use of this variable will amplify rounding errors in some elements of **U**. If all the elements of the solution vector that lie above the large one are also large there will be little cancellation, and the amplification of rounding errors will not be disproportionate. If any of these elements have small values, either cancellation occurs during the back substitution or the large values are multiplied by small coefficients. In this case, the coefficients are likely to have been found as differences between nearly equal numbers, and will have few significant figures. In either case, the element that is small will not be found very accurately.

From the equations

$$\left.\begin{aligned} 1{\cdot}00x_1 + 1{\cdot}11x_2 + 1{\cdot}01x_3 &= 1{\cdot}10, \\ 2{\cdot}00x_1 + 0{\cdot}98x_2 + 1{\cdot}10x_3 &= -1{\cdot}00, \\ 1{\cdot}87x_1 + 0{\cdot}46x_2 + 0{\cdot}33x_3 &= 1{\cdot}40, \end{aligned}\right\} \tag{7.5.16}$$

the following reduced equations may be derived:

$$\left.\begin{aligned} 1{\cdot}000x_1+1{\cdot}110x_2+1{\cdot}010x_3 &= 1{\cdot}100,\\ -1{\cdot}240x_2-0{\cdot}920x_3 &= -3{\cdot}200,\\ -0{\cdot}360x_3 &= 3{\cdot}603. \end{aligned}\right\} \tag{7.5.17}$$

From these it is found that

$$x_3 = -10{\cdot}0 \quad \text{and} \quad x_2 = 10{\cdot}0, \tag{7.5.18}$$

there being no apparent reason to expect an error of more than 0·5 in the last decimal place. For the evaluation of x_1, the first pivotal equation yields

$$\begin{aligned} x_1 &= 1{\cdot}100-(1{\cdot}010)(-10{\cdot}0)-(1{\cdot}110)(10{\cdot}0)\\ &= 1{\cdot}100+10{\cdot}10-11{\cdot}10\\ &= 0{\cdot}10. \end{aligned}$$

In this case, each product on the right of the equation may be in error by 0·05, supposing the coefficients to be exact in the original equations. Thus the value of x_1 might be out by as much as 0·1 in either direction, although one guard figure was kept in the calculation. If the possible error in the original coefficients is also 0·5 in the last quoted figure, that is 0·005, there is a total uncertainty of about 0·2 in either direction, and using more guard figures will not reduce this below 0·1.

The inverse of the matrix is, to four decimal places,

$$\mathbf{A}^{-1} = \begin{bmatrix} -0{\cdot}4091 & 0{\cdot}2202 & 0{\cdot}5180\\ 3{\cdot}1299 & -3{\cdot}4922 & 2{\cdot}0612\\ -2{\cdot}0446 & 3{\cdot}6199 & -2{\cdot}7781 \end{bmatrix}, \tag{7.5.19}$$

and postmultiplying this by the right-hand side column of equations (7.5.16) yields

$$x_1 = 0{\cdot}0550, \quad x_2 = 9{\cdot}8208, \quad x_3 = -9{\cdot}7583.$$

Thus it can be seen that the plausible values of x_2 and x_3 shown in (7.5.18) are also wrong by considerable proportions.

From the equations

$$\left.\begin{aligned} 1{\cdot}00x_1+1{\cdot}11x_2+1{\cdot}01x_3 &= 1{\cdot}20,\\ 2{\cdot}00x_1+0{\cdot}98x_2+1{\cdot}10x_3 &= -1{\cdot}00,\\ 1{\cdot}87x_1+0{\cdot}46x_2+0{\cdot}33x_3 &= 1{\cdot}49, \end{aligned}\right\} \tag{7.5.20}$$

may be derived the reduced equations (working to 3 decimal places)

$$\left.\begin{aligned} 1{\cdot}000x_1+1{\cdot}110x_2+1{\cdot}010x_3 &= 1{\cdot}200,\\ -1{\cdot}240x_2-0{\cdot}920x_3 &= -3{\cdot}400,\\ -0{\cdot}360x_3 &= 3{\cdot}676. \end{aligned}\right\} \tag{7.5.21}$$

From the last two equations it follows that

$$x_3 = -10{\cdot}21 \quad \text{and} \quad x_2 = 10{\cdot}32.$$

The value of x_3 has but three significant figures, since the pivot used in its evaluation $(-0{\cdot}360)$ has but three, and x_3 may be in error by as much as $\pm 0{\cdot}05$. The value of x_2 may be in error by about as much. When the first pivotal equation is used to yield x_1 we find

$$x_1 = 1{\cdot}20-(1{\cdot}01)\,(-10{\cdot}21)-(1{\cdot}11)\,(10{\cdot}32) = 0{\cdot}057.$$

Each product in this equation may be in error by $\pm 0{\cdot}05$, assuming the original coefficients to be exact, so that the error in x_1 may be as much as $0{\cdot}1$, which swamps the value found.

The exact solution may be found from the exact inverse of the matrix of coefficients, namely,

$$\mathbf{A}^{-1} = \frac{100}{446\,344}\begin{bmatrix} -1\,826 & 983 & 2\,312 \\ 13\,970 & -15\,587 & 9\,200 \\ -9\,126 & 16\,157 & -12\,400 \end{bmatrix} \qquad (7.5.22)$$

to be

$$x_1 = 27\,068/446\,344, \quad x_2 = 46\,505\,900/446\,344, \quad x_3 = -4\,558\,420/446\,344.$$

To four significant figures these values are

$$x_1 = 0{\cdot}06064, \quad x_2 = 10{\cdot}32, \quad x_3 = -10{\cdot}21,$$

and the value of x_1 was in error by about 6 per cent.

It is not easy to formulate rules which will enable a satisfactory estimate of the error in the solution to be made without an estimate of the inverse of the matrix being found, but for circumstances where the extra work is worth while, the change in the solution when the calculation is re-worked with several extra figures will indicate whether the solution is reliable or not.

A strict upper bound on the error has been derived by Wilkinson,† and it is shown in § 7.11 that the estimate of the size of the elements of the inverse matrix which is required in his method may be obtained by a simple device requiring n^2 additional operations.

7.6 Avoiding the difficulties. I

When a set of equations is well-conditioned, it may still happen that the technique of solution previously described leads to a very poor approximation to the solution. As was suggested in § 7.5, re-arranging the equations may avoid this difficulty. In fact, the best possible technique at present known seems to be a modification of the basic elimination process of § 7.2 in which re-arrangements are carried out as the work proceeds. It will be seen that this involves the loss of the reduction in recording given by the condensed elimination techniques, and unless recording is done with extra figures, an additional loss of accuracy will ensue. In § 7.8 a further modification will be described which is not such a safe way of obtaining solutions, but which has the advantages of being a condensed technique.

† Bounds on the errors due to rounding off have been obtained in two papers: Wilkinson, *J. Assn Comput. Mach.* **8** (1961), no. 3, pp. 281–330; J. H. Wilkinson, 'Rounding errors in algebraic processes', *Proc. Int. Conf. Information Processing* (UNESCO, 1959).

The modification to the basic technique which reduces the liability to trouble depends on the choice of the pivotal coefficients. In the basic technique, when a variable has been eliminated, the leading diagonal coefficient of the block of equations which have not yet been pivots is chosen as the new pivot. That is, the variables are eliminated in order of writing and the pivotal equations are taken as the equations in their given order. If, whenever a pivot has to be chosen, it is taken to be the largest coefficient in those equations which have not yet been pivotal, the best available technique is to hand. Since, to choose the largest coefficient, all the coefficients must be available for comparison, it is clear that the condensed elimination techniques, in which the coefficients of the non-pivotal equations are not computed until those equations become pivotal, cannot be used.

The 'strategy' described above will be applied to the set of equations

$$\left.\begin{aligned} 2x_1+3x_2+x_3 &= 13, \\ 3x_1-x_2-2x_3 &= 5, \\ -4x_1+6x_2+x_3 &= 1. \end{aligned}\right\} \tag{7.6.1}$$

The first pivot is '6', the pivotal variable being x_2 and the pivotal equation the last. The first two equations become

$$4{\cdot}0000x_1+0{\cdot}5000x_3 = 12{\cdot}5000,$$

$$2{\cdot}3332x_1-1{\cdot}8333x_3 = 5{\cdot}1667.$$

The next pivot is 4·0000, and x_1 is then eliminated from the second equation to yield

$$-2{\cdot}1250x_3 = -2{\cdot}1246.$$

The pivotal equations are

$$\left.\begin{aligned} -4{\cdot}0000x_1+6{\cdot}0000x_2+1{\cdot}0000x_3 &= 1{\cdot}0000, \\ 4{\cdot}0000x_1 \qquad\qquad +0{\cdot}5000x_3 &= 12{\cdot}5000, \\ -2{\cdot}1250x_3 &= -2{\cdot}1246, \end{aligned}\right\} \tag{7.6.2}$$

and the unknowns are found in turn to be

$$x_3 = 0{\cdot}9998,$$

$$x_1 = 3{\cdot}0000,$$

$$x_2 = 2{\cdot}0000.$$

Two advantages of the condensed processes have been lost, the amount of recording is now of order $n^3/3$, and scalar products cannot be accumulated double length, unless the recording is to double length accuracy. This is laborious for hand-working but may be quite possible when an automatic digital computer is used. The automatic computer is likely to be used when many small problems have to be solved, and then the double length storage required will not cause any difficulty. The computer will also be used when large problems have to be solved, and then its storage capacity may well be strained to the limit if numbers are stored only single length. For example, one machine in common use can store about 14 000 numbers in addition to a not-too-large programme. This is enough to solve a set of about 125 equations if all the coefficients are present, if single length recording is needed, but only about 85 equations if double length recording is necessary. This is the reason for describing the techniques of the next two sections.

The equations to be solved, if not pathological, will not have appreciably different solutions if they are modified in the following way. First each equation is multiplied by some number. Secondly, each variable is measured in a new unit, not necessarily the same unit for each variable. If the scaling factors this involves are powers of ten, the actual digits of each coefficient will not change and then the equations might be expected to yield exactly the same solutions as before scaling. If the scaling involves rounding-off, the solution may, in bad cases, be radically altered. Any method of solution which makes smaller errors in the solution than the changes due to such a scaling is a satisfactory method. Such a scaling could completely alter the choice of pivots in the process described above. Before the technique of picking the largest coefficient as pivot can be satisfactory, it is necessary to standardise the equations in some way. Theoretical considerations and practical experience suggest the following standardisation. The equations should be multiplied by such factors that the largest coefficient in each equation is about one. The variables should then be scaled so that the largest coefficient in each column is about one. Strictly, this should be done before each choice of a pivot, but in practice it is found that it is only the first standardisation which is of importance.

For the equations (7.6.1) the first step of standardisation is to multiply the first equation by $\frac{1}{3}$, the second by $\frac{1}{3}$ and the third by $\frac{1}{6}$, giving

$$\left.\begin{aligned} 0{\cdot}6667x_1+1{\cdot}0000x_2+0{\cdot}3333x_3 &= 4{\cdot}3333,\\ 1{\cdot}0000x_1-0{\cdot}3333x_2-0{\cdot}6667x_3 &= 1{\cdot}6667,\\ -0{\cdot}6667x_1+1{\cdot}0000x_2+0{\cdot}1667x_3 &= 0{\cdot}1667.\end{aligned}\right\} \tag{7.6.3}$$

The second step is to change the scale of x_3 by taking a new variable x_3^*, defined by

$$x_3 = 1{\cdot}5x_3^*,$$

which yields the equations

$$\left.\begin{aligned} 0{\cdot}6667x_1+1{\cdot}0000x_2+0{\cdot}5000x_3^* &= 4{\cdot}3333,\\ 1{\cdot}0000x_1-0{\cdot}3333x_2-1{\cdot}0000x_3^* &= 1{\cdot}6667,\\ -0{\cdot}6667x_1+1{\cdot}0000x_2+0{\cdot}2500x_3^* &= 0{\cdot}1667.\end{aligned}\right\} \tag{7.6.4}$$

Now any variable may be chosen as the first to eliminate. Choosing x_1 yields the two reduced equations

$$\begin{aligned} 1{\cdot}2222x_2+1{\cdot}1667x_3^* &= 3{\cdot}2221,\\ 0{\cdot}7778x_2-0{\cdot}4167x_3^* &= 1{\cdot}2779.\end{aligned}$$

The next pivot is 1·2222 and the last equation becomes

$$-1{\cdot}1592x_3^* = -0{\cdot}7726.$$

The pivotal equations are

$$\left.\begin{aligned} 1{\cdot}0000x_1 - 0{\cdot}3333x_2 - 1{\cdot}0000x_3^* &= 1{\cdot}6667, \\ 1{\cdot}2222x_2 + 1{\cdot}1667x_3^* &= 3{\cdot}2221, \\ -1{\cdot}1592x_3^* &= -0{\cdot}7726, \end{aligned}\right\} \quad (7.6.5)$$

from which the unknowns are found in turn to be

$$x_3^* = 0{\cdot}6665,$$

$$x_2 = 2{\cdot}0001,$$

$$x_1 = 2{\cdot}9998.$$

Rescaling the value of x_3^* gives $\quad x_3 = 0{\cdot}9998.$

These values are not quite as close to the true values as those of the previous example. What needs explanation, however, is not the discrepancies in this case, but the absence of discrepancy previously. That was due to a fortunate combination of rounding-off errors. In general, standardisation will improve matters if there is any real problem about solving the equations.

The difficulty present in equations (7.5.19) is not removed by this technique, and in fact is not avoided without the use of additional guard figures. The use of these will yield the solution vector corresponding to the given data, considered as exact, but the solution will still have a relatively high sensitivity to small changes in the right-hand sides, as far as the small element is concerned. There is no cure for this disease, nor is there any for the difficulty illustrated in equations (7.5.3), when the solution is very sensitive to small changes in the coefficients. These are properties of the physical situation represented by the equations, and if the physical system being investigated does not show these phenomena, the equations are a poor representation of the physics, and an alternative formulation should be tried.

EXAMPLES 7.6

1. Standardise and then solve equations (7.5.11) and (7.5.12).

2. (*a*) Write the coefficients in equations (7.5.1) as four-figure decimals and attempt to solve them by the straightforward method of §7.2. Verify that very small diagonal elements occur in **U**.

(*b*) Standardise equations (7.5.1) in the manner described in this section, write the coefficients as four-figure decimals, and then solve the equations. Verify that this leads to a much more accurate solution.

7.7 Avoiding the difficulties. II

At the cost of carrying out the computation twice, it is possible to reduce the amount of recording and retain the advantage of accumulating scalar products double length while still making use of interchanges. For hand working, where there is no difficulty about lack of storage space (although there is about having to write down too many results), it is practicable to run through the process described above, making a note of the positions of the successive pivotal coefficients, and then re-arranging the equations in such a way that these positions are the successive diagonal positions. When the equations have been re-arranged in this way the straightforward triangulation technique may be used; the pivots that will occur will be the same as those used in the first run through.

For equations (7.6.4), the pivots were chosen in order from the positions (2, 1), (1, 2) and (3, 3). Interchanging equations 1 and 2, which corresponds to premultiplication by the matrix

$$\mathbf{P} = \begin{bmatrix} 0 & 1 & 0 \\ 1 & 0 & 0 \\ 0 & 0 & 1 \end{bmatrix}$$

yields the equations

$$\left.\begin{aligned} 1{\cdot}0000x_1 - 0{\cdot}3333x_2 - 1{\cdot}0000x_3^* &= 1{\cdot}6667, \\ 0{\cdot}6667x_1 + 1{\cdot}0000x_2 + 0{\cdot}5000x_3^* &= 4{\cdot}3333, \\ -0{\cdot}6667x_1 + 1{\cdot}0000x_2 + 0{\cdot}2500x_3^* &= 0{\cdot}1667. \end{aligned}\right\} \qquad (7.7.1)$$

The pivots used in that example were derived from the elements which now stand in positions (1, 1), (2, 2) and (3, 3). The technique of triangular factorisation, described in § 7.2, can now be applied to yield the following results:

A			y	r	U			z	$\mathbf{r}^0$
1·0000	−0·3333	−1·0000	1·6667	1·3334	1·0000	−0·3333	−1·0000	1·6667	1·3334
0·6667	1·0000	0·5000	4·3333	6·5000	**L** 0·6667	1·2222	1·1667	3·2221	5·6110
−0·6667	1·0000	0·2500	0·1667	0·7500	−0·6667	0·6364	−1·1592	−0·7727	−1·9319
1·0000	1·6667	−0·2500			$\mathbf{c}^0$ 1·0000	1·6364			
					$\mathbf{x}^0$ 3·9999	3·0000	1·6666		
					x 2·9999	2·0000	0·6666		
					x_1 2·9999,	x_2 2·0000,	x_3 0·9999		

The pivots used this time are clearly the same as those used in § 7.6.

In general the variables will not be eliminated in natural order, and the re-arrangement will involve permuting the variables. This corresponds to post-multiplication of the matrix **A** by a square matrix **Q**, having one non-zero element, of value unity, in each row and column. If the equations are not used in the natural order then this corresponds to a premultiplication by a matrix **P** of the same type. If the equations are re-arranged so that the coefficient appearing in the rth diagonal position comes from position (i, j) then the rth row of **P** must be the ith row of the unit matrix, while the rth column of **Q** must be the jth column of the unit matrix. After the back substitution, the column of values of the permuted unknowns must be pre-multiplied by **Q** to yield the unknowns in their original order, and these must then be re-scaled to the original units.

If the single length storage required to carry out the preliminary elimination is too much, the original equations cannot be held in a computer for the second elimination. It may be convenient to re-calculate the equations' coefficients and right-hand sides, but in awkward cases this would be a very lengthy procedure. If sufficiently rapid output and input facilities are provided on a computer, it will be possible to punch out (or record on some other external medium) the original equations for re-use. If re-calculation and output/input are both to be avoided, it is possible to carry out the single and double length calculations at the same time. To do this, when a pivot has been chosen, the pivotal row is computed with double length accumulation of scalar products. The coefficients in the reduced equations, those not yet used as pivots, are calculated single length (if this is quicker than double length accumulation) and are compared as they are produced to find the largest, but are not stored. The largest indicates which row is next to be computed accurately. This process wastes much of the single length work, and involves some $n^4/12$ additional multiplications more than are needed for the elimination proper, whereas the two-stage process involves only some $2n^3/3$. Using the single-stage process will usually more than double the time to solve the equations.

EXAMPLES 7.7

1. The equations

$$\begin{aligned} 4x_1-6x_2-3x_3 &= 1,\\ -6x_1+10x_2-6x_3 &= 2,\\ -3x_1-6x_2-3x_3 &= 3, \end{aligned}$$

are dealt with in the arrangement

$$\begin{aligned} -3x_1-3x_3-6x_2 &= 3,\\ 4x_1-3x_3-6x_2 &= 1,\\ -6x_1-6x_3+10x_2 &= 2. \end{aligned}$$

Express the new matrix of coefficients, $\mathbf{A}^*$, in terms of the original one, $\mathbf{A}$, by means of an equation

$$\mathbf{A}^* = \mathbf{PAQ}.$$

2. Show that the matrices $\mathbf{P}$ and $\mathbf{Q}$ of the type described in this section are orthogonal, i.e. that

$$\mathbf{PP}' = \mathbf{I} = \mathbf{QQ}'.$$

Hence show that if
$$\mathbf{A}^* = \mathbf{PAQ},$$
then
$$\mathbf{A} = \mathbf{P}'\mathbf{A}^*\mathbf{Q}'.$$

7.8 Avoiding the difficulties. III

The technique to be described now is one that has been found to work in practice, but one which can fail if it is used for specially prepared equations. It is an economical process from the point of view of recording, and scalar products can be accumulated double-length.

In the straightforward process described in § 7.2 the first equation was used to eliminate x_1 from the remaining equations, and then the first of the modified equations so formed was used to eliminate x_2 from the remainder, and so on. In technical language, the first pivotal equation was the first equation and the second pivotal equation was the first of the modified equations, etc. In addition, the order in which the variables were eliminated, the order of the pivots, was the natural order $x_1, x_2, x_3, \ldots$.

In the process which will be described the order of the pivots (that is the order in which the variables are eliminated) will be the natural order, but the pivotal equations will be chosen in a specific way, not just as they happen to occur. The following criterion will be used. The pivotal equation at any stage will be chosen as the equation which has the largest coefficient of the next variable to be eliminated, that is, the largest coefficient in the next column. This is the major modification, but there is also another minor one, which is sometimes useful in the treatment of unpleasant matrices. This is to introduce square root factorisation similar to that described at the end of section 7.2.

The stages in the calculation are conveniently described with the aid of an auxiliary matrix **F**. The first column of **F** is the column of x_1-coefficients at the first stage, that is the first column of **A**. The second column of **F** is the column of x_2-coefficients of the partially reduced equations at the second stage, with a zero in the row which was used as the first pivot. Generally, the jth column of **F** is the column of x_j-coefficients of the partially reduced equations at the jth stage, with zeros in all the rows which have already been used as pivots.

In terms of matrix factorisation the method corresponds to a factorisation

$$\mathbf{A} = \mathbf{LU}, \tag{7.8.1}$$

where **U** is an upper triangular matrix and **L** may be obtained from a lower triangular matrix by row interchanges. In the straightforward process (that is, not using square roots) **U** may be taken to have unit diagonal elements and then the calculation proceeds as follows. The moduli of the elements of the first column of **F** (that is of **A**) are compared to find the largest. Suppose it is the sth, $f_{s1} = a_{s1}$. Then all the elements in the sth row of **L**, except the first, are taken to be zero. The first column of **L** is found from

$$l_{i1} = f_{i1} = a_{i1}, \tag{7.8.2}$$

and the remainder of the first row of **U** from

$$u_{1k} = a_{sk}/l_{s1} \quad (k = 2, 3, \ldots, n). \tag{7.8.3}$$

The equation that is used to eliminate x_1 is the sth of the original equations, which may be written

$$u_{11}x_1 + u_{12}x_2 + \ldots + u_{1n}x_n = y_s/l_{s1} = z_1. \tag{7.8.4}$$

The partially reduced system consists of this equation and

$$(a_{i2} - l_{i1}u_{12})\,x_2 + (a_{i3} - l_{i1}u_{13})\,x_3 + \ldots + (a_{in} - l_{in}u_{1n})\,x_n = y_i - l_{i1}z_1, \tag{7.8.5}$$

where $i = 1, 2, \ldots, s-1, s+1, \ldots, n$. Thus

$$f_{i2} = \begin{cases} a_{i2} - l_{i1}u_{12} & (i \neq s), \\ 0 & (i = s). \end{cases} \tag{7.8.6}$$

Now the largest element of this second column of **F** is found, say f_{t2}, and the tth row of **L** is taken to have non-zero elements only in its first two positions, and the following values are taken

$$\left.\begin{aligned} l_{i2} &= f_{i2}, \\ u_{2k} &= (a_{tk} - l_{t1}u_{1k})/l_{t2} \quad (k = 3, 4, \dots, n). \end{aligned}\right\} \tag{7.8.7}$$

At the next stage the tth modified equation,

$$u_{22}x_2 + u_{23}x_3 + \dots + u_{2n}x_n = (y_t - l_{t1}z_1)/l_{t2} = z_2 \tag{7.8.8}$$

is used to eliminate x_2 from the remaining $n-2$ equations which have not yet been used as pivots, and so the calculation proceeds.

Generally, if the ith row has not yet been pivotal then the ith element of the jth column of **F** is given by

$$f_{ij} = a_{ij} - l_{i1}u_{1j} - l_{i2}u_{2j} - \dots - l_{i,j-1}u_{j-1,j}. \tag{7.8.9}$$

If f_{rj} is the element of largest modulus in this column then the rth row of **L** has non-zero elements only in the first j positions, and the jth column of **L** and jth row of (**U**, **z**) are given by

$$\left.\begin{aligned} l_{ij} &= f_{ij}, \\ u_{jk} &= (a_{rk} - l_{r1}u_{1k} - l_{r2}u_{2k} - \dots - l_{r,j-1}u_{j-1,k})/l_{rj}, \\ z_j &= (y_r - l_{r1}z_1 - l_{r2}z_2 - \dots - l_{r,j-1}z_{j-1})/l_{rj}. \end{aligned}\right\} \tag{7.8.10}$$

It will be noted that the method does not involve any extra work, or any extra storage. To choose a pivot, only the next column of coefficients in the reduced equations has to be computed.

When square root factorisation is used the process becomes modified as follows. Instead of taking the diagonal elements of **U** (the pivots) to be unity we take them to be the square roots of the moduli of the greatest elements of the corresponding columns of **F**. Thus if $|f_{s1}| = |a_{s1}|$ is the greatest of the moduli of the elements of the first column of **F**, then

$$u_{11} = \sqrt{|f_{s1}|}. \tag{7.8.11}$$

This means that

$$\left.\begin{aligned} l_{i1} &= f_{i1}/u_{11}, \\ u_{1k} &= a_{sk}/l_{s1} \quad (k = 2, 3, \dots, n). \end{aligned}\right\} \tag{7.8.12}$$

The equations at the second stage corresponding to equations (7.8.7) are

$$\left.\begin{aligned} u_{22} &= \sqrt{|f_{t2}|}, \\ l_{i2} &= f_{i2}/u_{22}, \\ u_{2k} &= (a_{tk} - l_{t1}u_{1k})/l_{t2} \quad (k = 3, 4, \dots, n). \end{aligned}\right\} \tag{7.8.13}$$

For hand work, square root factorisation involves recording about $n^2/2$ extra elements, and the same number of extra multiplications to scale from **F** to **L**. In neither method is it convenient to pack **L** under the diagonal of **U**. For use in an automatic digital computer, **U** would be re-arranged so that the new row of **U** replaces the row of **A** from which it was derived. A note then has to be kept of the way in which **U** is re-arranged, that is, of the order in which the rows of **A** were used as pivots, so that the rows of **U** can be used in order of increasing length for back substitution. Row sums can be used to check the calculation of $(\mathbf{U}, \mathbf{z})$ column sums to check the calculation of **F** and **L**.

The purpose of square root factorisation is to reduce the effects of rounding off. The pivot need not be split between the elements of **L** and **U** exactly according to equations (7.8.11), (7.8.13), etc. As long as some number near the square root is used for u_{jj}, some of the ill-effects of rounding off will be reduced. The most convenient number for hand calculations is a power of ten near the square root, as this requires no additional recording or rounding off, only a change in the position of the decimal point. The process is described as 'elimination with row interchanges', the pivotal row being moved up so that its leading element is on the diagonal of **U**. The process is also known as one of 'partial pivoting for size'.

We shall apply square root factorisation to equations (7.6.3), namely,

$$0{\cdot}6667x_1 + 1{\cdot}0000x_2 + 0{\cdot}5000x_3^* = 4{\cdot}3333,$$

$$1{\cdot}0000x_1 - 0{\cdot}3333x_2 - 1{\cdot}0000x_3^* = 1{\cdot}6667,$$

$$-0{\cdot}6667x_1 + 1{\cdot}0000x_2 + 0{\cdot}2500x_3^* = 0{\cdot}1667.$$

The element of largest modulus in the first column, f_{21}, is $1{\cdot}0000$, so that $s = 2$ $u_{11} = 1{\cdot}0000$, $\mathbf{col}_1(\mathbf{L})$ is the same as $\mathbf{col}_1(\mathbf{F})$, and

$$\boldsymbol{\rho}_1(\mathbf{U}) = [1{\cdot}0000 \quad -0{\cdot}3333 \quad -1{\cdot}0000].$$

The second column of **F** is

$$\mathbf{col}_2(\mathbf{F}) = \{1{\cdot}2222 \quad 0 \quad 0{\cdot}7778\}.$$

The next pivot is therefore f_{12}, and

$$u_{22} = \sqrt{(1{\cdot}2222)} = 1{\cdot}1055.$$

Hence
$$\mathbf{col}_2(\mathbf{L}) = \{1{\cdot}1055 \quad 0 \quad 0{\cdot}7036\}$$

and
$$\boldsymbol{\rho}_2(\mathbf{U}) = [0 \quad 1{\cdot}1055 \quad 1{\cdot}0554].$$

The last element of this row is found as

$$u_{23} = (a_{13} - l_{11}u_{13})/l_{12}$$

$$= \{0{\cdot}5000 - (0{\cdot}6667)\,(-1.0000)\}/1{\cdot}1055.$$

The last column of **F** is then

$$\mathbf{col}_3(\mathbf{F}) = \{0 \quad 0 \quad -1{\cdot}1593\}$$

so that
$$u_{33} = 1{\cdot}0767,$$

and
$$\mathbf{col}_3(\mathbf{L}) = \{0 \quad 0 \quad -1{\cdot}0767\}.$$

Thus the matrix **A** has been factorised into the product **LU**, where

$$\left.\begin{aligned}\mathbf{L} &= \begin{bmatrix} 0{\cdot}6667 & 1{\cdot}1055 & 0 \\ 1{\cdot}0000 & 0 & 0 \\ -0{\cdot}6667 & 0{\cdot}7036 & -1{\cdot}0767 \end{bmatrix}, \\ \mathbf{U} &= \begin{bmatrix} 1{\cdot}0000 & -0{\cdot}3333 & -1{\cdot}0000 \\ 0 & 1{\cdot}1055 & 1{\cdot}0554 \\ 0 & 0 & 1{\cdot}0767 \end{bmatrix}.\end{aligned}\right\} \tag{7.8.14}$$

The modified right-hand sides are

$$\mathbf{z} = \begin{bmatrix} 1{\cdot}6667 \\ 2{\cdot}9146 \\ 0{\cdot}7178 \end{bmatrix}.$$

Solving the equations $\mathbf{Ux} = \mathbf{z}$ from the bottom up yields

$$\left.\begin{aligned} x_3^* &= 0{\cdot}6667, \\ x_2 &= 2{\cdot}0000, \\ x_1 &= 3{\cdot}0000, \end{aligned}\right\} \tag{7.8.15}$$

whence

$$x_3 = 1{\cdot}0000. \tag{7.8.16}$$

To obtain check values for the row sums of $(\mathbf{U}, \mathbf{z})$, the row sums of $(\mathbf{A}, \mathbf{y})$ are treated like $\mathbf{y}$, while to obtain check values for the column sums of **F**, the column sums of **A** are treated like any other row of **A**, it being remembered that the column sums of **L** are to be used in deriving the new column sum of **F**, since columns of **L** (*not* previous columns of **F**) are used in deriving the new column of **F**, see for example equation (7.8.6). The column sums for **L** are checked by scaling the column sums of **F** in the same way as the other elements in the columns of **F**. The calculation can be displayed thus:

	A		y	r	
0·6667	1·0000	0·5000	4·3333	6·5000	
1·0000	−0·3333	−1·0000	1·6667	1·3334	
−0·6667	1·0000	0·2500	0·1667	0·7500	
c 1·0000	1·6667	−0·2500	—	—	
	F			**L**	
0·6667	1·2222	0·0000	0·6667	1·1055	0·0000
1·0000	0·0000	0·0000	1·0000	0·0000	0·0000
−0·6667	0·7778	−1·1593	−0·6667	0·7036	−1·0767
$\mathbf{c}^0$ 1·0000	2·0000	−1·1593	$\mathbf{c}^1$ 1·0000	1·8091	−1·0767
	U		**z**	$\mathbf{r}^0$	
1·0000	−0·3333	−1·0000	1·6667	1·3334	
	1·1055	1·0554	2·9146	5·0755 (6)	
		1·0767	0·7178	1·7945	
x	$\mathbf{x}^0$				
3·0000	4·0000				
2·0000	3·0000				
0·6667	1·6667				

EXAMPLES 7.8

1. Find the matrices **L**, **U**, **z** which are obtained by applying the matrix factorisation of equations (7.8.1)–(7.8.10) to equations (7.6.1).

2. (i) Use matrix factorisation together with row interchanges to factorise the matrix

$$\mathbf{B} = \begin{bmatrix} 1 & -2 & 2\cdot5 \\ -2 & 8 & -15\cdot5 \\ 2\cdot5 & -15\cdot5 & 29 \end{bmatrix}.$$

(ii) Is it necessary to use 'square root factorisation'?

7.9 Several right-hand sides

All the techniques that have been described are applicable to the simultaneous solution of several sets of equations, the different sets having the same coefficients but different right-hand sides. Such a family of equations can be represented by the matrix equation

$$\mathbf{AX} = \mathbf{Y}, \tag{7.9.1}$$

where **X** and **Y** are now matrices conformable with **A**, that is with n rows, but with any number of columns. The only modification is that the matrix **Y** is written in place of a single right-hand side and its several columns are treated in the same way at the same time.

Consider the matrices **A** and **Y** which are defined by

$$\mathbf{A} = \begin{bmatrix} 4 & 3 & 2 \\ 2 & 5 & 1 \\ 1 & 2 & 3 \end{bmatrix}, \qquad \mathbf{Y} = \begin{bmatrix} 20 & 15 \\ 17 & 11 \\ 10 & 10 \end{bmatrix}. \tag{7.9.2}$$

The solution of equations (7.9.1) by the method of § 7.8 may be displayed as follows:

Scaling by rows yields

	$\bar{\mathbf{A}}$			$\bar{\mathbf{Y}}$		**r**
	1·0000	0·7500	0·5000	5·0000	3·7500	11·0000
	0·4000	1·0000	0·2000	3·4000	2·2000	7·2000
	0·3333	0·6667	1·0000	3·3333	3·3333	8·6667
c	1·7333	2·4167	1·7000			

Scaling the variables is not necessary, and elimination now yields

	F			**L**		
	1·0000	0	0	1·0000	0	0
	0·4000	0·7000	0	0·4000	0·7000	0
	0·3333	0·4167	0·8334	0·3333	0·4167	0·8334
$\mathbf{c}^0$	1·7333	1·1167	0·8334	1·7333	1·1167	0·8334

U			**Z**		$\mathbf{r}^0$
1·0000	0·7500	0·5000	5·0000	3·7500	11·0000
0	1·0000	0·0000	2·0000	1·0000	4·0000
0	0	1·0000	1·0000	1·9999	3·9999

Back substitution then yields the matrix **X** (found from the bottom up)

X		$\mathbf{x}^0$
3·0000	2·0000	6·0000
2·0000	1·0000	4·0000
1·0000	1·9999	3·9999

The numbers in the check column, $\mathbf{x}^0$, are, as they should be, one more than the row-sums of **X**.

In this example, the value of f_{sj} has been taken for l_{sj}, and hence the diagonal elements of **U** are ones. If there is no significant cancellation during the accumulation of the scalar product for f_{sj} (and there was none in the example) there is no need to take the square root. If cancellation does occur, but one of l_{sj} or u_{jj} is chosen to be unity, then the product $l_{sj}u_{jj}$ will be disproportionately different from f_{sj} if a fixed number of decimal places are kept in **L** and **U**. In that case, taking the square root, or values near it, gives both l_{sj} and u_{jj} the same number of significant figures, and makes the product agree as closely as possible with the scalar product accumulated for f_{sj}.

7.10 Inversion of a matrix

The inversion of a matrix is a problem of the type shown in equation (7.9.1), the right-hand side being the unit matrix,

$$\mathbf{AX} = \mathbf{I}. \tag{7.10.1}$$

These equations may be solved by any of the methods suggested in this chapter, and we illustrate one of these below.

We shall obtain an approximate inverse of the matrix

$$\mathbf{A} = \begin{bmatrix} 2 & 3 & 1 \\ 3 & -1 & -2 \\ -4 & 6 & 1 \end{bmatrix}, \tag{7.10.2}$$

using the method described in §§ 7.8 and 7.9. In this case it will be seen that all the pivotal products are between 1 and 10, so that **F** and **L** are identical (we choose the diagonal elements of **U** to be ones).

	A			I			R
	2·0000	3·0000	1·0000	1	0	0	7·0000
	3·0000	−1·0000	−2·0000	0	1	0	1·0000
	−4·0000	6·0000	1·0000	0	0	1	4·0000
c	1·0000	8·0000	0·0000				

	F(L)		
	2·0000	6·0000	0
	3·0000	3·5000	−2·1250
	−4·0000	0	0
$\mathbf{c}^0$	1·0000	9·5000	−2·1250

U			L^{-1}			R^0
1	−1·5000	−0·2500	0·0000	0·0000	−0·2500	−1·0000
0	1	0·2500	0·1667	0	0·0833	1·5000
0	0	1	0·2746	−0·4706	−0·2157	0·5883 (2)

X			X*l*	X^0
0·2158	0·0589	−0·0981	0·1766	1·1766
0·0981	0·1177	0·1372	0·3530	1·3529
0·2746	−0·4706	−0·2157	−0·4117	0·5883

As a check we may compute **AX** and we find that

$$10^4(\mathbf{AX}-\mathbf{I}) = \begin{bmatrix} 5 & 3 & -3 \\ 1 & 2 & -1 \\ 0 & 0 & -1 \end{bmatrix}. \tag{7.10.3}$$

7.11 Bounds on the error of a solution

The analysis of § 7.5 shows that the sensitivity of a computed solution to perturbation of the initial data, **A** and **y**, and to rounding off during the computation of the solution is essentially determined by the size of the elements of the inverse matrix $\mathbf{A}^{-1}$. The calculation of the inverse requires about $5n^3/6$ operations in addition to those needed for the factorisation. The solution for a single column on the right-hand side requires only about n^2 operations, so that the labour of finding the inverse is a considerable addition.

A more precise analysis† shows that the sensitivity of a set of equations may be expressed in terms of a fairly general measure of the size of the inverse, and that it is unnecessary to calculate the actual inverse. The measure that is used has to play a role in matrix theory similar to that played by the modulus of a complex number in ordinary algebra, and it is denoted by double modulus bars, $\|\mathbf{A}\|$. It has to be a function of a matrix which satisfies the following conditions:

$$\left.\begin{aligned} &(1)\quad \|\mathbf{A}\| > 0 \quad \text{if} \quad \mathbf{A} \neq \mathbf{0},\\ &(2)\quad \|\mathbf{0}\| = 0,\\ &(3)\quad \|\mathbf{A}+\mathbf{B}\| \leqslant \|\mathbf{A}\| + \|\mathbf{B}\|,\\ &(4)\quad \|\mathbf{AB}\| \leqslant \|\mathbf{A}\| \,.\, \|\mathbf{B}\|,\\ &(5)\quad \|\alpha\mathbf{A}\| = |\alpha| \,.\, \|\mathbf{A}\|, \text{ where } \alpha \text{ is any complex number.} \end{aligned}\right\} \tag{7.11.1}$$

A function of a matrix which satisfies these conditions is called a 'norm' of the matrix.

There are many different ways of defining a norm of a matrix. For example, we may take

$$\|\mathbf{A}\| = \surd\Big(\sum_{i,j} |a_{ij}|^2\Big); \tag{7.11.2}$$

† Wilkinson, *J. Assn Comput. Mach.* **8** (1961), no. 3, pp. 281–330.

This is called the Euclidean norm. Or we may take

$$\|\mathbf{A}\| = \sqrt{(\text{maximum eigenvalue of } \mathbf{A}'\mathbf{A})}. \tag{7.11.3}$$

These are often convenient in theoretical analysis, but a very useful norm in practice is the 'row-modulus-sum norm'

$$\|\mathbf{A}\| = \max_i.(\sum_j |a_{ij}|). \tag{7.11.4}$$

For a column vector, this norm reduces to the maximum modulus of the elements. This is the one we shall use later in this section, although most of the results we shall quote will be true for all norms.

We start by examining the perturbation of a solution of a set of equations due to perturbations of the coefficients and right-hand sides. Suppose the original equations to be

$$\mathbf{Ax} = \mathbf{y}, \tag{7.11.5}$$

and the perturbed equations

$$(\mathbf{A}+\boldsymbol{\delta}\mathbf{A})(\mathbf{x}+\boldsymbol{\delta}\mathbf{x}) = \mathbf{y}+\boldsymbol{\delta}\mathbf{y}. \tag{7.11.6}$$

Then we may expand the bracket in this equation, and subtract the previous equation to obtain

$$\mathbf{A}\boldsymbol{\delta}\mathbf{x}+\boldsymbol{\delta}\mathbf{A}(\mathbf{x}+\boldsymbol{\delta}\mathbf{x}) = \boldsymbol{\delta}\mathbf{y}, \tag{7.11.7}$$

whence

$$\boldsymbol{\delta}\mathbf{x} = \mathbf{A}^{-1}\{\boldsymbol{\delta}\mathbf{y}-\boldsymbol{\delta}\mathbf{A}(\mathbf{x}+\boldsymbol{\delta}\mathbf{x})\}. \tag{7.11.8}$$

This equation involves $\mathbf{A}^{-1}$, a matrix which we have no way of calculating. Even if we go through the labour of calculating the inverse, we do not get $\mathbf{A}^{-1}$, but a matrix $\mathbf{B}$ which satisfies the equation

$$(\mathbf{A}+\boldsymbol{\delta}\mathbf{A})\mathbf{B} = \mathbf{I}+\boldsymbol{\delta}\mathbf{I}, \tag{7.11.9}$$

where $\boldsymbol{\delta}\mathbf{I}$ is an effect of rounding off. A quantity that we *can* find simply is an upper bound to the row-modulus-sum of $(\mathbf{A}+\boldsymbol{\delta}\mathbf{A})^{-1}$, and we therefore rewrite equation (7.11.8) in terms of $(\mathbf{A}+\boldsymbol{\delta}\mathbf{A})^{-1}$. (The actual process of finding the upper bound will be described below.) We have

$$\begin{aligned}\mathbf{A}^{-1} &= \{(\mathbf{A}+\boldsymbol{\delta}\mathbf{A})-\boldsymbol{\delta}\mathbf{A}\}^{-1} = \{(\mathbf{A}+\boldsymbol{\delta}\mathbf{A})[\mathbf{I}-(\mathbf{A}+\boldsymbol{\delta}\mathbf{A})^{-1}\boldsymbol{\delta}\mathbf{A}]\}^{-1} \\ &= \{\mathbf{I}-(\mathbf{A}+\boldsymbol{\delta}\mathbf{A})^{-1}\boldsymbol{\delta}\mathbf{A}\}^{-1}(\mathbf{A}+\boldsymbol{\delta}\mathbf{A})^{-1}\end{aligned} \tag{7.11.10}$$

and substituting this in equation (7.11.8) we find

$$\boldsymbol{\delta}\mathbf{x} = \{\mathbf{I}-(\mathbf{A}+\boldsymbol{\delta}\mathbf{A})^{-1}\boldsymbol{\delta}\mathbf{A}\}^{-1}(\mathbf{A}+\boldsymbol{\delta}\mathbf{A})^{-1}\{\boldsymbol{\delta}\mathbf{y}-\boldsymbol{\delta}\mathbf{A}(\mathbf{x}+\boldsymbol{\delta}\mathbf{x})\}. \tag{7.11.11}$$

Applying the conditions satisfied by the norm we obtain

$$\|\boldsymbol{\delta}\mathbf{x}\| \leqslant \|\{\mathbf{I}-(\mathbf{A}+\boldsymbol{\delta}\mathbf{A})^{-1}\boldsymbol{\delta}\mathbf{A}\}^{-1}\|.\|(\mathbf{A}+\boldsymbol{\delta}\mathbf{A})^{-1}\|.\{\|\boldsymbol{\delta}\mathbf{y}\|+\|\boldsymbol{\delta}\mathbf{A}\|.\|\mathbf{x}+\boldsymbol{\delta}\mathbf{x}\|\}. \tag{7.11.12}$$

The first factor on the right-hand side of the last equation can be simplified. Let $\mathbf{C}$ be an arbitrary matrix such that $\|\mathbf{C}\| < 1$. Then

$$\mathbf{I} = (\mathbf{I}+\mathbf{C})(\mathbf{I}+\mathbf{C})^{-1} = (\mathbf{I}+\mathbf{C})^{-1}+\mathbf{C}(\mathbf{I}+\mathbf{C})^{-1}, \tag{7.11.13}$$

so that

$$\|(\mathbf{I}+\mathbf{C})^{-1}\| = \|\mathbf{I}-\mathbf{C}(\mathbf{I}+\mathbf{C})^{-1}\| \leqslant \|\mathbf{I}\|+\|\mathbf{C}\|.\|(\mathbf{I}+\mathbf{C})^{-1}\|, \tag{7.11.14}$$

and therefore

$$1 \geqslant (1-\|\mathbf{C}\|).\|(\mathbf{I}+\mathbf{C})^{-1}\|; \quad \|(\mathbf{I}+\mathbf{C})^{-1}\| \leqslant 1/(1-\|\mathbf{C}\|). \tag{7.11.15}$$

Applying this to the first factor of equation (7.11.12) we obtain

$$\|\{\mathbf{I}-(\mathbf{A}+\boldsymbol{\delta}\mathbf{A})^{-1}\boldsymbol{\delta}\mathbf{A}\}^{-1}\| \leqslant 1/\{1-\|(\mathbf{A}+\boldsymbol{\delta}\mathbf{A})^{-1}\boldsymbol{\delta}\mathbf{A}\|\}$$
$$\leqslant 1/\{1-\|(\mathbf{A}+\boldsymbol{\delta}\mathbf{A})^{-1}\|.\|\boldsymbol{\delta}\mathbf{A}\|\}, \qquad (7.11.16)$$

and therefore

$$\|\boldsymbol{\delta}\mathbf{x}\| \leqslant \|(\mathbf{A}+\boldsymbol{\delta}\mathbf{A})^{-1}\|.\{\|\boldsymbol{\delta}\mathbf{y}\|+\|\boldsymbol{\delta}\mathbf{A}\|.\|\mathbf{x}+\boldsymbol{\delta}\mathbf{x}\|\}/\{1-\|(\mathbf{A}+\boldsymbol{\delta}\mathbf{A})^{-1}\|.\|\boldsymbol{\delta}\mathbf{A}\|\}. \qquad (7.11.17)$$

We must now find estimates for the various terms occurring in this equation, namely $\|\boldsymbol{\delta}\mathbf{y}\|$, $\|\boldsymbol{\delta}\mathbf{A}\|$ and $\|(\mathbf{A}+\boldsymbol{\delta}\mathbf{A})^{-1}\|$.

Bounds for $\|\boldsymbol{\delta}\mathbf{y}\|$ and $\|\boldsymbol{\delta}\mathbf{A}\|$ have been given by Wilkinson,† and they may be expressed as follows. Suppose that the following quantities are known:

α = maximum recorded $|l_{ii}|$,
β = maximum recorded $|u_{ii}|$,
e = maximum rounding error when recording **L**, **U** and **z**,
f = maximum relative error in recording $\mathbf{x}+\boldsymbol{\delta}\mathbf{x}$. If **x** is recorded to t significant decimal figures, the relative error is $f = 0{\cdot}5\times 10^{-t}$, and $|\boldsymbol{\delta}\mathbf{x}| \leqslant f|\mathbf{x}|$.

Then $$\|\boldsymbol{\delta}\mathbf{A}\| \leqslant ne\max.(\alpha,\beta); \quad \|\boldsymbol{\delta}\mathbf{y}\| \leqslant e+\beta f\|\mathbf{L}\|.\|(\mathbf{x}+\boldsymbol{\delta}\mathbf{x})\|. \qquad (7.11.18)$$

To obtain a bound for $\|(\mathbf{A}+\boldsymbol{\delta}\mathbf{A})^{-1}\|$ we use a simple device which is usually regarded as giving a general idea of the sensitivity of a set of equations.‡ It involves treating an extra column of ones on the right-hand side. The treatment of the extra column parallels that of **y**, except that at every step all products are taken as positive and accumulated by addition, and all rounding off is done upwards. The largest number in the final column is then an upper bound for $\|(\mathbf{A}+\boldsymbol{\delta}\mathbf{A})^{-1}\|$. The formal definition of the quantities computed is

$$\left.\begin{aligned} g_i &= \{1+\sum_{j<i}|l_{ij}|\,g_j\}/|l_{ii}| \qquad (i=1,2,\ldots,n),\\ h_i &= \{g_i+\sum_{j>i}|u_{ij}|\,h_j\}/|u_{ii}| \quad (i=n,n-1,\ldots,2,1),\\ \gamma &= \max.(h_i) \geqslant \|(\mathbf{A}+\boldsymbol{\delta}\mathbf{A})^{-1}\|. \end{aligned}\right\} \qquad (7.11.19)$$

These equations parallel equations (7.2.17) and (7.2.18); in writing down the first equation it has not been assumed that $l_{ii}=1$, as in equation (7.2.17). Only a few figures need be kept in this part of the calculation.

As an example we consider the matrix of the last example in § 7.10. We form in succession

$$g_1 = 1/(1{\cdot}0000) = 1,$$
$$g_2 = \{1+(0{\cdot}5000)\,(1)\}/(0{\cdot}2886) = 5{\cdot}1975,$$

which we round to 5·2

$$g_3 = \{1+(0{\cdot}3333)\,(1)+(0{\cdot}2888)\,(5{\cdot}2)\}/(0{\cdot}0742) = 38{\cdot}3; \qquad (7.11.20)$$

† Wilkinson, *J. Assn Comput. Mach.* **8** (1961), no. 3.
‡ W. E. Milne, *Numerical Calculus* (Princeton University Press, 1949).

$$\left.\begin{aligned} h_3 &= 38{\cdot}3/(0{\cdot}0742) = 516{\cdot}2,\\ h_2 &= \{5{\cdot}2+(0{\cdot}2888)\,(516{\cdot}2)\}/(0{\cdot}2886) = 534{\cdot}6,\\ h_1 &= \{1+(0{\cdot}3333)\,(516{\cdot}2)+(0{\cdot}5000)\,(534{\cdot}6)\}/(1{\cdot}0000) = 440{\cdot}4.\end{aligned}\right\} \quad (7.11.21)$$

Finally $\gamma = 534{\cdot}6$ which we can round to 535 or 550. For the calculated inverse we see that the sums of the moduli of the elements in the three rows are respectively

$$75{\cdot}7759, \quad 412{\cdot}0118, \quad 393{\cdot}7308,$$

so that the method has given a reasonably close upper bound.

We shall now apply equation (7.11.17) to this matrix and find a bound for the errors in the first column of the inverse (although we have not really found it by back substitution). We have

$$\left.\begin{aligned} \alpha &= \beta = 1,\\ e &= f\,\|\mathbf{x}+\boldsymbol{\delta}\mathbf{x}\| = 0{\cdot}000\,05,\\ \|\mathbf{L}\| &= 1{\cdot}8333, \quad \|\mathbf{x}+\boldsymbol{\delta}\mathbf{x}\| = 37,\end{aligned}\right\} \quad (7.11.22)$$

so that

$$\left.\begin{aligned} \|\boldsymbol{\delta}\mathbf{y}\| &= 0{\cdot}000\,05+(2)\,(0{\cdot}000\,05) = 0{\cdot}000\,15,\\ \|\boldsymbol{\delta}\mathbf{A}\| &\leqslant (4)\,(1)\,(0{\cdot}000\,05) = 0{\cdot}0002,\\ \|\boldsymbol{\delta}\mathbf{x}\| &\leqslant (535)\{0{\cdot}000\,15+(0{\cdot}0002)\,(37)\}/\{1-(535)\,(0{\cdot}0002)\}\\ &= (4{\cdot}039)/0{\cdot}89\end{aligned}\right\} \quad (7.11.23)$$

$$= 4{\cdot}54. \quad (7.11.24)$$

Hence no element in the first column of the inverse is in error by more than $4{\cdot}54$.

The bound calculated in this way uses some information gathered from the computation and some general information which applies to all matrices. The more one uses the information contained in the computation, the more likely the estimate is to be near the actual error. One important piece of information that can be derived from the solution is the residual when it is substituted into the original equation. This provides both a means of bounding the error and of calculating a correction if the error bound is intolerably large. We define the residual $\mathbf{R}$ by

$$\mathbf{R} = \mathbf{y}-\mathbf{A}(\mathbf{x}+\boldsymbol{\delta}\mathbf{x}), \quad (7.11.25)$$

so that

$$\mathbf{R} = -\mathbf{A}\boldsymbol{\delta}\mathbf{x},$$

$$\boldsymbol{\delta}\mathbf{x} = -\mathbf{A}^{-1}\mathbf{R}. \quad (7.11.26)$$

This equation may be used to calculate a correction. It also provides a bound for the error, for

$$\|\boldsymbol{\delta}\mathbf{x}\| \leqslant \|\mathbf{A}^{-1}\|\,.\,\|\mathbf{R}\|, \quad (7.11.27)$$

and using equations (7.11.10) and (7.11.16) we find that

$$\|\mathbf{A}^{-1}\| \leqslant \gamma/(1-\gamma\,\|\boldsymbol{\delta}\mathbf{A}\|), \quad (7.11.28)$$

so that

$$\|\boldsymbol{\delta}\mathbf{x}\| \leqslant \gamma\,\|\mathbf{R}\|/(1-\gamma\,\|\boldsymbol{\delta}\mathbf{A}\|). \quad (7.11.29)$$

The values of the elements of **R** will be found after a great deal of cancellation of leading digits, and it is essential to accumulate scalar products full length in the accumulator if **R** is to be used to correct the approximate solution (as in equation (7.11.26)), but only one or two figures need be recorded if only an error bound is to be found from **R** (equation (7.11.29)).

For the first column of the inverse found in the last section we find

$$10^4\mathbf{R} = -\{0{\cdot}5530,\ -2{\cdot}0808,\ 2{\cdot}1226\}, \tag{7.11.30}$$

$$\|\mathbf{R}\| \doteqdot 0{\cdot}000\,48,\quad \gamma = 535,\quad \|\boldsymbol{\delta}\mathbf{A}\| \leqslant 0{\cdot}0002, \tag{7.11.31}$$

whence

$$\|\boldsymbol{\delta}\mathbf{x}\| \leqslant (535)\,(0{\cdot}000\,48)/\{1-(535)\,(0{\cdot}0002)\}$$

$$= 0{\cdot}2477 \tag{7.11.32}$$

and no element of the first column of the inverse is in error by more than this amount.

To improve the approximate solution if the bound on the error is too large, we solve, as best we can, the equation (7.11.22), that is we carry out a forward and backward substitution using the factors we have already found. We obtain

$$\left.\begin{aligned} 10^4\boldsymbol{\delta}\mathbf{z} &= \begin{bmatrix} 0{\cdot}5530 \\ -8{\cdot}1680 \\ 57{\cdot}9138 \end{bmatrix}, \\ 10^4\boldsymbol{\delta}\mathbf{x} &= \begin{bmatrix} 145{\cdot}0854 \\ -809{\cdot}3523 \\ 780{\cdot}5093 \end{bmatrix}, \end{aligned}\right\} \tag{7.11.33}$$

so that the new approximation is

$$\mathbf{x} = \begin{bmatrix} 9{\cdot}0699 \\ -36{\cdot}3650 \\ 30{\cdot}3410 \end{bmatrix} - \begin{bmatrix} 0{\cdot}0145 \\ -0{\cdot}0809 \\ 0{\cdot}0780 \end{bmatrix} = \begin{bmatrix} 9{\cdot}0554 \\ -36{\cdot}2841 \\ 30{\cdot}2630 \end{bmatrix}. \tag{7.11.34}$$

EXAMPLE 7.11

1. Estimate the error in the solution of the worked example of p. 268.

CHAPTER 8

ITERATIVE METHODS FOR CHARACTERISTIC VALUE PROBLEMS

There are one or two rules,
Half a dozen, may be,
That all family fools,
Of whatever degree,
Must observe if they love their profession.
The Yeomen of the Guard

8.1 Determination of the natural frequencies and principal modes of a system

The natural frequencies $\omega_1, \omega_2, \ldots, \omega_n$ of a system were defined in § 2.3 to be the roots of the frequency equation

$$|\mathbf{C} - \omega^2 \mathbf{A}| = 0. \tag{8.1.1}$$

The principal mode $\mathbf{\Psi}^{(r)}$ corresponding to ω_r satisfies the equation

$$(\mathbf{C} - \omega_r^2 \mathbf{A})\, \mathbf{\Psi}^{(r)} = \mathbf{0}. \tag{8.1.2}$$

If the system is able to move as a rigid body, the matrix $\mathbf{C}$ will be singular. If redundant coordinates are used in setting up the equations of motion then $\mathbf{A}$ will be singular (see §§ 2.11, 2.12, 4.7). While both of these circumstances may obtain, it will be assumed, unless otherwise stated, that at least one of $\mathbf{A}$ and $\mathbf{C}$ is positive definite (see § 2.1). As we shall see, a system for which this is the case can always be converted to the form

$$\mathbf{Bx} = \lambda \mathbf{x}, \tag{8.1.3}$$

in which $\mathbf{B}$ is symmetric, and the methods to be described are intended primarily for this problem—the characteristic value problem for a single symmetric matrix. The case in which $\mathbf{A}$ and $\mathbf{C}$ are both singular is discussed in § 2.12.

A few attempts to express the determinant of equation (8.1.1) as a polynomial for a system with, say, five degrees of freedom, will rapidly convince the reader that a systematic technique of solution is desirable. An attempt at a larger system will also indicate the desirability of avoiding the use of elementary methods of expanding determinants, and indeed of avoiding the expansion altogether. When only one or two frequencies and modes are desired this can well be done, and some of the iterative techniques for this are described in this chapter. When many of the modes are required, the so-called direct methods seem preferable. These are methods which amount to the computation of the characteristic polynomial (that is the expansion of the determinant $|\mathbf{B} - \lambda \mathbf{I}|$) if not in explicit polynomial form, at least in a form from which its value can readily be found for any chosen value of λ. Some of these methods are discussed in Chapter 9.

8.2 Reduction to standard form

The problem of equation (8.1.2) may be reduced to the form of equation (8.1.3) by a change of variables. The change to be made depends on the symmetric factorisation of **A** or **C** (see § 7.4). If both matrices are non-singular either may be chosen, that which is easier to factorise being the one to use. We shall suppose it to be **A**, and put

$$\mathbf{A} = \mathbf{LL'}, \quad \mathbf{x} = \mathbf{L'\Psi}, \quad \lambda = \omega^2, \tag{8.2.1}$$

whereupon equation (8.1.2) becomes

$$\mathbf{C}(\mathbf{L'})^{-1}\mathbf{x} = \lambda\mathbf{Lx}, \tag{8.2.2}$$

so that

$$\mathbf{L}^{-1}\mathbf{C}(\mathbf{L'})^{-1}\mathbf{x} = \lambda\mathbf{x}, \tag{8.2.3}$$

which is of the form required.

If we choose to factorise **C** instead of **A**, then the equations to use are

$$\mathbf{C} = \mathbf{LL'}, \quad \mathbf{x} = \mathbf{L'\Psi}, \quad \lambda = 1/\omega^2 \tag{8.2.4}$$

and

$$\mathbf{L}^{-1}\mathbf{A}(\mathbf{L'})^{-1}\mathbf{x} = \lambda\mathbf{x}. \tag{8.2.5}$$

If one of the matrices is singular there is some advantage in factorising that one, because thereby the problem may be reduced to one involving smaller matrices. If **A** is singular the characteristic polynomial is of degree less than n (see § 2.3) and the matrices can be reduced to have orders equal to the actual degree of the polynomial. When many fictitious coordinates have been used to simplify the formation of the equations of motion of the system this may result in a considerable saving of computing labour. If **C** is singular, then zero is a root of equation (8.1.1) and, considered as an equation in $1/\omega^2$, the polynomial involved is again of degree less than n and the order of the matrices can again be reduced to the degree of this polynomial.

Suppose **A** to be singular. Then it may be resolved into three factors (§ 7.4) of the form

$$\mathbf{A} = \mathbf{LDL'}, \tag{8.2.6}$$

where **L** (and hence **L'**) is not singular while **D** is diagonal with some zero elements. By changing the numbering of the variables we may bring the zero elements of **D** to lie in neighbouring places at the end of the diagonal, and by rescaling the variables we may take the diagonal elements to be ones. The transformation

$$\mathbf{x} = \mathbf{L'\Psi}, \quad \lambda = \omega^2 \tag{8.2.7}$$

then leads to

$$\mathbf{L}^{-1}\mathbf{C}(\mathbf{L'})^{-1}\mathbf{x} = \lambda\mathbf{Dx}. \tag{8.2.8}$$

The matrix **D** may then be partitioned symmetrically into four parts, thus

$$\mathbf{D} = \begin{bmatrix} \mathbf{I} & \mathbf{0} \\ \mathbf{0} & \mathbf{0} \end{bmatrix}, \tag{8.2.9}$$

while **x** and $\mathbf{L}^{-1}\mathbf{C}(\mathbf{L'})^{-1}$ may be similarly partitioned

$$\mathbf{x} = \begin{bmatrix} \mathbf{x}^0 \\ \mathbf{y} \end{bmatrix}, \quad \mathbf{L}^{-1}\mathbf{C}(\mathbf{L'})^{-1} = \begin{bmatrix} \mathbf{E} & \mathbf{F} \\ \mathbf{F'} & \mathbf{G} \end{bmatrix}. \tag{8.2.10}$$

Equation (8.2.8) may then be written

$$\left.\begin{aligned}\mathbf{E}\mathbf{x}^0+\mathbf{F}\mathbf{y} &= \lambda\mathbf{x}^0,\\ \mathbf{F}'\mathbf{x}^0+\mathbf{G}\mathbf{y} &= \mathbf{0}.\end{aligned}\right\} \tag{8.2.11}$$

The second equation gives $$\mathbf{y} = -\mathbf{G}^{-1}\mathbf{F}'\mathbf{x}^0 \tag{8.2.12}$$

and substitution of this in the first equation yields

$$(\mathbf{E}-\mathbf{F}\mathbf{G}^{-1}\mathbf{F}')\,\mathbf{x}^0 = \lambda\mathbf{x}^0, \tag{8.2.13}$$

which has the required form and order. We note that $\mathbf{G}$ is non-singular since $\mathbf{C}$ (and hence $\mathbf{L}^{-1}\mathbf{C}(\mathbf{L}')^{-1}$) is positive definite.

If $\mathbf{C}$ is singular (and $\mathbf{A}$ is non-singular) it may be factorised into three factors

$$\mathbf{C} = \mathbf{L}\mathbf{D}\mathbf{L}'. \tag{8.2.14}$$

The transformation $$\mathbf{x} = \mathbf{L}'\mathbf{\Psi}, \quad \lambda = 1/\omega^2 \tag{8.2.15}$$

then leads to $$\mathbf{L}^{-1}\mathbf{A}(\mathbf{L}')^{-1}\mathbf{x} = \lambda\mathbf{D}\mathbf{x}. \tag{8.2.16}$$

The matrix $\mathbf{D}$ may then be partitioned as before and $\mathbf{x}$ and $\mathbf{L}^{-1}\mathbf{A}(\mathbf{L}')^{-1}$ similarly, to yield an equation similar to (8.2.13).

It will be noticed that the above reduction may be applied in the case in which $\mathbf{A}$ and $\mathbf{C}$ are both singular, provided that $\mathbf{G}$ is non-singular. There are good reasons for believing that this is always so for physical systems; but even when $\mathbf{G}$ is singular the equations may be reduced to the standard form, although the process of reduction is more lengthy.

Consider the pair of matrices

$$\mathbf{C} = \begin{bmatrix} 5 & -2 & -1\\ -2 & 6 & -3\\ -1 & -3 & 7\end{bmatrix}, \quad \mathbf{A} = \begin{bmatrix} 1 & 2 & 3\\ 2 & 5 & 8\\ 3 & 8 & 13\end{bmatrix}. \tag{8.2.17}$$

The matrix $\mathbf{A}$, being positive definite, may be resolved into factors $\mathbf{L}$ and $\mathbf{L}'$, by one of the methods described in §7.4. It is found that

$$\mathbf{L} = \begin{bmatrix} 1 & 0 & 0\\ 2 & 1 & 0\\ 3 & 2 & 1\end{bmatrix}.$$

The matrix $\mathbf{L}^{-1}\mathbf{C}$ may be found by forward substitution in the equation

$$\mathbf{L}(\mathbf{L}^{-1}\mathbf{C}) = \mathbf{C}. \tag{8.2.18}$$

This statement is worth some comment. Suppose

$$\mathbf{L}^{-1}\mathbf{C} = \mathbf{D}, \tag{8.2.19}$$

then when $n = 3$ the general equation for **D** is

$$\begin{bmatrix} 1 & 0 & 0 \\ l_{21} & 1 & 0 \\ l_{31} & l_{32} & 1 \end{bmatrix} \begin{bmatrix} d_{11} & d_{12} & d_{13} \\ d_{21} & d_{22} & d_{23} \\ d_{31} & d_{32} & d_{33} \end{bmatrix} = \begin{bmatrix} c_{11} & c_{12} & c_{13} \\ c_{21} & c_{22} & c_{23} \\ c_{31} & c_{32} & c_{33} \end{bmatrix}, \tag{8.2.20}$$

from which the elements of **D** may be found as follows:

$$\left.\begin{aligned} d_{11} &= c_{11}, & d_{12} &= c_{12}, & d_{13} &= c_{13}; \\ d_{21} &= c_{21} - l_{21} d_{11}, & d_{22} &= c_{22} - l_{21} d_{12}, & d_{23} &= c_{23} - l_{21} d_{13}; \\ d_{31} &= c_{31} - l_{31} d_{11} - l_{32} d_{21}, & d_{32} &= c_{32} - l_{31} d_{12} - l_{32} d_{22}, & d_{33} &= c_{33} - l_{31} d_{13} - l_{32} d_{23}. \end{aligned}\right\} \tag{8.2.21}$$

In the present example it is found that

$$\mathbf{L}^{-1}\mathbf{C} = \begin{bmatrix} 5 & -2 & -1 \\ -12 & 10 & -1 \\ 8 & -17 & 12 \end{bmatrix}. \tag{8.2.22}$$

The matrix $\mathbf{L}^{-1}\mathbf{C}(\mathbf{L}')^{-1}$ may be found from the equation

$$\mathbf{L}^{-1}\mathbf{C}(\mathbf{L}')^{-1}\mathbf{L}' = \mathbf{L}^{-1}\mathbf{C} \tag{8.2.23}$$

by a process similar to that described above, which gives

$$\mathbf{L}^{-1}\mathbf{C}\mathbf{L}'^{-1} = \begin{bmatrix} 5 & -12 & 8 \\ -12 & 34 & -33 \\ 8 & -33 & 54 \end{bmatrix}. \tag{8.2.24}$$

The required frequency equation is thus

$$\begin{vmatrix} 5-\lambda & -12 & 8 \\ -12 & 34-\lambda & -33 \\ 8 & -33 & 54-\lambda \end{vmatrix} = 0. \tag{8.2.25}$$

The procedure for singular **A** is illustrated with the matrices

$$\mathbf{C} = \begin{bmatrix} 1 & 3 & 3 & 4 \\ 3 & 11 & 9 & 12 \\ 3 & 9 & 12 & 12 \\ 4 & 12 & 12 & 20 \end{bmatrix}, \quad \mathbf{A} = \begin{bmatrix} 1 & 2 & 3 & 4 \\ 2 & 5 & 8 & 11 \\ 3 & 8 & 13 & 18 \\ 4 & 11 & 18 & 25 \end{bmatrix}. \tag{8.2.26}$$

The singular matrix, **A**, may be resolved into a triple product by the method of equation (7.4.20) which gives

$$\mathbf{L} = \begin{bmatrix} 1 & 0 & 0 & 0 \\ 2 & 1 & 0 & 0 \\ 3 & 2 & 1 & 0 \\ 4 & 3 & 0 & 1 \end{bmatrix}, \quad \mathbf{D} = \begin{bmatrix} 1 & 0 & 0 & 0 \\ 0 & 1 & 0 & 0 \\ 0 & 0 & 0 & 0 \\ 0 & 0 & 0 & 0 \end{bmatrix}.$$

The evaluation of $\mathbf{L}^{-1}\mathbf{C}$ and $\mathbf{L}^{-1}\mathbf{C}(\mathbf{L}')^{-1}$ is as in the previous example, and

$$\mathbf{L}^{-1}\mathbf{C} = \begin{bmatrix} 1 & 3 & 3 & 4 \\ 1 & 5 & 3 & 4 \\ -2 & -10 & -3 & -8 \\ -3 & -15 & -9 & -8 \end{bmatrix}, \quad \mathbf{L}^{-1}\mathbf{C}(\mathbf{L}')^{-1} = \begin{bmatrix} 1 & 1 & -2 & -3 \\ 1 & 3 & -6 & -9 \\ -2 & -6 & 15 & 18 \\ -3 & -9 & 18 & 31 \end{bmatrix}.$$

Being of rank 2 the matrix $\mathbf{D}$ has to be partitioned into four 2×2 blocks, and the corresponding blocks of $\mathbf{L}^{-1}\mathbf{C}(\mathbf{L}')^{-1}$ are

$$\mathbf{E} = \begin{bmatrix} 1 & 1 \\ 1 & 3 \end{bmatrix}, \quad \mathbf{F} = \begin{bmatrix} -2 & -3 \\ -6 & -9 \end{bmatrix}, \quad \mathbf{G} = \begin{bmatrix} 15 & 18 \\ 18 & 31 \end{bmatrix}.$$

Then
$$\mathbf{G}^{-1} = \tfrac{1}{141}\begin{bmatrix} 31 & -18 \\ -18 & 15 \end{bmatrix}, \quad 141\mathbf{F}\mathbf{G}^{-1}\mathbf{F}' = \begin{bmatrix} 43 & 129 \\ 129 & 387 \end{bmatrix},$$

so that
$$\mathbf{E} - \mathbf{F}\mathbf{G}^{-1}\mathbf{F}' = \tfrac{1}{141}\begin{bmatrix} 98 & 12 \\ 12 & 36 \end{bmatrix}.$$

The characteristic equation for this last matrix is of degree two.

EXAMPLES 8.2

Reduce the following systems ($\omega^2\mathbf{A}\Psi = \mathbf{C}\Psi$) to the standard form ($\mathbf{B}\mathbf{x} = \lambda\mathbf{x}$):

1. $$\mathbf{A} = \begin{bmatrix} 2 & 0 & 0 & 0 \\ 0 & 3 & 0 & 0 \\ 0 & 0 & 1 & 0 \\ 0 & 0 & 0 & 5 \end{bmatrix}, \quad \mathbf{C} = \begin{bmatrix} 7 & -3 & -2 & -1 \\ -3 & 6 & -2 & -1 \\ -2 & -2 & 4 & 0 \\ -1 & -1 & 0 & 2 \end{bmatrix}.$$

2. $$\mathbf{A} = \begin{bmatrix} 4 & 2 & 2 \\ 2 & 2 & 3 \\ 2 & 3 & 6 \end{bmatrix}, \quad \mathbf{C} = \begin{bmatrix} 4 & -2 & -1 \\ -2 & 5 & -2 \\ -1 & -2 & 3 \end{bmatrix}.$$

3. $$\mathbf{A} = \begin{bmatrix} 1 & -3/2 & 1/3 & 1/12 \\ -3/2 & 7/3 & -7/12 & -2/15 \\ 1/3 & -7/12 & 1/5 & 1/30 \\ 1/12 & -2/15 & 1/30 & 1/105 \end{bmatrix}, \quad \mathbf{C} = \begin{bmatrix} 1 & -2 & 1 & 0 \\ -2 & 5 & -4 & 1 \\ 1 & -4 & 6 & -4 \\ 0 & 1 & -4 & 6 \end{bmatrix}.$$

4. $$\mathbf{A} = \begin{bmatrix} 4 & 2 & 1 & 1 \\ 2 & 4 & 2 & 1 \\ 1 & 2 & 3 & 2 \\ 1 & 1 & 2 & 3 \end{bmatrix}, \quad \mathbf{C} = \begin{bmatrix} 4 & -1 & 0 & -3 \\ -1 & 4 & -2 & -1 \\ 0 & -2 & 4 & -2 \\ -3 & -1 & -2 & 6 \end{bmatrix}.$$

8.3 Bounds for the characteristic values

Some simple bounds for the characteristic values of a matrix are given by Gershgorin.† They are couched in terms of the complex λ-plane because they apply to general (i.e. not necessarily symmetric) matrices which may have complex characteristic values.

Suppose that $\mathbf{x}$ is a characteristic vector of the matrix $\mathbf{B}$, and that $\mathbf{x}$ has been scaled so that its element of largest modulus, x_r, is unity. Then

$$\mathbf{B}\mathbf{x} = \lambda\mathbf{x}, \tag{8.3.1}$$

and

$$\boldsymbol{\rho}_r(\mathbf{B})\,\mathbf{x} = \lambda x_r = \lambda, \tag{8.3.2}$$

that is

$$\lambda = b_{r1}x_1 + \ldots + b_{rr}x_r + \ldots + b_{rn}x_n. \tag{8.3.3}$$

Hence

$$\lambda - b_{rr}x_r = b_{r1}x_1 + \ldots + b_{r,r-1}x_{r-1} + b_{r,r+1}x_{r+1} + \ldots + b_{rn}x_n. \tag{8.3.4}$$

Using the value of x_r, and the usual inequality for moduli, we find

$$|\lambda - b_{rr}| \leqslant \sum_{i \neq r} |b_{ri}x_i| \leqslant \sum_{i \neq r} |b_{ri}|. \tag{8.3.5}$$

This last inequality states that λ lies within the circle whose centre is at b_{rr} and whose radius is given by the right-hand side of the inequality. Unfortunately, until we have solved the characteristic vector problem, we do not know what the value of 'r' is. But this result tells us, before solving the problem, that if we consider all the circles defined in this way for $r = 1, 2, \ldots, n$, any characteristic value must lie in some one of them (or several if they overlap). Similarly, the roots lie inside the circles with the same centres but with radii defined by the sums taken down the columns, that is

$$|\lambda - b_{rr}| \leqslant \sum_{j \neq r} |b_{jr}|, \tag{8.3.6}$$

and only the smaller circle round each centre need be searched. Of course, when the matrix $\mathbf{B}$ is symmetric the bounds are both the same. When the characteristic values are known to be real (e.g. when the matrix is symmetric) these inequalities define a set of intervals on the real λ-axis within which all the characteristic values will lie.

For convenience, a simpler bound on the modulus of λ is often used. This is derived from the inequalities (8.3.5) and (8.3.6) by using the fact that

$$|\lambda - b_{rr}| \geqslant |\lambda| - |b_{rr}|. \tag{8.3.7}$$

This, when substituted into the equalities (8.3.5) and (8.3.6), yields

$$|\lambda| \leqslant \sum_{i} |b_{ri}|, \tag{8.3.8}$$

$$|\lambda| \leqslant \sum_{j} |b_{jr}|. \tag{8.3.9}$$

† S. Gershgorin, 'Über die Abgrenzung der Eigenwerte einer Matrix', *Izv. Akad. Nauk, SSSR*, **7** (1931), p. 249.

These bounds are also associated with the name of Gershgorin. They may be simplified, and coarsened, to yield

$$|\lambda| \leqslant n \max_{i,j}. |b_{ij}|. \tag{8.3.10}$$

For the symmetric matrix

$$\mathbf{B} = \begin{bmatrix} 1 & -5 & 0 & 0 \\ -5 & 70 & 10 & 0 \\ 0 & 10 & 4 & -1 \\ 0 & 0 & -1 & 1 \end{bmatrix}, \tag{8.3.11}$$

the inequalities (8.3.5) and (8.3.6) show that each characteristic value must lie in at least one of the intervals

$$-4 \leqslant \lambda \leqslant 6, \quad 55 \leqslant \lambda \leqslant 85, \quad -7 \leqslant \lambda \leqslant 15, \quad 0 \leqslant \lambda \leqslant 2.$$

Since the first and last intervals lie within the third, it is only the second and third which provide useful information. The inequalities (8.3.8) and (8.3.9) yield the bounds

$$-85 \leqslant \lambda \leqslant 85,$$

and the cruder inequality (8.3.10) yields

$$-280 \leqslant \lambda \leqslant 280.$$

For symmetric matrices a useful bound may be derived from Rayleigh's principle (§ 2.10). Suppose that $\mathbf{B}$ is a symmetric matrix, $\mathbf{x}$ is an arbitrary vector and

$$\mathbf{y} = \mathbf{Bx}. \tag{8.3.12}$$

Then

$$\frac{\mathbf{y}'\mathbf{y}}{\mathbf{x}'\mathbf{x}} = \frac{(\mathbf{Bx})'(\mathbf{Bx})}{\mathbf{x}'\mathbf{x}} = \frac{\mathbf{x}'(\mathbf{B}'\mathbf{B})\mathbf{x}}{\mathbf{x}'\mathbf{x}}. \tag{8.3.13}$$

The matrix $\mathbf{A}$ given by

$$\mathbf{A} = \mathbf{B}'\mathbf{B} \tag{8.3.14}$$

is symmetric and therefore, by Rayleigh's principle,

$$\frac{\mathbf{x}'\mathbf{A}\mathbf{x}}{\mathbf{x}'\mathbf{x}} > \mu_1, \tag{8.3.15}$$

where μ_1 is the least characteristic value of $\mathbf{A}$. Since $\mathbf{B}$ is symmetric, $\mathbf{B}'\mathbf{B}$ is the same as $\mathbf{B}^2$, and hence the characteristic values of $\mathbf{B}'\mathbf{B}$ (that is, $\mathbf{A}$) are the squares of those of $\mathbf{B}$. Therefore $\mathbf{B}$ has at least one characteristic value, λ, such that

$$\lambda^2 \leqslant \frac{\mathbf{y}'\mathbf{y}}{\mathbf{x}'\mathbf{x}}. \tag{8.3.16}$$

This bound may be applied to give an upper bound for the error in an estimate λ_0 of a characteristic value of $\mathbf{B}$, when an estimate, $\mathbf{x}_0$ of the corresponding characteristic vector is made. If

$$\mathbf{Bx}_0 - \lambda_0 \mathbf{x}_0 = \mathbf{y}_0, \tag{8.3.17}$$

then it follows from equation (8.3.16) that the matrix $\mathbf{B}-\lambda_0\mathbf{I}$ has a characteristic value, μ, such that

$$\mu^2 \leqslant \frac{\mathbf{y}_0'\mathbf{y}_0}{\mathbf{x}_0'\mathbf{x}_0}. \tag{8.3.18}$$

But if λ is a true characteristic value of $\mathbf{B}$ then the matrix $\mathbf{B}-\lambda_0\mathbf{I}$ has the characteristic value

$$\mu = \lambda-\lambda_0. \tag{8.3.19}$$

Therefore $\mathbf{B}$ must have a characteristic value λ such that

$$|\lambda-\lambda_0|^2 \leqslant \frac{\mathbf{y}_0'\mathbf{y}_0}{\mathbf{x}_0'\mathbf{x}_0}. \tag{8.3.20}$$

For the matrix

$$\mathbf{B} = \begin{bmatrix} 1 & 2 & 3 & 4 \\ 2 & 6 & 10 & 14 \\ 3 & 10 & 20 & 30 \\ 4 & 14 & 30 & 50 \end{bmatrix}, \tag{8.3.21}\dagger$$

an approximate characteristic vector is

$$\mathbf{x}_0 = \{3 \quad 10 \quad 21 \quad 33\}.$$

Rayleigh's quotient (applied to $\mathbf{B}$, not $\mathbf{B}'\mathbf{B}$) suggests the value

$$\lambda_0 = \frac{\mathbf{x}_0'\mathbf{B}\mathbf{x}_0}{\mathbf{x}_0'\mathbf{x}_0} = \frac{120\,189}{1639} = 73{\cdot}33.$$

The vector $\mathbf{y}_0$ of equation (8.3.17) is then

$$\mathbf{y}_0 = \{-1{\cdot}99 \quad 4{\cdot}70 \quad -20{\cdot}93 \quad 12{\cdot}11\}$$

so that

$$\frac{\mathbf{y}_0'\mathbf{y}_0}{\mathbf{x}_0'\mathbf{x}_0} = \frac{610{\cdot}767\,10}{1639} = 0{\cdot}372\,646.$$

We therefore deduce that there is a characteristic value, λ, of $\mathbf{B}$ such that

$$|\lambda-73{\cdot}33| \leqslant 0{\cdot}610\,447.$$

The inequality (8.3.20) yields a further bound, a bound on the disturbance caused to the characteristic values of a symmetric matrix $\mathbf{A}$ by the addition of a symmetric matrix $\mathbf{C}$. Let

$$\mathbf{B} = \mathbf{A}+\mathbf{C}, \tag{8.3.22}$$

and suppose that λ_0 is a characteristic value of $\mathbf{A}$ and $\mathbf{x}_0$ is the corresponding vector, so that

$$\mathbf{A}\mathbf{x}_0-\lambda\mathbf{x}_0 = \mathbf{0}. \tag{8.3.23}$$

Putting

$$\mathbf{y}_0 = \mathbf{B}\mathbf{x}_0-\lambda\mathbf{x}_0 = \mathbf{C}\mathbf{x}_0 \tag{8.3.24}$$

we deduce from equation (8.3.20) that $\mathbf{B}$ has a characteristic value, λ, such that

$$|\lambda-\lambda_0|^2 \leqslant \frac{\mathbf{y}_0'\mathbf{y}_0}{\mathbf{x}_0'\mathbf{x}_0} = \frac{\mathbf{x}_0'\mathbf{C}'\mathbf{C}\mathbf{x}_0}{\mathbf{x}_0'\mathbf{x}_0}. \tag{8.3.25}$$

† This matrix is used for a number of examples in this chapter and the next. A summary of information obtained about it is given in Appendix I at the end of the book.

The right-hand side of this inequality is less than the maximum characteristic value of $\mathbf{C}'\mathbf{C}$, that is, less than the square of the maximum characteristic value of $\mathbf{C}$. Therefore $\mathbf{B}$ has a characteristic value, λ, such that

$$|\lambda - \lambda_0| \leqslant \text{max.}\,|\text{characteristic value of } \mathbf{C}|. \tag{8.3.26}$$

Consider the matrices

$$\mathbf{A} = \begin{bmatrix} -2 & 1 & 0 \\ 1 & -2 & 1 \\ 0 & 1 & -2 \end{bmatrix}, \quad \mathbf{B} = \begin{bmatrix} -2{\cdot}00 & 1{\cdot}01 & 0{\cdot}02 \\ 1{\cdot}01 & -1{\cdot}99 & 1{\cdot}01 \\ 0{\cdot}02 & 1{\cdot}01 & -2{\cdot}01 \end{bmatrix},$$

$$\mathbf{C} = \begin{bmatrix} 0{\cdot}00 & 0{\cdot}01 & 0{\cdot}02 \\ 0{\cdot}01 & 0{\cdot}01 & 0{\cdot}01 \\ 0{\cdot}02 & 0{\cdot}01 & -0{\cdot}01 \end{bmatrix}.$$

The matrix $\mathbf{A}$ has a characteristic value

$$\lambda_0 = -4\sin^2(\tfrac{1}{4}\pi) = -2{\cdot}00.$$

The largest characteristic value of $\mathbf{C}$ is less than $0{\cdot}04$ in modulus (by equation (8.3.8) applied to the last row of $\mathbf{C}$) so that there is a characteristic value of $\mathbf{B}$ in the range

$$-2{\cdot}04 \leqslant \lambda \leqslant -1{\cdot}96.$$

A further bound on the characteristic values of a symmetric matrix has been given by Collatz in an enclosure theorem.† If $\mathbf{B}$ is a symmetric matrix and $\mathbf{x}$ and $\mathbf{y}$ are two vectors such that

$$\mathbf{Bx} = \mathbf{y}, \tag{8.3.27}$$

then the extreme ratios of corresponding elements of $\mathbf{y}$ and $\mathbf{x}$ enclose a characteristic value of $\mathbf{B}$.

For the matrix

$$\mathbf{B} = \begin{bmatrix} -2 & 1 & 0 \\ 1 & -2 & 1 \\ 0 & 1 & -2 \end{bmatrix},$$

we may choose the vector $\mathbf{x} = \{1 \quad 2 \quad 1\}$,

whence we derive $\mathbf{y} = \mathbf{Bx} = \{0 \quad -2 \quad 0\}$.

The ratios of corresponding elements of $\mathbf{y}$ and $\mathbf{x}$ are

$$\frac{0}{1}, \quad \frac{-2}{2} \quad \text{and} \quad \frac{0}{1},$$

or

$$0, \quad -1 \quad \text{and} \quad 0.$$

† L. Collazt, *Eigenwertaufgaben mit Technischen Anwendung* (Akademische Verlagsgesellschaft m.b.H., 1949), p. 289.

The extreme values of these are -1 and 0 so that there is a characteristic value of $\mathbf{B}$ in the range

$$-1 \leqslant \lambda \leqslant 0.$$

In fact this matrix has one characteristic value given by

$$\lambda = -4\sin^2\left(\tfrac{1}{8}\pi\right) = -0{\cdot}585\,786.$$

EXAMPLES 8.3

1. Apply Gershgorin's formulae to obtain bounds of the characteristic values of

$$\begin{bmatrix} 1 & 2 & 3 & -4 \\ 2 & -3 & 4 & -5 \\ 5 & -4 & 3 & 2 \\ 4 & 2 & -1 & 3 \end{bmatrix}.$$

2. Use Rayleigh's quotient to estimate a characteristic value of

$$\mathbf{B} = \begin{bmatrix} 1 & 1/2 & 1/3 & 1/4 \\ 1/2 & 1/3 & 1/4 & 1/5 \\ 1/3 & 1/4 & 1/5 & 1/6 \\ 1/4 & 1/5 & 1/6 & 1/7 \end{bmatrix}.$$

with the trial vector $\mathbf{x}_0 = \{1,\ 1,\ 1,\ 1\}.$

Deduce an upper bound on the error in this estimate and compare the resulting bounds with those derived from the enclosure theorem quoted in this section.

3. If the elements of the matrix of Ex. 8.3.2 are rounded to 6 decimal places, show that the maximum change this can make in any characteristic value is less than 0·7 in the sixth decimal place.

8.4 Further bounds on characteristic values and vectors of symmetric matrices

The formulae of § 8.3 may be used to estimate the error in a purported characteristic value. They require an approximate characteristic vector, but do not yield any information about the error in the vector. In this section we shall discuss some ways of estimating the error in a purported characteristic vector. We shall start our discussion by applying the bound (8.3.20) to the matrix $\mathbf{B}$ of equation (8.3.21).

For the matrix of equation (8.3.21) estimates for the characteristic value of least modulus and the corresponding vector are found in the table on page 340 and equation (9.7.7) and are†

$$\left.\begin{aligned} \lambda_0 &= 0{\cdot}209\,376, \\ \mathbf{x}_0 &= \{1 \quad -0{\cdot}790\,222 \quad 0{\cdot}377\,006 \quad -0{\cdot}085\,300\}. \end{aligned}\right\} \tag{8.4.1}$$

† See also the Appendix.

The vectors $\mathbf{Bx}_0$ and $\lambda_0\mathbf{x}_0$ are respectively

$$\left.\begin{aligned}\mathbf{Bx}_0 &= \{0{\cdot}209\,374 \quad -0{\cdot}165\,472 \quad 0{\cdot}078\,900 \quad -0{\cdot}017\,928\},\\ \lambda_0\mathbf{x}_0 &= \{0{\cdot}209\,376 \quad -0{\cdot}165\,454 \quad 0{\cdot}078\,936 \quad -0{\cdot}017\,860\},\end{aligned}\right\} \tag{8.4.2}$$

so that the residual vector $\mathbf{y}_0$ is given by

$$10^6\mathbf{y}_0 = \{-2 \quad -18 \quad -36 \quad -68\}. \tag{8.4.3}$$

In view of the possible rounding errors in each element of the vectors subtracted, each element of $\mathbf{y}_0$ may be too small (numerically) by 10^{-6}, and to be sure of having an upper bound on the error of λ_0 we increase the magnitude of each element by this amount before calculating

$$\mathbf{y}_0'\mathbf{y}_0 = 10^{-12} \times 6500.$$

Since $$\mathbf{x}_0'\mathbf{x}_0 = 1{\cdot}773\,860,$$

$\mathbf{B}$ has a characteristic value, λ, in the interval defined by

$$\begin{aligned}(\lambda-\lambda_0)^2 &\leqslant 10^{-12}\times 6500/1{\cdot}773\,860\\ &= 10^{-12}\times 3664,\end{aligned}$$

that is $$0{\cdot}209\,316 \leqslant \lambda \leqslant 0{\cdot}209\,436. \tag{8.4.4}$$

The interval found in this example is disappointingly long, for the value of λ is much nearer to λ_0 than this indicates. There does not seem to be any simple way of improving on this estimate of the error in λ_0. However, we shall now show that if we replace λ_0 by the Rayleigh quotient corresponding to $\mathbf{x}_0$, we may derive much closer bounds on the error of the estimated characteristic value, and also bound the error in $\mathbf{x}_0$. The estimates which follow are due to Wilkinson.†

We first note that $\mathbf{B}$ can be expressed in the form

$$\mathbf{B} = \mathbf{XDX}^{-1}, \tag{8.4.5}$$

where $\mathbf{X}$ is the matrix having the characteristic vectors of $\mathbf{B}$ as columns, and $\mathbf{D}$ is the diagonal matrix whose elements are the characteristic values of $\mathbf{B}$. This follows from the fact that the equation satisfied by $\mathbf{x}^{(r)}$, the rth column of $\mathbf{X}$ is

$$\mathbf{Bx}^{(r)} = \lambda_r\mathbf{x}^{(r)}, \tag{8.4.6}$$

and this is the rth column of the equation

$$\mathbf{BX} = \mathbf{DX}. \tag{8.4.7}$$

Using equation (8.4.5) we may write the Rayleigh quotient, μ, of $\mathbf{x}_0$ as

$$\mu = \frac{\mathbf{x}_0'\mathbf{Bx}_0}{\mathbf{x}_0'\mathbf{x}_0} = \frac{\mathbf{x}_0'\mathbf{XDX}^{-1}\mathbf{x}_0}{\mathbf{x}_0'\mathbf{x}_0}. \tag{8.4.8}$$

Now write $$\mathbf{X}^{-1}\mathbf{x}_0 = \boldsymbol{\xi} = \{\xi_1, \xi_2, \ldots, \xi_n\}, \tag{8.4.9}$$

† Wilkinson, 'Rigorous error bounds for computed eigensystems'. *Comput. J.* **4** (1961), pp. 230–40.

then
$$\mathbf{x}_0 = \xi_1\mathbf{x}^{(1)} + \xi_2\mathbf{x}^{(2)} + \dots + \xi_n\mathbf{x}^{(n)}. \tag{8.4.10}$$

This shows that $\xi_1, \xi_2, \dots, \xi_n$ are the components of $\mathbf{x}_0$ referred to the characteristic vectors of $\mathbf{B}$ as coordinate axes. In finding the vector corresponding to λ_r we try to make $\mathbf{x}_0$ have the direction of $\mathbf{x}^{(r)}$, that is to make $\xi_i = 0$ when $i \neq r$.

It is convenient to think of the characteristic vectors of $\mathbf{B}$ as being normalised to have length unity, so that
$$\mathbf{x}'^{(r)}\mathbf{x}^{(r)} = 1, \tag{8.4.11}$$
and then
$$\mathbf{X}^{-1} = \mathbf{X}'. \tag{8.4.12}$$
Now
$$\begin{aligned}\mathbf{B}\mathbf{x}_0 &= \xi_1\mathbf{B}\mathbf{x}^{(1)} + \xi_2\mathbf{B}\mathbf{x}^{(2)} + \dots + \xi_n\mathbf{B}\mathbf{x}^{(n)} \\ &= \xi_1\lambda_1\mathbf{x}^{(1)} + \xi_2\lambda_2\mathbf{x}^{(2)} + \dots + \xi_n\lambda_n\mathbf{x}^{(n)},\end{aligned} \tag{8.4.13}$$
and hence
$$\mathbf{x}_0'\mathbf{B}\mathbf{x}_0 = \lambda_1\xi_1^2 + \lambda_2\xi_2^2 + \dots + \lambda_n\xi_n^2. \tag{8.4.14}$$
Also
$$\mathbf{x}_0'\mathbf{x}_0 = \xi_1^2 + \xi_2^2 + \dots + \xi_n^2, \tag{8.4.15}$$
so that
$$\begin{aligned}0 &= \mathbf{x}_0'\mathbf{B}\mathbf{x}_0 - \mu\mathbf{x}_0'\mathbf{x}_0 \\ &= (\lambda_1 - \mu)\,\xi_1^2 + (\lambda_2 - \mu)\,\xi_2^2 + \dots + (\lambda_n - \mu)\,\xi_n^2.\end{aligned} \tag{8.4.16}$$

The residual vector, $\mathbf{y}_0$, may be expressed in terms of the ξ_i,
$$\begin{aligned}\mathbf{y}_0 &= \mathbf{B}\mathbf{x}_0 - \mu\mathbf{x}_0 \\ &= (\lambda_1 - \mu)\,\xi_1\mathbf{x}^{(1)} + (\lambda_2 - \mu)\,\xi_2\mathbf{x}^{(2)} + \dots + (\lambda_n - \mu)\,\xi_n\mathbf{x}^{(n)}\end{aligned} \tag{8.4.17}$$
and therefore
$$\mathbf{y}_0'\mathbf{y}_0 = (\lambda_1 - \mu)^2\xi_1^2 + (\lambda_2 - \mu)^2\xi_2^2 + \dots + (\lambda_n - \mu)^2\xi_n^2. \tag{8.4.18}$$

The inequality (8.3.20) states that there is at least one characteristic value of $\mathbf{B}$, λ_r say, such that
$$|\lambda_r - \mu|^2 \leqslant \frac{\mathbf{y}_0'\mathbf{y}_0}{\mathbf{x}_0'\mathbf{x}_0}. \tag{8.4.19}$$

We shall suppose this to be the only characteristic value in this interval around μ. We shall also suppose that the nearest of the other characteristic values is at a distance not less than δ from μ, i.e.
$$\min_{i \neq r}.\,|\lambda_i - \mu| \geqslant \delta. \tag{8.4.20}$$

If we drop the rth term on the right-hand side of equation (8.4.18), and replace the distances of the other λ_i from μ by their lower bound, δ, we obtain the inequality
$$\mathbf{y}_0'\mathbf{y}_0 \geqslant \delta^2 \sum_{i \neq r} \xi_i^2. \tag{8.4.21}$$

The sum on the right-hand side of this inequality is the length of that part of $\mathbf{x}_0$ which is not in the desired direction, along $\mathbf{x}^{(r)}$. The length of the wanted part of $\mathbf{x}_0$ is ξ_r, and equations (8.4.15) and (8.4.21) show that
$$\xi_r^2 \leqslant \mathbf{x}_0'\mathbf{x}_0 = \xi_r^2 + \sum_{i \neq r} \xi_i^2 \leqslant \xi_r^2 + \frac{1}{\delta^2}\mathbf{y}_0'\mathbf{y}_0., \tag{8.4.22}$$
which yields
$$\mathbf{x}_0'\mathbf{x}_0 - \frac{1}{\delta^2}\mathbf{y}_0'\mathbf{y}_0 \leqslant \xi_r^2 \leqslant \mathbf{x}_0'\mathbf{x}_0. \tag{8.4.23}$$

Re-arranging equation (8.4.16) and taking moduli we obtain

$$|\lambda_r-\mu|\,\xi_r^2 = \Big|\sum_{i\neq r}(\lambda_i-\mu)\,\xi_i^2\Big| \leqslant \sum_{i\neq r}|\lambda_i-\mu|\,\xi_i^2$$

$$\leqslant \sum_{i\neq r}\{(\lambda_i-\mu)^2\xi_i^2/|\lambda_i-\mu|\}. \tag{8.4.24}$$

We may now replace the denominators by their lower bound, δ, and use equation (8.4.18) (with the rth term dropped from the right-hand side) to obtain

$$|\lambda_r-\mu|\,\xi_r^2 \leqslant \frac{1}{\delta}\,\mathbf{y}_0'\mathbf{y}_0. \tag{8.4.25}$$

Finally we use the left-hand inequality of equation (8.4.23) for ξ_r to obtain

$$|\lambda_r-\mu| \leqslant \frac{(1/\delta)\,\mathbf{y}_0'\mathbf{y}_0}{\mathbf{x}_0'\mathbf{x}_0-(1/\delta^2)\,\mathbf{y}_0'\mathbf{y}_0}. \tag{8.4.26}$$

Since the left-hand side of this is $|\lambda_r-\mu|$ and not $(\lambda_r-\mu)^2$, this is a much closer bound than that provided by the inequality (8.3.19). Indeed the closeness of μ and λ_r is such that care must be exercised in the calculation of μ, not to introduce rounding errors which vitiate it.

We again treat the matrix of equation (8.3.21), using the approximate characteristic vector $\mathbf{x}_0$ given in the previous example. To calculate the Rayleigh quotient, μ, we use the equation

$$\mu = \lambda_0 + \frac{\mathbf{x}_0'(\mathbf{B}-\lambda_0\mathbf{I})\,\mathbf{x}_0}{\mathbf{x}_0'\mathbf{x}_0}. \tag{8.4.27}$$

This expresses μ as the sum of λ_0 and a small correction. The numerator of the correction term may be calculated by evaluating the scalar products involved double length, starting with the residual vector $\mathbf{y}_0$, namely

$$\mathbf{y}_0 = (\mathbf{B}-\lambda_0\mathbf{I})\,\mathbf{x}_0.$$

We have $$10^6\mathbf{y}_0 = \{-2 \quad -18{\cdot}478\,528 \quad -36{\cdot}008\,256 \quad -68{\cdot}227\,200\},$$

$$10^6\mathbf{x}_0'\mathbf{y}_0 = 4{\cdot}846\,591, \quad 10^6\mathbf{x}_0'\mathbf{y}_0/\mathbf{x}_0'\mathbf{x}_0 = 2{\cdot}732\,229,$$

$$\mu = 0{\cdot}209\,378\,732\,229.$$

This seems absurdly many figures to use, but most of them are justified, as may be seen by applying the inequality (8.4.26). Using this value of μ, and the same $\mathbf{x}_0$, we calculate a new residual vector $\mathbf{y}_0^*$, given by

$$\mathbf{y}_0^* = (\mathbf{B}-\mu\mathbf{I})\,\mathbf{x}_0,$$

and find

$$10^6\mathbf{y}_0^* = \{-4{\cdot}732\,229, \quad -16{\cdot}319\,461 \quad -37{\cdot}038\,323 \quad -67{\cdot}994\,141\}.$$

Before we can use the inequalities (8.4.23) and (8.4.26) we must find the quantity δ. In this case μ is near λ_1 and the nearest of the other characteristic values is λ_2. It may easily be shown (e.g. by calculating the Sturmian sign count for the triple diagonal matrix (9.3.34), as described in § 9.5) that λ_2 is greater than 0·5, so that we may take $\delta = 0{\cdot}29$. Then, since

$$\mathbf{y}_0'^{*}\mathbf{y}_0^{*} < 6248 \times 10^{-12},$$

we have
$$1{\cdot}773\,860\,348\,599 \leqslant \xi_1^2 \leqslant 1{\cdot}773\,860\,423\,320.$$

The square of the length of the unwanted component is

$$\sum_{i\neq 1} \xi_i^2 \leqslant 0{\cdot}000\,000\,074\,721. \tag{8.4.28}$$

Finally
$$|\lambda_1 - \mu| \leqslant \frac{(6284) \times 10^{-12}}{(0{\cdot}29)\,(1{\cdot}773\,860\,348\,599)}$$

$$< (12\,234) \times 10^{-12}$$

$$\doteqdot (1{\cdot}2) \times 10^{-8}.$$

Thus the first seven figures of μ are correct, and the eighth is very nearly right.

Since μ differs from λ_0 by only 2·7 in the last figure, we see that λ_0 is in error by less than 3 in the last place. This may be contrasted with the result found in equation (8.4.4), which suggested that the error might be as much as 61 in the last place. The length of the out-of-phase component of $\mathbf{x}_0$, given by equation (8.4.28), is less than 274×10^{-6}. This can be reduced by solving the equation

$$(\mathbf{B} - \mu\mathbf{I})\,\mathbf{x}_0^{*} = \mathbf{x}_0. \tag{8.4.29}$$

The Rayleigh quotient for $\mathbf{x}_0^{*}$ will then yield a much better estimate of λ_1. Thus there does not seem to be any way of improving the vector without a great deal of labour. Fortunately, it is not often necessary to do this.

The inequality (8.4.26) may be sharpened a little, the symmetric interval around the Rayleigh quotient μ being replaced by one which allows for the different distances to the nearest characteristic values on the two sides. The bounds are generalisations of one due to Temple,† and are due independently to Kohn‡ and Kato.§ If the two nearest characteristic values are λ_{r-1} and λ_{r+1}, and

$$\lambda_{r-1} \leqslant \lambda_r \leqslant \lambda_{r+1}, \quad \lambda_{r-1} \leqslant \mu \leqslant \lambda_{r+1}, \tag{8.4.30}$$

then
$$\mu - \frac{\mathbf{y}_0'\mathbf{y}_0}{(\lambda_{r+1} - \mu)\,\mathbf{x}_0'\mathbf{x}_0} \leqslant \lambda_r \leqslant \mu + \frac{\mathbf{y}_0'\mathbf{y}_0}{(\mu - \lambda_{r-1})\,\mathbf{x}_0'\mathbf{x}_0}. \tag{8.4.31}$$

The inequalities may be derived from the fact that, for all values of i,

$$(\lambda_i - \lambda_r)\,(\lambda_i - \lambda_{r+1}) \geqslant 0. \tag{8.4.32}$$

The right-hand side of this inequality may be rearranged to give

$$(\lambda_i - \mu)^2 + (2\mu - \lambda_r - \lambda_{r+1})\,(\lambda_i - \mu) - (\mu - \lambda_r)\,(\lambda_{r+1} - \mu) \geqslant 0. \tag{8.4.33}$$

† G. Temple, 'The computation of characteristic numbers and characteristic functions', *Proc. Lond. Math. Soc.* **29** (1928), pp. 257–80.

‡ W. Kohn, 'A note on Weinstein's variational method', *Phys. Rev.* **71** (1947), pp. 902–4.

§ T. Kato, 'On the upper and lower bounds of eigenvalues', *J. Phys. Soc. Japan*, **4** (1949), p. 334.

Multiplying this by x_i^2, summing over i, and using equations (8.4.15), (8.4.16) and (8.4.18) we obtain

$$\mathbf{y}_0'\mathbf{y}_0 - (\mu-\lambda_r)(\lambda_{r+1}-\mu)\,\mathbf{x}_0'\mathbf{x}_0 \geqslant 0, \tag{8.4.34}$$

which yields the first half of the inequality; the second may be obtained in a similar manner. The estimates, λ^*_{r-1}, λ^*_{r+1}, of the two neighbouring characteristic values that are used in the inequality (8.4.31) may be in error, so long as the errors bring them nearer to λ_r, that is so long as

$$\lambda_{r-1} \leqslant \lambda^*_{r-1}, \quad \lambda^*_{r+1} \leqslant \lambda_{r+1}. \tag{8.4.35}$$

It will be noticed that the first half of the inequality (8.4.31) breaks down when $r = n$, and the second half when $r = 1$. However, in these cases the inequalities which break down may be replaced by the sharper ones

$$\mu \leqslant \lambda_n, \quad \lambda_1 \leqslant \mu, \tag{8.4.36}$$

respectively.

It may be shown that an approximate characteristic value for the matrix†

$$\mathbf{B} = \begin{bmatrix} -2 & 1 & 0 & 0 \\ 1 & -2 & 1 & 0 \\ 0 & 1 & -2 & 1 \\ 0 & 0 & 1 & -2 \end{bmatrix} \tag{8.4.37}$$

is

$$\lambda_2 = -2{\cdot}619, \tag{8.4.38}$$

and a corresponding characteristic vector is

$$\mathbf{x}_0 = \{1 \quad -0{\cdot}617 \quad -0{\cdot}616 \quad 1\}. \tag{8.4.39}$$

For this vector the Rayleigh quotient is found to be

$$\mu = -2{\cdot}618\,031. \tag{8.4.40}$$

The residual vector $\mathbf{y}_0$ is

$$\begin{aligned}\mathbf{y}_0 &= (\mathbf{B}-\mu\mathbf{I})\,\mathbf{x}_0 \\ &= \{0{\cdot}001\,031 \quad 0{\cdot}002\,675 \quad 0{\cdot}002\,293 \quad 0{\cdot}002\,031\},\end{aligned}$$

and the two neighbouring characteristic values are about $-1{\cdot}4$ and $-3{\cdot}6$. Thus

$$\mathbf{y}_0'\mathbf{y}_0 = 10^{-6}\times 17{\cdot}601, \quad \mathbf{y}_0'\mathbf{y}_0/\mathbf{x}_0'\mathbf{x}_0 = 10^{-6}\times 6{\cdot}377,$$

$$\lambda_3-\mu = 1{\cdot}218\,031, \quad \mu-\lambda_1 = 0{\cdot}981\,969,$$

and this gives the bounds

$$-2{\cdot}618\,031 - 5{\cdot}3\times 10^{-6} \leqslant \lambda_2 \leqslant -2{\cdot}618\,031 + 6{\cdot}5\times 10^{-6},$$

that is

$$-2{\cdot}618\,036 \leqslant \lambda_2 \leqslant -2{\cdot}618\,024. \tag{8.4.41}$$

In practice, one would use the Kohn–Kato bounds for the error in Rayleigh's quotient and Wilkinson's bounds for the error in the vector.

† See the Appendix for the list of examples on this matrix.

EXAMPLES 8.4

1. The following estimates are given for data referring to the matrix (8.3.21):

$$\lambda_2 = 0{\cdot}535, \quad \lambda_3 = 2{\cdot}919\,381, \quad \lambda_4 = 73{\cdot}335\,849,$$

$$\mathbf{x}^{(3)} = \{0{\cdot}486\,186, \quad 1, \quad 0{\cdot}637\,905, \quad -0{\cdot}745\,146\},$$

$$\mathbf{x}^{(4)} = \{0{\cdot}089\,578, \quad 0{\cdot}303\,300, \quad 0{\cdot}624\,378, \quad 1\}.$$

(i) Find bounds for the errors in λ_3 and λ_4 by finding the Rayleigh quotients corresponding to the given approximate vectors and using equation (8.4.26).

(ii) Find bounds for the lengths of the unwanted components in $\mathbf{x}^{(3)}$ and $\mathbf{x}^{(4)}$.

2. Use the data of the last example to find the bounds for λ_3 given by equation (8.4.31), that is, the Kohn–Kato bounds.

8.5 Iterative processes for symmetric matrices

Most of the iterative processes that are used are based on the behaviour of powers of a matrix. This behaviour is most easily described by using the representation

$$\mathbf{B} = \mathbf{XDX}', \tag{8.5.1}$$

where $\mathbf{X}$ is the matrix whose columns are the characteristic vectors $\mathbf{x}^{(i)}$, normalised so that

$$\mathbf{x}'^{(i)}\mathbf{x}^{(i)} = 1, \tag{8.5.2}$$

and

$$\mathbf{D} = \mathbf{diag}\ (\lambda_1 \quad \lambda_2 \quad \dots \quad \lambda_n). \tag{8.5.3}$$

(The processes described in this section may be applied, with slight modifications, to unsymmetric matrices also; this is discussed briefly in section 8.7.)

If $\mathbf{B}$ is symmetric the matrix $\mathbf{X}$ will satisfy

$$\mathbf{XX}' = \mathbf{X}'\mathbf{X} = \mathbf{I}, \quad \mathbf{X}' = \mathbf{X}^{-1}. \tag{8.5.4}$$

If $\mathbf{B}$ is given by equation (8.5.1) then

$$\mathbf{B}^2 = (\mathbf{XDX}')\,(\mathbf{XDX}') = \mathbf{XD}(\mathbf{X}'\mathbf{X})\,\mathbf{DX}' = \mathbf{XD}^2\mathbf{X}', \tag{8.5.5}$$

and generally

$$\mathbf{B}^r = \mathbf{XD}^r\mathbf{X}'. \tag{8.5.6}$$

Equations (8.5.1) and (8.5.6) may be expanded to give

$$\mathbf{B} = \sum_{i=1}^{n} \lambda_i \mathbf{x}^{(i)}\mathbf{x}'^{(i)}. \tag{8.5.7}$$

$$\mathbf{B}^r = \sum_{i=1}^{n} \lambda_i^r \mathbf{x}^{(i)}\mathbf{x}'^{(i)}. \tag{8.5.8}$$

If one characteristic value, say λ_n, is larger in modulus than all the others, then, as r increases the term $\lambda_n^r\mathbf{x}^{(n)}\mathbf{x}'^{(n)}$ will predominate in $\mathbf{B}^r$, so that $\mathbf{B}^r$ will become more and more like a multiple of $\mathbf{x}^{(n)}\mathbf{x}'^{(n)}$, and the ratio of successive matrices will yield the value of λ_n. This property is not of great use directly, because the number of multiplications needed in multiplying two matrices is n^3, so that the work of forming $\mathbf{B}^r$ increases as n^{3r}. It is not necessary, however, to form the powers of $\mathbf{B}$. The repeated multiplication of an arbitrary vector by $\mathbf{B}$ will show similar behaviour; if we take a vector, $\mathbf{v}$, then we can form the sequence

$$\mathbf{Bv},\quad \mathbf{B}^2\mathbf{v},\quad \ldots,\quad \mathbf{B}^r\mathbf{v},\quad \ldots,$$

with only n^2 multiplications per step. Using equation (8.5.8) we may write

$$\mathbf{B}^r\mathbf{v} = \sum_{i=1}^{n} \lambda_i^r \mathbf{x}^{(i)}(\mathbf{x}'^{(i)}\mathbf{v}), \tag{8.5.9}$$

and as r increases this is dominated by the term

$$\lambda_n^r(\mathbf{x}'^{(n)}\mathbf{v})\,\mathbf{x}^{(n)},$$

so that the vectors of the sequence come to lie more and more along the characteristic vector $\mathbf{x}^{(n)}$. The corresponding elements of successive vectors grow in a ratio which approaches λ_n.

Suppose

$$\mathbf{B} = \begin{bmatrix} 1 & 2 & 3 & 4 \\ 2 & 6 & 10 & 14 \\ 3 & 10 & 20 & 30 \\ 4 & 14 & 30 & 50 \end{bmatrix}. \tag{8.5.10}$$

As an initial vector we take

$$\mathbf{v} = \{1 \quad 1 \quad 1 \quad 1\}.$$

Then

$$\mathbf{Bv} = \{10 \quad 32 \quad 63 \quad 98\},$$

$$\mathbf{B}^2\mathbf{v} = \{655 \quad 2214 \quad 4550 \quad 7278\}.$$

It becomes difficult to compare the vectors unless they are standardised in some way, and we shall do this by making their largest components unity. The sequence of vectors then becomes

$$\begin{aligned}
\mathbf{v}^{(0)} &= \{\,1 \qquad 1 \qquad 1 \qquad 1\},\\
\mathbf{Bv}^{(0)} &= \{10 \qquad 32 \qquad 63 \qquad 98\},\\
\mathbf{v}^{(1)} &= \{\,0{\cdot}102 \quad 0{\cdot}327 \quad 0{\cdot}643 \quad 1{\cdot}000\},\\
\mathbf{Bv}^{(1)} &= \{\,6{\cdot}685 \quad 22{\cdot}596 \quad 46{\cdot}436 \quad 74{\cdot}276\},\\
\mathbf{v}^{(2)} &= \{\,0{\cdot}090 \quad 0{\cdot}304 \quad 0{\cdot}625 \quad 1{\cdot}000\},\\
\mathbf{Bv}^{(2)} &= \{\,6{\cdot}573 \quad 22{\cdot}254 \quad 45{\cdot}810 \quad 73{\cdot}366\}.
\end{aligned}$$

The difference between $\mathbf{Bv}^{(1)}$ and $\mathbf{Bv}^{(2)}$ is not very great, so we now use more figures in recording the standardised vector $\mathbf{v}^{(3)}$, and in the further working.

$$\begin{aligned}
\mathbf{v}^{(3)} &= \{0{\cdot}089\,592 \quad 0{\cdot}303\,329 \quad 0{\cdot}624\,404 \quad 1{\cdot}000\,000\},\\
\mathbf{Bv}^{(3)} &= \{6{\cdot}569\,462 \quad 22{\cdot}243\,198 \quad 45{\cdot}790\,146 \quad 73{\cdot}337\,094\},\\
\mathbf{v}^{(4)} &= \{0{\cdot}089\,579 \quad 0{\cdot}303\,301 \quad 0{\cdot}624\,379 \quad 1{\cdot}000\,000\},\\
\mathbf{Bv}^{(4)} &= \{6{\cdot}569\,318 \quad 22{\cdot}242\,754 \quad 45{\cdot}789\,327 \quad 73{\cdot}335\,900\},\\
\mathbf{v}^{(5)} &= \{0{\cdot}089\,578 \quad 0{\cdot}303\,300 \quad 0{\cdot}624\,378 \quad 1{\cdot}000\,000\}.
\end{aligned}$$

The difference between $\mathbf{v}^{(4)}$ and $\mathbf{v}^{(5)}$ will change $\mathbf{B}\mathbf{v}^{(5)}$ by at most 48 in the last element, as compared with $\mathbf{B}\mathbf{v}^{(4)}$, and hence will affect at most the rounding off when the vector is normalised. The best estimate this yields for λ_4 is

$$\lambda_4 = 73{\cdot}335\,900.$$

This can be improved by calculating the Rayleigh's quotients for the vectors. We display below the Rayleigh's quotients for the sequence of vectors $\mathbf{v}^{(0)}$ to $\mathbf{v}^{(3)}$

$$50{\cdot}75, \quad 73{\cdot}299\,209, \quad 73{\cdot}335\,812, \quad 73{\cdot}335\,849\,04.$$

The ratio of corresponding elements of successive vectors is not a very good estimate of λ_4 compared with that found from the Rayleigh quotient. The value of λ_4 to six decimal places is

$$\lambda_4 = 73{\cdot}335\,849. \tag{8.5.11}$$

This may be obtained by two more iterations, using nine decimal places.

The smallest characteristic value of the matrix may be found by iterating with the inverse of the matrix. The characteristic values of the inverse are the reciprocals of those of the matrix, and therefore the smallest characteristic value of the latter corresponds to the greatest value of the former. To iterate with the inverse we use the equation

$$\mathbf{v}^{(r+1)} = \mathbf{B}^{-1}\mathbf{v}^{(r)}. \tag{8.5.12}$$

This can be re-written as an equation to be solved for $\mathbf{v}^{(r+1)}$,

$$\mathbf{B}\mathbf{v}^{(r+1)} = \mathbf{v}^{(r)}, \tag{8.5.13}$$

and this may be solved by any convenient method. The matrix factorisation method (§ 7.2) is usually the most convenient, as the same factors $\mathbf{L}$ and $\mathbf{U}$ may be used at every step of the iteration.

For the matrix $\mathbf{B}$ of the last example we can find the factors exactly, and they are

$$\mathbf{L} = \begin{bmatrix} 1 & 0 & 0 & 0 \\ 2 & 1 & 0 & 0 \\ 3 & 2 & 1 & 0 \\ 4 & 3 & 2 & 1 \end{bmatrix}, \quad \mathbf{U} = \begin{bmatrix} 1 & 2 & 3 & 4 \\ 0 & 2 & 4 & 6 \\ 0 & 0 & 3 & 6 \\ 0 & 0 & 0 & 4 \end{bmatrix}. \tag{8.5.14}$$

Starting from the same $\mathbf{v}_0$ as in the last example we solve the equation

$$\mathbf{B}\mathbf{w}^{(0)} = \mathbf{v}^{(0)}$$

by writing it as

$$\mathbf{L}\mathbf{z}^{(0)} = \mathbf{v}^{(0)}, \quad \mathbf{U}\mathbf{w}^{(0)} = \mathbf{z}^{(0)},$$

and solving first for $\mathbf{z}^{(0)}$ and then for $\mathbf{w}^{(0)}$. We then take $\mathbf{w}^{(0)}$, normalised to have its largest element unity, as the vector $\mathbf{v}^{(1)}$, and repeat the process with $\mathbf{v}^{(1)}$, and so on. The details of the calculation are as follows:

$$\begin{aligned} \mathbf{v}^{(0)} &= \{1 \quad\quad 1 \quad\quad 1 \quad\quad 1\}, \\ \mathbf{z}^{(0)} &= \{1 \quad\quad -1 \quad\quad 0 \quad\quad 0\}, \\ \mathbf{w}^{(0)} &= \{2 \quad\quad -0{\cdot}5 \quad\quad 0 \quad\quad 0\}, \end{aligned}$$

$$\mathbf{v}^{(1)} = \{1 \quad -0{\cdot}25 \quad 0 \quad 0\},$$
$$\mathbf{z}^{(1)} = \{1 \quad -2{\cdot}25 \quad 1{\cdot}5 \quad -0{\cdot}25\},$$
$$\mathbf{w}^{(1)} = \{3{\cdot}75 \quad -2{\cdot}187\,5 \quad 0{\cdot}625 \quad -0{\cdot}062\,5\},$$

$$\mathbf{v}^{(2)} = \{1 \quad -0{\cdot}583\,333 \quad 0{\cdot}166\,667 \quad -0{\cdot}016\,667\},$$
$$\mathbf{z}^{(2)} = \{1 \quad -2{\cdot}583\,333 \quad 2{\cdot}333\,334 \quad -0{\cdot}933\,336\},$$
$$\mathbf{w}^{(2)} = \{4{\cdot}361\,111 \quad -3{\cdot}080\,556 \quad 1{\cdot}244\,446 \quad -0{\cdot}233\,334\},$$

$$\mathbf{v}^{(3)} = \{1 \quad -0{\cdot}706\,370 \quad 0{\cdot}285\,350 \quad -0{\cdot}053\,503\},$$
$$\mathbf{z}^{(3)} = \{1 \quad -2{\cdot}706\,370 \quad 2{\cdot}698\,090 \quad -1{\cdot}330\,573\},$$
$$\mathbf{w}^{(3)} = \{4{\cdot}606\,233 \quad -3{\cdot}485\,554 \quad 1{\cdot}565\,149 \quad -0{\cdot}332\,643\},$$

$$\mathbf{v}^{(4)} = \{1 \quad -0{\cdot}756\,704 \quad 0{\cdot}339\,789 \quad -0{\cdot}072\,216\},$$
$$\mathbf{z}^{(4)} = \{1 \quad -2{\cdot}756\,704 \quad 2{\cdot}853\,197 \quad -1{\cdot}508\,498\},$$
$$\mathbf{w}^{(4)} = \{4{\cdot}707\,773 \quad -3{\cdot}657\,608 \quad 1{\cdot}705\,314 \quad -0{\cdot}377\,124\}.$$

It is clear that this is not converging as well as the process used in the last example, but the difference is not due to the process itself, but to the difference in the spacing of the characteristic values of the matrix at the two ends of the range. Rayleigh's quotient for $\mathbf{v}^{(r)}$ is

$$\frac{\mathbf{v}'^{(r)}\mathbf{B}^{-1}\mathbf{v}^{(r)}}{\mathbf{v}'^{(r)}\mathbf{v}^{(r)}} = \frac{\mathbf{w}'^{(r)}\mathbf{v}^{(r)}}{\mathbf{v}'^{(r)}\mathbf{v}^{(r)}},$$

and again the quotients give better estimates for the characteristic value than do the ratios of corresponding elements of successive vectors. The Rayleigh quotients for $\mathbf{v}^{(0)}$ to $\mathbf{v}^{(4)}$ are found to be

$$0{\cdot}375, \quad 4{\cdot}044\,118, \quad 4{\cdot}654\,860, \quad 4{\cdot}757\,780, \quad 4{\cdot}773\,112.$$

The number derived in this way is the largest characteristic value of $\mathbf{B}^{-1}$, and its reciprocal is the value we seek; the last quotient gives

$$\lambda_1 = 0{\cdot}209\,507. \tag{8.5.15}$$

It is shown in § 8.4 that the correct value is

$$\lambda_1 = 0{\cdot}209\,379 \tag{8.5.16}$$

to six decimal places.

The vector $\mathbf{v}^{(r)}$, after scaling, has unwanted components whose coefficients are of the form $c(\lambda_i/\lambda_n)^r$ and the rate at which the vectors converge to a characteristic value is determined by the ratio $(\lambda_{n-1}/\lambda_n)$.

For the matrix of the last two examples the characteristic values are approximately

$$0{\cdot}2, \quad 0{\cdot}5, \quad 2{\cdot}9, \quad 73{\cdot}3.$$

The convergence found with the matrix (8.5.10) is like that of a geometric series with common ratio $(2{\cdot}9)/(73{\cdot}3) = 0{\cdot}04$, while that for the matrix (8.5.14) is like that of a series with common ratio $(1/0{\cdot}5)/(1/0{\cdot}2) = 0{\cdot}4$. The former would require

4 steps to reduce the unwanted components by a factor of about 0·000 002, while the latter would require about 14 steps.

The amount of work involved in factorising **B** to use the iteration for its smallest characteristic value is about half that required for Householder's reduction to triple-diagonal form described in Chapter 9. However, if more than two distinct factorisations have to be performed, it is probably preferable to rotate to triple-diagonal form, except when the matrix has a large proportion of zeros so arranged that the factorisation is made less laborious.

The technique described above for the calculation of the smallest characteristic value, and corresponding vector of a matrix, is a special case of a more general process. The inverse of the matrix, $\mathbf{B}^{-1}$, was used because this made the desired characteristic value dominate the others. In order to find any characteristic value we may perform transformations on the matrix to make that value dominate. One such transformation is as follows. Suppose that α is an approximation to a characteristic value λ of a matrix **B**. The approximation may be quite crude, so long as α is nearer to λ than to any other characteristic value. If this is the case then the dominant characteristic value of $(\mathbf{B}-\alpha\mathbf{I})^{-1}$ will be $(\lambda-\alpha)^{-1}$. However, the convergence ratio may not be very good. Other transformations may be used,† but unless **B** has many non-zero elements, the labour of solving the resulting equations may be excessive.

EXAMPLES 8.5

1.‡ (*a*) Calculate approximations to the characteristic value of greatest modulus, and the corresponding vector, for the matrix of Ex. 8.3.2. First round the elements of the matrix to six decimal places. Then, starting from the trial vector

$$\mathbf{v}^{(0)} = \{1 \quad 1 \quad 1 \quad 1\}$$

iterate until $\mathbf{v}^{(5)}$.

(*b*) Calculate the Rayleigh quotient corresponding to $\mathbf{v}^{(5)}$ and estimate the error in this by means of equations (8.3.20) and (8.3.25).

2.‡ Use an iterative method to calculate the characteristic value of greatest modulus of the matrix of Ex. 7.8.2, correct to four places of decimals; find an approximation to the characteristic vector also.

3.‡ Use the factorisation found in Ex. 7.8.2 to find the numerically smallest characteristic value, and the corresponding characteristic vector for the matrix of Ex. 7.8.2. Start from the vector $\mathbf{v}^{(0)} = \{1, 1, 1, 1\}$ and find values which are correct to three places of decimals.

8.6 Aitken's method of accelerating the convergence of iterates

The approximate characteristic values and vectors which occur as iterates in the processes described in this chapter all have the form

$$x^{(r)} = x + ab^r + \text{terms small compared with } b^r. \tag{8.6.1}$$

† H. Wielandt, 'Das Iterationsverfahren bei Nichtselbstadjugierte Eigenwertprobleme', *Math. Z.* **50** (1944).

‡ The calculation involved in the solution should be retained for use in Examples 8.6.

Usually $b\ (<1)$ is the ratio $(\lambda_{n-1}/\lambda_n)$. Aitken† showed that such iterates may be made more rapidly convergent by means of the following device. If we ignore the small terms we have the equations

$$\left.\begin{aligned} x^{(r-1)} &= x+ab^{r-1}, \\ x^{(r)} &= x+ab^{r}, \\ x^{(r+1)} &= x+ab^{r+1}. \end{aligned}\right\} \tag{8.6.2}$$

After eliminating a and b we obtain

$$x = \bar{x}^{(r+1)} = x^{(r+1)} - \frac{(x^{(r+1)}-x^{(r)})^2}{x^{(r-1)}-2x^{(r)}+x^{(r+1)}} \tag{8.6.3}$$

as the improved estimate. Because the denominator of the correction term is the second central difference of $x^{(r)}$, denoted by $\delta^2 x^{(r)}$, this process is often called the δ^2 process.

We illustrate the method by applying it to the iteration for the dominant characteristic value of the unsymmetric matrix

$$\mathbf{B} = \begin{bmatrix} 1 & 0 & 0 & -1 \\ -1 & 1 & 1 & -3 \\ -2 & 3 & 0 & -4 \\ -7 & 3 & 4 & -7 \end{bmatrix}. \tag{8.6.4}‡$$

With the same starting vector as before, the sequence of vectors is, to three decimal places,

$$\begin{aligned}
\mathbf{v}^{(0)} &= \{1 \quad 1 \quad 1 \quad 1\}, \\
\mathbf{v}^{(1)} &= \{0 \quad 0{\cdot}286 \quad 0{\cdot}428 \quad 1\}, \\
\mathbf{v}^{(2)} &= \{0{\cdot}226 \quad 0{\cdot}516 \quad 0{\cdot}709 \quad 1\}, \\
\mathbf{v}^{(3)} &= \{0{\cdot}184 \quad 0{\cdot}477 \quad 0{\cdot}692 \quad 1\}, \\
\mathbf{v}^{(4)} &= \{0{\cdot}200 \quad 0{\cdot}493 \quad 0{\cdot}718 \quad 1\}, \\
\mathbf{v}^{(5)} &= \{0{\cdot}198 \quad 0{\cdot}491 \quad 0{\cdot}721 \quad 1\}, \\
\mathbf{v}^{(6)} &= \{0{\cdot}199 \quad 0{\cdot}493 \quad 0{\cdot}725 \quad 1\}, \\
\mathbf{v}^{(7)} &= \{0{\cdot}200 \quad 0{\cdot}494 \quad 0{\cdot}727 \quad 1\}.
\end{aligned}$$

The scaling factors at each stage, found as $(\mathbf{B}\mathbf{v}^{(r-1)})_4$, are shown in the left-hand column of the table below:

r	$x^{(r)}$	$(x^{(r+1)}-x^{(r)})^2$	$x^{(r+1)}-2x^{(r)}+x^{(r-1)}$	$\bar{x}^{(r+1)}$
2	−4·430	—	—	—
3	−4·198	0·011881	−0·123	−3·992
4	−4·089	0·001600	−0·069	−4·026
5	−4·049	0·000400	−0·020	−4·009
6	−4·029	0·000225	−0·005	−3·969
7	−4·014	—	—	—

It is clear that the values of $x^{(r)}$ are too crude to be treated by the process. We

† A. C. Aitken, 'The evaluation of the latent roots and latent vectors of a matrix', *Proc. Roy. Soc. Edinb.* **62** (1937), pp. 269–304.

‡ The justification for this iterative process for an unsymmetric matrix is given in section 8.7.

derive some better values by carrying on the iteration with six decimal places, for three steps, re-computing $\mathbf{v}^{(7)}$ from $\mathbf{B}\mathbf{v}^{(6)}$ as a starting vector. We find

$$\mathbf{v}^{(7)} = \{0{\cdot}199\,552 \quad 0{\cdot}493\,523 \quad 0{\cdot}727\,205 \quad 1\},$$
$$\mathbf{v}^{(8)} = \{0{\cdot}199\,739 \quad 0{\cdot}493\,783 \quad 0{\cdot}728\,273 \quad 1\},$$
$$\mathbf{v}^{(9)} = \{0{\cdot}199\,879 \quad 0{\cdot}493\,960 \quad 0{\cdot}728\,853 \quad 1\},$$
$$\mathbf{v}^{(10)} = \{0{\cdot}199\,937 \quad 0{\cdot}494\,036 \quad 0{\cdot}729\,130 \quad 1\}.$$

We can now derive the following table, in which $x^{(r)} = (\mathbf{B}\mathbf{v}^{(r-1)})_4$:

r	$x^{(r)}$	$(x^{(r+1)}-x^{(r)})^2$	$x^{(r+1)}-2x^{(r)}+x^{(r-1)}$	$\bar{x}^{(r+1)}$
7	$-4{\cdot}014\,000$	—	—	—
8	$-4{\cdot}007\,475$	$0{\cdot}000\,014\,010\,049$	$-0{\cdot}002\,782$	$-3{\cdot}998\,696$
9	$-4{\cdot}003\,732$	$0{\cdot}000\,003\,485\,689$	$-0{\cdot}001\,876$	$-4{\cdot}000\,007$
10	$-4{\cdot}001\,865$	—	—	—

The true value of λ_4 is -4. It is to be noticed that the use of the poor value $x^{(7)}$ does not yield a very great improvement, while the use of three six-figure values does.

In principle it is possible to improve the vector in the same way, applying equation (8.6.3) to successive estimates of each component in turn, but this will often lead to quite insignificant changes in the latest estimate of the vector. However, in the present case we find

$$\bar{\mathbf{v}}^{(10)} - \mathbf{v}^{(10)} = \{0{\cdot}000\,041 \quad 0{\cdot}000\,057 \quad 0{\cdot}000\,257 \quad 0{\cdot}000\,000\}$$

so that
$$\bar{\mathbf{v}}^{(10)} = \{0{\cdot}199\,978 \quad 0{\cdot}494\,093 \quad 0{\cdot}729\,387 \quad 1\}. \tag{8.6.5}$$

The true value of the characteristic vector, to six decimal places, is

$$\mathbf{v} = \{0{\cdot}2 \quad 0{\cdot}494\,118 \quad 0{\cdot}729\,412 \quad 1\}. \tag{8.6.6}$$

EXAMPLES 8.6

1. Find a more exact characteristic vector by applying Aitken's δ^2 process to the iterates

$$\mathbf{v}^{(3)} = \{1 \quad 0{\cdot}570\,740 \quad 0{\cdot}407\,381 \quad 0{\cdot}318\,706\},$$
$$\mathbf{v}^{(4)} = \{1 \quad 0{\cdot}570\,236 \quad 0{\cdot}406\,847 \quad 0{\cdot}318\,204\},$$
$$\mathbf{v}^{(5)} = \{1 \quad 0{\cdot}570\,179 \quad 0{\cdot}406\,787 \quad 0{\cdot}318\,148\},$$

of Ex. 8.5.1.

2. Starting from the vector

$$\mathbf{v}^{(3)} = \{0{\cdot}097\,676 \quad -0{\cdot}532\,184 \quad 1\}$$

of Ex. 8.5.2, and working with nine decimal places, find more exact vectors $\mathbf{v}^{(4)}$ and $\mathbf{v}^{(5)}$. Apply Aitken's δ^2 process to the three vectors to find a better approximation to the characteristic vector, and use this approximation to find the characteristic value correct to six decimal places. Give reasons for thinking that the desired accuracy has been obtained.

3. Apply Aitken's δ^2 process to the working of Ex. 8.5.3 to find the characteristic value and vector correct to six places of decimals.

8.7 Iteration to find a non-dominant characteristic value

For a symmetric matrix, the characteristic vector for any characteristic value is orthogonal to the vectors corresponding to any other characteristic value. Thus if we start with a vector orthogonal to the vector $\mathbf{x}^{(n)}$, corresponding to the largest characteristic value, and apply the iteration technique of § 8.5, we shall obtain the characteristic vector corresponding to the characteristic value of second largest modulus—that is, we should if we could perform exact arithmetic. Unfortunately, the inevitable rounding-off will almost always introduce some component in the direction of the dominant vector, and the iteration will converge to this vector. The deleterious effect of rounding-off can be eliminated by making the successive iterates orthogonal to the dominant vector.

Thus, starting from any vector $\mathbf{v}^{(0)}$ such that

$$\mathbf{x}'^{(n)}\mathbf{v}^{(0)} = 0, \tag{8.7.1}$$

we construct the sequence defined by

$$\mathbf{w}^{(r+1)} = \mathbf{B}\mathbf{v}^{(r)} - \alpha_r \mathbf{x}^{(n)}, \tag{8.7.2}$$

$$\mathbf{v}^{(r+1)} = c_r \mathbf{w}^{(r+1)}, \tag{8.7.3}$$

where α_r is chosen so that

$$\mathbf{x}'^{(n)}\mathbf{w}^{(r+1)} = 0, \tag{8.7.4}$$

and the scale factor c_r is chosen so that the largest element in $\mathbf{v}^{(r+1)}$ is unity. Equation (8.7.2) shows that

$$\alpha_r = \frac{\mathbf{x}'^{(n)}\mathbf{B}\mathbf{v}^{(r)}}{\mathbf{x}'^{(n)}\mathbf{x}^{(n)}}. \tag{8.7.5}$$

Once more we make use of the symmetric matrix of equation (8.5.10) for which an estimate of the dominant characteristic vector is

$$\mathbf{x}^{(4)} = \{0{\cdot}089\,578 \quad 0{\cdot}303\,300 \quad 0{\cdot}624\,378 \quad 1\}.$$

A simple vector which is orthogonal to this is

$$\mathbf{v}^{(0)} = \{0 \quad 0 \quad 1 \quad -0{\cdot}624\,378\}.$$

We then find

$$\begin{aligned}
\mathbf{B}\mathbf{v}^{(0)} &= \{0{\cdot}502\,488 \quad 1{\cdot}258\,708 \quad 1{\cdot}268\,660 \quad -1{\cdot}218\,900\},\\
\mathbf{x}'^{(4)}\mathbf{B}\mathbf{v}^{(0)} &= 0{\cdot}000\,001 \qquad \alpha_0 = 0,\\
\mathbf{v}^{(1)} &= \{0{\cdot}396\,078 \quad 0{\cdot}992\,156 \quad 1 \quad -0{\cdot}960\,778\},\\
\mathbf{B}\mathbf{v}^{(1)} &= \{1{\cdot}537\,278 \quad 3{\cdot}294\,200 \quad 2{\cdot}286\,454 \quad -2{\cdot}564\,404\},\\
\mathbf{x}'^{(4)}\mathbf{B}\mathbf{v}^{(1)} &= 0{\cdot}000\,044\,7 \qquad \alpha_1 = 0{\cdot}000\,030,\\
\mathbf{v}^{(2)} &= \{0{\cdot}466\,662 \quad 1 \quad 0{\cdot}694\,081 \quad -0{\cdot}778\,472\},\\
\mathbf{B}\mathbf{v}^{(2)} &= \{1{\cdot}435\,017 \quad 2{\cdot}975\,526 \quad 1{\cdot}927\,446 \quad -2{\cdot}234\,522\},
\end{aligned}$$

$$\mathbf{x}'^{(4)}\mathbf{B}\mathbf{v}^{(2)} = -0{\cdot}000\,044\,1 \qquad \alpha_2 = -0{\cdot}000\,030,$$

$$\mathbf{v}^{(3)} = \{0{\cdot}482\,273 \quad 1 \quad 0{\cdot}647\,771 \quad -0{\cdot}750\,955\},$$

$$\mathbf{B}\mathbf{v}^{(3)} = \{1{\cdot}421\,766 \quad 2{\cdot}928\,886 \quad 1{\cdot}873\,589 \quad -2{\cdot}185\,528\},$$

$$\mathbf{x}'^{(4)}\mathbf{B}\mathbf{v}^{(3)} = -0{\cdot}000\,010\,2 \qquad \alpha_3 = -0{\cdot}000\,007,$$

$$\mathbf{v}^{(4)} = \{0{\cdot}485\,429 \quad 1 \quad 0{\cdot}639\,694 \quad -0{\cdot}746\,195\},$$

$$\mathbf{B}\mathbf{v}^{(4)} = \{1{\cdot}419\,731 \quad 2{\cdot}921\,068 \quad 1{\cdot}864\,317 \quad -2{\cdot}177\,214\}.$$

The successive scale factors c_r are

$$1{\cdot}268\,660 \quad 3{\cdot}294\,191 \quad 2{\cdot}975\,535 \quad 2{\cdot}928\,888.$$

Applying Aitken's δ^2 process to the last three of these we obtain

$$2{\cdot}928\,888 - (0{\cdot}046\,647)^2/0{\cdot}272\,009 = 2{\cdot}920\,888.$$

We may also calculate the Rayleigh quotients for the last three vectors; they are

$$2{\cdot}915\,807 \quad 2{\cdot}919\,261 \quad 2{\cdot}919\,377$$

and Aitken's δ^2 process applied to these gives

$$\lambda_3 = 2{\cdot}919\,381. \tag{8.7.6}$$

When several characteristic vectors are known for a matrix, the process is essentially similar to that described, except that the initial vector must be orthogonal to all the known vectors, and the successive vectors $\mathbf{B}\mathbf{v}^{(r)}$ must be orthogonalised to them also, so that equations (8.7.2), (8.7.3) and (8.7.5) are replaced by

$$\mathbf{w}^{(r+1)} = \mathbf{B}\mathbf{v}^{(r)} - \alpha_r \mathbf{x}^{(n)} - \beta_r \mathbf{x}^{(n-1)} - \ldots, \quad \mathbf{v}^{(r+1)} = c_r \mathbf{w}^{(r+1)}, \tag{8.7.7}$$

where

$$\alpha_r = \frac{\mathbf{x}'^{(n)}\mathbf{B}\mathbf{v}^{(r)}}{\mathbf{x}'^{(n)}\mathbf{x}^{(n)}}, \quad \beta_r = \frac{\mathbf{x}'^{(n-1)}\mathbf{B}\mathbf{v}^{(r)}}{\mathbf{x}'^{(n-1)}\mathbf{x}^{(n-1)}}, \quad \ldots, \tag{8.7.8}$$

and c_r is chosen so that the largest element of $\mathbf{v}^{(r+1)}$ is unity. The amount of work in each step of the iteration is much increased as more characteristic values are required.

An unsymmetric matrix not only has characteristic columns $\mathbf{x}^{(r)}$ satisfying equation (8.4.6), but also characteristic rows $\mathbf{y}'^{(r)}$ satisfying

$$\mathbf{y}'^{(r)}\mathbf{B} = \lambda_r \mathbf{y}'^{(r)}. \tag{8.7.9}$$

The characteristic columns of an unsymmetric matrix are not mutually orthogonal, but form half a bi-orthogonal system, the other half being formed by the characteristic rows. In fact, provided it satisfies certain general conditions, the matrix will have a total of $2n$ characteristic vectors which may be normalised so that

$$\mathbf{y}'^{(r)}\mathbf{x}^{(s)} = \delta_{rs}, \quad (r, s = 1, 2, \ldots, n). \tag{8.7.10}$$

This means that if

$$\mathbf{X} = [\mathbf{x}^{(1)} \quad \mathbf{x}^{(2)} \quad \ldots \quad \mathbf{x}^{(n)}], \quad \mathbf{Y}' = [\mathbf{y}^{(1)} \quad \mathbf{y}^{(2)} \quad \ldots \quad \mathbf{y}^{(n)}] \tag{8.7.11}$$

then

$$\mathbf{Y}'\mathbf{X} = \mathbf{I}, \quad \mathbf{B} = \mathbf{X}\mathbf{D}\mathbf{Y}' = \mathbf{X}\mathbf{D}\mathbf{X}^{-1}. \tag{8.7.12}$$

Using this result we may write $\mathbf{B}$ in the form

$$\mathbf{B} = \sum_{i=1}^{n} \lambda_i \mathbf{x}^{(i)} \mathbf{y}'^{(i)} \tag{8.7.13}$$

so that

$$\mathbf{B}\mathbf{v} = \sum_{i=1}^{n} \lambda_i (\mathbf{y}'^{(i)}\mathbf{v})\, \mathbf{x}^{(i)}, \quad \mathbf{w}'\mathbf{B} = \sum_{i=1}^{n} \lambda_i (\mathbf{w}'\,\mathbf{x}^{(i)})\, \mathbf{y}'^{(i)}. \tag{8.7.14}$$

This iterative premultiplication by $\mathbf{B}$ will lead to the eigenvalue λ_n and the characteristic column $\mathbf{x}^{(n)}$, while iterative postmultiplication by $\mathbf{B}$ will lead to the characteristic row $\mathbf{y}'^{(n)}$. The re-orthogonalisation process for sub-dominant characteristic values may still be used provided that the column vectors and the row vectors are found at the same time. The equations (8.7.8) become

$$\alpha_r = \frac{\mathbf{y}'^{(n)}\mathbf{B}\mathbf{v}^{(r)}}{\mathbf{y}'^{(n)}\mathbf{x}^{(n)}}, \quad \beta_r = \frac{\mathbf{y}'^{(n-1)}\mathbf{B}\mathbf{v}^{(r)}}{\mathbf{y}'^{(n-1)}\mathbf{x}^{(n-1)}}, \quad \ldots. \tag{8.7.15}$$

The matrix of equation (8.6.4) has the following characteristic values and vectors:

$$\lambda_4 = -4, \quad \mathbf{x}^{(4)} = \{0{\cdot}2 \quad 0{\cdot}494 \quad 0{\cdot}729 \quad 1\},$$

$$\mathbf{y}'^{(4)} = \{2{\cdot}125 \quad 0 \quad -2{\cdot}125 \quad 2{\cdot}125\},$$

$$\lambda_3 = -2, \quad \mathbf{x}^{(3)} = \{0{\cdot}25 \quad 0{\cdot}5 \quad 1 \quad 0{\cdot}75\},$$

$$\mathbf{y}'^{(3)} = \{-1 \quad -2 \quad 3 \quad -1\}.$$

The row vectors have been scaled so that $\mathbf{y}'^{(4)}\mathbf{x}^{(4)} = 1 = \mathbf{y}'^{(3)}\mathbf{x}^{(3)}$, and so

$$\alpha_r = \mathbf{y}'^{(4)}\mathbf{B}\mathbf{v}^{(r)}, \quad \beta_r = \mathbf{y}'^{(3)}\mathbf{B}\mathbf{v}^{(r)}.$$

We start with an arbitrary vector and make it orthogonal to $\mathbf{y}'^{(4)}$ and $\mathbf{y}'^{(3)}$,

$$\mathbf{v}^{(0)} = \{1 \quad 1 \quad 1 \quad 1\},$$

$$\mathbf{y}'^{(4)}\mathbf{v}^{(0)} = 2{\cdot}125, \quad \mathbf{y}'^{(3)}\mathbf{v}^{(0)} = -1.$$

Thus

$$\mathbf{w}^{(1)} = \{\mathbf{v}^{(0)} \quad -2{\cdot}125\mathbf{x}^{(4)} \quad +1.\mathbf{x}^{(3)}\},$$

$$= \{0{\cdot}825 \quad 0{\cdot}450 \quad 0{\cdot}451 \quad -0{\cdot}375\},$$

and scaling this to make the largest element unity we find

$$\mathbf{v}^{(1)} = \{1 \quad 0{\cdot}545 \quad 0{\cdot}547 \quad -0{\cdot}455\}.$$

We then find

$$\mathbf{B}\mathbf{v}^{(1)} = \{0{\cdot}545 \quad 1{\cdot}457 \quad 1{\cdot}455 \quad 0{\cdot}008\},$$

$$\alpha_1 = -1{\cdot}912, \qquad \beta_1 = 0{\cdot}898,$$

$$\mathbf{w}^{(2)} = \mathbf{B}\mathbf{v}^{(1)} - \alpha_1\mathbf{x}^{(4)} - \beta_1\mathbf{x}^{(3)}$$

$$= \{0{\cdot}703 \quad 1{\cdot}952 \quad 1{\cdot}951 \quad 1{\cdot}246\},$$

and scaling this we find

$$\mathbf{v}^{(2)} = \{0{\cdot}360 \quad 1 \quad 0{\cdot}999 \quad 0{\cdot}638\}.$$

The process continues, giving

$$\mathbf{Bv}^{(2)} = \{-0{\cdot}278 \quad -0{\cdot}275 \quad -0{\cdot}272 \quad 0{\cdot}010\},$$

$$\alpha_2 = 0{\cdot}008, \qquad \beta_2 = 0{\cdot}002,$$

$$\mathbf{w}^{(3)} = \{-0{\cdot}280 \quad -0{\cdot}280 \quad -0{\cdot}280 \quad 0{\cdot}000\},$$

$$\mathbf{v}^{(3)} = \{\ 1{\cdot}000 \quad 1{\cdot}000 \quad 1{\cdot}000 \quad 0{\cdot}000\},$$

$$\mathbf{Bv}^{(3)} = \{\ 1{\cdot}000 \quad 1{\cdot}000 \quad 1{\cdot}000 \quad 0{\cdot}000\}.$$

Hence the second characteristic value is

$$\lambda_2 = 1, \tag{8.7.16}$$

and the corresponding characteristic vector is

$$\mathbf{x}^{(2)} = \{1{\cdot}000 \quad 1{\cdot}000 \quad 1{\cdot}000 \quad 0{\cdot}000\}. \tag{8.7.17}$$

It is possible to eliminate the effects of dominant characteristic values from a matrix in such a way that simple iteration produces sub-dominant values. Techniques for doing this are known as 'deflation' of the matrix. We shall describe one method which reduces the order of the matrix by one for each value eliminated. There are several other methods.† Suppose that the dominant characteristic value and column vector are λ and $\mathbf{x}$, and that the largest element of $\mathbf{x}$ is unity, in position s. We form a modified matrix $\mathbf{C}$ such that

$$\mathbf{C} = \mathbf{B} - \mathbf{x}\boldsymbol{\rho}_s(\mathbf{B}). \tag{8.7.18}$$

Then

$$\begin{aligned}\mathbf{Cx} &= \mathbf{Bx} - \mathbf{x}\boldsymbol{\rho}_s(\mathbf{B})\,\mathbf{x} \\ &= \mathbf{Bx} - \mathbf{x}\boldsymbol{\rho}_s(\mathbf{Bx}).\end{aligned} \tag{8.7.19}$$

But

$$\begin{aligned}\boldsymbol{\rho}_s(\mathbf{Bx}) &= \lambda\boldsymbol{\rho}_s(\mathbf{x}) \\ &= \lambda,\end{aligned} \tag{8.7.20}$$

hence

$$\begin{aligned}\mathbf{Cx} &= \mathbf{Bx} - \mathbf{x}\lambda \\ &= \lambda\mathbf{x} - \mathbf{x}\lambda \\ &= \mathbf{0}.\end{aligned} \tag{8.7.21}$$

Thus the component of a trial vector in the direction of $\mathbf{x}$ will be annihilated at the first step of the iteration. In the matrix $\mathbf{C}$, the sth row is zero, as is the sth element of any product of the form $\mathbf{Cv}^{(r)}$, so that all the characteristic vectors of $\mathbf{C}$ have zero sth element, and hence the sth column of $\mathbf{C}$ is irrelevant to the determination of the characteristic vectors of $\mathbf{C}$. We delete the sth row and column of $\mathbf{C}$, and obtain a matrix of order $n-1$ by $n-1$, which we shall denote by $\mathbf{D}$. Iteration with this matrix will yield a characteristic vector corresponding to a dominant value. Suppose $\mathbf{y}$ and μ to be such a pair. Then μ is a characteristic value of $\mathbf{B}$, and we can reverse the deflation to yield a characteristic vector of $\mathbf{B}$. We denote by $\mathbf{z}$ the vector

† Wilkinson, 'The calculation of the latent roots and vectors of matrices on the pilot model of the ACE', *Proc. Camb. Phil. Soc.* **50** (1954), pp. 536–66; Crandall, *Engineering Analysis*, p. 100.

formed from **y** by inserting a zero element between the $(s-1)$th and sth elements. Then the elements, other than the sth, of **Cz** are the same as those of **Dy**, so that

$$\mathbf{Cz} = \mu\mathbf{z}, \tag{8.7.22}$$

$$\begin{aligned}\mathbf{Bz} &= \mathbf{Cz}+\mathbf{x}\boldsymbol{\rho}_s(\mathbf{B})\,\mathbf{z}\\ &= \mu\mathbf{z}+\{\boldsymbol{\rho}_s(\mathbf{B})\,\mathbf{z}\}\mathbf{x}.\end{aligned} \tag{8.7.23}$$

The last step can be made because $\boldsymbol{\rho}_s(\mathbf{B})\,\mathbf{z}$ is a scalar. We may now modify **z** by adding a suitable multiple of **x** to it, and form a characteristic vector of **B**. We put

$$\mathbf{u} = \mathbf{z}+\{\boldsymbol{\rho}_s(\mathbf{B})\,\mathbf{z}\}\mathbf{x}/(\mu-\lambda), \tag{8.7.24}$$

and then

$$\begin{aligned}\mathbf{Bu} &= \mathbf{Bz}+\{\boldsymbol{\rho}_s(\mathbf{B})\,\mathbf{z}\}\mathbf{Bx}/(\mu-\lambda)\\ &= \mu\mathbf{z}+\{\boldsymbol{\rho}_s(\mathbf{B})\,\mathbf{z}\}\mathbf{x}+\{\boldsymbol{\rho}_s(\mathbf{B})\,\mathbf{z}\}\lambda\mathbf{x}/(\mu-\lambda)\\ &= \mu\mathbf{z}+\mu\{\boldsymbol{\rho}_s(\mathbf{B})\,\mathbf{z}\}\mathbf{x}/(\mu-\lambda)\\ &= \mu\mathbf{u}.\end{aligned} \tag{8.7.25}$$

This process can be applied to unsymmetric matrices, although we shall exemplify it only for a symmetric matrix. It is, to some extent, similar to the re-orthogonalisation process in that the effect of multiplication by **C** is to multiply by **B** and then subtract a multiple of **x** from the product. It has the advantage of not requiring a knowledge of the row-vectors, and of leading to the use of smaller and smaller matrices, but it is probably somewhat more sensitive to errors in the estimated characteristic vectors.

For the symmetric matrix of equation (8.5.10), estimates for the dominant characteristic value and vector are

$$\lambda = 73{\cdot}335\,849,$$

$$\mathbf{x} = \{0{\cdot}089\,578 \quad 0{\cdot}303\,300 \quad 0{\cdot}624\,378 \quad 1\}. \tag{8.7.26}$$

The largest element is unity, in position 4, so that the value of s is 4. The deflated matrix **C** is given by

$$c_{ij} = b_{ij} - x_{i4}b_{sj},$$

whence

$$\mathbf{C} = \begin{bmatrix} 0{\cdot}641\,688 & 0{\cdot}745\,908 & 0{\cdot}312\,660 & -0{\cdot}478\,900\\ 0{\cdot}786\,800 & 1{\cdot}753\,800 & 0{\cdot}901\,000 & -1{\cdot}165\,000\\ 0{\cdot}502\,488 & 1{\cdot}258\,708 & 1{\cdot}268\,660 & -1{\cdot}218\,900\\ 0 & 0 & 0 & 0 \end{bmatrix}. \tag{8.7.27}$$

We work with the three-by-three matrix formed from the first three rows and columns of this, namely

$$\mathbf{D} = \begin{bmatrix} 0{\cdot}641\,688 & 0{\cdot}745\,908 & 0{\cdot}312\,660\\ 0{\cdot}786\,800 & 1{\cdot}753\,800 & 0{\cdot}901\,000\\ 0{\cdot}502\,488 & 1{\cdot}258\,708 & 1{\cdot}268\,660 \end{bmatrix}. \tag{8.7.28}$$

We derive the following sequence of vectors:

$$
\begin{aligned}
\mathbf{v}^{(0)} &= \{1 \quad 1 \quad 1\},\\
\mathbf{D}\mathbf{v}^{(0)} &= \{1{\cdot}700\,256 \quad 3{\cdot}441\,600 \quad 3{\cdot}029\,856\},\\
\mathbf{v}^{(1)} &= \{0{\cdot}494\,031 \quad 1{\cdot}000\,000 \quad 0{\cdot}880\,363\},\\
\mathbf{D}\mathbf{v}^{(1)} &= \{1{\cdot}338\,176 \quad 2{\cdot}935\,711 \quad 2{\cdot}623\,834\},\\
\mathbf{v}^{(2)} &= \{0{\cdot}455\,827 \quad 1{\cdot}000\,000 \quad 0{\cdot}893\,764\},\\
\mathbf{D}\mathbf{v}^{(2)} &= \{1{\cdot}317\,851 \quad 2{\cdot}917\,726 \quad 2{\cdot}621\,638\},\\
\mathbf{v}^{(3)} &= \{0{\cdot}451\,671 \quad 1{\cdot}000\,000 \quad 0{\cdot}898\,521\},\\
\mathbf{D}\mathbf{v}^{(3)} &= \{1{\cdot}316\,671 \quad 2{\cdot}918\,742 \quad 2{\cdot}625\,585\},\\
\mathbf{v}^{(4)} &= \{0{\cdot}451\,109 \quad 1{\cdot}000\,000 \quad 0{\cdot}899\,560\},\\
\mathbf{D}\mathbf{v}^{(4)} &= \{1{\cdot}316\,636 \quad 2{\cdot}919\,236 \quad 2{\cdot}626\,621\},\\
\mathbf{v}^{(5)} &= \{0{\cdot}451\,021 \quad 1{\cdot}000\,000 \quad 0{\cdot}899\,763\},\\
\mathbf{D}\mathbf{v}^{(5)} &= \{1{\cdot}316\,643 \quad 2{\cdot}919\,350 \quad 2{\cdot}626\,834\},\\
\mathbf{v}^{(6)} &= \{0{\cdot}451\,006 \quad 1{\cdot}000\,000 \quad 0{\cdot}899\,801\}.
\end{aligned}
$$

We take $\mathbf{v}^{(6)}$ as $\mathbf{y}$ and form $\mathbf{z}$, $\boldsymbol{\rho}_4(\mathbf{B})\,\mathbf{z}$ and $\mathbf{u}$. For our estimate of μ we take the last three scale factors and apply Aitken's process, which yields

$$
\begin{aligned}
\mu &= 2{\cdot}919\,350 - (0{\cdot}000\,114)^2/(-0{\cdot}000\,380)\\
&= 2{\cdot}919\,384.
\end{aligned}
$$

Then

$$
\left.
\begin{aligned}
\mathbf{z} &= \{0{\cdot}451\,006 \quad 1{\cdot}000\,000 \quad 0{\cdot}899\,801 \quad 0{\cdot}000\,000\},\\
\boldsymbol{\rho}_4(\mathbf{B})\,\mathbf{z} &= 42{\cdot}798\,054,\\
\boldsymbol{\rho}_4(\mathbf{B})\,\mathbf{z}/(\mu-\lambda) &= -0{\cdot}607\,785,\\
\mathbf{u} &= \{0{\cdot}396\,562 \quad 0{\cdot}815\,659 \quad 0{\cdot}520\,313 \quad -0{\cdot}607\,785\}.
\end{aligned}
\right\}
\qquad (8.7.29)
$$

A detailed account of a number of variants of the methods described in this section is given by Wilkinson.† This account was written before the methods which depend on rotation to triple diagonal form, which are described in the next chapter, were known. It includes descriptions of techniques for treating characteristic values of equal modulus, and also for dealing with those unsymmetric matrices in which the dominant values form a complex pair.

EXAMPLES 8.7

1. Use the characteristic vector

$$\mathbf{x} = \{0{\cdot}097\,672 \quad -0{\cdot}532\,174 \quad 1\}$$

to find the remaining characteristic value and vector of the matrix of Ex. 7.8.2. Use iteration with repeated orthogonalisation together with Aitken's δ^2 process.

† Wilkinson, *Proc. Camb. Phil. Soc.* **50** (1954), pp. 536–66.

2. Use the characteristic value and vector

$$\lambda = 1{\cdot}500214, \qquad \mathbf{x} = \{1 \quad 0{\cdot}570172 \quad 0{\cdot}406779 \quad 0{\cdot}318141\}$$

to find the next characteristic value and vector of the matrix of Ex. 8.3.2. Use deflation, iteration and Aitken's δ^2 process. (To avoid loss of accuracy due to cancellation of leading digits work to seven places of decimals when forming $\mathbf{B}$ and $\mathbf{x}\boldsymbol{\rho}_1(\mathbf{B})$.)

3. Find the next characteristic value and vector of the matrix of Ex. 8.3.2 by deflating the matrix $\mathbf{C}_3$ obtained in the previous example. This is an example on the deflation of an unsymmetric matrix. Notice that the two characteristic vectors which are obtained for the deflated matrix are not orthogonal. Why is this?

CHAPTER 9

DIRECT METHODS FOR CHARACTERISTIC VALUE PROBLEMS

I should show you in a moment how to grapple with the question,
And you'd really be astonished at the force of my suggestion. *Ruddigore*

9.1 Introduction

A direct method for finding characteristic values is one in which values, which are exact apart from the inevitable rounding off errors, are found after a certain finite number of steps. Direct methods may thus be contrasted with iterative methods, which proceed by means of a sequence of approximate values. In this chapter we present one direct method by which the characteristic values of a symmetric matrix may be found, and another method for finding the characteristic values of an unsymmetric matrix. Thus the chapter is not, and is not intended to be, a compendium of methods; for this the reader must look elsewhere.†

Both of the methods presented depend on the reduction of the matrix to 'triple-diagonal form'. The technique applicable to symmetric matrices is described in § 9.3, and that applicable to unsymmetric matrices in § 9.8. When a symmetric matrix has been reduced to a symmetric triple diagonal form its characteristic values may be found by using certain convenient properties of what are called 'Sturm's sequences'. These are described in §§ 9.4 and 9.5. A method for finding the characteristic *vectors* of the reduced, triple-diagonal matrix, and the corresponding vectors of the original matrix, is described in §§ 9.6 and 9.7.

The triple-diagonal matrix obtained by reducing an unsymmetric matrix does not always have the properties necessary for the use of Sturm's sequences, but the reduction to triple-diagonal form still provides a convenient way of finding the characteristic equation of the original matrix. A method for finding the roots of any polynomial equation, and in particular the characteristic equation of an unsymmetric matrix, is described in § 9.9.

9.2 Permissible transformations—strict equivalence

The set of all matrices of the form $\mathbf{C}-\lambda\mathbf{A}$ is called the matrix 'pencil' $\mathbf{C}-\lambda\mathbf{A}$. The 'characteristic polynomial' of this pencil is the polynomial obtained by expanding the determinant $|\mathbf{C}-\lambda\mathbf{A}|$. The characteristic polynomial will not be changed in essence if it is multiplied by a non-zero constant, unless its zeros are very

† See, for example, E. Bodewig, *Matrix Calculus* (North Holland Publishing Co., 2nd edn, 1959); V. N. Faddeeva, *Computational Methods of Linear Algebra* (Dover, 1959); U.S. National Bureau of Standards, Appl. Math. Ser., nos. 29, 39, 49; *Modern Computing Methods* (H.M.S.O., 1959).

sensitive to small changes in its coefficients, such as rounding off might cause. (Polynomials which show such sensitivity are called 'ill-conditioned' and the same term is applied to those particular zeros which are disproportionately changed.) In consequence, multiplying the matrix pencil on the left or the right by matrices whose determinants are non-zero constants (that is, independent of λ) does not essentially change the characteristic polynomial. Thus if

$$|\mathbf{P}| \neq 0 \quad \text{and} \quad |\mathbf{Q}| \neq 0, \tag{9.2.1}$$

then

$$|\mathbf{PCQ} - \lambda\mathbf{PAQ}| = |\mathbf{P}|\,.\,|\mathbf{C} - \lambda\mathbf{A}|\,.\,|\mathbf{Q}|, \tag{9.2.2}$$

and the two sides vanish for the same values of λ. Such a transformation was used in § 8.2 to reduce the pencil to the standard form $\mathbf{B} - \lambda\mathbf{I}$.

A transformation of the type

$$\mathbf{S} - \lambda\mathbf{M} = \mathbf{P}(\mathbf{C} - \lambda\mathbf{A})\,\mathbf{Q} \tag{9.2.3}$$

in which $\mathbf{P}$ and $\mathbf{Q}$ may contain λ but have constant non-zero determinants is called an 'equivalence' transformation. If $\mathbf{P}$ and $\mathbf{Q}$ do not involve λ, they define a 'strict equivalence'.

Although the characteristic values of a matrix pencil are unchanged by equivalences, the characteristic vectors are changed. In fact, if $\mathbf{y}$ is a characteristic vector of $\mathbf{S} - \lambda\mathbf{M}$, so that

$$(\mathbf{S} - \lambda\mathbf{M})\,\mathbf{y} = \mathbf{0} = \mathbf{P}(\mathbf{C} - \lambda\mathbf{A})\,\mathbf{Q}\mathbf{y}, \tag{9.2.4}$$

then

$$(\mathbf{C} - \lambda\mathbf{A})\,(\mathbf{Q}\mathbf{y}) = \mathbf{0}, \tag{9.2.5}$$

and therefore

$$\mathbf{x} = \mathbf{Q}\mathbf{y} \tag{9.2.6}$$

is a characteristic vector of $\mathbf{C} - \lambda\mathbf{A}$. In applications it is essential to have the transformation $\mathbf{Q}$ linking $\mathbf{x}$ and $\mathbf{y}$ independent of λ, and for this to hold it is necessary to use strict equivalence transformations.

The application of a strict equivalence transformation to the matrix pencil $\mathbf{B} - \lambda\mathbf{I}$ will, in general, change it to the form $\mathbf{C} - \lambda\mathbf{A}$, since

$$\mathbf{P}(\mathbf{B} - \lambda\mathbf{I})\,\mathbf{Q} = (\mathbf{PBQ}) - \lambda(\mathbf{PQ}). \tag{9.2.7}$$

This is undesirable, if only because it increases the storage requirements. This complication does not occur if

$$\mathbf{PQ} = \mathbf{I}, \quad \text{that is} \quad \mathbf{P} = \mathbf{Q}^{-1}. \tag{9.2.8}$$

Strict equivalences which satisfy this equation are called 'similarity transformations', and may be written

$$\mathbf{S} - \lambda\mathbf{M} = \mathbf{Q}^{-1}(\mathbf{C} - \lambda\mathbf{A})\,\mathbf{Q}. \tag{9.2.9}$$

For a similarity transformation

$$|\mathbf{S} - \lambda\mathbf{M}| = |\mathbf{C} - \lambda\mathbf{A}|. \tag{9.2.10}$$

In general, an equivalence transformation changes a symmetric pencil into an unsymmetric pencil. Those which preserve symmetry satisfy the equation

$$\mathbf{P} = \mathbf{Q}'. \tag{9.2.11}$$

For such transformations the transform of $\mathbf{C}-\lambda\mathbf{A}$ is

$$\mathbf{S}-\lambda\mathbf{M} = \mathbf{Q}'(\mathbf{C}-\lambda\mathbf{A})\mathbf{Q}. \tag{9.2.12}$$

To preserve both the symmetry and the simplicity of the characteristic value problem for a single matrix, transformations must be used which satisfy both equation (9.2.8) and equation (9.2.11) that is

$$\mathbf{P} = \mathbf{Q}^{-1} = \mathbf{Q}', \quad \text{so that} \quad \mathbf{Q}'\mathbf{Q} = \mathbf{I}. \tag{9.2.13}$$

Matrices $\mathbf{Q}$ for which this relation holds are called 'rotation' matrices, the transformations of space which they represent being rotations. Such matrices are also called 'orthogonal' (see § 3.7).

The transformations which are of most use in simplifying the characteristic value problem for a single matrix are rotations, and it will be shown in the next section that any symmetric matrix may be rotated to triple-diagonal form. The corresponding reduction for unsymmetric matrices will be discussed in § 9.8.

EXAMPLES 9.2

1. (*a*) Show that the product of any number of rotation matrices is another rotation matrix.

(*b*) Show that the determinant of a rotation matrix is of value ± 1.

2. Show that any matrix $\mathbf{Q}$ obtained from the unit matrix $\mathbf{I}_n$ by altering the four elements in positions $i, i; i, j; j, i; j, j$ according to the prescription

$$q_{ii} = \cos\theta \quad q_{ij} = -\sin\theta,$$
$$q_{ji} = \sin\theta \quad q_{jj} = \cos\theta,$$

is a rotation matrix.

3. If

$$\mathbf{B} = \begin{bmatrix} 1 & -2 & 2{\cdot}5 \\ -2 & 8 & -15{\cdot}5 \\ 2{\cdot}5 & -15{\cdot}5 & 29 \end{bmatrix}, \quad \mathbf{Q} = \begin{bmatrix} 0 & 0 & 0 \\ 0 & \cos\theta & -\sin\theta \\ 0 & \sin\theta & \cos\theta \end{bmatrix},$$

find the angle θ which is such that the matrix $\mathbf{Q}'\mathbf{B}\mathbf{Q}$ has zeros in position 2,3 or 3,2.

Show that the resulting matrix $\mathbf{Q}'\mathbf{B}\mathbf{Q}$ is 'more diagonal' than $\mathbf{B}$ in the sense that the sum of the squares of its off-diagonal elements is smaller than that for $\mathbf{B}$.

The matrices $\mathbf{Q}$ introduced in Ex. 9.2.2 form the basis of the method of 'diagonalisation by successive rotations' in which the technique exemplified in Ex. 9.2.3 is applied again and again to eliminate all the off-diagonal elements. The practical application of this method is dealt with elsewhere.†

9.3 Householder's reduction to triple-diagonal form

Any symmetric matrix may be rotated to triple diagonal form, but there is no simple technique for calculating the elements of the rotation matrix. It has been shown by Givens‡ that a suitable matrix could be built up as the product of plane

† Crandall, *Engineering Analysis*, p. 118.

‡ W. Givens, 'Numerical computation of the characteristic values of a real symmetric matrix', *Oak Ridge Nat. Lab. Rep.* 1574 (1954).

rotations, each transformation being chosen to eliminate one pair of the elements not on the three diagonals to be retained. This construction was shortened by Householder and Bauer,† who devised a simple rotation matrix which reduces a whole row and column in one step. This is the process we describe in detail (see also Wilkinson).‡

Each rotation matrix used is of the simple form

$$\mathbf{R} = \mathbf{I} - 2\mu\mathbf{z}\mathbf{z}', \tag{9.3.1}$$

where

$$\mu = 1/(\mathbf{z}'\mathbf{z}). \tag{9.3.2}$$

For such a matrix

$$\mathbf{R}' = \mathbf{R} \tag{9.3.3}$$

and

$$\begin{aligned}\mathbf{R}'\mathbf{R} &= \mathbf{I} - 4\mu\mathbf{z}\mathbf{z}' + 4\mu^2\mathbf{z}\mathbf{z}'\mathbf{z}\mathbf{z}' \\ &= \mathbf{I} - 4\mu\mathbf{z}\mathbf{z}' + 4\mu^2(\mathbf{z}'\mathbf{z})\,\mathbf{z}\mathbf{z}' \\ &= \mathbf{I}.\end{aligned} \tag{9.3.4}$$

The first rotation is used to reduce the first row and column to triple diagonal form. The post-multiplication by $\mathbf{R}$ reduces the row, the pre-multiplication by $\mathbf{R}'$ the column.

Now

$$\mathbf{B}\mathbf{R} = \mathbf{B} - 2\mu\mathbf{B}\mathbf{z}\mathbf{z}', \tag{9.3.5}$$

and

$$\boldsymbol{\rho}_1(\mathbf{B}\mathbf{R}) = \boldsymbol{\rho}_1(\mathbf{B}) - 2\mu(\boldsymbol{\rho}_1(\mathbf{B})\,\mathbf{z})\,\mathbf{z}'. \tag{9.3.6}$$

The matrix $\mathbf{R}$ is to be chosen in such a way that the third and subsequent elements of $\boldsymbol{\rho}_1(\mathbf{B}\mathbf{R})$ are zero. Therefore the third and subsequent elements of $\mathbf{z}'$ must be proportional to the corresponding elements of the first row of $\mathbf{B}$. The constant of proportionality may be taken as unity, so that

$$z_j = b_{ij} \quad \text{for} \quad j \geqslant 3. \tag{9.3.7}$$

The corresponding elements of $\boldsymbol{\rho}_1(\mathbf{B}\mathbf{R})$ will then be zero if

$$1 = 2\mu\boldsymbol{\rho}_1(\mathbf{B})\,\mathbf{z}. \tag{9.3.8}$$

This is one equation for the remaining unknowns z_1 and z_2.

The matrix $\mathbf{B}\mathbf{R}$ has the form

$$\mathbf{B}\mathbf{R} = \mathbf{C} = \begin{bmatrix} c_{11} & c_{12} & 0 & \cdots & 0 \\ c_{21} & c_{22} & c_{23} & \cdots & c_{2n} \\ \cdots & \cdots & \cdots & \cdots & \cdots \\ c_{n1} & c_{n2} & c_{n3} & \cdots & c_{nn} \end{bmatrix}, \tag{9.3.9}$$

and the matrix $\mathbf{R}'\mathbf{B}\mathbf{R}$ is

$$\mathbf{R}'\mathbf{B}\mathbf{R} = \mathbf{R}'\mathbf{C} = \mathbf{C} - 2\mu\mathbf{z}(\mathbf{z}'\mathbf{C}), \tag{9.3.10}$$

so that

$$\boldsymbol{\rho}_1(\mathbf{R}'\mathbf{B}\mathbf{R}) = \boldsymbol{\rho}_1(\mathbf{C}) - 2\mu z_1(\mathbf{z}'\mathbf{C}). \tag{9.3.11}$$

† A. S. Householder and F. L. Bauer, 'On certain methods for expanding the characteristic polynomial', *Num. Math.* **1** (1959), pp. 29–37.

‡ J. H. Wilkinson, 'Householder's method for the solution of the algebraic eigenproblem', *Comput. J.* **3** (1960), pp. 23–7.

Therefore in order that the premultiplication by $\mathbf{R}'$ shall not replace elements in the first row, we must choose

$$z_1 = 0. \tag{9.3.12}$$

The unknown z_2 may now be determined from equation (9.3.8). This may be written

$$2\boldsymbol{\rho}_1(\mathbf{B})\,\mathbf{z} = \mathbf{z}'\mathbf{z}. \tag{9.3.13}$$

Substituting for the z's from equation (9.3.7) we find

$$2(b_{12}z_2 + b_{13}^2 + \ldots + b_{1n}^2) = z_2^2 + b_{13}^2 + \ldots + b_{1n}^2, \tag{9.3.14}$$

whence

$$z_2 = b_{12} \pm \sqrt{(b_{12}^2 + b_{13}^2 + \ldots + b_{1n}^2)}. \tag{9.3.15}$$

We define

$$S^2 = \sum_{j=2}^{n} b_{1j}^2, \tag{9.3.16}$$

and then

$$\mathbf{z}' = [0, \quad b_{12}+S, \quad b_{13}, \ldots, b_{1n}], \tag{9.3.17}$$

$$\mu = 1/(2S(S+b_{12})). \tag{9.3.18}$$

The transformed matrix $\overline{\mathbf{B}}$ is given by

$$\overline{\mathbf{B}} = \mathbf{B} - 2\mu(\mathbf{Bz})\,\mathbf{z}' - 2\mu\mathbf{z}(\mathbf{z}'\mathbf{B}) + 4\mu^2\mathbf{z}(\mathbf{z}'\mathbf{Bz})\,\mathbf{z}'. \tag{9.3.19}$$

This formula may be simplified by putting

$$\left.\begin{aligned} \mathbf{y} &= \mathbf{Bz}, \\ q &= \mathbf{z}'\mathbf{y}, \\ \mathbf{w} &= 2\mu(\mathbf{y} - \mu q\mathbf{z}), \end{aligned}\right\} \tag{9.3.20}$$

when equation (9.3.19) becomes

$$\overline{\mathbf{B}} = \mathbf{B} - \mathbf{wz}' - \mathbf{zw}'. \tag{9.3.21}$$

In terms of the elements of the matrices this is

$$\bar{b}_{ij} = b_{ij} - w_i z_j - z_i w_j. \tag{9.3.22}$$

We notice that

$$\bar{b}_{11} = b_{11}. \tag{9.3.23}$$

We take for $\mathbf{B}$ the matrix of equation (8.2.24), namely

$$\mathbf{B} = \begin{bmatrix} 5 & -12 & 8 \\ -12 & 34 & -33 \\ 8 & -33 & 54 \end{bmatrix}.$$

Then

$$\begin{aligned}
S^2 &= (-12)^2 + 8^2 = 208, \quad S = 14{\cdot}422\,205, \\
\mathbf{z} &= \{0 \quad 2{\cdot}422\,205 \quad 8\}, \\
\mathbf{y} &= \{34{\cdot}933\,540 \quad -181{\cdot}645\,030 \quad 352{\cdot}067\,235\}, \\
\mu &= 0{\cdot}014\,312\,894, \quad q = 2376{\cdot}556\,380, \quad \mu q = 34{\cdot}015\,400, \\
\mathbf{w} &= \{1{\cdot}000\,000 \quad -7{\cdot}558\,276 \quad 2{\cdot}288\,461\}, \\
\bar{b}_{11} &= 5 - 0 - 0 = 5, \\
\bar{b}_{12} &= -12 - (1{\cdot}000\,000)\,(2{\cdot}422\,205) - 0(-7{\cdot}558\,276) \\
&= -14{\cdot}422\,205 = -S, \\
\bar{b}_{13} &= 8 - (1{\cdot}000\,000)\,(8) = 0,
\end{aligned}$$

and so on, to yield

$$\overline{\mathbf{B}} = \begin{bmatrix} 5 & -14{\cdot}422\,205 & 0 \\ -14{\cdot}422\,205 & 70{\cdot}615\,388 & 21{\cdot}923\,086 \\ 0 & 21{\cdot}923\,086 & 17{\cdot}384\,624 \end{bmatrix}.$$

A check is afforded by the equations

$$w_1 = 1, \quad \bar{b}_{12} = -S.$$

In general, when the first row and column have been reduced, the elements of the second row and column have to be eliminated in the same way. At this stage, to avoid spoiling the reduced first row and column, the vector **z** must be chosen to have its first element zero. The first row and column of **R** are then those of the unit matrix, and we effectively operate with the last $(n-1)$ rows and columns only. We may, therefore, at this stage, extract the trailing principal minor of $\overline{\mathbf{B}}$ and work with an $(n-1)$-dimensional space, treating the first row and column of the smaller matrix as we did those of the original matrix. The amount of work thus decreases as we treat later rows and columns, amounting to about $2(n-r)^2$ multiplications to clear out the rth row and column, in all about $n^3/3$ multiplications. It should be noted that only the first $(n-2)$ rows and columns need to be reduced, since a triple diagonal matrix has a two-by-two square in its bottom right-hand corner.

The equations used at the rth stage are

$$S^2 = \sum_{j=r+1}^{n} b_{rj}^2, \tag{9.3.24}$$

$$z_j = \begin{cases} 0 & \text{for } j < r+1, \\ b_{r,r+1} + S & \text{for } j = r+1, \\ b_{rj} & \text{for } j > r+1. \end{cases} \tag{9.3.25}$$

$$\mu = 1/(2S(S + b_{r,r+1})). \tag{9.3.26}$$

$$\left.\begin{aligned} \mathbf{y} &= \mathbf{Bz}, \\ q &= \mathbf{z}'\mathbf{y}, \\ \mathbf{w} &= 2\mu(\mathbf{y} - \mu q\mathbf{z}), \\ \overline{\mathbf{B}} &= \mathbf{B} - \mathbf{wz}' - \mathbf{zw}'. \end{aligned}\right\} \tag{9.3.27}$$

Here, **B** stands for the partially reduced matrix and $\overline{\mathbf{B}}$ for the matrix reduced one stage further. In order to avoid loss of accuracy by cancellation of leading digits when $b_{r,r+1}$ dominates its partial row, it is advisable to choose S to have the same sign as $b_{r,r+1}$ in such cases, and it is then easier to make this the working rule in all cases.

Before exemplifying the calculation as a whole we shall describe a method of checking the steps of the calculation. We again denote the column of ones by $\boldsymbol{l}$, so that

$$\boldsymbol{l}' = [1, 1, \ldots, 1]. \tag{9.3.28}$$

The associativity of matrix multiplications then enables us to write the following equation connecting column sums,

$$\begin{aligned} \boldsymbol{l}'\mathbf{y} &= \boldsymbol{l}'(\mathbf{Bz}) \\ &= \boldsymbol{l}'\mathbf{Bz} \\ &= (\boldsymbol{l}'\mathbf{B})\,\mathbf{z}. \end{aligned} \tag{9.3.29}$$

This expresses the sum of the elements of the column $\mathbf{y}$ in terms of the sums of the elements down the columns of $\mathbf{B}$, and enables us to check the formation of $\mathbf{y}$. The formation of $\mathbf{w}$ can be checked by using the equation

$$\boldsymbol{l}'\mathbf{w} = 2\mu(\boldsymbol{l}'\mathbf{y} - \mu q \boldsymbol{l}'\mathbf{z}). \tag{9.3.30}$$

The sums down the columns of $\overline{\mathbf{B}}$, namely $\boldsymbol{l}'\overline{\mathbf{B}}$, should satisfy the equation

$$\boldsymbol{l}'\overline{\mathbf{B}} = \boldsymbol{l}'\mathbf{B} - (\boldsymbol{l}'\mathbf{w})\,\mathbf{z} - (\boldsymbol{l}'\mathbf{z})\,\mathbf{w}. \tag{9.3.31}$$

The right-hand side of this should be formed before $\overline{\mathbf{B}}$ and used to check $\overline{\mathbf{B}}$ as it is formed. To check the right-hand equation of (9.3.31), which is a row matrix, the sum of its elements can be checked by using the equation

$$\{\boldsymbol{l}'\mathbf{B} - (\boldsymbol{l}'\mathbf{w})\,\mathbf{z} - (\boldsymbol{l}'\mathbf{z})\,\mathbf{w}\}\,\boldsymbol{l} = \boldsymbol{l}'\mathbf{B}\boldsymbol{l} - 2(\boldsymbol{l}'\mathbf{z})\,(\boldsymbol{l}'\mathbf{w}). \tag{9.3.32}$$

The example on p. 330 includes these checks and shows a practicable layout for the calculation when desk calculators are used. Only half of each symmetric matrix is recorded, and vectors are shown as being of length $n-r+1$ at the rth stage.

For the matrix of equation (8.3.21)† the calculation would be recorded as shown on p. 330.

In calculating the column sums of a symmetric matrix such as $\mathbf{B}$, it must be remembered that only half the matrix is recorded, and when the diagonal is encountered while proceeding down a column, the path must continue along the row to the right. For example, in forming the sum of the elements in the second column, the elements used come from the positions b_{12}, b_{22}, b_{23}, b_{24}. In checking the first partially reduced matrix, four columns need to be checked. Of these four columns only three are relevant to the next step, and since there is a non-zero element at the top of the second column, a new column sum must be prepared and used in the checking of the next step of reduction. The extra column sum is written below the sum used in checking the first transformation. The part of $\overline{\mathbf{B}}$, and the check sums used in the next step are enclosed in rules.

The triple-diagonal matrix which has the same characteristic values as $\mathbf{B}$ is therefore

$$\overline{\overline{\mathbf{B}}} = \begin{bmatrix} 1 & -5{\cdot}385\,165 & 0 & 0 \\ -5{\cdot}385\,165 & 71{\cdot}310\,352 & 10{\cdot}599\,618 & 0 \\ 0 & 10{\cdot}599\,618 & 4{\cdot}176\,061 & -0{\cdot}462\,331 \\ 0 & 0 & -0{\cdot}462\,331 & 0{\cdot}513\,586 \end{bmatrix}. \tag{9.3.33}$$

† See the Appendix for the list of examples on this matrix.

	B				z	y	w		
	1	2	3	4	0	39·770 310	1	$S^2 = 29$	$\mu q = 41{\cdot}456\,689$
		6	10	14	7·385 165	130·310 990	−4·421 726	$S = 5{\cdot}385\,165$	$(l'\mathbf{B})\,\mathbf{z} = 817{\cdot}325\,280$
			20	30	3	253·851 650	3·255 733	$2\mu = 0{\cdot}025\,144\,373$	$2\mu(l'\mathbf{y}-\mu q l'\mathbf{z}) = 5{\cdot}556\,000$
				50	4	393·392 310	5·721 993	$q = 3297{\cdot}492\,352$	$l'\mathbf{B}l-2(l'\mathbf{z})\,(l'\mathbf{w}) = 43{\cdot}152\,047$
		$l'\mathbf{B}$						$l'\mathbf{B}-(l'\mathbf{z})\,\mathbf{w}-(l'\mathbf{w})\,\mathbf{z}$	
	10	32	63	98	14·385 165	817·325 280	5·556 000	−4·385 165 54·575 281	−0·502 256 −6·535 813
	1	−5·385 165	0	0				$S^2 = 112{\cdot}351\,911$	$\mu q = 2{\cdot}018\,328$
		71·310 352	−0·778 947	−10·570 959	0	120·608 452	1	$S = -10{\cdot}599\,618$	$(l'\overline{\mathbf{B}})\,\mathbf{z} = 74{\cdot}804\,757$
			0·465 602	−0·188 911	−11·378 565	−3·300 913	0·163 046	$2\mu = 0{\cdot}008\,291\,294$	$2\mu(l'\mathbf{y}-\mu q l'\mathbf{z}) = 0{\cdot}987\,544$
				4·224 056	−10·570 958	−42·502 782	−0·175 503	$q = 486{\cdot}854\,829$	$l'\overline{\mathbf{B}}l-2(l'\mathbf{z})\,(l'\mathbf{w}) = 96{\cdot}274\,574$
			$l'\overline{\mathbf{B}}$					$l'\overline{\mathbf{B}}-(l'\mathbf{z})\,\mathbf{w}-(l'\mathbf{w})\,\mathbf{z}$	
†	−4·385 165	54·575 281	−0·502 256	−6·535 813					
‡		59·960 447	−0·502 256	−6·353 813	−21·949 523	74·804 757	0·987 543	81·909 970 14·313 348	0·051 255
		71·310 352	10·599 618	0					
			4·176 061	−0·462 331					
				0·513 586					
		81·909 971	14·313 349	0·051 255					

† This level is check sum of first stage. ‡ This level is for second stage.

Wilkinson† has obtained bounds on the error that is introduced by rounding off. The effect of rounding off at any stage of the reduction to triple-diagonal form will be the same as the addition of a small perturbing matrix to the matrix currently being treated. By the nature of the method being used the perturbing matrix will be symmetric and therefore a bound on the effect on the characteristic value may be obtained using the inequality (8.3.26). By deriving upper bounds on the values of the elements being found, together with some simple estimates of the largest characteristic value of the perturbing matrix, we may show that the effect of rounding off during one stage of the process will change the characteristic value by not more than

$$\{(10\sqrt{\mu}+1)\sqrt{n}+4\mu+1\}e.$$

Here e is the maximum rounding error in an operation, namely 0·5 in the last decimal place recorded. The addition of quantities like this, one for each stage, gives an upper bound to disturbances caused to the characteristic values.

For the matrix of the last example the values of μ on the two stages were

$$\mu = 0{\cdot}0126 \quad \text{and} \quad \mu = 0{\cdot}00415,$$

and the two contributions to the error estimate are

$$[\{(10)\,(0{\cdot}11)+1\}\,2+0{\cdot}05+4]\,(0{\cdot}5)\,(10^{-6}) = (4)\,10^{-6},$$

$$[\{(10)\,(0{\cdot}065)+1\}\,2+0{\cdot}02+4]\,(0{\cdot}5)\,(10^{-6}) = (3{\cdot}5)\,10^{-6}.$$

The total bound on the perturbation due to the rotation is eight in the last decimal place.

EXAMPLES 9.3

1. Use Householder's method to reduce to triple-diagonal form the matrix **B** of Ex. 8.2.2, namely

$$\mathbf{B} = \begin{bmatrix} 1 & -2 & 2{\cdot}5 \\ -2 & 8 & -15{\cdot}5 \\ 2{\cdot}5 & -15{\cdot}5 & 29 \end{bmatrix}.$$

2. Use Householder's method to reduce to triple-diagonal form the matrix **B** of Ex. 8.3.2, namely

$$\mathbf{B} = \begin{bmatrix} 1{\cdot}000000 & 0{\cdot}500000 & 0{\cdot}333333 & 0{\cdot}250000 \\ 0{\cdot}500000 & 0{\cdot}333333 & 0{\cdot}250000 & 0{\cdot}200000 \\ 0{\cdot}333333 & 0{\cdot}250000 & 0{\cdot}200000 & 0{\cdot}166667 \\ 0{\cdot}250000 & 0{\cdot}200000 & 0{\cdot}166667 & 0{\cdot}142857 \end{bmatrix}.$$

Cancellation of leading digits occurs at the second stage of the reduction, and extra figures must be used to avoid loss of accuracy.)

† Wilkinson, 'Error analysis of eigenvalue techniques based on orthogonal transformations', *J. Soc. Indust. Appl. Math.* **10** (1962), pp. 162–95.

9.4 The characteristic polynomials of triple-diagonal matrices—Sturm's sequences

The typical triple diagonal matrix pencil has the form

$$\mathbf{B}-\lambda\mathbf{I} = \begin{bmatrix} b_1-\lambda & c_1 & 0 & 0 & \dots & 0 \\ a_2 & b_2-\lambda & c_2 & 0 & \dots & 0 \\ 0 & a_3 & b_3-\lambda & c_3 & \dots & 0 \\ \dots & \dots & \dots & \dots & \dots & \dots \\ 0 & 0 & 0 & 0 & \dots & b_n-\lambda \end{bmatrix}. \tag{9.4.1}$$

We denote by P_r the determinant formed from the first r rows and r columns, the leading principal minor of order r. Expanding this in terms of its last column we find

$$P_r = (b_r-\lambda)P_{r-1} - c_{r-1}\begin{vmatrix} b_1-\lambda & c_1 & 0 & \dots & 0 \\ a_2 & b_2-\lambda & c_2 & \dots & 0 \\ \dots & \dots & \dots & \dots & \dots \\ 0 & 0 & 0 & \dots & c_{r-2} \\ 0 & 0 & 0 & \dots & a_r \end{vmatrix}. \tag{9.4.2}$$

The last determinant can be expanded in terms of its last row and its value is $a_r P_{r-2}$. There is thus a linear relation between any three consecutive leading principal minors, namely

$$P_r = (b_r-\lambda)P_{r-1} - a_r c_{r-1} P_{r-2}. \tag{9.4.3}$$

It is convenient to define a minor of order zero, P_0, with value unity, and the sequence of minors then starts with

$$P_0 = 1, \quad P_1 = b_1-\lambda. \tag{9.4.4}$$

There is thus a sequence of polynomials which is formed of the leading principal minors, namely

$$\left.\begin{aligned} &P_0 = 1, \quad P_1 = b_1-\lambda, \\ &P_r = (b_r-\lambda)P_{r-1} - a_r c_{r-1} P_{r-2} \quad (r = 2, 3, \dots, n). \end{aligned}\right\} \tag{9.4.5}$$

When the product $a_r c_{r-1}$ is positive, as it is for a symmetric matrix, the sequence of polynomials has properties which make it valuable in locating the zeros of P_n (the characteristic values of $\mathbf{B}$). This will be discussed below.

Suppose

$$\mathbf{B} = \begin{bmatrix} 2 & -1 & 0 & 0 \\ -1 & 2 & -1 & 0 \\ 0 & -1 & 2 & -1 \\ 0 & 0 & -1 & 2 \end{bmatrix}. \tag{9.4.6}$$

The sequence of principal minors is determined from

$$P_0 = 1, \quad P_1 = 2-\lambda,$$

$$P_r = (2-\lambda)P_{r-1} - P_{r-2},$$

to be

$$\begin{aligned} P_0 &= 1, \\ P_1 &= 2-\lambda, \\ P_2 &= 3-4\lambda+\lambda^2, \\ P_3 &= 4-10\lambda+6\lambda^2-\lambda^3, \\ P_4 &= 5-20\lambda+21\lambda^2-8\lambda^3+\lambda^4. \end{aligned}$$

The derivatives of the polynomials (9.4.5) with respect to λ can also be found readily, from the relations

$$\left.\begin{aligned} &P'_0 = 0, \quad P'_1 = -1, \\ &P'_r = (b_r-\lambda)\,P'_{r-1} - a_r c_{r-1} P'_{r-2} - P_{r-1}. \end{aligned}\right\} \tag{9.4.7}$$

For the matrix (9.4.6), equations (9.4.7) yield

$$\left.\begin{aligned} P'_0 &= 0, \\ P'_1 &= -1, \\ P'_2 &= (2-\lambda)(-1) - 1(0) - (2-\lambda) \\ &= -4+2\lambda, \\ P'_3 &= (2-\lambda)(-4+2\lambda) - 1(-1) - (3-4\lambda+\lambda^2) \\ &= -10+12\lambda-3\lambda^2, \\ P'_4 &= (2-\lambda)(-10+12\lambda-3\lambda^2) \\ &\quad -1(-4+2\lambda) - (4-10\lambda+6\lambda^2-\lambda^3) \\ &= -20+42\lambda-24\lambda^2+4\lambda^3. \end{aligned}\right\} \tag{9.4.8}$$

With any sequence of polynomials it is possible to associate an integer-valued function, the number of changes of sign in the sequence of numerical values obtained for each value of the argument, λ, of the polynomials. We shall denote this function by $s(\lambda)$.

Consider the sequence of polynomials

$$\left.\begin{aligned} P_0(\lambda) &= 1, \\ P_1(\lambda) &= 2-\lambda, \\ P_2(\lambda) &= 3-4\lambda+\lambda^2, \\ P_3(\lambda) &= 4-10\lambda+6\lambda^2-\lambda^3. \end{aligned}\right\} \tag{9.4.9}$$

Then for $\lambda = 0$ the sequence of values is

$$1,\ 2,\ 3,\ 4$$

and, since there is no change of sign in this sequence,

$$s(0) = 0.$$

For $\lambda = 1$ the sequence of values is

$$1,\ 1,\ 0,\ -1, \tag{9.4.10}$$

and therefore $s(1) = 1.$

For $\lambda = 4$ the sequence is $1, -2, 3, -4,$

and $s(4) = 3.$

When one of the polynomials vanishes, this may make the value of $s(\lambda)$ indeterminate, the value depending upon whether zero is counted as positive or negative. This indeterminacy can only arise if the two values which are on either side of the zero value have the same sign. If they have opposite signs, the sign count is independent of the sign assigned to zero, as in the sequence (9.4.10) where the signs may be taken to be

$$+, +, +, -$$

or

$$+, +, -, -$$

without altering the number of changes of sign.

A 'Sturms sequence' is a sequence of polynomials for which $s(\lambda)$ has the following properties:

(*a*) $s(\lambda)$ alters only as λ passes through a zero of the last polynomial, P_n;

(*b*) there it always alters in the same sense as λ increases (that is, it is a monotonic function of λ); and

(*c*) it alters by the order of the zero.

The fundamental property of a Sturm's sequence is that the number of zeros of $P_n(\lambda)$ between two values of λ is equal to the change in $s(\lambda)$ from one value to the other.

Sufficient conditions for a sequence $P_r(\lambda)$ $(r = 0, 1, 2, \ldots, n)$ to be a Sturm's sequence are:

(i) P_r is exactly of degree r in λ, and $P_0 \neq 0$;

(ii) the leading coefficients of the polynomials (i.e. the coefficients of λ^0 in P_0, λ in P_1, λ^2 in P_2, etc.) alternate in sign;

(iii) when P_r vanishes, the values of P_{r-1} and P_{r+1} are of opposite sign.

These conditions are satisfied by any sequence of the type (9.4.5) for which $a_r c_{r-1}$ is positive.

We shall now sketch a proof that the sequence (9.4.5) does have the properties of a Sturm's sequence. First of all it is clear that $s(\lambda)$ can alter only when λ passes through a zero of one of the polynomials. Therefore, to prove (*a*), it is sufficient to prove that $s(\lambda)$ does not alter when λ passes through a zero of one of the intermediate polynomials $P_r(\lambda)$ with $r < n$. This we may prove as follows. Suppose λ_0 is a zero of $P_r(\lambda)$, then neither $P_{r-1}(\lambda_0)$ nor $P_{r+1}(\lambda_0)$ are zero (for if two adjacent members of the sequence were zero for the same value of λ, all the members would be zero, and this is impossible since $P_0 \neq 0$) and, by (iii), the signs of $P_{r-1}(\lambda_0)$ and $P_{r+1}(\lambda_0)$ are different. Therefore the signs of P_{r-1}, P_r, P_{r+1} at λ_0 are

$$\text{either} \quad +0- \quad \text{or} \quad -0+.$$

Suppose the first to be the case and suppose that P_r increases as it passes through λ_0

(the other possibilities may be handled similarly). Then for values of λ sufficiently close to λ_0 and less than λ_0 the signs would be

$$+ \quad - \quad -,$$

while for sufficiently close values of λ greater than λ_0 the signs will be

$$+ \quad + \quad -.$$

Thus, whether λ is greater or less than λ_0 there is just one change of sign in the triplet of values of P_{r-1}, P_r and P_{r+1}. In other words, this triplet of polynomials will not contribute any change to $s(\lambda)$ as λ passes through λ_0. But no other members of the sequence will contribute any change to $s(\lambda)$ as λ passes through λ_0, and therefore $s(\lambda)$ will not change at all. This proves property (*a*).

To prove property (*b*) we first prove that, for $r = 1, 2, \ldots, n$, the function Q_r defined by

$$Q_r(\lambda) = \frac{P_r(\lambda)}{P_{r-1}(\lambda)} \tag{9.4.11}$$

is a steadily decreasing function. We may prove this by induction. When $r = 1$

$$Q_1(\lambda) = P_1(\lambda) = b_1 - \lambda, \tag{9.4.12}$$

so that

$$Q_1'(\lambda) = -1. \tag{9.4.13}$$

Suppose $Q_{r-1}(\lambda)$ is a decreasing function, then

$$Q_{r-1}'(\lambda) < 0 \tag{9.4.14}$$

(wherever it exists). But equation (9.4.5) shows that

$$Q_r(\lambda) = \frac{P_r(\lambda)}{P_{r-1}(\lambda)} = b_r - \lambda - a_r c_{r-1} \frac{P_{r-2}(\lambda)}{P_{r-1}(\lambda)} = b_r - \lambda - \frac{a_r c_{r-1}}{Q_{r-1}(\lambda)}, \tag{9.4.15}$$

so that

$$Q_r'(\lambda) = -1 + a_r c_{r-1} \frac{Q_{r-1}'(\lambda)}{Q_{r-1}^2(\lambda)}, \tag{9.4.16}$$

and if $Q_{r-1}'(\lambda)$ is negative then so will $Q_r'(\lambda)$ be. This proves the result.

Having proved this we may now prove (*b*). Suppose λ_0 is a simple zero of $P_n(\lambda)$. Then there will be a neighbourhood of λ_0 in which $P_{n-1}(\lambda)$ is non-zero. Suppose first that $P_{n-1}(\lambda)$ is positive. Then, by the result just proved, $P_n(\lambda)$ must change from positive below λ_0 to negative above. Thus the signs of P_{n-1} and P_n will change from

$$+ \; + \quad \text{to} \quad + \; -,$$

and so $s(\lambda)$ will increase by one as λ passes through λ_0. If $P_{n-1}(\lambda)$ is negative in the neighbourhood of λ_0 then $P_n(\lambda)$ will change from negative to positive, so that the signs will change from

$$- \; - \quad \text{to} \quad - \; +,$$

and again $s(\lambda)$ will increase by one. This proves (*b*) and the part of (*c*) referring to simple roots. Multiple roots may be handled by a slight modification of the argument just used. The fundamental property now follows as a simple corollary.

The sequence considered in the last example is Sturmian because it consists of the first four of the polynomials derived from the matrix (9.4.6). For this sequence we have the following table of values:

λ	0	0·5	1	1·5	2	2·5	3	3·5
P_0	1	1·000	1	1·000	1	1·000	1	1·000
P_1	2	1·500	1	0·500	0	$-0·500$	-1	$-1·500$
P_2	3	1·250	0	$-0·750$	-1	$-0·750$	0	1·250
P_3	4	0·375	-1	$-0·875$	0	0·875	1	$-0·375$
$s(\lambda)$	0	0	1	1		2	2	3

This shows that $P_3(\lambda)$ has roots between 0·5 and 1; at 2; and between 3 and 3·5. The roots of $P_3(\lambda)$ are in fact at 0·585 785, 2 and 3·414 217. We note that when the polynomial of highest degree vanishes there is no need to assign a value to $s(\lambda)$.

EXAMPLES 9.4

1. Expand into explicit polynomial form the leading principal minors of the matrices

$$(a)\ \begin{bmatrix} 1-\lambda & 2 & 0 \\ 1 & 2-\lambda & 3 \\ 0 & 2 & 3-\lambda \end{bmatrix}, \qquad (b)\ \begin{bmatrix} 1-\lambda & -2 & 0 \\ 1 & 2-\lambda & 3 \\ 0 & -2 & 3-\lambda \end{bmatrix},$$

$$(c)\ \begin{bmatrix} -2-\lambda & -2 & 0 & 0 \\ 1 & -2-\lambda & 1 & 0 \\ 0 & 1 & -2-\lambda & 1 \\ 0 & 0 & 1 & -2-\lambda \end{bmatrix}, \qquad (d)\ \begin{bmatrix} 4-\lambda & 2 & 0 & 0 \\ 1 & 4-\lambda & 1 & 0 \\ 0 & 2 & 4-\lambda & 2 \\ 0 & 0 & 1 & 4-\lambda \end{bmatrix}.$$

2. Do the principal minors of the matrix of Ex. 9.4.1 (*b*) form a Sturmian sequence? Test the sequence to see whether it has the three fundamental properties (*a*), (*b*), (*c*).

3. (*a*) Determine the number of negative roots of the determinant of the matrix of Ex. 9.4.1 (*d*).

(*b*) Find the roots of the determinant of Ex. 9.4.1 (*d*) correct to the nearest whole numbers.

9.5 Location of the characteristic values of a triple-diagonal matrix

In this section we shall describe a technique for calculating the characteristic values of a triple-diagonal matrix based on the fact that when $a_r c_{r-1}$ is positive the leading principal minors of $\mathbf{B}-\lambda\mathbf{I}$ form a Sturmian sequence.

Denote the zeros of $P_n(\lambda)$, in increasing order, by $\lambda_1, \lambda_2, \ldots, \lambda_n$, and denote $a_r c_{r-1}$ by β_r. Then, as in equation (9.4.5),

$$\left.\begin{aligned} P_0 &= 1, \quad P_1 = b_1-\lambda, \\ P_r &= (b_r-\lambda)P_{r-1}-\beta_r P_{r-2}. \end{aligned}\right\} \tag{9.5.1}$$

We use the fact that the sign count function $s(\lambda)$ indicates the number of characteristic values (zeros of $P_n(\lambda)$) to the left of λ.

Suppose we wish to determine an interval, of length less than some pre-assigned amount, e, within which the ith zero, λ_i, lies. The technique consists of the following steps:

(i) Determine a_0 and b_0 such that λ_i is between them, for example by using the inequality (8.3.8).

(ii) Calculate $s(x_1)$ for $x_1 = (a_0+b_0)/2$.

(iii) If $s(x_1) < i$, which means that λ_i lies to the right of x_1, put

$$a_1 = x_1, \quad b_1 = b_0,$$

otherwise put

$$a_1 = a_0, \quad b_1 = x_1.$$

(iv) If $|b_1 - a_1| < e$, stop, otherwise repeat step (ii) using a_1 and b_1 instead of a_0 and b_0, forming new values a_2 and b_2 which are tested in step (iii) and so on.

This technique will eventually yield an interval of length less than e within which λ_i lies.

For the matrix **B** we take

$$\mathbf{B} = \begin{bmatrix} -2 & 1 & 0 & 0 \\ 1 & -2 & 1 & 0 \\ 0 & 1 & -2 & 1 \\ 0 & 0 & 1 & -2 \end{bmatrix}. \tag{9.5.2}$$

Then

$$\beta_2 = \beta_3 = \beta_4 = 1,$$

and the leading principal minors may be calculated from

$$\begin{aligned} P_0 &= 1, \\ P_1 &= -2-\lambda, \\ P_2 &= (-2-\lambda)P_1 - P_0, \\ P_3 &= (-2-\lambda)P_2 - P_1, \\ P_4 &= (-2-\lambda)P_3 - P_2. \end{aligned}$$

We choose to determine λ_2 with an error of at most 0·05; by taking the mid-point of the final interval as the best available estimate of λ_2 we may take

$$e = 0{\cdot}1.$$

The steps are then:

(i) From the inequality (8.3.8) it follows that

$$|\lambda| \leqslant 4,$$

and therefore we may put

$$a_0 = -4, \quad b_0 = 4.$$

(ii) $x_1 = 0$ and at this point

$$P_0 = 1, \quad P_1 = -2, \quad P_2 = (-2)P_1 - P_0 = 3,$$
$$P_3 = (-2)P_2 - P_1 = -4, \quad P_4 = (-2)P_3 - P_2 = 5,$$

so that $$s(x_1) = 4.$$

(iii) $s(x_1) > 2$, therefore $$a_1 = -4, \quad b_1 = 0.$$

(iv) The length of the interval known to contain λ_2 is

$$|b_1 - a_1| = 4 > e,$$

so we repeat the steps from (ii) onwards. The calculation may be displayed thus:

r	0	1	2	3	4	5	—	—
x_r	—	0	-2	-3	$-2{\cdot}5$	$-2{\cdot}75$	$-2{\cdot}625$	$-2{\cdot}5625$
$P_0(x_r)$	—	1	1	1	1	1	1	1
$P_1(x_r)$	—	-2	0	1	$0{\cdot}5$	$0{\cdot}75$	$0{\cdot}625$	$0{\cdot}5625$
$P_2(x_r)$	—	3	-1	0	$-0{\cdot}75$	$-0{\cdot}4375$	$-0{\cdot}609375$	$-0{\cdot}683594$
$P_3(x_r)$	—	-4	0	-1	$-0{\cdot}875$	$-1{\cdot}078125$	$-1{\cdot}005859$	$-0{\cdot}947022$
$P_4(x_r)$	—	5	1	-1	$0{\cdot}3125$	$-0{\cdot}371094$	$-0{\cdot}019287$	$0{\cdot}150894$
$s(x_r)$	—	4	2	1	2	1	1	2
a_r	-4	-4	-4	-3	-3	$-2{\cdot}75$	$-2{\cdot}625$	$-2{\cdot}625$
b_r	4	0	-2	-2	$-2{\cdot}5$	$-2{\cdot}5$	$-2{\cdot}5$	$-2{\cdot}5625$

The length of the last interval is 0·0625, which is less than e. The mid-point of this interval is $-2{\cdot}59375$, which is within 0·05 of λ_2. A better estimate than this can be found from the two latest values of P_4. Taking the curve of P_4 against λ to be linear over the last interval we can estimate the position of the zero to be

$$\lambda = -2{\cdot}5625 - \frac{(0{\cdot}0625)(0{\cdot}150894)}{(0{\cdot}150894 + 0{\cdot}019287)}$$

$$= -2{\cdot}617917.$$

The value of $s(x)$ is then $$s(-2{\cdot}617917) = 2,$$

and the zero lies between this and $-2{\cdot}625$.

When a zero of P_n has been isolated, that is, when it is known that it is the only zero between a and b, it may be located more rapidly than by bisection if the values of P_n have been calculated reliably. Linear interpolation, as used at the end of the last example, is useful, and so is the Newton–Raphson formula. This states that if x is a close enough estimate of a zero of

$$f(x) = 0, \tag{9.5.3}$$

then a closer estimate will be $$y = x - f(x)/f'(x), \tag{9.5.4}$$

unless the curve of $f(x)$ against x has an inflection at the zero being sought. The derivative of P_n may be found by using equations (9.4.7).

For the polynomials of the last example, and for $\lambda = -2{\cdot}625$, equation (9.4.7) yields

$$\begin{aligned} P_0' &= 0, \\ P_1' &= -1, \\ P_2' &= (0{\cdot}625)(-1) - 0 - 0{\cdot}625 = -1{\cdot}25, \\ P_3' &= (0{\cdot}625)(-1{\cdot}25) - (-1) - (-0{\cdot}609\,375) = 0{\cdot}828\,125, \\ P_4' &= (0{\cdot}625)(0{\cdot}828\,125) - (-1{\cdot}25) - (-1{\cdot}005\,859) = 2{\cdot}773\,437, \\ y &= -2{\cdot}625 + (0{\cdot}019\,287)/2{\cdot}773\,437 = -2{\cdot}618\,046. \end{aligned}$$

The correct value is
$$\lambda_2 = -4\sin^2(3\pi/10) = -2{\cdot}618\,034.$$

Unless great accuracy is needed, linear interpolation will usually give an adequate approximation, but care must be taken to ensure that the interpolation is always carried out between values of P_n which are of opposite signs. For machine computation it is easier to programme the bisection process than a mixed process.

We shall determine the characteristic value of smallest modulus of the matrix of equation (9.3.33), namely

$$\bar{\bar{\mathbf{B}}} = \begin{bmatrix} 1 & -5{\cdot}385\,165 & 0 & 0 \\ -5{\cdot}385\,165 & 71{\cdot}310\,352 & 10{\cdot}599\,618 & 0 \\ 0 & 10{\cdot}599\,618 & 4{\cdot}176\,061 & -0{\cdot}462\,331 \\ 0 & 0 & -0{\cdot}462\,331 & 0{\cdot}513\,586 \end{bmatrix}.$$

First we find the ordinal numbers of the characteristic values nearest to zero on the positive and negative sides. The number of the negative value is equal to $s(o)$, the sign count of the sequence of principal minors of the matrix, while the positive characteristic value is the next one. We then locate these two values and choose that nearest to zero.

The leading principal minors of $\bar{\bar{\mathbf{B}}} - \lambda\mathbf{I}$ satisfy

$$\begin{aligned} P_0 &= 1, \\ P_1 &= 1 - \lambda, \\ P_2 &= (71{\cdot}310\,352 - \lambda)P_1 - 29{\cdot}000\,002, \\ P_3 &= (4{\cdot}176\,061 - \lambda)P_2 - 112{\cdot}351\,902 P_1, \\ P_4 &= (0{\cdot}513\,586 - \lambda)P_3 - 0{\cdot}213\,750 P_2. \end{aligned}$$

For $\lambda = 0$ these yield

$$P_0 = 1, \quad P_1 = 1, \quad P_2 = 42{\cdot}310\,350, \quad P_3 = 64{\cdot}338\,700,$$
$$P_4 = 23{\cdot}999\,618, \quad s(o) = 0.$$

The matrix therefore has no negative characteristic values, and the value of least modulus is λ_1. The bound (8.3.8) suggests that this may lie in the range

$$0 \leqslant \lambda \leqslant 7.$$

Since the matrix has no negative characteristic values, $P_4(\lambda)$ will be convex towards the λ-axis for $\lambda < \lambda_1$, and therefore Newton's formula should provide a sequence of approximations which increases steadily to λ_1. To use this formula we need to calculate the derivative, $P_4'(\lambda)$, and we use equations (9.4.7), namely

$$\begin{aligned} P_0' &= 0, \\ P_1' &= -1, \\ P_2' &= (71{\cdot}310\,352-\lambda)\,P_1'-P_1, \\ P_3' &= (4{\cdot}176\,061-\lambda)\,P_2'-112{\cdot}351\,902P_1'-P_2, \\ P_4' &= (0{\cdot}513\,586-\lambda)\,P_3'-0{\cdot}213\,750P_2'-P_3. \end{aligned}$$

We then derive the correction, $\delta\lambda$, from

$$\delta\lambda = -P_4/P_4'.$$

For $\lambda = 0$, these equations yield

$$P_0' = 0,\quad P_1' = -1,\quad P_2' = 72{\cdot}310\,352,\quad P_3' = -231{\cdot}930\,889,$$

$$P_4' = -167{\cdot}998\,820,\quad \delta\lambda = 0{\cdot}142\,856.$$

The subsequent tabulation can then be presented as follows:

	$\lambda = 0{\cdot}142856$		$\lambda = 0{\cdot}198568$	
r	P_r	P_r'	P_r	P_r'
0	1	0	1	0
1	0·857144	−1	0·801432	−1
2	32·000790	−72·024640	27·991257	−71·913216
3	32·763988	−210·067026	21·292619	−201·673068
4	5·306424	−95·246870	0·724427	−69·452005
	$\delta\lambda = 0{\cdot}055712$		$\delta\lambda = 0{\cdot}010431$	

	$\lambda = 0{\cdot}208999$		$\lambda = 0{\cdot}209376$	
r	P_r	P_r'	P_r	P_r'
0	1	0	1	—
1	0·791001	−1	0·790624	—
2	27·241239	−71·892354	27·214136	—
3	19·197217	−200·090763	19·121795	—
4	0·024408	−64·775272	0·000020	—
	$\delta\lambda = 0{\cdot}000377$		—	

The value of P_4' is clearly not changing much at the last step and is therefore about −64, which indicates a change of about 1/3 in the last figure of λ_1. A further check is furnished by the fact that consecutive changes in λ, say δ_1 and δ_2, approximately satisfy the equation

$$\delta_2 = 3(\delta_1)^2.$$

The quadratic law is correct for Newton's method of successive improvement, and the rough constancy of the multiplier indicates that the process is working reliably.

Wilkinson,† using a simple analysis, has shown that if e is the maximum rounding error in an operation, namely 0·5 in the last decimal place recorded, then the effect of rounding errors in the process will be to displace the characteristic value λ by at most

$$2 \max_{r}. \left(|a_r| + |b_r| + |c_r| + |\lambda| \right).$$

The effect of rounding off in the working of the last example is to cause an error which is bounded by

$$2(87{\cdot}3 + 0{\cdot}21)\, e \leqslant (88)\, 10^{-6}.$$

If we combine this with the error due to rounding off in the formation of the triple diagonal matrix, which we found at the end of § 9.4 to be (8) 10^{-6}, we obtain (96) 10^{-6} as an overall estimate of the error in the characteristic value. However, in § 8·4 we found that the value 0·209 376 is in error by less than (3) 10^{-6}, and therefore the present bound is very crude.

EXAMPLES 9.5

1. Use the triple-diagonal matrix obtained in Ex. 9.3.1 to obtain the characteristic values of the matrix **B** given there. First locate the roots roughly using $s(\lambda)$ and then use the Newton–Raphson process.

2. Find the (algebraically) greatest characteristic value of the matrix of equation (9.5.2), correct to six decimal places, using linear interpolation or the Newton–Raphson process, or both.

3. Use the triple-diagonal form obtained in Ex. 9.3.2 to find the third (i.e. the second largest) characteristic value of the matrix of Ex. 8.3.2.

Compare the result, and the time taken to get it, with that of Ex. 8.7.2. Once the matrix has been reduced to triple-diagonal form it is relatively easy to find (all) its characteristic values.

4. Given that it lies between 1 and 2, find the least root of the determinant of Ex. 9.4.1 (d) correct to two decimal places by using the bisection process alone. Round the bounds a_r, b_r to three decimal places.

5. How many steps are needed in the method of 'iteration by bisection' to obtain a value for a root having its first four figures correct if it is known initially that the root is in a certain interval of length l?

9.6 Determination of the characteristic vectors of a triple-diagonal matrix

When approximate values of characteristic values have been determined, it is often necessary to find approximations to the characteristic vectors. If the exact value of a characteristic value, λ, were known, it would be possible to determine the ratios of the elements of the characteristic vector from the equation

$$\mathbf{Bx} - \lambda\mathbf{x} = \mathbf{0}. \qquad (9.6.1)$$

But

$$|\mathbf{B} - \lambda\mathbf{I}| \neq 0, \qquad (9.6.2)$$

† Wilkinson, 'Error analysis of floating-point computation', *Numer. Math.* **2** (1960), pp. 319–40.

unless λ is an exact characteristic value, and the only solution of the equation is the zero vector. It might be thought that since the equations represented by the matrix equation (9.6.1) are very nearly linearly dependent, approximate values of the ratios of the elements of the characteristic vector could be found from $n-1$ of the equations, leaving the other equation as a check. This is not so in general, and quite small sets of equations (8 or so) often yield completely wrong results when treated in this way, without being in any way pathological.

The usual method for finding the characteristic vectors makes use of the fact that $\mathbf{B}-\lambda\mathbf{I}$ is not singular. Since the matrix is not singular it has an inverse, and the effect of multiplying any column vector by the inverse can be very simply studied in terms of the spectral representation of $\mathbf{B}$. As in § 8.4 we write

$$\mathbf{B} = \mathbf{X}\mathbf{D}\mathbf{X}^{-1}, \tag{9.6.3}$$

where

$$\mathbf{D} = \begin{bmatrix} \lambda_1 & 0 & \dots & 0 \\ 0 & \lambda_2 & \dots & 0 \\ \dots & \dots & \dots & \dots \\ 0 & 0 & \dots & \lambda_n \end{bmatrix}, \tag{9.6.4}$$

and $\mathbf{X}$ is the matrix whose columns are the characteristic vectors $\mathbf{x}^{(r)}$. Since

$$\mathbf{B}-\lambda\mathbf{I} = \mathbf{X}(\mathbf{D}-\lambda\mathbf{I})\,\mathbf{X}^{-1}, \tag{9.6.5}$$

the inverse of $\mathbf{B}-\lambda\mathbf{I}$ is

$$(\mathbf{B}-\lambda\mathbf{I})^{-1} = \mathbf{X}(\mathbf{D}-\lambda\mathbf{I})^{-1}\mathbf{X}^{-1}. \tag{9.6.6}$$

Hence if $\mathbf{y}$ is any vector and

$$(\mathbf{B}-\lambda\mathbf{I})\,\mathbf{x} = \mathbf{y}, \tag{9.6.7}$$

then

$$\mathbf{x} = (\mathbf{B}-\lambda\mathbf{I})^{-1}\mathbf{y} = \mathbf{X}(\mathbf{D}-\lambda\mathbf{I})^{-1}\mathbf{X}^{-1}\mathbf{y}. \tag{9.6.8}$$

If we write

$$\mathbf{X}^{-1}\mathbf{y} = \boldsymbol{\eta} = \{\eta_1, \eta_2, \dots, \eta_n\}, \tag{9.6.9}$$

then

$$\mathbf{y} = \mathbf{X}\boldsymbol{\eta} = \eta_1\mathbf{x}^{(1)} + \eta_2\mathbf{x}^{(2)} + \dots + \eta_n\mathbf{x}^{(n)}. \tag{9.6.10}$$

In other words, the η's are the components of $\mathbf{y}$ referred to the characteristic vectors $\mathbf{x}^{(r)}$ as coordinate axes.

Substituting from equation (9.6.10) into equation (9.6.8) we find

$$\mathbf{x} = \sum_{t=1}^{n} \frac{\eta_t}{\lambda_t - \lambda}\mathbf{x}^{(t)}. \tag{9.6.11}$$

Thus the effect of solving equation (9.6.7) for $\mathbf{x}$ is to resolve the vector $\mathbf{y}$ into components along the characteristic vectors, multiplying the components by appropriate scaling factors $(\lambda_t-\lambda)^{-1}$, and then to recombine them to give $\mathbf{x}$. If λ is nearer to λ_r than to any other characteristic value, the scaling factor in the direction of $\mathbf{x}^{(r)}$ will be much larger than the scaling factor in any other direction, and as long as η_r, the component of $\mathbf{y}$ along $\mathbf{x}^{(r)}$ is not unduly small, $\mathbf{x}$ will be largely in the direction of $\mathbf{x}^{(r)}$. Thus, if we can find a vector $\mathbf{y}$ which has an appreciable component in the direction of $\mathbf{x}^{(r)}$, the method provides a means of determining $\mathbf{x}^{(r)}$.

We take the matrix **B** of equation (9.5.2) for which an approximate characteristic value is $-2{\cdot}619$. Then

$$\mathbf{B}-\lambda\mathbf{I} = \begin{bmatrix} 0{\cdot}619 & 1 & 0 & 0 \\ 1 & 0{\cdot}619 & 1 & 0 \\ 0 & 1 & 0{\cdot}619 & 1 \\ 0 & 0 & 1 & 0{\cdot}619 \end{bmatrix}.$$

This may be factorised by the method of § 7.8, that is, row-interchanges (without using square roots) to yield

$$\mathbf{U} = \begin{bmatrix} 1 & 0{\cdot}619 & 1 & 0 \\ 0 & 1 & 0{\cdot}619 & 1 \\ 0 & 0 & -1{\cdot}000\,823 & -0{\cdot}616\,839 \\ 0 & 0 & 0 & 0{\cdot}002\,668 \end{bmatrix},$$

and

$$\mathbf{L} = \begin{bmatrix} 0{\cdot}619 & 0{\cdot}616\,839 & 1 & 0 \\ 1 & 0 & 0 & 0 \\ 0 & 1 & 0 & 0 \\ 0 & 0 & -0{\cdot}999\,178 & 1 \end{bmatrix}.$$

As a trial right-hand side we choose

$$\mathbf{y} = \{1 \quad 1 \quad 1 \quad 1\}, \tag{9.6.12}$$

and we solve

$$\mathbf{Lz} = \mathbf{y}, \tag{9.6.13}$$

finding the modified right-hand side

$$\mathbf{z} = \{1 \quad 1 \quad -0{\cdot}235\,839 \quad 0{\cdot}764\,355\}. \tag{9.6.14}$$

We then solve

$$\mathbf{Ux} = \mathbf{z}, \tag{9.6.15}$$

and find

$$\mathbf{x} = \{286{\cdot}490 \quad -176{\cdot}337 \quad -176{\cdot}337 \quad 286{\cdot}490\}.$$

Scaling this vector so that its largest coefficient is unity, we obtain

$$\mathbf{x} = \{1 \quad -0{\cdot}615\,508 \quad -0{\cdot}615\,508 \quad 1\}. \tag{9.6.16}$$

The scaling factor, which is roughly $(\lambda_2-\lambda)^{-1}$, is $286{\cdot}490$, which suggests that the error in the trial value, $-2{\cdot}619$, is about $0{\cdot}003$.

This process may be repeated to magnify the desired component further. We take the vector **x** above to be the new **y** and calculate **z** and **x** as before, and find

$$\mathbf{z} = \{-0{\cdot}615\,508 \quad -0{\cdot}615\,508 \quad 1{\cdot}760\,669 \quad 2{\cdot}759\,222\},$$

$$\mathbf{x} = \{1034{\cdot}191 \quad -639{\cdot}164 \quad -639{\cdot}164 \quad 1034{\cdot}191\},$$

and scaling this we have

$$\mathbf{x} = \{1 \quad -0{\cdot}618\,033 \quad -0{\cdot}618\,033 \quad 1\}. \tag{9.6.17}$$

The scaling factor this time is 1034·191, which suggests that the error in our trial value of λ was 0·001. More exactly it gives

$$\lambda_2 - (-2{\cdot}619) = (1034{\cdot}191)^{-1} = 0{\cdot}000\,967,$$

or

$$\lambda_2 = -2{\cdot}618\,033.$$

The exact characteristic vector of the matrix **B** is

$$\mathbf{x} = \{\sin 72^0 \quad -\sin 36^0 \quad -\sin 36^0 \quad \sin 72^0\}, \tag{9.6.18}$$

which on scaling agrees with the last approximation, (9.6.17), for **x**.

When the matrix has two characteristic values close to a trial λ, the convergence of this iteration is very slow. If the wanted vector corresponds to λ_r and the next nearest is λ_s, the proportion of **x** which corresponds to unwanted components is reduced by a factor

$$\rho = (\lambda - \lambda_r)/(\lambda - \lambda_s) \tag{9.6.19}$$

at each stage. This is called the 'convergence ratio'. It often occurs in practice that the convergence ratio is nearly one, so that many steps are required to purify **x**. Under these circumstances it is important to choose an initial vector which has a large proportion of the desired characteristic vector. Engineering intuition will often provide a good estimate of the amplitude ratios in a normal mode, and hence a good starting vector. When intuition fails, the suggestion of Wilkinson makes a good substitute. He suggests starting with a modified right-hand side, **z**, given by

$$\mathbf{z} = \{1, 1, 1, \ldots, 1\} \tag{9.6.20}$$

adducing sound reasons for expecting this to behave as though it had been found from an initial right-hand side **y**, which had a large component in the desired direction.

We use the matrix, approximate characteristic value and factors of the last example. Instead of using equations (9.6.12) and (9.6.13) to determine the **z** of equation (9.6.14), we choose **z** as given by equation (9.6.20) and solve equation (9.6.15), finding

$$\mathbf{x} = \{374{\cdot}813, \quad -232{\cdot}008, \quad -230{\cdot}200, \quad 375{\cdot}502\}.$$

When this is scaled it yields

$$\mathbf{x} = \{0{\cdot}998\,165, \quad -0{\cdot}617\,861, \quad -0{\cdot}613\,046, \quad 1\}.$$

This is not much worse than the estimate of equation (9.6.16). The rest of the improvement process proceeds as before.

A more realistic specimen is provided by the matrix of equation (9.3.33) and the characteristic value 0·209 376 obtained in § 9.5 (p. 340). For these, we have

$$\mathbf{B} - \lambda\mathbf{I} = \begin{bmatrix} 0{\cdot}790\,624 & -5{\cdot}385\,165 & 0 & 0 \\ & 71{\cdot}100\,976 & 10{\cdot}599\,618 & 0 \\ & & 3{\cdot}966\,685 & -0{\cdot}462\,331 \\ & & & 0{\cdot}304\,210 \end{bmatrix}, \tag{9.6.21}$$

where only the upper half of the symmetric matrix is displayed. It must be remembered that all the matrix must be taken into account when forming check sums. Using row-interchanges, as in § 7.8, this matrix may be factorised into the product **LU**, where

$$\mathbf{L} = \begin{bmatrix} 0{\cdot}790\,624 & 5{\cdot}053\,538 & -0{\cdot}334\,995 & 0{\cdot}001\,091 \\ -5{\cdot}385\,165 & 0 & 0 & 0 \\ 0 & 10{\cdot}599\,618 & 0 & 0 \\ 0 & 0 & -0{\cdot}462\,331 & 0 \end{bmatrix},$$

$$\mathbf{U} = \begin{bmatrix} 1 & -13{\cdot}203\,119 & -1{\cdot}968\,300 & 0 \\ 0 & 1 & 0{\cdot}374\,229 & -0{\cdot}043\,618 \\ 0 & 0 & 1 & -0{\cdot}657\,992 \\ 0 & 0 & 0 & 0{\cdot}001\,091 \end{bmatrix}.$$

Since none of the pivots but the last ($l_{14}u_{44}$) was small, square roots were not used until the last pivot was treated. The product that had to be split between **L** and **U** was then 0·000 001 190 444, and it is essential to use double-length working in forming this.

The iteration then proceeds from **z** to **x** to **y** to **z** to $\underline{\mathbf{z}}$ to **x**, and so on. The various vectors are scaled to keep their largest elements unity, before being used in further calculation—this is the difference between a vector **x** and the **y** used in the next step, and between **z** and $\underline{\mathbf{z}}$. The magnification factor of a step is then the product of the scale factors used:

z	x	y	z	$\underline{\mathbf{z}}$
1	−1253·767 382	1	−0·027 414	−0·000 017
1	−185·095 300	0·147 631	−0·045 458	−0·000 028
1	604·109 074	−0·481 835	1·581 268	0·000 969
1	916·590 284	−0·731 069	1632·552 704	1

x	y	z	$\underline{\mathbf{z}}$
−1265·001 070	1	−0·027 263	−0·000 017
−185·721 461	0·146 815	−0·044 980	−0·000 028
603·110 043	−0·476 766	1·567 226	0·000 964
916·590 284	−0·724 577	1625·175 069	1

The difference between this last vector $\underline{\mathbf{z}}$ and the previous version is too small to affect the vector **x**, and the estimate we have for the characteristic vector (we take **y** so that the largest element is unity) is

$$\mathbf{x} = \{1 \quad 0{\cdot}146\,815 \quad -0{\cdot}476\,766 \quad -0{\cdot}724\,577\}. \tag{9.6.22}$$

It is always advisable to check the calculation by forming the product **Bx** and comparing it with $\lambda\mathbf{x}$. We find

$$\mathbf{Bx} = \{0{\cdot}209\,377 \quad 0{\cdot}030\,727 \quad -0{\cdot}099\,827 \quad -0{\cdot}151\,709\},$$

$$\lambda\mathbf{x} = \{0{\cdot}209\,376 \quad 0{\cdot}030\,740 \quad -0{\cdot}099\,823 \quad -0{\cdot}151\,709\}.$$

The largest discrepancy is 13 in the sixth decimal place, in the second component. This comes from the second row of **B**, which has as its diagonal element 71·310352. The sum of the moduli of the elements of this row is about 86, so that rounding off the elements of **y** might easily cause a larger discrepancy. This agreement is therefore satisfactory.

EXAMPLES 9.6

1. Find the characteristic vector of the matrix of equation (9.5.2) corresponding to the characteristic value $\lambda = -0{\cdot}381966$ found in Ex. 9.5.2. Use the following method: Take $\lambda = -0{\cdot}382$ and factorise $\mathbf{B}-\lambda\mathbf{I}$ into factors **L**, **U** using the row interchange factorisation of §7.8 with square root factorisation for the last pivot. Then proceed as in the text.

2. Find the characteristic vector of the triple-diagonal matrix $\overline{\mathbf{B}}$ found in Ex. 9.3.1 corresponding to the characteristic value $\lambda_2 = 1{\cdot}013642$. (Take $\lambda = 1{\cdot}0137$ and factorise $\overline{\mathbf{B}}-\lambda\mathbf{I}$.)

3. The method described in this section can be used to find the characteristic value and vector of any (not necessarily triple-diagonal) matrix, once an approximation to the corresponding characteristic value has been found. Apply it to the matrix of equation (8.3.21) and the approximation $\lambda = 0{\cdot}209$ and check the characteristic value and vector so found against those found elsewhere (λ_2 in §8.4, $\mathbf{x}^{(2)}$ in §9.7).

9.7 Reversing Householder's process

To each characteristic vector of the triple-diagonal matrix there corresponds a vector of the original matrix. The original matrix was transformed to triple-diagonal form by a succession of rotations, $n-2$ in all. These transformations of the matrix can be considered to be induced by a change of coordinate system in the space of the vectors. To regain the vectors of the original matrix we have to reverse the change of coordinates.

Each step of the reduction involves a matrix **R** as in equation (9.3.1) and the transformation of **B** by this matrix (a strict equivalence) corresponds to a rotation of the axes in the vector space, such that if **v** is a vector referred to the old coordinates and **y** is the vector referred to the new coordinates, then

$$\mathbf{v} = \mathbf{R}\mathbf{y}. \tag{9.7.1}$$

If **v** satisfies

$$\mathbf{B}\mathbf{v} = \lambda\mathbf{v}, \tag{9.7.2}$$

then

$$\mathbf{B}\mathbf{R}\mathbf{y} = \lambda\mathbf{R}\mathbf{y}, \tag{9.7.3}$$

and hence

$$\mathbf{R}^{-1}\mathbf{B}\mathbf{R}\mathbf{y} = \lambda\mathbf{y}. \tag{9.7.4}$$

If we know **y** such that equation (9.7.4) is satisfied, we may use equation (9.7.1) to determine **v** so that equation (9.7.2) is satisfied.

The matrix **R** is given by

$$\mathbf{R} = \mathbf{I} - 2\mu\mathbf{z}\mathbf{z}', \tag{9.7.5}$$

so that

$$\mathbf{v} = \mathbf{y} - 2\mu(\mathbf{z}'\mathbf{y})\,\mathbf{z}. \tag{9.7.6}$$

This equation is used $n-2$ times, starting with the last rotation used to transform **B** and working back to the first.

We shall now derive the characteristic vector of the matrix of equation (8.3.21) from the vector found in equation (9.6.23) to be a characteristic vector of the equivalent triple-diagonal matrix. The vector $\mathbf{y}$ of equation (9.7.4) is taken to be

$$\mathbf{y} = \{1, \quad 0{\cdot}146\,815, \quad -0{\cdot}476\,766, \quad -0{\cdot}724\,577\}.$$

The *last* rotation applied to the original matrix, when it was reduced to the triple-diagonal form in § 9.3 (p. 330), is associated with the values

$$\mathbf{z} = \{0, \quad 0, \quad -11{\cdot}378\,565, \quad -10{\cdot}570\,958\},$$
$$2\mu = 0{\cdot}008\,291\,294.$$

Applying equation (9.7.6) we find

$$2\mu\mathbf{z}'\mathbf{y} = 0{\cdot}108\,486,$$
$$\mathbf{v} = \{1, \quad 0{\cdot}146\,815, \quad 0{\cdot}757\,649, \quad 0{\cdot}422\,224\}.$$

The rotation before last (the first applied in this case) is associated with

$$\mathbf{z} = \{0, \quad 7{\cdot}385\,165, \quad 3, \quad 4\},$$
$$2\mu = 0{\cdot}025\,144\,373.$$

Applying this rotation to the vector $\mathbf{v}$ we have

$$2\mu\mathbf{z}'\mathbf{v} = 0{\cdot}126\,881,$$
$$\mathbf{x} = \{1, \quad -0{\cdot}790\,222, \quad 0{\cdot}377\,006, \quad -0{\cdot}085\,300\}. \tag{9.7.7}$$

As a check on the arithmetic we can, at any stage, test that the appropriate partially reduced matrix, $\mathbf{C}$, and partially rotated vector, $\mathbf{v}$, satisfy

$$\mathbf{Cv} = \lambda\mathbf{v}. \tag{9.7.8}$$

We shall check the final vector of this simple example. We find, for the original matrix $\mathbf{B}$ of equation (8.2.24) that

$$\mathbf{Bx} = \{0{\cdot}209\,374, \quad -0{\cdot}165\,472, \quad 0{\cdot}078\,900, \quad -0{\cdot}017\,928\}.$$

Also $$\lambda\mathbf{x} = \{0{\cdot}209\,376, \quad -0{\cdot}165\,454, \quad 0{\cdot}078\,936, \quad -0{\cdot}017\,860\}.$$

In view of the size of the elements of $\mathbf{B}$, the discrepancies are not unduly large, and there is no indication of any error.

EXAMPLE 9.7

1. Use the characteristic vector of the triple-diagonal matrix of Ex. 9.3.1 which was found in Ex. 9.6.2 to find the second characteristic vector of the original matrix of Ex. 9.3.1.

9.8 Lanczos's reduction of an unsymmetric matrix to triple-diagonal form

Although an unsymmetric triple-diagonal matrix need not be Sturmian, the simplicity and stability of the recurrence relations (9.4.5) is such that they are a valuable way of evaluating the characteristic polynomial of any matrix which can be reduced to this form. It was shown by Lanczos† that an arbitrary matrix could be reduced to triple-diagonal form by a similarity, and he devised an elegant algorithm for the step by step determination of the transformation. It has been shown by Wilkinson‡ that the practical application of this algorithm is liable to encounter certain difficulties, and the modifications needed to overcome these have been described by him. We shall describe and exemplify the process in its practical form.

The original matrix is denoted by **B**, the triple-diagonal matrix by **T** and the transforming matrix by **Q**, so that

$$\mathbf{T} = \mathbf{Q}^{-1}\mathbf{B}\mathbf{Q}. \tag{9.8.1}$$

In the course of the calculation, **Q** is found a column at a time, and $\mathbf{Q}^{-1}$ is found simultaneously, a row at a time. To be more precise, we determine rows of a matrix $\mathbf{P}'$, such that $\mathbf{P}'\mathbf{Q}$ is diagonal, so that the rows of $\mathbf{Q}^{-1}$ are proportional to the rows of $\mathbf{P}'$. We shall denote the columns of **Q** by $\mathbf{q}^{(j)}$, and the rows of $\mathbf{P}'$ by $\mathbf{p}'^{(i)}$, so that

$$\mathbf{p}'^{(i)}\mathbf{q}^{(j)} = 0 \quad (i \neq j). \tag{9.8.2}$$

The elements of **T** are given by

$$\mathbf{T} = \begin{bmatrix} b_1 & c_1 & 0 & \dots & 0 \\ a_2 & b_2 & c_2 & \dots & 0 \\ 0 & a_3 & b_3 & \dots & 0 \\ \dots & \dots & \dots & \dots & \dots \\ \dots & \dots & \dots & \dots & \dots \\ 0 & 0 & 0 & \dots & b_n \end{bmatrix}. \tag{9.8.3}$$

Equation (9.8.1) can be rewritten as

$$\mathbf{QT} = \mathbf{BQ}, \tag{9.8.4}$$

and in terms of column vectors this is

$$c_{j-1}\mathbf{q}^{(j-1)} + b_j\mathbf{q}^{(j)} + a_{j+1}\mathbf{q}^{(j+1)} = \mathbf{B}\mathbf{q}^{(j)}. \tag{9.8.5}$$

Equation (9.8.2) requires that

$$\left.\begin{aligned} \mathbf{p}'^{(j-1)}\mathbf{q}^{(j)} &= 0 = \mathbf{p}'^{(j-1)}\mathbf{q}^{(j+1)}, \\ \mathbf{p}'^{(j)}\mathbf{q}^{(j-1)} &= 0 = \mathbf{p}'^{(j)}\mathbf{q}^{(j+1)}, \end{aligned}\right\} \tag{9.8.6}$$

and these relations used in conjunction with equation (9.8.5) give the values of c_{j-1} and b_j, namely

$$c_{j-1} = \frac{\mathbf{p}'^{(j-1)}\mathbf{B}\mathbf{q}^{(j)}}{\mathbf{p}'^{(j-1)}\mathbf{q}^{(j-1)}}, \quad b_j = \frac{\mathbf{p}'^{(j)}\mathbf{B}\mathbf{q}^{(j)}}{\mathbf{p}'^{(j)}\mathbf{q}^{(j)}}. \tag{9.8.7}$$

The value of a_{j+1} remains at our disposal, and provides a means of scaling the vectors $\mathbf{q}^{(j)}$ so that they have elements of some convenient size. Provided that the

† C. Lanczos, 'An iteration method for the solution of the eigenvalue problem of linear differential and integral operators', *J. Res. Nat. Bur. Standards*, **45** (1950), pp. 255–82.

‡ Wilkinson, 'The calculation of eigenvectors by the method of Lanczos', *Comput. J.* **1** (1958), pp. 148–52.

vectors $\mathbf{p}'^{(j)}$ can be found at the same time, so that equations (9.8.7) can be used to determine the elements of $\mathbf{T}$, equation (9.8.5) provides a means of finding all the vectors $\mathbf{q}^{(j)}$, once $\mathbf{q}^{(1)}$ has been chosen.

In order to find the $\mathbf{p}$'s we notice that they satisfy a relation similar to (9.8.5). If the diagonal matrix $\mathbf{P}'\mathbf{Q}$ is denoted by

$$\mathbf{P}'\mathbf{Q} = \mathbf{D} = \begin{bmatrix} d_{11} & 0 & \dots & 0 \\ 0 & d_{22} & \dots & 0 \\ \dots & \dots & \dots & \dots \\ 0 & 0 & \dots & d_{nn} \end{bmatrix}, \tag{9.8.8}$$

then equation (9.8.1) can be rewritten

$$(\mathbf{D}\mathbf{T}\mathbf{D}^{-1})\mathbf{P}' = \mathbf{P}'\mathbf{B}, \tag{9.8.9}$$

or

$$\mathbf{S}\mathbf{P}' = \mathbf{P}'\mathbf{B}, \tag{9.8.10}$$

where

$$\mathbf{S} = \mathbf{D}\mathbf{T}\mathbf{D}^{-1}. \tag{9.8.11}$$

Since $\mathbf{D}$ is diagonal, $\mathbf{S}$ like $\mathbf{T}$, will be triple-diagonal, and have elements which are simply related to those of $\mathbf{T}$. In fact, if

$$\mathbf{S} = \begin{bmatrix} e_1 & f_1 & 0 & \dots & 0 \\ d_2 & e_2 & f_2 & \dots & 0 \\ 0 & d_3 & e_3 & \dots & \dots \\ \dots & \dots & \dots & \dots & \dots \\ \dots & \dots & \dots & \dots & \dots \\ 0 & 0 & 0 & \dots & e_n \end{bmatrix}, \tag{9.8.12}$$

then

$$d_i = \frac{a_i d_{ii}}{d_{i-1,\,i-1}}, \quad e_i = b_i, \quad f_i = \frac{c_i d_{ii}}{d_{i+1,\,i+1}}. \tag{9.8.13}$$

In terms of rows, equation (9.8.10) is

$$d_j \mathbf{p}'^{(j-1)} + e_j \mathbf{p}'^{(j)} + f_j \mathbf{p}'^{(j+1)} = \mathbf{p}'^{(j)}\mathbf{B} \tag{9.8.14}$$

and postmultiplication by $\mathbf{q}^{(j-1)}$ and $\mathbf{q}^{(j)}$, and use of equation (9.8.2) yields

$$d_j = \frac{\mathbf{p}'^{(j)}\mathbf{B}\mathbf{q}^{(j-1)}}{\mathbf{p}'^{(j-1)}\mathbf{q}^{(j-1)}}, \quad e_j = \frac{\mathbf{p}'^{(j)}\mathbf{B}\mathbf{q}^{(j)}}{\mathbf{p}'^{(j)}\mathbf{q}^{(j)}}. \tag{9.8.15}$$

This time it is f_j which remains at our disposal, and this may be used to scale $\mathbf{p}'^{(j+1)}$.

The elements of $\mathbf{S}$ and $\mathbf{T}$ have been found by using the condition (9.8.2). However, in every case, the suffices i and j have differed only by one or two, as in equations (9.8.6). We must now investigate whether equation (9.8.2) is satisfied when i and j differ by more than two. We may use a kind of induction. Suppose that the condition is satisfied for the products $\mathbf{p}'^{(1)}\mathbf{q}^{(2)}$, $\mathbf{p}'^{(1)}\mathbf{q}^{(3)}$, $\mathbf{p}'^{(2)}\mathbf{q}^{(3)}$; we shall prove that it holds for $\mathbf{p}'^{(1)}\mathbf{q}^{(4)}$. We have

$$\begin{aligned} a_4\mathbf{p}'^{(1)}\mathbf{q}^{(4)} &= \mathbf{p}'^{(1)}(\mathbf{B}\mathbf{q}^{(3)} - c_2\mathbf{q}^{(2)} - b_2\mathbf{q}^{(3)}) \\ &= \mathbf{p}'^{(1)}\mathbf{B}\mathbf{q}^{(3)} \\ &= (e_1\mathbf{p}'^{(1)} + f_1\mathbf{p}'^{(2)})\,\mathbf{q}^{(3)} \\ &= 0. \end{aligned} \tag{9.8.16}$$

The general result may be proved in the same way. We suppose that

$$\mathbf{p}'^{(i)}\mathbf{q}^{(j)} = 0 \quad (i \neq j), \tag{9.8.17}$$

when $i \leqslant k$ and $j \leqslant k$ and we examine the next vector $\mathbf{q}^{(k+1)}$. Since we know that the result holds for $i = k, j = k+1$ we may take $i \leqslant k-1$ and then

$$\begin{aligned} a_{k+1}\mathbf{p}'^{(i)}\mathbf{q}^{(k+1)} &= \mathbf{p}'^{(i)}(\mathbf{B}\mathbf{q}^{(k)} - c_{k-1}\mathbf{q}^{(k-1)} - b_k\mathbf{q}^{(k)}) \\ &= \mathbf{p}'^{(i)}\mathbf{B}\mathbf{q}^{(k)} \end{aligned} \tag{9.8.18}$$

by hypothesis, and now

$$\begin{aligned} \mathbf{p}'^{(i)}\mathbf{B}\mathbf{q}^{(k)} &= (d_i\mathbf{p}'^{(i+1)} + e_i\mathbf{p}'^{(i)} + f_i\mathbf{p}'^{(i+1)})\,\mathbf{q}^{(k)} \\ &= 0 \end{aligned} \tag{9.8.19}$$

by the same hypothesis. This proves the result.

The sequence of calculations is thus

(i) Choose $\mathbf{p}'^{(1)}$ and $\mathbf{q}^{(1)}$ so that $\mathbf{p}'^{(1)}\mathbf{q}^{(1)} \neq 0$.

(ii) Calculate $\mathbf{B}\mathbf{q}^{(1)}$ and $\mathbf{p}'^{(1)}\mathbf{B}$.

(iii) Calculate

$$b_1 = e_1 = \frac{\mathbf{p}'^{(1)}\mathbf{B}\mathbf{q}^{(1)}}{\mathbf{p}'^{(1)}\mathbf{q}^{(1)}}.$$

(iv) Calculate

$$\mathbf{B}\mathbf{q}^{(1)} - \mathbf{b}_1\mathbf{q}^{(1)} = a_2\mathbf{q}^{(2)},$$

$$\mathbf{p}'^{(1)}\mathbf{B} - e_1\mathbf{p}'^{(1)} = f_1\mathbf{p}'^{(2)}.$$

(v) Choose a_2 and f_1 and scale the vectors found in (iv) to give $\mathbf{q}^{(2)}$ and $\mathbf{p}'^{(2)}$.

(vi) Calculate $\mathbf{B}\mathbf{q}^{(2)}$ and $\mathbf{p}'^{(2)}\mathbf{B}$.

(vii) Calculate $\mathbf{p}'^{(1)}\mathbf{B}\mathbf{q}^{(2)}$, $\mathbf{p}'^{(2)}\mathbf{B}\mathbf{q}^{(1)}$, $\mathbf{p}'^{(2)}\mathbf{B}\mathbf{q}^{(2)}$ and hence

$$c_1 = \frac{\mathbf{p}'^{(1)}\mathbf{B}\mathbf{q}^{(2)}}{\mathbf{p}'^{(1)}\mathbf{q}^{(1)}}, \quad d_2 = \frac{\mathbf{p}'^{(2)}\mathbf{B}\mathbf{q}^{(1)}}{\mathbf{p}'^{(1)}\mathbf{q}^{(1)}},$$

$$b_2 = e_2 = \frac{\mathbf{p}'^{(2)}\mathbf{B}\mathbf{q}^{(2)}}{\mathbf{p}'^{(2)}\mathbf{q}^{(2)}}.$$

(viii) Calculate

$$\mathbf{B}\mathbf{q}^{(2)} - c_1\mathbf{q}^{(1)} - b_2\mathbf{q}^{(2)} = a_3\mathbf{q}^{(3)},$$

$$\mathbf{p}'^{(2)}\mathbf{B} - d_2\mathbf{p}'^{(1)} - e_2\mathbf{p}'^{(2)} = f_2\mathbf{p}'^{(3)}.$$

(ix) Choose a_3 and f_2 and scale the vectors given in step (viii) to give $\mathbf{q}^{(3)}$ and $\mathbf{p}'^{(3)}$.

Continue the process as in steps (vi)–(ix) until a_n, b_n, d_n and e_n are found.

If the arithmetic in the process as described above could be carried out exactly we should find that

$$\left.\begin{aligned}\mathbf{p}'^{(j-1)}\mathbf{B}\mathbf{q}^{(j)} &= f_{j-1}\mathbf{p}'^{(j)}\mathbf{q}^{(j)},\\ \mathbf{p}'^{(j)}\mathbf{B}\mathbf{q}^{(j-1)} &= a_j\mathbf{p}'^{(j)}\mathbf{q}^{(j)}.\end{aligned}\right\} \tag{9.8.20}$$

Thus if we choose the arbitrary elements (the a's and f's) to be equal, to unity say, we shall have

$$c_{j-1} = d_j. \tag{9.8.21}$$

But, since we always have

$$b_j = e_j, \tag{9.8.22}$$

it follows that, with this choice of the arbitrary elements,

$$\mathbf{S} = \mathbf{T}', \tag{9.8.23}$$

and we need compute only one set of multipliers for the elements of the triple-diagonal matrix.

We apply the method described above, taking a_j and f_j to be unity, to the matrix of equation (8.6.4), namely

$$\mathbf{B} = \begin{bmatrix} 1 & 0 & 0 & -1 \\ -1 & 1 & 1 & -3 \\ -2 & 3 & 0 & -4 \\ -7 & 3 & 4 & -7 \end{bmatrix}.$$

As starting vectors we choose

$$\mathbf{q}^{(1)} = \{1 \quad 0 \quad 0 \quad 0\}, \quad \mathbf{p}'^{(1)} = [1 \quad 0 \quad 0 \quad 0],$$

so that

$$\mathbf{p}'^{(1)}\mathbf{q}^{(1)} = 1.$$

Then

$$\mathbf{B}\mathbf{q}^{(1)} = \{1 \quad -1 \quad -2 \quad -7\}, \quad \mathbf{p}'^{(1)}\mathbf{B} = [1 \quad 0 \quad 0 \quad -1]$$

and

$$\mathbf{p}'^{(1)}(\mathbf{B}\mathbf{q}^{(1)}) = (\mathbf{p}'^{(1)}\mathbf{B})\,\mathbf{q}^{(1)} = 1.$$

Hence

$$b_1 = e_1 = \mathbf{p}'^{(1)}\mathbf{B}\mathbf{q}^{(1)}/\mathbf{p}'^{(1)}\mathbf{q}^{(1)} = 1,$$

and

$$\mathbf{q}^{(2)} = \mathbf{B}\mathbf{q}^{(1)} - b_1\mathbf{q}^{(1)} = \{0 \quad -1 \quad -2 \quad -7\},$$

$$\mathbf{p}'^{(2)} = \mathbf{p}'^{(1)}\mathbf{B} - e_1\mathbf{p}'^{(1)} = [0 \quad 0 \quad 0 \quad -1],$$

and

$$\mathbf{p}'^{(2)}\mathbf{q}^{(2)} = 7.$$

We now find

$$\mathbf{B}\mathbf{q}^{(2)} = \{7 \quad 18 \quad 25 \quad 38\}, \quad \mathbf{p}'^{(2)}\mathbf{B} = [7 \quad -3 \quad -4 \quad 7],$$

$$\mathbf{p}'^{(1)}(\mathbf{B}\mathbf{q}^{(2)}) = (\mathbf{p}'^{(2)}\mathbf{B})\,\mathbf{q}^{(1)} = 7, \quad \mathbf{p}'^{(2)}(\mathbf{B}\mathbf{q}^{(2)}) = (\mathbf{p}'^{(2)}\mathbf{B})\,\mathbf{q}^{(2)} = -38,$$

whence

$$c_1 = d_2 = \mathbf{p}'^{(1)}\mathbf{B}\mathbf{q}^{(2)}/\mathbf{p}'^{(1)}\mathbf{q}^{(1)} = 7,$$

$$b_2 = e_2 = \mathbf{p}'^{(2)}\mathbf{B}\mathbf{q}^{(2)}/\mathbf{p}'^{(2)}\mathbf{q}^{(2)} = -5{\cdot}428\,571,$$

so that

$$\begin{aligned}\mathbf{q}^{(3)} &= \mathbf{B}\mathbf{q}^{(2)} - c_1\mathbf{q}^{(1)} - b_2\mathbf{q}^{(2)}\\ &= \{0 \quad 12{\cdot}571\,429 \quad 14{\cdot}142\,858 \quad 0{\cdot}000\,003\},\end{aligned}$$

$$\begin{aligned}\mathbf{p}'^{(3)} &= \mathbf{p}'^{(2)}\mathbf{B} - d_2\mathbf{p}'^{(1)} - e_2\mathbf{p}'^{(2)}\\ &= [0 \quad -3 \quad -4 \quad 1{\cdot}571\,429],\end{aligned}$$

and
$$\mathbf{p}'^{(3)}\mathbf{q}^{(3)} = -94{\cdot}285\,714.$$

Then
$$\mathbf{Bq}^{(3)} = \{-0{\cdot}000\,003 \quad 26{\cdot}714\,278 \quad 37{\cdot}714\,275 \quad 94{\cdot}285\,698\},$$
$$\mathbf{p}'^{(3)}\mathbf{B} = [-0{\cdot}000\,003 \quad -10{\cdot}285\,713 \quad 3{\cdot}285\,716 \quad 13{\cdot}999\,997],$$
$$\mathbf{p}'^{(2)}(\mathbf{Bq}^{(3)}) = (\mathbf{p}'^{(2)}\mathbf{B})\,\mathbf{q}^{(3)} = -94{\cdot}285\,698,$$
$$\mathbf{p}'^{(3)}(\mathbf{Bq}^{(3)}) = (\mathbf{p}'^{(3)}\mathbf{B})\,\mathbf{q}^{(3)} = -82{\cdot}836\,654,$$

so that
$$c_2 = d_3 = \mathbf{p}'^{(2)}\mathbf{Bq}^{(3)}/\mathbf{p}'^{(2)}\mathbf{q}^{(2)} = -13{\cdot}469\,385,$$
$$b_3 = e_3 = \mathbf{p}'^{(3)}\mathbf{Bq}^{(3)}/\mathbf{p}'^{(3)}\mathbf{q}^{(3)} = 0{\cdot}878\,571,$$
$$\mathbf{q}^{(4)} = \{-0{\cdot}000\,003 \quad 2{\cdot}200\,000 \quad -1{\cdot}650\,000 \quad 0{\cdot}000\,000\},$$
$$\mathbf{p}'^{(4)} = [0{\cdot}000\,003 \quad -7{\cdot}650\,000 \quad 6{\cdot}800\,000 \quad -0{\cdot}850\,000],$$
$$\mathbf{p}'^{(4)}\mathbf{q}^{(4)} = -28{\cdot}050\,000.$$

Lastly,
$$\mathbf{Bq}^{(4)} = \{-0{\cdot}000\,003 \quad 0{\cdot}550\,003 \quad 6{\cdot}600\,006 \quad 0{\cdot}000\,021\},$$
$$\mathbf{p}'^{(4)}\mathbf{B} = [-0{\cdot}000\,003 \quad 10{\cdot}200\,000 \quad -11{\cdot}050\,000 \quad 1{\cdot}700\,003].$$
$$\mathbf{p}'^{(3)}(\mathbf{Bq}^{(4)}) = (\mathbf{p}'^{(4)}\mathbf{B})\,\mathbf{q}^{(3)} = -28{\cdot}050\,000,$$
$$\mathbf{p}'^{(4)}(\mathbf{Bq}^{(4)}) = (\mathbf{p}'^{(4)}\mathbf{B})\,\mathbf{q}^{(4)} = 40{\cdot}672\,500,$$

so that
$$c_3 = d_4 = \mathbf{p}'^{(3)}\mathbf{Bq}^{(4)}/\mathbf{p}'^{(3)}\mathbf{q}^{(3)} = 0{\cdot}297\,500,$$
$$b_4 = e_4 = \mathbf{p}'^{(4)}\mathbf{Bq}^{(4)}/\mathbf{p}'^{(4)}\mathbf{q}^{(4)} = -1{\cdot}450\,000.$$

We now have the matrices

$$\mathbf{P}' = \begin{bmatrix} 1 & 0 & 0 & 0 \\ 0 & 0 & 0 & -1 \\ 0 & -3 & -4 & 1{\cdot}571\,429 \\ -0{\cdot}000\,003 & -7{\cdot}650\,000 & 6{\cdot}800\,000 & -0{\cdot}850\,000 \end{bmatrix}, \tag{9.8.24}$$

$$\mathbf{Q} = \begin{bmatrix} 1 & 0 & 0 & -0{\cdot}000\,003 \\ 0 & -1 & 12{\cdot}571\,429 & 2{\cdot}200\,000 \\ 0 & -2 & 14{\cdot}142\,858 & -1{\cdot}650\,000 \\ 0 & -7 & 0{\cdot}000\,003 & 0 \end{bmatrix}. \tag{9.8.25}$$

The product $\mathbf{P}'\mathbf{Q}$ is not quite diagonal, but we extract the diagonal elements to form the matrix

$$\mathbf{D} = \begin{bmatrix} 1 & 0 & 0 & 0 \\ 0 & 7 & 0 & 0 \\ 0 & 0 & -94{\cdot}285\,714 & 0 \\ 0 & 0 & 0 & -28{\cdot}050\,000 \end{bmatrix}. \tag{9.8.26}$$

The approximate inverse of **Q** is then

$$\mathbf{R}' = \mathbf{D}^{-1}\mathbf{P}' = \begin{bmatrix} 1 & 0 & 0 & 0 \\ 0 & 0 & 0 & -0{\cdot}142\,857 \\ 0 & 0{\cdot}031\,818 & 0{\cdot}042\,424 & -0{\cdot}016\,667 \\ 0 & 0{\cdot}272\,727 & -0{\cdot}242\,424 & 0{\cdot}030\,303 \end{bmatrix}. \tag{9.8.27}$$

As a check that the transformation we have employed is a similarity, we calculate **R**′**Q**, and find

$$\mathbf{R}'\mathbf{Q} = \begin{bmatrix} 1 & 0 & 0 & 0 \\ 0 & 0{\cdot}999\,999 & 0 & 0 \\ 0 & 0{\cdot}000\,003 & 0{\cdot}999\,994 & 0 \\ 0 & 0 & 0 & 0{\cdot}999\,999 \end{bmatrix}. \tag{9.8.28}$$

The triple-diagonal matrix **T** is

$$\mathbf{T} = \begin{bmatrix} 1 & 7 & 0 & 0 \\ 1 & -5{\cdot}428\,571 & -13{\cdot}469\,385 & 0 \\ 0 & 1 & 0{\cdot}878\,571 & 0{\cdot}297\,500 \\ 0 & 0 & 1 & -1{\cdot}450\,000 \end{bmatrix}. \tag{9.8.29}$$

This is the matrix we use to find the characteristic values of **B**.

As a measure of the effects of rounding off, we may calculate the matrix **B*** which would have given rise to the triple-diagonal matrix **T** if we had applied the transformation $\mathbf{Q}^{-1}\mathbf{B}^*\mathbf{Q}$ without rounding off. This is the matrix $\mathbf{QTQ}^{-1}$, and we find (on working to 9 decimal places)

$$10^9[(\mathbf{QTQ}^{-1}) - \mathbf{B}] = \begin{bmatrix} 0 & 2007 & -1783 & 223 \\ 0 & -773 & 679 & -95 \\ 0 & -1552 & 1370 & -187 \\ 0 & -5463 & 4811 & -635 \end{bmatrix}, \tag{9.8.30}$$

where we have used the matrix (9.8.25) for $\mathbf{Q}^{-1}$.

It may well happen that $\mathbf{QTQ}^{-1}$ differs from **B** in the first decimal place, or, in other words, that $\mathbf{Q}^{-1}\mathbf{BQ}$ differs from **T** by amounts of order unity. Such changes would not be negligible, and the algorithm must be made to avoid this difficulty. Two distinct steps must be taken to avoid this. First, the arbitrary elements a_j and f_j may be chosen so that no element is either very big or very small. Secondly, we must not rely upon the new vector being orthogonal to *all* the previous vectors, we must re-orthogonalise it explicitly.

To allow for these changes we may work as follows. *Previously* we started with arbitrary vectors $\mathbf{p}'^{(1)}$ and $\mathbf{q}^{(1)}$ such that $\mathbf{p}'^{(1)}\mathbf{q}^{(1)} \neq 0$ and at the $(j+1)$th stage determined $\mathbf{q}^{(j+1)}$ and $\mathbf{p}'^{(j+1)}$ from the equations

$$c_{j-1}\mathbf{q}^{(j-1)} + b_j\mathbf{q}^{(j)} + a_{j+1}\mathbf{q}^{(j+1)} = \mathbf{B}\mathbf{q}^{(j)}, \tag{9.8.31}$$

$$d_j\mathbf{p}'^{(j-1)} + e_j\mathbf{p}'^{(j)} + f_j\mathbf{p}'^{(j+1)} = \mathbf{p}'^{(j)}\mathbf{B}, \tag{9.8.32}$$

taking the values of c_{j-1} and b_j to be

$$c_{j-1} = \frac{\mathbf{p}'^{(j-1)}(\mathbf{B}\mathbf{q}^{(j)})}{\mathbf{p}'^{(j-1)}\mathbf{q}^{(j-1)}}, \quad b_j = \frac{\mathbf{p}'^{(j)}(\mathbf{B}\mathbf{q}^{(j)})}{\mathbf{p}'^{(j)}\mathbf{q}^{(j)}}, \tag{9.8.33}$$

those of d_j and e_j to be

$$d_j = \frac{(\mathbf{p}'^{(j)}\mathbf{B})\,\mathbf{q}^{(j-1)}}{\mathbf{p}'^{(j-1)}\mathbf{q}^{(j-1)}}, \quad e_j = \frac{(\mathbf{p}'^{(j)}\mathbf{B})\,\mathbf{q}^{(j)}}{\mathbf{p}'^{(j)}\mathbf{q}^{(j)}}, \tag{9.8.34}$$

and a_{j+1} and f_j to be unity. This ensured that $\mathbf{q}^{(j+1)}$ was orthogonal to $\mathbf{p}'^{(j-1)}$ and $\mathbf{p}^{(j)}$, and $\mathbf{p}'^{(j+1)}$ was orthogonal to $\mathbf{q}^{(j-1)}$ and $\mathbf{q}^{(j)}$. *Now* we start with arbitrary $\mathbf{p}'^{(1)}$ and $\mathbf{q}^{(1)}$ such that $\mathbf{p}'^{(1)}\mathbf{q}^{(1)} \neq 0$ and at the $(j+1)$th stage we take the values of c_{j-1}, b_j, d_j and e_j given in equations (9.8.26) and (9.8.27) to form vectors

$$\boldsymbol{l}^{(j)} = \mathbf{B}\mathbf{q}^{(j)} - c_{j-1}\mathbf{q}^{(j-1)} - b_j\mathbf{q}^{(j)}, \tag{9.8.35}$$

$$\mathbf{k}'^{(j)} = \mathbf{p}'^{(j)}\mathbf{B} - d_j\mathbf{p}'^{(j-1)} - e_j\mathbf{p}'^{(j)}, \tag{9.8.36}$$

being careful to find the products involved in equations (9.8.33) and (9.8.34) in the ways indicated by the brackets. Now we choose a_{j+1} and f_j so that the scaled vectors

$$\mathbf{n}^{(j)} = \frac{\boldsymbol{l}^{(j)}}{a_{j+1}}, \quad \mathbf{m}'^{(j)} = \frac{\mathbf{k}'^{(j)}}{f_j} \tag{9.8.37}$$

have their largest elements of order unity. It will be convenient to choose a_{j+1} and f_j to be powers of ten when calculations are performed by hand, and powers of two when they are performed by an automatic digital computer (with a few exceptions). We now take

$$\mathbf{q}^{(j+1)} = \mathbf{n}^{(j)} - u_{jj}\mathbf{q}^{(j)} - u_{j-1,j}\,\mathbf{q}^{(j-1)} - \ldots - u_{1j}\mathbf{q}^{(1)}, \tag{9.8.38}$$

$$\mathbf{p}'^{(j+1)} = \mathbf{m}'^{(j)} - v_{jj}\mathbf{p}'^{(j)} - v_{j,j-1}\mathbf{p}'^{(j-1)} - \ldots - v_{j1}\mathbf{p}'^{(1)}, \tag{9.8.39}$$

and choose the values of the u's and v's to make $\mathbf{q}^{(j+1)}$ orthogonal to all the previous $\mathbf{p}$'s, and $\mathbf{p}'^{(j+1)}$ orthogonal to all the previous $\mathbf{q}$'s; the values required for this are

$$u_{ij} = \frac{\mathbf{p}'^{(i)}\mathbf{n}^{(j)}}{\mathbf{p}'^{(i)}\mathbf{q}^{(i)}}, \quad v_{ji} = \frac{\mathbf{m}'^{(j)}\mathbf{q}^{(i)}}{\mathbf{p}'^{(i)}\mathbf{q}^{(i)}}. \tag{9.8.40}$$

Now equation (9.8.1) will be true to very nearly full working accuracy in almost all cases. The numbers u_{ij} and v_{ij} may be considered as elements of matrices $\mathbf{U}$ and $\mathbf{V}$; $\mathbf{U}$ will be upper triangular and $\mathbf{V}$ will be lower triangular.

We apply the modified algorithm to the matrix of the last example. Only slight improvement will result, since there is not room for much. We choose the same starting vectors as before, namely

$$\mathbf{q}^{(1)} = \{1 \quad 0 \quad 0 \quad 0\}, \quad \mathbf{p}'^{(1)} = [1 \quad 0 \quad 0 \quad 0],$$

for which

$$\mathbf{p}'^{(1)}\mathbf{q}^{(1)} = 1.$$

Then

$$\mathbf{B}\mathbf{q}^{(1)} = \{1 \quad -1 \quad -2 \quad -7\}, \quad \mathbf{p}'^{(1)}\mathbf{B} = [1 \quad 0 \quad 0 \quad -1];$$

$$\mathbf{p}'^{(1)}(\mathbf{B}\mathbf{q}^{(1)}) = 1, \quad b_1 = 1; \quad (\mathbf{p}'^{(1)}\mathbf{B})\,\mathbf{q}^{(1)} = 1, \quad e_1 = 1;$$

$$\boldsymbol{l}^{(1)} = \{0 \quad -1 \quad -2 \quad -7\}, \quad \mathbf{k}'^{(1)} = [0 \quad 0 \quad 0 \quad -1].$$

We now choose $$a_2 = 10, \quad f_1 = 1,$$

and then $$\mathbf{n}^{(1)} = \{0 \quad -0{\cdot}1 \quad -0{\cdot}2 \quad -0{\cdot}7\}, \quad \mathbf{m}'^{(1)} = [0 \quad 0 \quad 0 \quad -1];$$

$$\mathbf{p}'^{(1)}\mathbf{n}^{(1)} = 0, \quad u_{11} = 0; \quad \mathbf{m}'^{(1)}\mathbf{q}^{(1)} = 0, \quad v_{11} = 0;$$

$$\mathbf{q}^{(2)} = \mathbf{n}^{(1)} - u_{11}\mathbf{q}^{(1)} = \{0 \quad -0{\cdot}1 \quad -0{\cdot}2 \quad -0{\cdot}7\},$$

$$\mathbf{p}'^{(2)} = \mathbf{m}'^{(1)} - v_{11}\mathbf{p}'^{(1)} = [0 \quad 0 \quad 0 \quad -1];$$

$$\mathbf{p}'^{(2)}\mathbf{q}^{(2)} = 0{\cdot}7.$$

Then, in the third stage,

$$\mathbf{B}\mathbf{q}^{(2)} = \{0{\cdot}7 \quad 1{\cdot}8 \quad 2{\cdot}5 \quad 3{\cdot}8\}, \quad \mathbf{p}'^{(2)}\mathbf{B} = [7 \quad -3 \quad -4 \quad 7];$$

$$\mathbf{p}'^{(2)}(\mathbf{B}\mathbf{q}^{(2)}) = -3{\cdot}8, \quad b_2 = -5{\cdot}428\,571; \quad \mathbf{p}'^{(1)}(\mathbf{B}\mathbf{q}^{(2)}) = 0{\cdot}7, \quad c_1 = 0{\cdot}7;$$

$$(\mathbf{p}'^{(2)}\mathbf{B})\,\mathbf{q}^{(2)} = -3{\cdot}8, \quad e_2 = -5{\cdot}428\,571; \quad (\mathbf{p}'^{(2)}\mathbf{B})\,\mathbf{q}^{(1)} = 7, \quad d_2 = 7;$$

$$\boldsymbol{l}^{(2)} = \{0 \quad 1{\cdot}257\,143 \quad 1{\cdot}414\,286 \quad 0\},$$

$$\mathbf{k}'^{(2)} = [0 \quad -3 \quad -4 \quad 1{\cdot}571\,429].$$

We now choose $$a_3 = 1, \quad f_2 = 10,$$

so that $$\mathbf{n}^{(2)} = \{0 \quad 1{\cdot}257\,143 \quad 1{\cdot}414\,286 \quad 0\},$$

$$\mathbf{m}'^{(2)} = [0 \quad -0{\cdot}3 \quad -0{\cdot}4 \quad 0{\cdot}157\,143];$$

$$\mathbf{p}'^{(2)}\mathbf{n}^{(2)} = 0, \quad u_{22} = 0; \quad \mathbf{p}'^{(1)}\mathbf{n}^{(2)} = 0, \quad u_{12} = 0;$$

$$\mathbf{m}'^{(2)}\mathbf{q}^{(2)} = 0, \quad v_{22} = 0; \quad \mathbf{m}'^{(2)}\mathbf{q}^{(1)} = 0, \quad v_{21} = 0;$$

$$\mathbf{q}^{(3)} = \mathbf{n}^{(2)} - u_{22}\mathbf{q}^{(2)} - u_{12}\mathbf{q}^{(1)} = \{0 \quad 1{\cdot}257\,143 \quad 1{\cdot}414\,286 \quad 0\}$$

$$\mathbf{p}'^{(3)} = \mathbf{m}'^{(2)} - v_{22}\mathbf{p}'^{(2)} - v_{21}\mathbf{p}'^{(1)} = [0 \quad -0{\cdot}3 \quad -0{\cdot}4 \quad 0{\cdot}157\,143]$$

$$\mathbf{p}'^{(3)}\mathbf{q}^{(3)} = -0{\cdot}942\,857.$$

For the next stage,

$$\mathbf{B}\mathbf{q}^{(3)} = \{0 \quad 0{\cdot}267\,429 \quad 3{\cdot}771\,429 \quad 9{\cdot}428\,573\},$$

$$\mathbf{p}'^{(3)}\mathbf{B} = [-0{\cdot}000\,001 \quad -1{\cdot}028\,571 \quad 0{\cdot}328\,572 \quad 1{\cdot}399\,999];$$

$$\mathbf{p}'^{(3)}(\mathbf{B}\mathbf{q}^{(3)}) = -0{\cdot}828\,366, \quad b_3 = 0{\cdot}878\,570;$$

$$\mathbf{p}'^{(2)}(\mathbf{B}\mathbf{q}^{(3)}) = -9{\cdot}428\,573, \quad c_2 = -13{\cdot}469\,390;$$

$$(\mathbf{p}'^{(3)}\mathbf{B})\,\mathbf{q}^{(3)} = -0{\cdot}828\,366, \quad e_3 = 0{\cdot}878\,570;$$

$$(\mathbf{p}'^{(3)}\mathbf{B})\,\mathbf{q}^{(2)} = -0{\cdot}942\,857, \quad d_3 = -1{\cdot}346\,939;$$

$$\boldsymbol{l}^{(3)} = \{0 \quad 0{\cdot}220\,002 \quad -0{\cdot}164\,998 \quad 0\},$$

$$\mathbf{k}'^{(3)} = [-0{\cdot}000\,001 \quad -0{\cdot}765\,000 \quad 0{\cdot}680\,000 \quad -0{\cdot}085\,001].$$

Choosing $$a_4 = 1, \quad f_3 = 1,$$

we have
$$\mathbf{n}^{(3)} = \boldsymbol{l}^{(3)}, \quad \mathbf{m}'^{(3)} = \mathbf{k}'^{(3)}$$

$$\mathbf{p}'^{(3)}\mathbf{n}^{(3)} = -0{\cdot}000\,001, \quad u_{33} = 0{\cdot}000\,001;$$

$$\mathbf{p}'^{(2)}\mathbf{n}^{(3)} = 0, \quad u_{23} = 0; \quad \mathbf{p}'^{(1)}\mathbf{n}^{(3)} = 0, \quad u_{13} = 0;$$

$$\mathbf{m}'^{(3)}\mathbf{q}^{(3)} = 0, \quad v_{33} = 0;$$

$$\mathbf{m}'^{(3)}\mathbf{q}^{(2)} = 0{\cdot}000\,001, \quad v_{32} = 0{\cdot}000\,001;$$

$$\mathbf{m}'^{(3)}\mathbf{q}^{(1)} = -0{\cdot}000\,001, \quad v_{31} = -0{\cdot}000\,001;$$

$$\begin{aligned}\mathbf{q}^{(4)} &= \mathbf{n}^{(3)} - u_{33}\mathbf{q}^{(3)} - u_{23}\mathbf{q}^{(2)} - u_{13}\mathbf{q}^{(1)} \\ &= \{0 \quad 0{\cdot}220\,001 \quad -0{\cdot}164\,999 \quad 0\},\end{aligned}$$

$$\begin{aligned}\mathbf{p}'^{(4)} &= \mathbf{m}'^{(3)} - v_{33}\mathbf{p}'^{(3)} - v_{32}\mathbf{p}'^{(2)} - v_{31}\mathbf{p}'^{(1)} \\ &= [0 \quad -0{\cdot}765\,000 \quad 0{\cdot}680\,000 \quad -0{\cdot}085\,000];\end{aligned}$$

$$\mathbf{p}'^{(4)}\mathbf{q}^{(4)} = -0{\cdot}280\,500.$$

Finally
$$\begin{aligned}\mathbf{B}\mathbf{q}^{(4)} &= \{0 \quad 0{\cdot}055\,002 \quad 0{\cdot}660\,003 \quad 0{\cdot}000\,007\}, \\ \mathbf{p}'^{(4)}\mathbf{B} &= [0 \quad 1{\cdot}020\,000 \quad -1{\cdot}105\,000 \quad 0{\cdot}170\,000];\end{aligned}$$

$$\mathbf{p}'^{(4)}(\mathbf{B}\mathbf{q}^{(4)}) = 0{\cdot}406\,725, \quad b_4 = -1{\cdot}450\,000;$$

$$\mathbf{p}'^{(3)}(\mathbf{B}\mathbf{q}^{(4)}) = -0{\cdot}280\,501, \quad c_3 = 0{\cdot}297\,501;$$

$$(\mathbf{p}'^{(4)}\mathbf{B})\,\mathbf{q}^{(4)} = 0{\cdot}406\,725, \quad e_4 = -1{\cdot}450\,000;$$

$$(\mathbf{p}'^{(4)}\mathbf{B})\,\mathbf{q}^{(3)} = -0{\cdot}280\,500, \quad d_4 = 0{\cdot}297\,500;$$

$$\boldsymbol{l}^{(4)} = \{0 \quad 0{\cdot}000\,002 \quad 0{\cdot}000\,003 \quad 0{\cdot}000\,007\},$$

$$\mathbf{m}'^{(4)} = [0 \quad 0 \quad 0 \quad 0].$$

These last two vectors should be orthogonal to four of the preceding vectors, and, since they have only four elements, they should vanish. The extent to which they do not is a measure of the imprecision of our calculations.

The matrices found by this process are

$$\mathbf{Q} = \begin{bmatrix} 1{\cdot}000\,000 & 0{\cdot}000\,000 & 0{\cdot}000\,000 & 0{\cdot}000\,000 \\ 0{\cdot}000\,000 & -0{\cdot}100\,000 & 1{\cdot}257\,143 & 0{\cdot}220\,001 \\ 0{\cdot}000\,000 & -0{\cdot}200\,000 & 1{\cdot}414\,286 & -0{\cdot}164\,999 \\ 0{\cdot}000\,000 & -0{\cdot}700\,000 & 0{\cdot}000\,000 & 0{\cdot}000\,000 \end{bmatrix}, \qquad (9.8.41)$$

$$\mathbf{P}' = \begin{bmatrix} 1{\cdot}000\,000 & 0{\cdot}000\,000 & 0{\cdot}000\,000 & 0{\cdot}000\,000 \\ 0{\cdot}000\,000 & 0{\cdot}000\,000 & 0{\cdot}000\,000 & -1{\cdot}000\,000 \\ 0{\cdot}000\,000 & -0{\cdot}300\,000 & -0{\cdot}400\,000 & 0{\cdot}157\,143 \\ 0{\cdot}000\,000 & -0{\cdot}765\,000 & 0{\cdot}680\,000 & -0{\cdot}085\,000 \end{bmatrix}, \qquad (9.8.42)$$

$$\mathbf{T} = \begin{bmatrix} 1{\cdot}000\,000 & 0{\cdot}700\,000 & 0 & 0 \\ 10{\cdot}000\,000 & -5{\cdot}428\,571 & -13{\cdot}469\,390 & 0 \\ 0 & 1{\cdot}000\,000 & 0{\cdot}878\,570 & 0{\cdot}297\,501 \\ 0 & 0 & 1{\cdot}000\,000 & -1{\cdot}450\,000 \end{bmatrix}, \tag{9.8.43}$$

$$\mathbf{S} = \begin{bmatrix} 1{\cdot}000\,000 & 1{\cdot}000\,000 & 0 & 0 \\ 7{\cdot}000\,000 & -5{\cdot}428\,571 & 10{\cdot}000\,000 & 0 \\ 0 & -1{\cdot}346\,939 & 0{\cdot}878\,570 & 1{\cdot}000\,000 \\ 0 & 0 & 0{\cdot}297\,500 & -1{\cdot}450\,000 \end{bmatrix}. \tag{9.8.44}$$

Ve shall not make use of the matrices **U** and **V** and shall not record them; we ıote only that all their elements are small. The matrix

$$\mathbf{D} = \mathbf{P}'\mathbf{Q}$$

s diagonal, and is

$$\mathbf{D} = \begin{bmatrix} 1 & 0 & 0 & 0 \\ 0 & 0{\cdot}7 & 0 & 0 \\ 0 & 0 & -0{\cdot}942\,857 & 0 \\ 0 & 0 & 0 & -0{\cdot}280\,500 \end{bmatrix}.$$

The approximate inverse of **Q**, namely

$$\mathbf{R}' = \mathbf{D}^{-1}\mathbf{P}'$$

s

$$\mathbf{R}' = \begin{bmatrix} 1{\cdot}000\,000 & 0{\cdot}000\,000 & 0{\cdot}000\,000 & 0{\cdot}000\,000 \\ 0{\cdot}000\,000 & 0{\cdot}000\,000 & 0{\cdot}000\,000 & -1{\cdot}428\,571 \\ 0{\cdot}000\,000 & 0{\cdot}318\,182 & 0{\cdot}424\,242 & -0{\cdot}166\,667 \\ 0{\cdot}000\,000 & 2{\cdot}727\,273 & -2{\cdot}424\,242 & 0{\cdot}303\,030 \end{bmatrix}.$$

Ve find that

$$\mathbf{R}'\mathbf{Q} = \begin{bmatrix} 1{\cdot}000\,000 & 0{\cdot}000\,000 & 0{\cdot}000\,000 & 0{\cdot}000\,000 \\ 0{\cdot}000\,000 & 1{\cdot}000\,000 & 0{\cdot}000\,000 & 0{\cdot}000\,000 \\ 0{\cdot}000\,000 & 0{\cdot}000\,000 & 1{\cdot}000\,000 & 0{\cdot}000\,001 \\ 0{\cdot}000\,000 & 0{\cdot}000\,000 & 0{\cdot}000\,001 & 1{\cdot}000\,000 \end{bmatrix},$$

o that the transformation is very close to a true similarity. As a check on the ccuracy of the transformation we calculate

$$\mathbf{QTQ}^{-1} = \begin{bmatrix} 1{\cdot}000\,000 & 0{\cdot}000\,000 & 0{\cdot}000\,000 & -1{\cdot}000\,000 \\ -1{\cdot}000\,000 & 1{\cdot}000\,000 & 0{\cdot}999\,999 & -3{\cdot}000\,000 \\ -2{\cdot}000\,000 & 3{\cdot}000\,001 & -0{\cdot}000\,002 & -4{\cdot}000\,000 \\ -7{\cdot}000\,000 & 3{\cdot}000\,002 & 3{\cdot}999\,997 & -7{\cdot}000\,002 \end{bmatrix}, \tag{9.8.45}$$

ısing $\mathbf{R}'$ for $\mathbf{Q}^{-1}$; this differs from **B** by at most three in the last figure.

The importance of the modified algorithm is that it prevents the accumulation of rounding off errors, so that $\mathbf{QTQ}^{-1}$ will give $\mathbf{B}$ to an accuracy comparable to that found in the above example, even when the simple process first described would lead to large disagreement.

The process, either in its simple or in its modified form, may break down because for some value of j the product $\mathbf{p}'^{(j)}\mathbf{q}^{(j)}$ vanishes. This may happen even when neither of the vectors $\mathbf{p}'^{(j)}$ and $\mathbf{q}^{(j)}$ vanishes, and when the matrix $\mathbf{B}$ is quite well-behaved. In this case it is necessary to start again with new vectors $\mathbf{p}'^{(1)}$ and $\mathbf{q}^{(1)}$. The product may also vanish because one (or both) of the vectors $\mathbf{p}'^{(j)}$ and $\mathbf{q}^{(j)}$ vanishes. In this case, the process can be continued if the vanishing vector is replaced by any other vector which is orthogonal to all the previous vectors of the alternate set. Suppose that the process breaks down at the step which would have determined $\mathbf{q}^{(j+1)}$—that is, suppose that $\boldsymbol{l}^{(j)}$ given by equation (9.8.35) is zero. A new $\mathbf{q}^{(j+1)}$ may be determined by choosing a new, non-zero, vector $\mathbf{n}^{(j)}$ and using equation (9.8.37). But if $\mathbf{n}^{(j)}$ is non-zero then the equation (9.8.37), namely

$$a_{j+1}\mathbf{n}^{(j)} = \boldsymbol{l}^{(j)}, \tag{9.8.46}$$

can be satisfied only by putting

$$a_{j+1} = 0. \tag{9.8.47}$$

Also we have

$$d_{j+1} = \frac{(\mathbf{p}'^{(j+1)}\mathbf{B})\,\mathbf{q}^{(j)}}{\mathbf{p}'^{(j)}\mathbf{q}^{(j)}} = \frac{\mathbf{p}'^{(j+1)}(\mathbf{B}\mathbf{q}^{(j)})}{\mathbf{p}'^{(j)}\mathbf{q}^{(j)}} = \frac{\mathbf{p}'^{(j+1)}(\boldsymbol{l}^{(j)} + c_{j-1}\mathbf{q}^{(j-1)} + b_j\mathbf{q}^{(j)})}{\mathbf{p}'^{(j)}\mathbf{q}^{(j)}}, \tag{9.8.48}$$

and the first product in the numerator vanishes because $\boldsymbol{l}^{(j)}$ does, while the last vanishes because of the orthogonality relations: therefore

$$d_{j+1} = 0. \tag{9.8.49}$$

Similarly, if $\mathbf{p}'^{(j+1)}$ vanishes, we may replace it by any vector orthogonal to all the previous $\mathbf{q}^{(i)}$, and set

$$c_j = 0 = f_{j+1}. \tag{9.8.50}$$

If the process breaks down because $\mathbf{q}^{(j+1)}$ or $\mathbf{p}'^{(j+1)}$ vanishes the matrix $\mathbf{T}$ has an advantageous form. It has a zero co-diagonal element, and its characteristic polynomial can be written as the product of two factors, the minors before and after the zero elements. Thus there are two polynomial equations of lower degree to solve instead of one of the full degree.

For the matrix

$$\mathbf{T} = \begin{bmatrix} -2 & 1 & 0 & 0 & 0 \\ 1 & -2 & 1 & 0 & 0 \\ 0 & 0 & -2 & 1 & 0 \\ 0 & 0 & 1 & -2 & 1 \\ 0 & 0 & 0 & 1 & -2 \end{bmatrix},$$

for example, we have

$$|\mathbf{T}-\lambda\mathbf{I}| = \begin{vmatrix} -2-\lambda & 1 \\ 1 & -2-\lambda \end{vmatrix} \cdot \begin{vmatrix} -2-\lambda & 1 & 0 \\ 1 & -2-\lambda & 1 \\ 0 & 1 & -2-\lambda \end{vmatrix}.$$

The characteristic vectors also have special forms. Suppose that the zero appears below the diagonal. Then if λ is a characteristic value of the upper minor the corresponding column vector $\mathbf{q}$ will have zeros in the positions multiplied by the lower minor. Similarly, if λ is a characteristic value of the lower minor the corresponding row vector $\mathbf{p}'$ will have zeros for its upper elements.

For the matrix of the above example $\lambda = -1$ is a characteristic value of the upper minor. The corresponding characteristic column vector is

$$\mathbf{q} = \{1 \quad -1 \quad 0 \quad 0 \quad 0\}.$$

A characteristic value of the lower minor is $\lambda = -0{\cdot}292\,893$ and the corresponding row vector is

$$\mathbf{p}' = [0 \quad 0 \quad \sqrt{2}/2 \quad 1 \quad \sqrt{2}/2].$$

EXAMPLE 9.8

1. Use Lanczos's method in its modified form to transform to triple-diagonal form the matrix

$$\begin{bmatrix} 1{\cdot}000\,000 & 0{\cdot}500\,000 & 0{\cdot}166\,667 & 0{\cdot}083\,333 \\ 0{\cdot}500\,000 & 0{\cdot}333\,333 & 0{\cdot}250\,000 & 0{\cdot}066\,667 \\ 0{\cdot}666\,667 & 0{\cdot}500\,000 & 0{\cdot}200\,000 & 0{\cdot}111\,111 \\ 0{\cdot}750\,000 & 0{\cdot}600\,000 & 0{\cdot}250\,000 & 0{\cdot}142\,857 \end{bmatrix}.$$

9.9 Muller's method of solving equations

The triple diagonal matrix formed by Lanczos's process may have the Sturmian property; that is, its symmetrically placed co-diagonal elements may have the same sign. If this is so, the method of § 9.5 may be used to locate its characteristic values to any desired accuracy. If the matrix does not have this property it becomes necessary to make use of more general processes for solving equations. One such process is based upon the Newton–Raphson formula (9.5.4), the characteristic determinant and its derivative being evaluated as in the example on page 339, using equation (9.4.7). This will often involve arithmetic using complex numbers, and the labour of using two recursions for each step of the iteration is quite considerable. Only one function need be evaluated for each step of Muller's method,† while the convergence is as good as it is with the Newton–Raphson process.

Suppose that the equation to be solved is

$$f(x) = 0. \tag{9.9.1}$$

If we know three values of f we may fit a quadratic polynomial to these values and take one of its zeros as an approximate root of the equation. If we denote the three arguments by x_1, x_2 and x_3, the three values will be

$$f_1 = f(x_1), \quad f_2 = f(x_2) \quad \text{and} \quad f_3 = f(x_3). \tag{9.9.2}$$

† D. Muller, 'A method for solving algebraic equations using an automatic computer', *Math. Tables Aids Comput.* **10** (1956), pp. 208–15.

We choose the zero of the quadratic nearer to x_3, and use this as a new argument, x_4. We then calculate

$$f_4 = f(x_4), \tag{9.9.3}$$

fit a quadratic to f_2, f_3 and f_4 and take the zero of this quadratic nearer to x_4 as the new argument, x_5. This process may be continued until convergence is achieved.

There are many ways of deriving the coefficients of the interpolating quadratic, but perhaps the most straightforward is to solve the three equations which express the fact that the polynomial takes the three prescribed values for the given values of x. The resulting formulae are

$$q(x) = (q_0 x^2 + q_1 x + q_2)/(x_1 - x_2)(x_2 - x_3)(x_3 - x_1), \tag{9.9.4}$$

$$q_0 = -f_1(x_2 - x_3) - f_2(x_3 - x_1) - f_3(x_1 - x_2), \tag{9.9.5}$$

$$q_1 = f_1(x_2 - x_3)(x_2 + x_3) + f_2(x_3 - x_1)(x_3 + x_1) + f_3(x_1 - x_2)(x_1 + x_2), \tag{9.9.6}$$

$$q_2 = -f_1(x_2 - x_3)x_2 x_3 - f_2(x_3 - x_1)x_3 x_1 - f_3(x_1 - x_2)x_1 x_2. \tag{9.9.7}$$

The denominator of equation (9.9.4) may be ignored, since it does not affect the zeros of the polynomial. The zeros of the quadratic are readily found from the elementary formula

$$x = [-q_1 \pm \sqrt{(q_1^2 - 4q_0 q_2)}]/2q_0, \tag{9.9.8}$$

but care must be taken in evaluating this. One zero may involve much cancellation of leading digits, and then it should be found by dividing the other zero into q_2. The zero nearer to x_3 is chosen for x_4.

At the rth step, equations (9.9.5) to (9.9.7) are to be used with all the suffices on the right-hand sides advanced by $r-1$.

We shall apply the method to the polynomial equation

$$f(x) = (0{\cdot}5 - x)^3 + (0{\cdot}5 - x) \tag{9.9.9}$$

showing in detail the calculations of the first step, and quoting the results of the later steps. Let us seek a zero near $x = 0$, taking as our first three estimates

$$x_1 = i, \quad x_2 = 1, \quad x_3 = 0. \tag{9.9.10}$$

From these we derive

$$f_1 = -0{\cdot}875 - 0{\cdot}750i; \quad f_2 = -0{\cdot}625; \quad f_3 = 0{\cdot}625.$$

Then

$$-f_1(x_2 - x_3) = (0{\cdot}875 + 0{\cdot}75i)\,.\,1 = 0{\cdot}875 + 0{\cdot}75i,$$

$$-f_2(x_3 - x_1) = -0{\cdot}625\,.\,i = -0{\cdot}625i,$$

$$-f_3(x_1 - x_2) = -0{\cdot}625(i - 1) = 0{\cdot}625 - 0{\cdot}625i.$$

Hence

$$q_0 = 1{\cdot}5 - 0{\cdot}5i. \tag{9.9.11}$$

We also need the following quantities

$$x_2 + x_3 = 1, \quad x_3 + x_1 = i, \quad x_1 + x_2 = 1 + i,$$

$$x_2 x_3 = 0, \quad x_3 x_1 = 0, \quad x_1 x_2 = i,$$

and from these we find

$$q_1 = (-0{\cdot}875-0{\cdot}75i)\,(1)+(0{\cdot}625i)\,(i)+(-0{\cdot}625+0{\cdot}625i)(1+i)$$
$$= -2{\cdot}75-0{\cdot}75i, \tag{9.9.12}$$

$$q_2 = (0{\cdot}875+0{\cdot}75i)\,(0)+(-0{\cdot}625i)\,(0)+(0{\cdot}625-0{\cdot}625i)\,(i)$$
$$= 0{\cdot}625+0{\cdot}625i. \tag{9.9.13}$$

The following quadratic equation has now to be solved

$$(1{\cdot}5-0{\cdot}5i)\,x^2+(-2{\cdot}75-0{\cdot}75i)\,x+(0{\cdot}625+0{\cdot}625i) = 0 \tag{9.9.14}$$

and, for this, we use the usual elementary formula

$$x = \frac{-q_1 \pm \sqrt{(q_1^2-4q_0q_2)}}{2q_0}. \tag{9.9.15}$$

The numerical values involved are

$$q_1^2 = 7{\cdot}000+4{\cdot}125i, \quad q_0q_2 = 1{\cdot}250+0{\cdot}625i,$$

so that
$$q_1^2-4q_0q_2 = 2{\cdot}000+1{\cdot}625i.$$

The easiest way to find the square root is to convert to modulus and phase angle form. We find

$$|q_1^2-4q_0q_2|^2 = 6{\cdot}640\,625, \quad |q_1^2-4q_0q_2| = 2{\cdot}576\,941,$$

$$\sqrt{|q_1^2-4q_0q_2|} = 1{\cdot}605\,285,$$

$$\tan\phi = 0{\cdot}8125, \quad \phi = 39^\circ\ 5{\cdot}63', \quad \tfrac{1}{2}\phi = 19^\circ\ 32{\cdot}82',$$

$$\cos(\tfrac{1}{2}\phi) = 0{\cdot}942\,367, \quad \sin(\tfrac{1}{2}\phi) = 0{\cdot}334\,580.$$

Hence
$$\sqrt{(q_1^2-4q_0q_2)} = 1{\cdot}512\,768+0{\cdot}537\,096i,$$
$$-q_1+\sqrt{(q_1^2-4q_0q_2)} = 4{\cdot}262\,768+1{\cdot}287\,096i,$$
$$-q_1-\sqrt{(q_1^2-4q_0q_2)} = 1{\cdot}237\,232+0{\cdot}212\,904i.$$

Now
$$1/(2q_0) = 1/(3-i) = (3+i)/10 = 0{\cdot}3+0{\cdot}1i,$$

hence the two roots of the equation are

$$1{\cdot}150\,121+0{\cdot}812\,406i,$$
$$0{\cdot}349\,879+0{\cdot}187\,594i.$$

We choose the value nearer x_3, that is 0, for x_4, so that

$$x_4 = 0{\cdot}349\,879+0{\cdot}187\,594i, \tag{9.9.16}$$

and then
$$f_4 = f(x_4) = 0{\cdot}137\,655-0{\cdot}193\,676i.$$

The process continues, using x_2, x_3, x_4, f_2, f_3 and f_4 in place of x_1, x_2, x_3, f_1, f_2 and f_3 in equations (9.9.5)–(9.9.7). When x_5 has been found, x_3, x_4, x_5 and the corresponding function values are used in the following step. The process is continued in this way until convergence is achieved. The following table gives the values of the successive approximations and the corresponding values of the function:

$$\left.\begin{array}{cc} x & f(x) \\ 0{\cdot}349\,879+0{\cdot}187\,594i & 0{\cdot}137\,655-0{\cdot}193\,676i \\ 0{\cdot}469\,579+0{\cdot}037\,316i & 0{\cdot}030\,322-0{\cdot}037\,368i \\ 0{\cdot}497\,852-0{\cdot}005\,881i & 0{\cdot}002\,148+0{\cdot}005\,881i \\ 0{\cdot}500\,061-0{\cdot}000\,039i & -0{\cdot}000\,061+0{\cdot}000\,039i \\ 0{\cdot}500\,000+0{\cdot}000\,000i & 0{\cdot}000\,000+0{\cdot}000\,000i \end{array}\right\} \tag{9.9.17}$$

There is no hard and fast rule for the choice of the initial values x_1, x_2 and x_3. If they can be chosen near to a zero of the function, convergence will be that much quicker, but the process will usually converge to some zero, and if it is not in the region where the desired zero is, the process can be repeated with new starting values and a further zero found, this being done again and again until the desired zero is found.

In finding several zeros it is advisable to have a technique which prevents convergence upon a zero that has already been found. If the function values, computed from the original function, are modified by dividing them by factors which remove the effects of the known zeros, the process will yield a known zero only if it is a multiple zero, and then will reveal it as many times as it is multiple, so that all the linear factors of a polynomial may be found with the right multiplicities. To ensure this it is advisable to restart the process after a zero has been found using three values of x near that zero, when it may be expected that the next zero found will be the same one if it is multiple. The modification to the function values is made according to the equation

$$f_j = f(x_j)/(x_j - a_1)\,(x_j - a_2)\ldots(x_j - a_r). \tag{9.9.18}$$

Here, $a_1, a_2, \ldots, a_r$ are the zeros which have already been found, and x_j is the approximate zero for which a function value is needed. As can be seen, cancellation of leading digits will occur if a repeated zero is being found for the second or later time, and extra precision may be necessary in the working.

For the polynomial of the last example, which has one zero at $x = a_1 = 0{\cdot}5$, we choose for the three starting values

$$x_1 = 0{\cdot}5 + 0{\cdot}100\,000i, \quad x_2 = 0{\cdot}6, \quad x_3 = 0{\cdot}5005, \tag{9.9.19}$$

the last being close to the known zero.

We then calculate

$$\begin{aligned} f_1 &= ((0{\cdot}5 - x_1)^3 + (0{\cdot}5 - x_1))/(x_1 - 0{\cdot}5) \\ &= -0{\cdot}099i/0{\cdot}1i = -0{\cdot}99, \\ f_2 &= ((0{\cdot}5 - 0{\cdot}6)^3 + (0{\cdot}5 - 0{\cdot}6))/(0{\cdot}6 - 0{\cdot}5) \\ &= -0{\cdot}101/0{\cdot}1 = -1{\cdot}01, \end{aligned}$$

and
$$f_3 = ((0{\cdot}5 - 0{\cdot}500\,5)^3 + (0{\cdot}5 - 0{\cdot}500\,5))/(0{\cdot}500\,5 - 0{\cdot}5)$$
$$= -0{\cdot}000\,500\,000/0{\cdot}000\,500$$
$$= -1{\cdot}000\,000.$$

Then
$$-f_1(x_2 - x_3) = -(-0{\cdot}99)(+0{\cdot}0995) = +0{\cdot}098\,505,$$
$$-f_2(x_3 - x_1) = -(-1{\cdot}01)(0{\cdot}000\,500 - 0{\cdot}100\,000i)$$
$$= 0{\cdot}000\,505 - 0{\cdot}101\,000i,$$
$$-f_3(x_1 - x_2) = -(-1{\cdot}000\,000)(-0{\cdot}100\,000 + 0{\cdot}100\,000i)$$
$$= -0{\cdot}100\,000 + 0{\cdot}100\,000i.$$

Hence
$$q_0 = -0{\cdot}000\,990 - 0{\cdot}001\,000i. \tag{9.9.20}$$

Also
$$x_2 + x_3 = 1{\cdot}100\,500, \qquad x_2x_3 = 0{\cdot}300\,300,$$
$$x_3 + x_1 = 1{\cdot}000\,500 + 0{\cdot}100\,000i, \quad x_3x_1 = 0{\cdot}250\,250 + 0{\cdot}050\,050i,$$
$$x_1 + x_2 = 1{\cdot}100\,000 + 0{\cdot}100\,000i, \quad x_1x_2 = 0{\cdot}300\,000 + 0{\cdot}600\,000i.$$

Hence
$$q_1 = (-0{\cdot}098\,505)(1{\cdot}100\,500)$$
$$-(0{\cdot}000\,505 - 0{\cdot}101\,000i)(1{\cdot}000\,500 + 0{\cdot}100\,000i)$$
$$-(-0{\cdot}100\,000 + 0{\cdot}100\,000i)(1{\cdot}100\,000 + 0{\cdot}100\,000i)$$
$$= 0{\cdot}000\,990 + 0{\cdot}001\,000i, \tag{9.9.21}$$
$$q_2 = (0{\cdot}098\,505)(0{\cdot}300\,300)$$
$$+(0{\cdot}000\,505 - 0{\cdot}101\,000i)(0{\cdot}250\,250 + 0{\cdot}050\,050i)$$
$$+(-0{\cdot}100\,000 + 0{\cdot}100\,000i)(0{\cdot}300\,000 + 0{\cdot}060\,000i)$$
$$= -0{\cdot}001\,138 - 0{\cdot}001\,250i. \tag{9.9.22}$$

We solve the quadratic equation in the same way as in the last example, recording each complex number with about six significant figures in either its real or imaginary part (or both). We find
$$q_0q_2 = 10^{-6}(-0{\cdot}024\,380 + 2{\cdot}475\,500i),$$
$$q_1^2 = 10^{-6}(-0{\cdot}019\,900 + 1{\cdot}980\,000i),$$
$$q_1^2 - 4q_0q_2 = 10^{-6}(0{\cdot}077\,620 - 7{\cdot}922\,000i).$$

Then
$$|q_1^2 - 4q_0q_2|^2 = 10^{-12}(62{\cdot}764\,109), \quad |q_1^2 - 4q_0q_2| = 10^{-6}(7{\cdot}922\,380),$$
$$\sqrt{|(q_1^2 - 4q_0q_2)|} = 10^{-3}(2{\cdot}814\,672).$$
$$\tan\phi = -102{\cdot}061, \quad \phi = -89^\circ\ 26{\cdot}32', \quad \tfrac{1}{2}\phi = -44^\circ\ 43{\cdot}16'.$$
$$\cos(\tfrac{1}{2}\phi) = 0{\cdot}710\,562, \quad \sin(\tfrac{1}{2}\phi) = -0{\cdot}703\,634.$$

Hence
$$\sqrt{(q_1^2 - 4q_0q_2)} = 10^{-3}(1{\cdot}999\,999 - 1{\cdot}980\,499i),$$
$$-q_1 + \sqrt{(q_1^2 - 4q_0q_2)} = 10^{-3}(1{\cdot}009\,999 - 2{\cdot}980\,499i),$$
$$-q_1 - \sqrt{(q_1^2 - 4q_0q_2)} = 10^{-3}(-2{\cdot}989\,999 + 0{\cdot}980\,499i).$$

Now
$$1/(2q_0) = 1/(-0{\cdot}001\,980 - 0{\cdot}002\,000i)$$
$$= (-0{\cdot}001\,980 + 0{\cdot}002\,000i)/[(0{\cdot}001\,980)^2 + (0{\cdot}002\,000)^2]$$
$$= -249{\cdot}987 + 252{\cdot}512i,$$

so that the two roots of the quadratic equation are

$$0{\cdot}500\,125 + 1{\cdot}000\,123i$$

$$0{\cdot}499\,873 - 1{\cdot}000\,123i.$$

Of these, the former is nearer to x_3, so that

$$x_4 = 0{\cdot}500\,125 + 1{\cdot}000\,123i, \tag{9.9.23}$$

and
$$f(x_4) = 10^{-3}(0{\cdot}246\,000 - 0{\cdot}250\,031i).$$

The next two steps yield the approximations

$$0{\cdot}500\,015 + 1{\cdot}000\,004i,$$

$$0{\cdot}500\,000 + 1{\cdot}000\,000i, \tag{9.9.24}$$

At the last step, the discarded approximation to the zero was

$$0{\cdot}500\,000 - 1{\cdot}000\,001i.$$

and at the last stage, when the function to which the quadratic polynomial is being fitted is effectively a quadratic itself, the discarded approximation will be very nearly the right value for the last zero to be found, and should certainly be used as one of the three trial approximations at the last stage. At the last stage, the fitted quadratic may have a very small coefficient of x^2, compared with the coefficient of x. If it is found that the ratio q_0/q_1 is very much smaller than q_1/q_2, then the equation may be treated as linear, and the last stage is then essentially one of repeated application of the rule of false position (linear extrapolation).

As can be seen from this example, the arithmetic of this process is tedious in the extreme. It is not easy to check, and is very liable to error, because of the process's irregular structure. However, it is very simple to write a programme for almost any digital computer, to make it carry out this process, and such programmes are already available for most machines, so that the practical application of the process is almost trivial.

In particular, the process may be applied directly to the solution of the equation for the normal frequencies of a damped system

$$|\mathbf{A}\lambda^2 + \mathbf{B}\lambda + \mathbf{C}| = 0.$$

For a trial value of λ, this determinant may be found by factorising the matrix $\mathbf{A}\lambda^2 + \mathbf{B}\lambda + \mathbf{C}$, and taking the product of the diagonal elements of the factors. Since the values of λ will normally be complex, this will involve heavy arithmetic which makes it highly unsuitable for hand computation. Nevertheless, it can be mechanised successfully. This problem can also be reduced to that treated in § 9.8, as long as either **A** or **C** is non-singular. Taking new variables **y** defined by

$$\mathbf{y} = \lambda\mathbf{x},$$

we have

$$\lambda\mathbf{x} = \mathbf{y}$$

$$\lambda\mathbf{y} = -\mathbf{A}^{-1}\mathbf{C}\mathbf{x} - \mathbf{A}^{-1}\mathbf{B}\mathbf{y}.$$

The characteristic equation is

$$\begin{bmatrix} \mathbf{0} & \mathbf{I} \\ -\mathbf{A}^{-1}\mathbf{B} & -\mathbf{A}^{-1}\mathbf{C} \end{bmatrix} \begin{bmatrix} \mathbf{x} \\ \mathbf{y} \end{bmatrix} = \lambda \begin{bmatrix} \mathbf{x} \\ \mathbf{y} \end{bmatrix},$$

where the matrix to be treated by the method of § 9.8 is of order $2n$.

APPENDIX

Most of the worked examples in Chapters 8 and 9 are based on one or other of two matrices. We summarise below the information which is obtained about these matrices. Best values are marked with an asterisk.

1. $$\mathbf{B} = \begin{bmatrix} 1 & 2 & 3 & 4 \\ 2 & 6 & 10 & 14 \\ 3 & 10 & 20 & 30 \\ 4 & 14 & 30 & 50 \end{bmatrix}.$$

Page 301. Crude bounds for highest λ, λ_4:

$$|\lambda_4 - 73{\cdot}33| < 0{\cdot}610447.$$

Page 304. Bounds for least λ, λ_1, using data from §§9.5 and 9.7:

$$0{\cdot}209316 \leqslant \lambda \leqslant 0{\cdot}209436.$$

Page 307. Better bounds for λ_1 (Wilkinson):

* $$|\lambda_1 - 0{\cdot}209378732229| \leqslant (0{\cdot}012234)10^{-6}.$$

Page 310. Iterative method for λ_4 and $\mathbf{x}^{(4)}$.
Estimate from iteration above: $\lambda_4 = 73{\cdot}335900$.
* Estimate using Rayleigh's quotient: $\lambda_4 = 73{\cdot}335849$.
The latter is correct to 6 decimal places.

* $$\mathbf{x}^{(4)} = \{0{\cdot}089578 \quad 0{\cdot}303300 \quad 0{\cdot}624378 \quad 1\}.$$

Page 312. Factorisation into triangular factors and iteration for λ_1 and $\mathbf{x}^{(1)}$.
Estimate from iteration and Rayleigh's quotient:

$$\lambda_1 = 0{\cdot}209507,$$

$$\mathbf{x}^{(1)} = \{1 \quad -0{\cdot}756704 \quad 0{\cdot}339789 \quad -0{\cdot}072216\}.$$

Page 312. Convergence ratios for λ_1 and λ_4.

Page 317. Iteration (by orthogonalisation) for λ_3 and $\mathbf{x}^{(3)}$ using $\mathbf{x}^{(4)}$ from page 310.

Best estimate of λ_3 from:

	(*a*) iteration alone	2·928888,
	(*b*) Aitken's δ^2 process applied to (*a*)	2·920888,
	(*c*) Rayleigh's quotients	2·919377,
*	(*d*) Aitken's δ^2 process applied to (*c*)	2·919381,

$$\mathbf{x}^{(3)} = \{0{\cdot}485429 \quad 1 \quad 0{\cdot}639694 \quad -0{\cdot}746195\}.$$

Page 321. λ_3 and $\mathbf{x}^{(3)}$ by deflation using data of §8.5.
Estimates: $\lambda_3 = 2{\cdot}919384$.

$$\mathbf{x}^{(3)} = \{0{\cdot}396562 \quad 0{\cdot}815659 \quad 0{\cdot}520313 \quad -0{\cdot}607785\}.$$

Page 329. Reduction to triple diagonal form—see equation (9.3.33).

Page 331. Estimate of error in characteristic values due to rotation.

Page 340. Calculation of λ_1 using Sturm' sequences.
Estimate: $\lambda_1 = 0{\cdot}209376$.

Page 341. Estimate of error in λ_1 due to rounding off.

Page 345. Calculation of characteristic vector of reduced matrix of page 344 corresponding to λ_1 of page 340; see equation (9.6.22).

Page 347. Calculation of $\mathbf{x}^{(1)}$ from equation (9.6.22).

* Estimate: $\mathbf{x}^{(1)} = \{1 \quad -0{\cdot}790222 \quad 0{\cdot}377006 \quad -0{\cdot}085300\}$.

2.

$$\mathbf{B} = \begin{bmatrix} -2 & 1 & 0 & 0 \\ 1 & -2 & 1 & 0 \\ 0 & 1 & -2 & 1 \\ 0 & 0 & 1 & -2 \end{bmatrix}.$$

Page 308. Bounds for λ_2 (Kohn–Kato)

$$-2{\cdot}618036 \leqslant \lambda_2 \leqslant -2{\cdot}618024.$$

Page 333. Sturm's polynomials $P_r(\lambda)$.

Page 333. Derivatives $P_r'(\lambda)$.

Page 336. Sign count for $P_3(\lambda)$.

Page 338. Approximate location of the roots of $P_3(\lambda)$.

Page 338. Approximate location of λ_2 using linear interpolation $\lambda_2 = -2{\cdot}617917$.

Page 339. Approximate location of λ_2 using the Newton–Raphson formula.

* Estimate: $\lambda_2 = -2{\cdot}618046$.

Correct value: $\lambda_2 = -2{\cdot}618034$.

ANSWERS TO EXAMPLES

1.1.1 (*a*) $\mathbf{A}(1, 4)$. (*b*) $\mathbf{A}(3, 3)$, or 'of order 3'; symmetric.
(*c*) $\mathbf{A}(4, 1)$. (*d*) $\mathbf{A}(3, 4)$. (*e*) $\mathbf{A}(3, 3)$ or 'of order 3'; skew.

1.1.2 (*a*) $\{0 \quad -1 \quad 4 \quad 3\}$, $\mathbf{A}(4, 1)$. (*b*) $\begin{bmatrix} 3 & 2 & 5 \\ 2 & 4 & 9 \\ 5 & 9 & -16 \end{bmatrix}$, $\mathbf{A}(3, 3)$, symmetric.

(*c*) $[1 \quad 9 \quad -4 \quad 2]$, $\mathbf{A}(1, 4)$. (*d*) $\begin{bmatrix} 0 & -4 & 1 \\ 4 & 0 & -2 \\ -1 & 2 & 0 \\ 2 & 0 & 0 \end{bmatrix}$, $\mathbf{A}(4, 3)$.

(*e*) $\begin{bmatrix} -1 & 2 & 9 \\ -2 & 3 & -14 \\ -9 & 14 & 4 \end{bmatrix}$, $\mathbf{A}(3, 3)$; skew.

1.1.3 (*a*) $n(n+1)/2$. (*b*) n^2. (*c*) n^2-n.

1.2.1 $\mathbf{A}+\mathbf{B} = \begin{bmatrix} 7 & 6 & 17 \\ 0 & 0 & 0 \\ -1 & 9 & -1 \end{bmatrix}$; $\mathbf{A}-2\mathbf{B} = \begin{bmatrix} -11 & 3 & -7 \\ 6 & 18 & 9 \\ -1 & -6 & -7 \end{bmatrix}$.

1.2.2 (*a*) No. (*b*) No. (*c*) Only if **A** has 4 columns and **B** has 2 rows.

1.2.3 If $\mathbf{A} = \mathbf{B}+\mathbf{C}$ where $\mathbf{B} = \mathbf{B}'$ and $\mathbf{C} = -\mathbf{C}'$ then

$$\mathbf{A}' = \mathbf{B}'+\mathbf{C}' = \mathbf{B}-\mathbf{C}$$

and therefore
$$\mathbf{A}+\mathbf{A}' = 2\mathbf{B}, \quad \mathbf{A}-\mathbf{A}' = 2\mathbf{C}.$$

$$\mathbf{A} = \tfrac{1}{2}(\mathbf{A}+\mathbf{A}')+\tfrac{1}{2}(\mathbf{A}-\mathbf{A}')$$
$$= \begin{bmatrix} 1 & \frac{7}{2} & 4 \\ \frac{7}{2} & 6 & \frac{7}{2} \\ 4 & \frac{7}{2} & -3 \end{bmatrix} + \begin{bmatrix} 0 & \frac{3}{2} & 5 \\ -\frac{3}{2} & 0 & -\frac{1}{2} \\ -5 & \frac{1}{2} & 0 \end{bmatrix}.$$

1.3.1 $\mathbf{AB} = \begin{bmatrix} 46 & -70 & 21 \\ 87 & -118 & 74 \\ -131 & 241 & 17 \end{bmatrix}$, $\mathbf{BA} = \begin{bmatrix} -52 & -91 & 121 \\ 82 & 142 & -187 \\ 19 & -2 & -145 \end{bmatrix}$.

1.3.2 (*a*) $\mathbf{A}+\mathbf{B}$ only. (*b*) **AB** only. (*c*) **AB** and **BA** only.
(*d*) None. When **A** and **B** are square and have the same order.

1.3.3
$$\begin{bmatrix} 1 & 7 & -3 & 0 \\ 0 & 1 & 1 & 0 \\ 3 & -4 & 11 & 1 \\ 5 & 0 & 0 & 4 \end{bmatrix} \begin{bmatrix} x \\ y \\ z \\ t \end{bmatrix} = \begin{bmatrix} 4 \\ 2 \\ 9 \\ 12 \end{bmatrix}.$$

1.3.4 $$\mathbf{A}(\theta)\,\mathbf{A}(\phi) = \begin{bmatrix} \cos\theta\cos\phi - \sin\theta\sin\phi, & -\cos\theta\sin\phi - \sin\theta\cos\phi \\ \sin\theta\cos\phi + \cos\theta\sin\phi, & -\sin\theta\sin\phi + \cos\theta\cos\phi \end{bmatrix}$$
$$= \begin{bmatrix} \cos(\theta+\phi) & -\sin(\theta+\phi) \\ \sin(\theta+\phi) & \cos(\theta+\phi) \end{bmatrix} = \mathbf{A}(\theta+\phi).$$

1.4.1 $$(\mathbf{AB})\mathbf{C} = \begin{bmatrix} -192 & -91 & -449 \\ -237 & -192 & -660 \\ 867 & 224 & 1759 \end{bmatrix}, \quad \mathbf{BC} = \begin{bmatrix} -27 & 7 & -37 \\ 42 & -11 & 56 \\ -39 & -18 & -90 \end{bmatrix}.$$

1.4.2
$$[x, \ y, \ z] \begin{bmatrix} a & h & g \\ h & b & f \\ g & f & c \end{bmatrix} \begin{bmatrix} x \\ y \\ z \end{bmatrix}.$$

1.4.3 Proof by induction. The required result holds for $r = 1$. Assume it true for $r = n$. Then
$$(\mathbf{A}(\theta))^n = \mathbf{A}(n\theta),$$
and
$$(\mathbf{A}(\theta))^{n+1} = (\mathbf{A}(\theta))^n.\mathbf{A}(\theta) = \mathbf{A}(n\theta).\mathbf{A}(\theta) = \mathbf{A}((n+1)\theta). \qquad \text{Q.E.D.}$$

1.4.4 A sufficient condition is $\mathbf{AB} = \mathbf{BA}$. For then
$$(\mathbf{AB})^2 = \mathbf{AB}.\mathbf{AB} = \mathbf{A}.\mathbf{BA}.\mathbf{B} = \mathbf{A}.\mathbf{AB}.\mathbf{B} = \mathbf{A}^2\mathbf{B}^2,$$
$$(\mathbf{AB})^3 = \mathbf{AB}.\mathbf{AB}.\mathbf{AB} = \mathbf{A}.\mathbf{BA}.\mathbf{BA}.\mathbf{B} = \mathbf{A}.\mathbf{AB}.\mathbf{AB}.\mathbf{B}$$
$$= \mathbf{A}^2.\mathbf{BA}.\mathbf{B}^2 = \mathbf{A}^2.\mathbf{AB}.\mathbf{B}^2 = \mathbf{A}^3\mathbf{B}^3 \quad \text{and so on.}$$

1.5.4 $\mathbf{A} = -\mathbf{A}'$ and therefore $|\mathbf{A}| = |-\mathbf{A}'| = (-1)^n|\mathbf{A}'| = -|\mathbf{A}'|$. But $|\mathbf{A}| = |\mathbf{A}'|$ for any matrix $\mathbf{A}$, and therefore $|\mathbf{A}| = 0$.

1.5.6 (*a*) Even. (*b*) Even. (*c*) Odd. (*d*) Odd.

1.5.7 A determinant $|\mathbf{A}|$ may be expressed as the sum of all the possible determinants having zeros everywhere except for one place in each row and each column, where it has the appropriate element of $|\mathbf{A}|$. The determinant, whose only non-zero terms are $a_{i1}, a_{j2}, \ldots, a_{tn}$ in their respective positions, will have the value $\pm a_{i1}, a_{j2}, \ldots a_{tn}$, where the sign may be obtained by interchanging two rows of the determinant successively until all the terms lie on the leading diagonal.

1.6.1 $$\mathbf{A}^{-1} = \frac{1}{39}\begin{bmatrix} -147 & 126 & -36 & 60 \\ 25 & -14 & 4 & -11 \\ -23 & 16 & 1 & 7 \\ 43 & -35 & 10 & -8 \end{bmatrix}.$$

1.6.2 $x = \dfrac{-861}{39}$, $y = \dfrac{113}{39}$, $z = \dfrac{-118}{39}$, $t = \dfrac{263}{39}$.

1.6.3 If $\mathbf{A} = \mathbf{diag}\,(a_1, a_2, \ldots, a_n) = \begin{bmatrix} a_1 & 0 & \ldots & 0 \\ 0 & a_2 & \ldots & 0 \\ \ldots & \ldots & \ldots & \ldots \\ 0 & 0 & \ldots & a_n \end{bmatrix}$,

then $\mathbf{A}^{-1} = \mathbf{diag}\,(a_1^{-1}, a_2^{-1}, \ldots, a_n^{-1})$ provided that the a's are all non-zero.

1.6.4 If $\mathbf{AR} = \mathbf{I}$ then $\mathbf{A}$ must be non-singular, and therefore must have an inverse, $\mathbf{A}^{-1}$. Therefore

$$\mathbf{A}^{-1}(\mathbf{AR}) = (\mathbf{A}^{-1}\mathbf{A})\,\mathbf{R} = \mathbf{IR} = \mathbf{R},$$

$$\mathbf{A}^{-1}(\mathbf{AR}) = \mathbf{A}^{-1}\mathbf{I} = \mathbf{A}^{-1}.$$

Therefore $$\mathbf{R} = \mathbf{A}^{-1} \quad \text{and} \quad \mathbf{RA} = \mathbf{A}^{-1}\mathbf{A} = \mathbf{I}.$$

1.6.5 For suppose $\hat{\mathbf{A}}$ were non-singular. It would have an inverse $\mathbf{S}$ say, such that

$$\mathbf{S}\hat{\mathbf{A}} = \hat{\mathbf{A}}\mathbf{S} = \mathbf{I}.$$

But if $\mathbf{A}$ is singular then $$\mathbf{A}\hat{\mathbf{A}} = \mathbf{0},$$

and therefore $$\mathbf{A}\hat{\mathbf{A}}\mathbf{S} = \mathbf{0}.\mathbf{S} = \mathbf{0}.$$

But $\mathbf{A}\hat{\mathbf{A}}\mathbf{S} = \mathbf{A}(\hat{\mathbf{A}}\mathbf{S}) = \mathbf{AI} = \mathbf{A}$ so that $\mathbf{A} = \mathbf{0}$. But if $\mathbf{A} = \mathbf{0}$ then so is $\hat{\mathbf{A}}$. Therefore $\hat{\mathbf{A}}$ is singular. Contradiction.

1.6.6 $$\mathbf{A}\hat{\mathbf{A}} = |\mathbf{A}|.\mathbf{I}.$$

$$|\mathbf{A}\hat{\mathbf{A}}| = |\mathbf{A}|.|\hat{\mathbf{A}}| = ||\mathbf{A}|.\mathbf{I}'| = |\mathbf{A}|^n.$$

Therefore $$|\hat{\mathbf{A}}| = |\mathbf{A}|^{n-1}.$$

The adjoint of $\hat{\mathbf{A}}$ satisfies

$$\hat{\hat{\mathbf{A}}}\hat{\mathbf{A}} = |\hat{\mathbf{A}}|\,\mathbf{I} = |\mathbf{A}|^{n-1}\mathbf{I}.$$

But $\{|\mathbf{A}|^{n-2}\mathbf{A}\}\,\hat{\mathbf{A}} = |\mathbf{A}|^{n-1}\mathbf{I}$.
Therefore $$\{|\mathbf{A}|^{n-2}\mathbf{A} - \hat{\hat{\mathbf{A}}}\}\,\hat{\mathbf{A}} = \mathbf{0},$$

and therefore $\hat{\hat{\mathbf{A}}} = |\mathbf{A}|^{n-2}\mathbf{A}$, because $\hat{\mathbf{A}}$ is non-singular.

1.7.1 There are an infinity of such trios, but for all such trios $\mathbf{A}$ must be singular.

1.7.2 $\mathbf{AB} = \mathbf{BA}$. Therefore

$$\mathbf{B}^{-1}(\mathbf{AB})\,\mathbf{B}^{-1} = (\mathbf{B}^{-1}\mathbf{A})\,(\mathbf{BB}^{-1}) = \mathbf{B}^{-1}\mathbf{A},$$

and $$\mathbf{B}^{-1}(\mathbf{BA})\,\mathbf{B}^{-1} = (\mathbf{B}^{-1}\mathbf{B})\,(\mathbf{AB}^{-1}) = \mathbf{AB}^{-1}.$$

1.7.3 $\mathbf{AA}^{-1} = \mathbf{I}$. Therefore $(\mathbf{AA}^{-1})' = \mathbf{I}$. But $(\mathbf{AA}^{-1})' = (\mathbf{A}^{-1})'\mathbf{A}'$. Therefore $(\mathbf{A}^{-1})'$ is an inverse for $\mathbf{A}'$ and since there is only one such we must have

$$(\mathbf{A}^{-1})' = (\mathbf{A}')^{-1}.$$

1.8.1 $s:t:u:v = 7:5:-27:-7.$

1.8.2 $\mathbf{A}(\theta\mathbf{x}^{(1)}+\phi\mathbf{x}^{(2)}) = \theta\mathbf{A}\mathbf{x}^{(1)}+\phi\mathbf{A}\mathbf{x}^{(2)} = \theta.\mathbf{0}+\phi.\mathbf{0} = \mathbf{0}.$

1.8.3 $s:t:u:v = 0:1:0:-2$

$s:t:u:v = 7:0:-27:3$

are solutions, and any combination of these is a solution. All the cofactors of the elements of the determinant of coefficients are zero.

2.1.1 (*a*) $\mathbf{A} = m\begin{bmatrix} 2 & -4 & 5 \\ -4 & 26 & -34 \\ 5 & -34 & \frac{111}{2} \end{bmatrix}$, $\mathbf{C} = mg/l\begin{bmatrix} 20 & 2 & 6 \\ 2 & 4 & -5 \\ 6 & -5 & \frac{29}{2} \end{bmatrix}$.

Both **A** and **C** are positive definite.

(*b*) $\mathbf{A} = ml^2\begin{bmatrix} 9 & 0 & 0 \\ 0 & 7 & 0 \\ 0 & 0 & 2 \end{bmatrix}$, $\mathbf{C} = mgl\begin{bmatrix} 1 & \frac{1}{2} & -\frac{7}{2} \\ \frac{1}{2} & 1 & \frac{1}{2} \\ -\frac{7}{2} & \frac{1}{2} & 19 \end{bmatrix}$.

A is positive definite; **C** is positive semi-definite. There is one rigid-body mode, viz.

$$\theta:\phi:\psi = 5:-3:1.$$

2.1.2 (*a*)
$$T = \tfrac{1}{2}ml^2\{4\dot{\theta}^2+12\dot{\phi}^2+9\dot{\psi}^2+12\dot{\theta}\dot{\phi}+6\dot{\theta}\dot{\psi}+12\dot{\phi}\dot{\psi}\},$$
$$V = \tfrac{1}{2}mgl\{4\theta^2+6\phi^2+3\psi^2\}.$$

(*b*) $\mathbf{A} = ml^2\begin{bmatrix} 4 & 6 & 3 \\ 6 & 12 & 6 \\ 3 & 6 & 9 \end{bmatrix}$, $\mathbf{C} = mgl\begin{bmatrix} 4 & 0 & 0 \\ 0 & 6 & 0 \\ 0 & 0 & 3 \end{bmatrix}$.

(*c*)
$$4\ddot{\theta}+6\ddot{\phi}+3\ddot{\psi}+\frac{4g}{l}\theta = \frac{P}{ml},$$
$$\ddot{\theta}+2\ddot{\phi}+\ddot{\psi}+\frac{g}{l}\phi = \frac{P}{3ml},$$
$$\ddot{\theta}+2\ddot{\phi}+3\ddot{\psi}+\frac{g}{l}\psi = \frac{P}{ml}.$$

2.1.3 (*a*) *Necessity*. Suppose $U(x, y, z)$ is positive definite

$$U(1, 0, 0) = a > 0,$$
$$U(h, -a, 0) = a(ab-h^2) > 0, \text{ therefore } ab-h^2 > 0,$$
$$U(hf-bg, hg-af, ab-h^2) = (ab-h^2)\,\Delta,$$

herefore
$$\Delta = \begin{vmatrix} a & h & g \\ h & b & f \\ g & f & c \end{vmatrix} > 0.$$

(*b*) *Sufficiency*. Suppose the condition holds

$$U(x, y, z) = \frac{1}{a}(ax+hy+gz)^2+\frac{1}{a(ab-h^2)}\{(ab-h^2)y+(af-hg)z\}^2+\frac{\Delta}{(ab-h^2)}z^2,$$

herefore
$$U(x, y, z) = 0 \quad \text{only if} \quad x = 0 = y = z.$$

(*c*) If the condition holds, then $U(x, y, z)$ is positive definite and therefore

$$b = U(0, 1, 0) > 0, \text{ etc.}$$

2.2.1 $\alpha = (1/\Delta)\begin{bmatrix} 54\sigma^4-135\sigma^2+57 & -6\sigma^2+5 & -18\sigma^2+9 \\ -6\sigma^2+5 & 6\sigma^4-23\sigma^2+18 & 3\sigma^2-4 \\ -18\sigma^2+9 & 3\sigma^2-4 & 9\sigma^4-24\sigma^2+11 \end{bmatrix}$,

where $\Delta = (mg/l)\{-54\sigma^6+243\sigma^4-285\sigma^2+91\}$ and $\sigma = \omega\surd(l/g)$.

2.2.2 $\theta = \dfrac{(F/mg)\sin\omega t}{-4\sigma^6+27\sigma^4-24\sigma^2+4}$,

where $\sigma = \omega\surd(l/g)$.

2.2.3 $\alpha_{12} = \dfrac{\tau^4-32\tau^2+151}{k\{\tau^6-26\tau^4+169\tau^2-144\}}$,

where $\tau = \omega\surd(I/k)$.

2.3.1 The equation for $\sigma \equiv \omega\surd(l/g)$ is

$$(\sigma^2-1)\,(2\sigma^4-19\sigma^2+5) = 0,$$

$$\omega_1 = 0{\cdot}5206\surd(g/l),\quad \omega_2 = \surd(g/l),\quad \omega_3 = 3{\cdot}038\surd(g/l).$$

2.3.2 $\omega_1 = 0{\cdot}468\surd(g/l),\quad \omega_2 = 0{\cdot}893\surd(g/l),\quad \omega_3 = 2{\cdot}39\surd(g/l).$

$$\mathbf{\Psi}^{(1)} = \begin{bmatrix} 1 \\ 1{\cdot}26 \\ 2{\cdot}24 \end{bmatrix},\quad \mathbf{\Psi}^{(2)} = \begin{bmatrix} 1 \\ 1{\cdot}07 \\ -1{\cdot}80 \end{bmatrix},\quad \mathbf{\Psi}^{(3)} = \begin{bmatrix} 1 \\ -0{\cdot}578 \\ 0{\cdot}055 \end{bmatrix}.$$

2.3.3 Three sets of equations, each involving the elements of one row of **A**, may be found by using equation (2.3.10).

$$\mathbf{A} = (I/5184)\begin{bmatrix} 2476 & 3794 & 4098 \\ 3794 & 32659 & 15387 \\ 4098 & 15387 & 11619 \end{bmatrix}.$$

2.4.1

$$\omega_1 = \surd(g/l) \quad x:y:z = 4:1:-3,$$
$$\omega_2 = 2\surd(g/l) \quad x:y:z = 4:-1:-1,$$
$$\omega_3 = 5\surd(g/l) \quad x:y:z = 2:1:-1.$$

$$\mathbf{X} = \begin{bmatrix} 4 & -4 & 2 \\ 1 & 1 & 1 \\ -3 & 1 & -1 \end{bmatrix},\quad \begin{aligned} x &= 4p_1-4p_2+2p_3, \\ y &= p_1+p_2+p_3, \\ z &= -3p_1+p_2-p_3, \end{aligned}$$

$$p_1 = -\tfrac{1}{4}(x+y+3z),\quad p_2 = -\tfrac{1}{4}(x-y+z),\quad p_3 = \tfrac{1}{2}(x+2y+2z).$$

2.4.2

$$p_s^* = p_s/\lambda_s.$$

The equation for the p_s^* is

$$\begin{bmatrix} 0 \\ 2 \\ 0 \end{bmatrix} = \begin{bmatrix} 2 & 2 & 2 \\ 2 & 5 & 1/5 \\ 2 & -9 & -1 \end{bmatrix}\begin{bmatrix} p_1^* \\ p_2^* \\ p_3^* \end{bmatrix}.$$

To solve this multiply by (the new) $\mathbf{X'C}$ to give

$$p_1^* = \tfrac{5}{9}, \quad p_2^* = \tfrac{5}{24}, \quad p_3^* = -\tfrac{55}{72}.$$

2.5.2 (*a*) Inconsistent. $\mathbf{\Psi}'^{(2)}\mathbf{A}\mathbf{\Psi}^{(3)} \neq 0.$

(*b*) $c_{12} = c_{21} = 2(mg/l)$; $\mathbf{\Psi}^{(2)} = \begin{bmatrix} 4 \\ -3 \\ 2 \end{bmatrix}$; $\mathbf{A} = m\begin{bmatrix} 2 & 2 & -1 \\ 2 & 6 & 0 \\ -1 & 0 & 7 \end{bmatrix}.$

(*c*) $a_{13} = -3I, \quad a_{23} = 8I; \quad a_{12} = -I; \quad \mathbf{\Psi}^{(3)} = \{1 \quad 3 \quad -1\},$

$c_{12} = c_{21} = -2k, \quad c_{13} = c_{31} = -20k, \quad c_{23} = c_{32} = 55k,$

$\omega_1 = \sqrt{(k/I)}, \quad \omega_2 = 3\sqrt{(k/I)}, \quad \omega_3 = 4\sqrt{(k/I)}.$

The quantities may most conveniently be obtained in the order in which they are given here.

2.5.3 First calculate $\mathbf{X'CX} = \mathbf{N}$, then $\mathbf{L} = \mathbf{N\Omega}^{-1}$ and $\mathbf{A} = (\mathbf{X}^{-1})'\mathbf{LX}^{-1}$. N.B. This method may be used to advantage in Ex. 2.5.2 (*b*).

2.6.1 $\alpha = \sqrt{15}/15.$

2.6.2 $p_1 = \sqrt{(m)}\,(x+2y+z),$

$p_2 = -\sqrt{(2m)}\,(y-z),$

$p_3 = -\sqrt{(m)}\,(x-2z).$

Yes. By choosing the $\mathbf{\Psi}$'s so that $\mathbf{\Psi}'^{(r)}\mathbf{C}\mathbf{\Psi}^{(r)} = \omega_r^2.$

2.6.3 (*a*) $1/(\omega_1^2-\omega^2).$ (*b*) 0.

2.7.2 (*a*) $p_1 = p_2 = \tfrac{1}{3}, \quad p_3 = 0.$

(*b*) $p_1 = 0 = p_3, \quad p_2 = 1.$

2.8.1 $\alpha_{12} = -\dfrac{1}{3I}\left\{\dfrac{1}{\omega_1^2-\omega^2}+\dfrac{1}{\omega_2^2-\omega^2}-\dfrac{1}{\omega_3^2-\omega^2}\right\}.$

2.8.2 $\alpha_{11} = \dfrac{1}{I}\left\{\dfrac{0{\cdot}108}{\omega_1^2-\omega^2}+\dfrac{0{\cdot}543}{\omega_2^2-\omega^2}+\dfrac{0{\cdot}349}{\omega_3^2-\omega^2}\right\}.$

2.8.3 (*a*) $\alpha_{11} = \dfrac{1}{m}\left\{\dfrac{1}{\omega_1^2-\omega^2}+\dfrac{4}{\omega_2^2-\omega^2}+\dfrac{49}{\omega_3^2-\omega^2}+\dfrac{1}{\omega_4^2-\omega^2}\right\}.$

(*b*) $\alpha_{xy} = \dfrac{1}{m}\left\{\dfrac{12}{\omega_1^2-\omega^2}+\dfrac{9}{\omega_2^2-\omega^2}+\dfrac{48}{\omega_3^2-\omega^2}-\dfrac{18}{\omega_4^2-\omega^2}\right\}.$

2.8.4 (*a*) Expand $\mathbf{X\zeta}^{-1}\mathbf{X}'$ using equation (2.4.10).

(*b*) If the scale of measurement of p_r is changed by a factor θ_r, then the rth column $\mathbf{\Psi}^{(r)}$ in $\mathbf{X}$ becomes $\theta_r\mathbf{\Psi}^{(r)}$ so that $\mathbf{\Psi}^{(r)}\mathbf{\Psi}'^{(r)}$ in equation (2.8.20) is multiplied by θ_r^2. But equation (2.5.12) shows that a_r is also multiplied by θ_r^2, so that (by (2.8.16)) α_r is divided by θ_r^2. Therefore the rth term in the expression (2.8.20) is unaffected by the change of scale.

2.9.1 $\boldsymbol{\Phi}^{(1)} = 2I \begin{bmatrix} 1 \\ 5 \\ 3 \end{bmatrix}, \quad \boldsymbol{\Phi}^{(2)} = (5I/4) \begin{bmatrix} -2 \\ 5 \\ -3 \end{bmatrix}, \quad \boldsymbol{\Phi}^{(3)} = (I/9) \begin{bmatrix} 14 \\ -11 \\ -3 \end{bmatrix}.$

Yes. $\boldsymbol{\Phi}^{(r)} = \omega_r^{-2}\mathbf{C}\boldsymbol{\Psi}^{(r)}$.

2.9.2 (*a*) $(\mathbf{A}-\omega_r^{-2}\mathbf{C})\,\boldsymbol{\Psi}^{(r)} = \mathbf{0}$, therefore $\boldsymbol{\Phi}^{(r)}-\omega_r^{-2}\mathbf{C}\boldsymbol{\Psi}^{(r)} = \mathbf{0}$, therefore $\mathbf{C}^{-1}\boldsymbol{\Phi}^{(r)}-\omega_r^{-2}\boldsymbol{\Psi}^{(r)} = \mathbf{0}$, therefore $(\mathbf{C}^{-1}-\omega_r^{-2}\mathbf{A}^{-1})\,\boldsymbol{\Phi}^{(r)} = \mathbf{0}$.

(*b*) This follows direct from equation (2.9.13) which holds for all values of ω.

2.9.3 $$\boldsymbol{\Phi} = (F/3I)\,\{-4\boldsymbol{\Phi}^{(1)}-2\boldsymbol{\Phi}^{(2)}+5\boldsymbol{\Phi}^{(3)}\}.$$

2.10.1 $\omega_1 \doteqdot 0{\cdot}486\sqrt{(g/l)}$ $\quad \boldsymbol{\Psi}^{(1)} \doteqdot \{1 \quad 1{\cdot}25 \quad 1{\cdot}80\}$,

$\omega_1 \doteqdot 0{\cdot}470\sqrt{(g/l)}$ $\quad \boldsymbol{\Psi}^{(1)} \doteqdot \{1 \quad 1{\cdot}26 \quad 2{\cdot}19\}$.

2.10.2 For stationary values

$$\frac{c_{11}x^2+2c_{12}x+c_{22}}{a_{11}x^2+2a_{12}x+a_{22}} = \frac{c_{11}x+c_{12}}{a_{11}x+a_{12}} = \lambda^2 \text{ (say).}$$

Clearly $$\frac{c_{12}x+c_{22}}{a_{12}x+a_{22}} = \lambda^2 \text{ also, and therefore}$$

$$(c_{11}-\lambda^2 a_{11})\,x+c_{12}-\lambda^2 a_{12} = 0,$$

$$(c_{12}-\lambda^2 a_{12})\,x+c_{22}-\lambda^2 a_{22} = 0.$$

In other words $$[\mathbf{C}-\lambda^2\mathbf{A}]\begin{bmatrix} x \\ 1 \end{bmatrix} = \mathbf{0}.$$ Q.E.D.

$$\omega_1^2 = \left(1-\frac{2\sqrt{7}}{7}\right)\frac{g}{l}, \quad x = \frac{-5-2\sqrt{7}}{3},$$

$$\omega_2^2 = \left(1+\frac{2\sqrt{7}}{7}\right)\frac{g}{l}, \quad x = \frac{-5+2\sqrt{7}}{3}.$$

2.11.1 $2T = ml^2[4\dot{x}^2+2\dot{\theta}^2+\dot{\phi}^2+4\dot{x}\dot{\theta}+2\dot{x}\dot{\phi}+2\dot{\theta}\dot{\phi}]$,

$2V = mgl[2\theta^2+\phi^2]$;

x is ignorable and $$\frac{\partial T}{\partial \dot{x}} = ml^2[4\dot{x}+2\dot{\theta}+\dot{\phi}] = \text{const.} = a.$$

therefore $$4x+2\theta+\phi = at+b,$$

and $a = 0 = b$ in small motion. The relation expresses the fact that the centre of mass of the system remains directly below O.

2.11.2 $\omega_1 = 0, \quad x:\theta:\phi = 1:0:0;$

$\omega_2 = \sqrt{(g/l)}, \quad x:\theta:\phi = -1:1:2;$

$\omega_3 = 2\sqrt{(g/l)}, \quad x:\theta:\phi = 1:-4:4.$

2.11.4 (*a*) $\mathbf{A} = M\begin{bmatrix} 5 & 2 & 1 & 2 \\ 2 & 2 & 0 & 0 \\ 1 & 0 & 1 & 0 \\ 2 & 0 & 0 & 2 \end{bmatrix}$, $\mathbf{C} = \begin{bmatrix} 0 & 0 & 0 & 0 \\ 0 & 3k+\lambda & -k & -\lambda \\ 0 & -k & 2k & -k \\ 0 & -k & -k & k+\lambda \end{bmatrix}$.

(*b*) $\omega_1^2 = 0;\quad q_1\!:\!q_2\!:\!q_3\!:\!q_4 = 1\!:\!0\!:\!0\!:\!0.$

(*c*) $5q_1+2q_2+q_3+2q_4 = 0.$

(*d*) $\mathbf{\Psi}^{(4)} = \{-1, 1, 1, 1\},\quad q_2 = 0.$

(*e*) q_2 and q_3 are the simplest to eliminate; corresponding to $\{q_1 \quad q_4\}$ we have

$$\mathbf{A}_1 = M\begin{bmatrix} 20 & 10 \\ 10 & 6 \end{bmatrix},\quad \mathbf{C}_1 = \begin{bmatrix} 50k & 25k \\ 25k & 13k+\lambda \end{bmatrix}.$$

(*f*) $|\mathbf{C}_1-\omega^2\mathbf{A}_1| = 0$ gives

$$\omega_2^2 = \frac{k+2\lambda}{2M},\quad q_1\!:\!q_4 = 1\!:\!-2,\quad \text{or}$$

$$q_1\!:\!q_2\!:\!q_3\!:\!q_4 = 1\!:\!0\!:\!-1\!:\!-2;$$

$$\omega_3^2 = \frac{5k}{2M},\quad q_1\!:\!q_4 = 1\!:\!0,\quad \text{or}$$

$$q_1\!:\!q_2\!:\!q_3\!:\!q_4 = 1\!:\!0\!:\!-5\!:\!0.$$

The complete set of principal modes can therefore be set out in semi-graphical form as follows, the arrows indicating absolute displacements of the three masses:

$$\overrightarrow{1} \quad \overrightarrow{1} \quad \overrightarrow{1},\qquad \omega_1^2 = 0;$$

$$\overrightarrow{1} \quad 0 \quad \overleftarrow{1},\qquad \omega_2^2 = \frac{k+2\lambda}{2M};$$

$$\overrightarrow{1} \quad \overleftarrow{4} \quad \overrightarrow{1},\qquad \omega_3^2 = \frac{5k}{2M}.$$

Note that it is implicitly assumed throughout that $\lambda \neq 2k$.

2.12.1 (*a*)
$$\mathbf{L} = \mathbf{X'AX} = M\begin{bmatrix} 5 & 0 & 0 \\ 0 & 4 & 0 \\ 0 & 0 & 20 \end{bmatrix}.$$

(*b*)
$$\mathbf{N} = \mathbf{L\Omega} = \begin{bmatrix} 0 & 0 & 0 \\ 0 & 2k+4\lambda & 0 \\ 0 & 0 & 50k \end{bmatrix}.$$

(*c*)
$$\boldsymbol{\xi} \equiv \mathbf{X}^{-1} = \mathbf{L'X'A} = \begin{bmatrix} \frac{2}{5} & \frac{1}{5} & \frac{2}{5} \\ \frac{1}{2} & 0 & -\frac{1}{2} \\ \frac{1}{10} & -\frac{1}{5} & \frac{1}{10} \end{bmatrix}.$$

(*d*) Find $\mathbf{C}$ as $\boldsymbol{\xi}'\mathbf{N}\boldsymbol{\xi}$

$$\mathbf{C} = \begin{bmatrix} k+\lambda & -k & -\lambda \\ -k & 2k & -k \\ -\lambda & -k & k+\lambda \end{bmatrix}.$$

(*e*) $\bar{p}_1 = \sqrt{(5M)}p_1$, $\bar{p}_2 = 2\sqrt{(M)}p_2$, $\bar{p}_3 = 2\sqrt{(5M)}p_3$, so that $\bar{\mathbf{p}} = \bar{\boldsymbol{\xi}}\mathbf{q}$ gives

$$\bar{\boldsymbol{\xi}} = \sqrt{M}\begin{bmatrix} \frac{2}{\sqrt{5}} & \frac{1}{\sqrt{5}} & \frac{2}{\sqrt{5}} \\ 1 & 0 & -1 \\ \frac{1}{\sqrt{5}} & -\frac{2}{\sqrt{5}} & \frac{1}{\sqrt{5}} \end{bmatrix},$$

$$\bar{\mathbf{X}} = (\bar{\boldsymbol{\xi}})^{-1} = \frac{1}{\sqrt{M}}\begin{bmatrix} \frac{1}{\sqrt{5}} & \frac{1}{2} & \frac{1}{2\sqrt{5}} \\ \frac{1}{\sqrt{5}} & 0 & -\frac{2}{\sqrt{5}} \\ \frac{1}{\sqrt{5}} & -\frac{1}{2} & \frac{1}{2\sqrt{5}} \end{bmatrix}.$$

2.12.2

$$\lambda_1^2 = \frac{2M}{3k}, \quad q_1{:}q_2{:}q_3 = 1{:}0{:}-2;$$

$$\lambda_2^2 = \frac{2M}{5k}, \quad q_1{:}q_2{:}q_3 = 1{:}0{:}0;$$

$$\lambda_3^2 = 0, \quad q_1{:}q_2{:}q_3 = 1{:}-1{:}-1.$$

2.12.3 $\omega_1^2 = 0$; $\boldsymbol{\Psi}^{(1)} = \{1 \quad 0 \quad 0 \quad 0\}$ whence

$$5q_1+2q_2+q_3+2q_4 = 0.$$

This gives inertia and stability matrices $\mathbf{A}_1$ and $\mathbf{C}_1$ which are identical with those of Ex. 2.12.2 except that they now refer to $\{q_1 \quad q_2 \quad q_4\}$. Using previous results, therefore

$$\omega_2^2 = \frac{3k}{2M}, \quad \boldsymbol{\Psi}^{(2)} = \{1 \quad 0 \quad -1 \quad -2\};$$

$$\omega_3^2 = \frac{5k}{2M}, \quad \boldsymbol{\Psi}^{(3)} = \{1 \quad 0 \quad -5 \quad 0\};$$

$$\omega_4^2 = \infty; \quad \boldsymbol{\Psi}^{(4)} = \{1 \quad -1 \quad -1 \quad -1\}.$$

The last mode represents vibration of the system in which only the left-hand spring moves; it is the limiting case in which a very small mass is attached at q_1.

3.1.1 The vector $-\mathbf{b}$ is represented by OQ', the line equal in magnitude and opposite in direction to OQ. Therefore $\mathbf{a}-\mathbf{b} = \mathbf{a}+(-\mathbf{b})$ is represented by the diagonal OR' of the parallelogram $OQ'R'P$ and this diagonal is equal and opposite to QP.

The vector equal and parallel to AC is the sum of the vectors equal and parallel to BC and AB. Therefore

$$-\mathbf{b} = \mathbf{a}+\mathbf{c}.$$

3.1.2
$$\begin{aligned} \mathbf{A}_1\mathbf{A}_3 &= \mathbf{A}_1\mathbf{A}_2+\mathbf{A}_2\mathbf{A}_3 = \mathbf{a}^{(1)}+\mathbf{a}^{(2)},\\ \mathbf{A}_1\mathbf{A}_4 &= \mathbf{A}_1\mathbf{A}_3+\mathbf{A}_3\mathbf{A}_4 = \mathbf{a}^{(1)}+\mathbf{a}^{(2)}+\mathbf{a}^{(3)},\\ &\cdots\cdots\cdots\cdots\cdots\cdots\\ \mathbf{A}_1\mathbf{A}_n &= \mathbf{A}_1\mathbf{A}_{n-1}+\mathbf{A}_{n-1}\mathbf{A}_n = \mathbf{a}^{(1)}+\mathbf{a}^{(2)}+\ldots+\mathbf{a}^{(n-1)}\\ &= -\mathbf{A}_n\mathbf{A}_1 = -\mathbf{a}^{(n)}.\end{aligned}$$

Therefore $$\mathbf{a}^{(1)}+\mathbf{a}^{(2)}+\ldots+\mathbf{a}^{(n)} = \mathbf{0}.$$

3.2.1 (*a*) More vectors than elements in each. Therefore linearly dependent.
(*b*) Linearly dependent. $3\mathbf{A}^{(1)}+2\mathbf{A}^{(2)}-\mathbf{A}^{(3)} = \mathbf{0}$.
(*c*) Linearly independent.

3.2.2 Its columns are linearly dependent and therefore it is zero.

3.2.3 (*a*) Yes. (*b*) No. (*c*) No.

3.2.5 $a = 1, 8$.

3.3.1 (*a*) 1. (*b*) 2. (*c*) 3. (*d*) 2.

3.3.2 (*a*) Yes. (*b*) No. (*c*) No. (*d*) Yes.

3.3.3 (*a*) Not possible by selection of a and b. (*b*) $a = 1$, $b = 5$ or $a = 2$, $b = 5$. (*c*) Otherwise. [For $\mathbf{A}$ to have rank less than 3 we must have

(i) $a^2-3a+2 = 0$, i.e. $a = 1$ or $a = 2$;
(ii) $a(b-5) = 0$, i.e. $a = 0$ or $b = 5$;
(iii) $(2b-5)a^2-(b+10)a+10 = 0$.

For $a = 1$ or $a = 2$ (iii) gives $b = 5$ and for $b = 5$, (iii) gives $a = 1$ or $a = 2$.]

3.4.1 $a = 1, \dfrac{-39}{17}$.

3.4.2 (*a*) 3. (*b*) 4. (*c*) 3.

3.4.3 (*a*) 2. (*b*) 3.

3.5.1 (*a*) None. (*b*) $\{38 \quad -39 \quad 5 \quad 0\}$ and $\{-54 \quad 67 \quad 0 \quad 5\}$ are two linearly independent solutions.

(*c*) $\{-1 \quad 7 \quad 39 \quad 0 \quad 0\}$, $\{-12 \quad -7 \quad 0 \quad 13 \quad 0\}$, $\{-2 \quad 1 \quad 0 \quad 0 \quad 13\}$

are three linearly independent solutions.

In cases (*b*) and (*c*) any combination of the given solutions is a solution, and vice versa.

3.5.2 (*a*) $a = 1$, $b = 4$; 3 solutions:

$$\{-2 \quad 1 \quad 0 \quad 0\},\ \{1 \quad 0 \quad 1 \quad 0\},\ \{-4 \quad 0 \quad 0 \quad 1\}.$$

(*b*) $a \neq 1$, $b = -4$; 2 solutions:

$$\{-1 \quad \tfrac{1}{2}(a+1) \quad 1 \quad 0\},\ \{0 \quad -2 \quad 0 \quad 1\}.$$

(*c*) $a = -1$, $b = 4$; 2 solutions:

$$\{-1 \quad 0 \quad 1 \quad 0\},\ \{4 \quad 0 \quad 0 \quad 1\}.$$

In cases (*a*), (*b*) and (*c*) any combination of the given solutions is a solution, and vice versa.

(*d*) Otherwise; 1 solution:

$$\{b-4a \quad 0 \quad ab-4 \quad a^2-1\}.$$

3.5.3 (*a*) $\mathbf{U}$ is the same as $\mathbf{I}_m$ except that row 1 is multiplied by k.

(*b*) $\mathbf{V}$ is the same as $\mathbf{I}_n$ except that column r is the sum of columns 1 and r of $\mathbf{I}_n$.

(*c*) Postmultiply by the $n \times n$ matrix obtained from $\mathbf{I}_n$ by interchanging columns 1 and 2.

3.6.1 (*a*) $\frac{1}{101}\{-20 \quad 197 \quad 119\}$.

(*b*) $\{x \quad y \quad z\} = \frac{1}{23}\{13 \quad 2 \quad 0\} + \lambda\{7 \quad 109 \quad 23\}$.

(*c*) None.

3.6.2
$$\begin{aligned}\{x \quad y \quad z \quad t \quad u\} &= \tfrac{1}{10}\{116 \quad 27 \quad 0 \quad -20 \quad 0\}\\ &\quad +\lambda\{8 \quad 11 \quad 10 \quad 0 \quad 0\}\\ &\quad +\mu\{378 \quad 51 \quad 0 \quad -130 \quad 40\}.\end{aligned}$$

3.6.3 4. There are an infinity of such sets; one example is

$$\begin{aligned}\mathbf{x}^{(1)} &= \tfrac{1}{7}\{22 \quad 35 \quad 0 \quad 0 \quad 16 \quad 0\},\\ \mathbf{x}^{(2)} &= \tfrac{1}{77}\{445 \quad 420 \quad 77 \quad 0 \quad 176 \quad 0\},\\ \mathbf{x}^{(3)} &= \tfrac{1}{77}\{-199 \quad 378 \quad 0 \quad 77 \quad 176 \quad 0\},\\ \mathbf{x}^{(4)} &= \tfrac{1}{77}\{214 \quad -196 \quad 0 \quad 0 \quad -132 \quad 77\}.\end{aligned}$$

All the solutions are given by

$$\begin{aligned}\mathbf{x} &= \tfrac{1}{7}\{22 \quad 35 \quad 0 \quad 0 \quad 16 \quad 0\} + \lambda_1\{29 \quad 5 \quad 11 \quad 0 \quad 0 \quad 0\}\\ &\quad + \lambda_2\{63 \quad 1 \quad 0 \quad -11 \quad 0 \quad 0\} + \lambda_3\{4 \quad 83 \quad 0 \quad 0 \quad 44 \quad -11\}.\end{aligned}$$

3.7.1 $t = 1, \frac{1}{3}$.

3.7.2 $\{x_1 \quad x_2 \quad x_3\} = \lambda\{11 \quad 23 \quad -27\}$; $\lambda = \sqrt{1379}$.

3.7.3 There are infinitely many possible sets, but each set has just two vectors; one set is

$$\mathbf{u}^{(1)} = 1/\sqrt{14}\{-1 \quad 3 \quad 2\}, \quad \mathbf{u}^{(2)} = 1/\sqrt{2982}\{37 \quad -13 \quad 38\}.$$

4.1.1 x, y must satisfy $10 - 6x + 49y = 0$. Two possible vectors are $\{1 \quad 18 \quad 2\}$ and $\{1 \quad -31 \quad -4\}$; any other is a linear combination of these.

4.1.2 The normalised modes are

$$\mathbf{u}^{(1)} = \mathbf{\Psi}^{(1)}/4\sqrt{(2m)}, \quad \mathbf{u}^{(2)} = \mathbf{\Psi}^{(2)}/4\sqrt{m}, \quad \mathbf{u}^{(3)} = \mathbf{\Psi}^{(3)}/2\sqrt{m}.$$

4.1.3 $\mathbf{u}^{(1)} = \mathbf{\Psi}^{(1)}/3\sqrt{I}, \quad \mathbf{u}^{(2)} = \mathbf{\Psi}^{(2)}/3\sqrt{(2I)}, \quad \mathbf{u}^{(3)} = \{1 \quad 3 \quad -1\}/3\sqrt{I}.$

4.2.1
$$\begin{aligned}2T &= m\{\dot{q}_1^2 + \dot{q}_2^2 + \dot{q}_3^2\},\\ 2V &= (S/l)\{q_1^2 + (q_2 - q_1)^2 + (q_3 - q_2)^2 + q_3^2\},\\ 2T &= m\{\dot{\bar{q}}_1^2 + (\dot{\bar{q}}_1 + \dot{\bar{q}}_2)^2 + (\dot{\bar{q}}_1 + \dot{\bar{q}}_2 + \dot{\bar{q}}_3)^2\},\\ 2V &= (S/l)\{\bar{q}_1^2 + \bar{q}_2^2 + \bar{q}_3^2 + (\bar{q}_1 + \bar{q}_2 + \bar{q}_3)^2\}.\end{aligned}$$

$$\mathbf{Z} = \begin{bmatrix}1 & 0 & 0\\ 1 & 1 & 0\\ 1 & 1 & 1\end{bmatrix}.$$

4.3.1 $\omega_1 = 1{\cdot}140\surd(g/l)$, $\omega_2 = 2{\cdot}586\surd(g/l)$, $\omega_3 = 5{\cdot}315\surd(g/l)$.

4.3.2 $\omega_1^* = 0$, $\omega_2^* = \surd[4(2+k^2\ g/(4+4k+3k^2)\,l]$;

$k = 2$, $x:\theta:\phi = -1:1:2$;

$k = -1$, $x:\theta:\phi = 1:-4:4$.

These are the second and third modes, respectively, of the unconstrained system. (See Ex. 2.11.2.)

4.3.3 $b_1 = 0$, $b_2 = \surd(2)\ (k-2)$, $b_3 = -4(k+1)$.

4.4.1 (*a*) $\{1 \quad 0 \quad -1\}/\surd(2m)$. (*b*) $\surd(2S/ml)$.

(*c*) $\bar{\mathbf{q}} = \surd(m)\ \{(q_1-q_3)/\surd 2 \quad q_2 \quad (q_1+q_3)\}$.

(*d*) $\mathbf{A}_2 = \begin{bmatrix} 1 & 0 \\ 0 & \frac{1}{2} \end{bmatrix}$, $\mathbf{C}_2 = (S/ml)\begin{bmatrix} 2 & -1 \\ -1 & 1 \end{bmatrix}$.

(*e*) $\omega^2 = (2-\surd 2)\,S/ml$ giving $\{1 \quad \surd 2 \quad 1\}/2\surd m$;

$\omega^2 = (2+\surd 2)\,S/ml$ giving $\{-1 \quad \surd 2 \quad -1\}/2\surd m$.

4.4.2 $\omega_1 = \omega_2 = \surd(2k/m)$. In free vibration $x = A\cos(\omega_1 t+\theta_1)$, $y = B\cos(\omega_1 t+\theta_2)$. Possible motions are:

(*a*) $x = A\cos(\omega_1 t+\theta_1)$, $y = 0$; (motion along x-axis).

(*b*) $x = 0$, $y = B\cos(\omega_1 t+\theta_2)$; (along y-axis).

(*c*) $\theta_2 = \theta_1$, then $y = (B/A)\,x$.

(*d*) $\theta_2 = \theta_1 \pm \pi/2$, $B = A$; (motion in circle $x^2+y^2 = A^2$).

(*e*) $\theta_2 = \theta_1 \pm \pi/2$; (motion in ellipse $x^2/A^2+y^2/B^2 = 1$).

4.4.3 The vertical displacement of the pendulum is small and is approximately

$$z_1 = (1/2l)\ [(x-x_1)^2+(y-y_1)^2],$$

$$2T = m(\dot{x}^2+\dot{y}^2)+M(\dot{x}_1^2+\dot{y}_2^2),$$

$$2V = 2k(x^2+y^2)+(Mg/l)\ [(x-x_1)^2+(y-y_1)^2].$$

4.4.4 $\omega_1 = \omega_2 = 2\surd 3/3$, $\omega_3 = \omega_4 = 4$.

A set of mutually orthogonal modes giving the ratios $x:y:x_1:y_1$ is

$$\boldsymbol{\Psi}^{(1)} = \begin{bmatrix} 2 \\ 0 \\ 3 \\ 0 \end{bmatrix},\quad \boldsymbol{\Psi}^{(2)} = \begin{bmatrix} 0 \\ 2 \\ 0 \\ 3 \end{bmatrix},\quad \boldsymbol{\Psi}^{(3)} = \begin{bmatrix} -3 \\ 0 \\ 1 \\ 0 \end{bmatrix},\quad \boldsymbol{\Psi}^{(4)} = \begin{bmatrix} 0 \\ -3 \\ 0 \\ 1 \end{bmatrix}.$$

4.5.2 (*a*) $\tilde{\omega}_1 = 0{\cdot}766\surd(g/l)$, $\tilde{\omega}_2 = 1{\cdot}85\surd(g/l)$.

(*b*) $\tilde{\omega}_1 = 0{\cdot}654\surd(g/l)$, $\tilde{\omega}_2 = 2{\cdot}498\surd(g/l)$.

(*c*) $\tilde{\omega}_1 = 0{\cdot}919\surd(g/l)$, $\tilde{\omega}_2 = 1{\cdot}776\surd(g/l)$.

4.5.3 (*a*) The frequency equation

$$\begin{vmatrix} 3-3\sigma^2 & -2\sigma^2 & -\sigma^2 & 1 & 0 \\ -2\sigma^2 & 2-2\sigma^2 & -\sigma^2 & -1 & 1 \\ -\sigma^2 & -\sigma^2 & 1-\sigma^2 & 0 & -1 \\ 1 & -1 & 0 & 0 & 0 \\ 0 & 1 & -1 & 0 & 0 \end{vmatrix} = 0$$

(where $\sigma^2 = l\omega^2/g$) gives $\sigma^2 = \frac{3}{7}$, or

$$\tilde{\omega}_1 = \sqrt{\left(\frac{3g}{7l}\right)}.$$

(*b*)

$$\begin{vmatrix} \mathbf{C}-\omega^2\mathbf{A} & \mathbf{D} \\ \mathbf{D}' & \mathbf{0} \end{vmatrix} = 0,$$

where $\mathbf{D}(n, m)$ is a matrix whose columns are the vectors $\mathbf{d}^{(i)}$.

5.1.1 (*a*) $2\begin{bmatrix} 1 & 1 \\ 1 & 1 \end{bmatrix}.$

(*b*) $2\begin{bmatrix} 1 & e^{-2i\theta} \\ e^{2i\theta} & 1 \end{bmatrix} = 2\begin{bmatrix} 1 & \cos 2\theta \\ \cos 2\theta & 1 \end{bmatrix} + 2i\begin{bmatrix} 0 & -\sin 2\theta \\ \sin 2\theta & 0 \end{bmatrix}.$

(*c*) $\begin{bmatrix} 1+e^{2i\theta} & 1+e^{2i\theta} \\ 1+e^{-2i\theta} & 1+e^{-2i\theta} \end{bmatrix} = (1+\cos 2\theta)\begin{bmatrix} 1 & 1 \\ 1 & 1 \end{bmatrix} + i\sin 2\theta \begin{bmatrix} 1 & 1 \\ -1 & -1 \end{bmatrix}.$

5.1.2 (*a*) Real; $\mathbf{A} = \bar{\mathbf{A}}$; $a_{ij} = \bar{a}_{ij}$; all elements real.

Imaginary; $\mathbf{A} = -\bar{\mathbf{A}}$; $a_{ij} = -\bar{a}_{ij}$; all elements imaginary.

Hermitian; $\mathbf{A} = \mathbf{A}^*$; $a_{ij} = \bar{a}_{ji}$; a_{ii} real.

Skew Hermitian; $\mathbf{A} = -\mathbf{A}^*$; $a_{ij} = -\bar{a}_{ji}$; a_{ii} imaginary.

Orthogonal; $\mathbf{AA}' = \mathbf{A}'\mathbf{A} = \mathbf{I}$; $\mathbf{A}^{-1} = \mathbf{A}'$;

$$\sum_{k=1}^{n} a_{ik}a_{jk} = \sum_{k=1}^{n} a_{ki}a_{kj} = \delta_{ij}.$$

Unitary; $\mathbf{AA}^* = \mathbf{A}^*\mathbf{A} = \mathbf{I}$; $\mathbf{A}^{-1} = \mathbf{A}^*$;

$$\sum_{k=1}^{n} a_{ik}\bar{a}_{jk} = \sum_{k=1}^{n} \bar{a}_{ki}a_{kj} = \delta_{ij}.$$

Generally 'unitary' plays the part for complex matrices that 'orthogonal' plays for real matrices.

(*b*) Consider an Hermitian matrix $\mathbf{H} = \mathbf{A}+i\mathbf{B}$, where $\mathbf{A}$ and $\mathbf{B}$ are real; $\mathbf{H}^* = \mathbf{A}'-i\mathbf{B}'$; but $\mathbf{H}^* = \mathbf{H}$ so that $\mathbf{A}' = \mathbf{A}$, $\mathbf{B}' = -\mathbf{B}$.

(*c*) If $\mathbf{A}$ is real then $\mathbf{A}' = \mathbf{A}^*$, so that if $\mathbf{AA}' = \mathbf{I}$ then $\mathbf{AA}^* = \mathbf{I}$.

5.1.3 (*a*) If $\mathbf{G} = \mathbf{AB}$, then by forming g_{ij} we see that $\bar{\mathbf{G}} = \bar{\mathbf{A}}\bar{\mathbf{B}}$. Therefore if $\mathbf{F} = \mathbf{ABC} = \mathbf{GC}$ then $\bar{\mathbf{F}} = \bar{\mathbf{G}}\bar{\mathbf{C}} = \bar{\mathbf{A}}\bar{\mathbf{B}}\bar{\mathbf{C}}$.

(*b*) $\bar{\mathbf{F}} = \bar{\mathbf{A}}\bar{\mathbf{B}}\bar{\mathbf{C}}$ so that $\mathbf{F}^* = \bar{\mathbf{F}}' = \bar{\mathbf{C}}'\bar{\mathbf{B}}'\bar{\mathbf{A}}' = \mathbf{C}^*\mathbf{B}^*\mathbf{A}^*$.

5.2.1 If the right-hand side of equation (5.2.8) is replaced by nought, the meaning of the symbol ω on the left-hand side is left open. If a particular value is assigned to ω, such as $\omega = \omega_1$, on the left-hand side, the system reduces to one with viscous damping.

5.2.2 (*a*) If $\tan\eta = \mu/(1-\beta^2)$, then

$$\sin\eta = \frac{\mu}{\sqrt{[(1-\beta^2)^2+\mu^2]}}$$

so that, through equation (5.2.13), y may be written in the form stated.

(*b*) Taking axes OY in the direction of F and OX in the perpendicular direction to the right, we have

$$x = \frac{F\sin^2\eta}{h}; \quad y = \frac{F\sin\eta\cos\eta}{h}.$$

On eliminating the quantity η, it is found that

$$\left(x-\frac{F}{2h}\right)^2+y^2 = \left(\frac{F}{2h}\right)^2.$$

The centre of this circle lies on OX and the radius is $F/2h$.

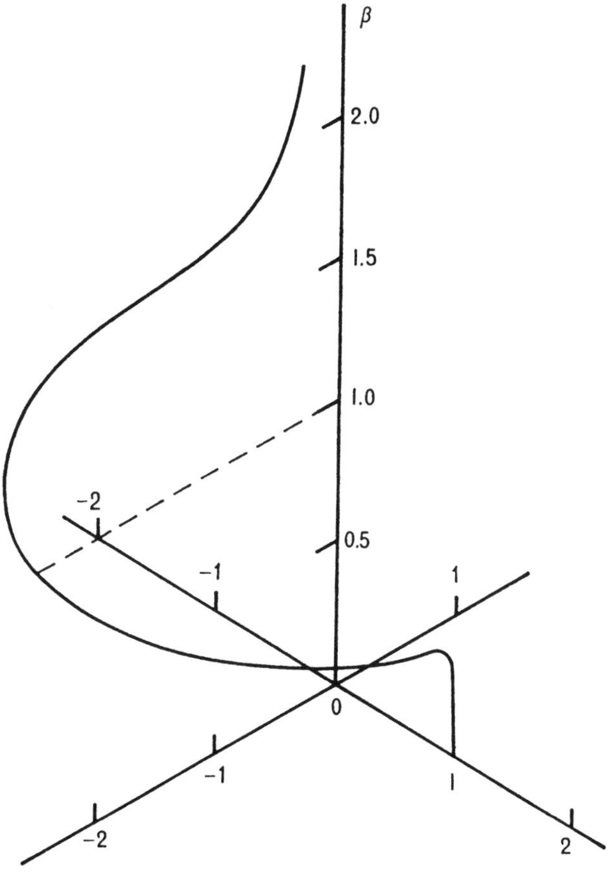

(*a*)

Ans. to Ex. 5.2.3.

5.2.3 See the diagrams. (The curves are drawn isometrically for the particular value $\mu = 0{\cdot}4$ and the 'horizontal' scales are in units of kY/F.)

5.3.1 The response $\mathbf{q} = \Pi_r \mathbf{\Psi}^{(r)} \sin(\omega t - \theta)$, when substituted in the equation

$$\mathbf{A}\ddot{\mathbf{q}} + \mathbf{B}\dot{\mathbf{q}} + \mathbf{C}\mathbf{q} = \mathbf{Q}$$

gives
$$\mathbf{Q} = \Pi_r\{\sin(\omega t - \theta).(\mathbf{C} - \omega^2\mathbf{A})\,\mathbf{\Psi}^{(r)} + \omega\cos(\omega t - \theta)\,.\,\mathbf{B}\mathbf{\Psi}^{(r)}\}.$$

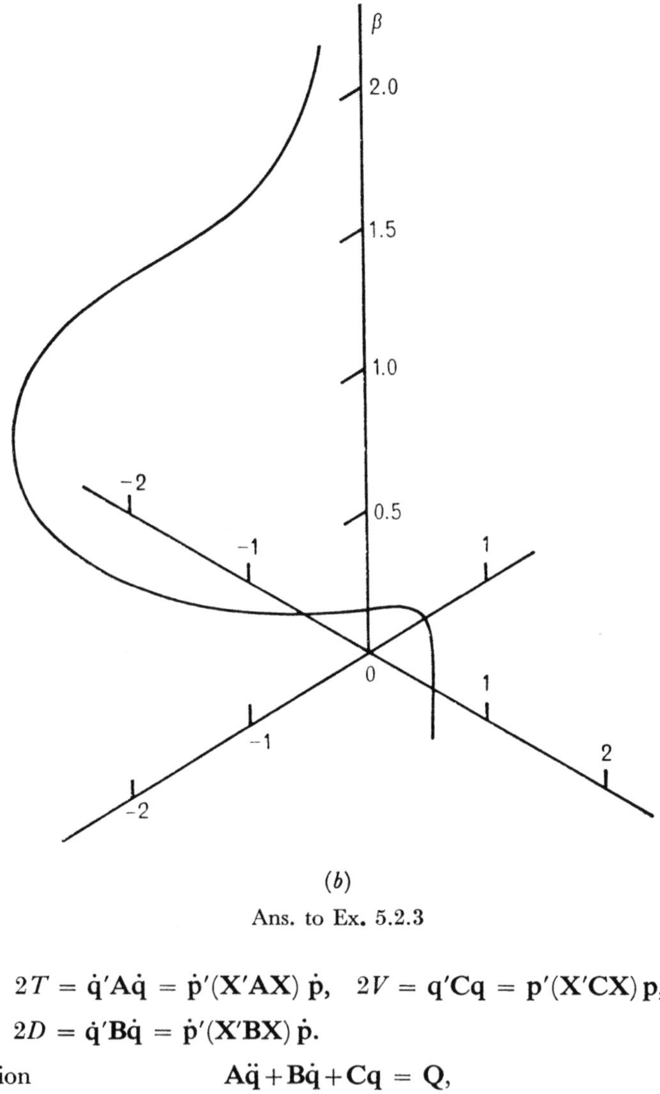

(*b*)

Ans. to Ex. 5.2.3

5.3.2 (*a*) $2T = \dot{\mathbf{q}}'\mathbf{A}\dot{\mathbf{q}} = \dot{\mathbf{p}}'(\mathbf{X}'\mathbf{A}\mathbf{X})\,\dot{\mathbf{p}}, \quad 2V = \mathbf{q}'\mathbf{C}\mathbf{q} = \mathbf{p}'(\mathbf{X}'\mathbf{C}\mathbf{X})\,\mathbf{p},$

$$2D = \dot{\mathbf{q}}'\mathbf{B}\dot{\mathbf{q}} = \dot{\mathbf{p}}'(\mathbf{X}'\mathbf{B}\mathbf{X})\,\dot{\mathbf{p}}.$$

In the equation
$$\mathbf{A}\ddot{\mathbf{q}} + \mathbf{B}\dot{\mathbf{q}} + \mathbf{C}\mathbf{q} = \mathbf{Q},$$

substitute for $\mathbf{q}$ and premultiply by $\mathbf{X}'$ to obtain

$$\mathbf{L}\ddot{\mathbf{p}} + \mathbf{M}\dot{\mathbf{p}} + \mathbf{N}\mathbf{p} = \mathbf{X}'\mathbf{Q} = \mathbf{P}.$$

(*b*) If $\mathbf{q} = \Pi_r \mathbf{\Psi}^{(r)} \sin \omega t$, then $p_s = 0$ if $s \neq r$ and $p_r = \Pi_r \sin \omega t$. The forcing required is given by $P_s = 0$ if $s \neq r$, $P_r = a_r\ddot{p}_r + b_r\dot{p}_r + c_r p_r$. This reduces to

$$P_r = \Pi_r a_r \omega_r^2 \sqrt{[(1-\beta_r^2)^2 + \mu_r^2]}\sin(\omega t + \theta_r),$$

where
$$\tan\theta_r = \frac{\mu_r}{1-\beta_r^2}.$$

Now $$\mathbf{Q} = (\mathbf{X}')^{-1}\mathbf{P} = (\mathbf{AXL}^{-1})\,\mathbf{P} = \mathbf{A}\boldsymbol{\Psi}^{(r)}a_r^{-1}P_r$$

since only P_r is non-zero. Therefore

$$\mathbf{Q} = \Pi_r\omega_r^2\sqrt{[(1-\beta_r^2)^2+\mu_r^2]}\,\mathbf{A}\boldsymbol{\Psi}^{(r)}\sin(\omega t+\theta_r).$$

5.3.3 The equation for θ corresponding to equation (5.3.16) is

$$|\tan\theta(\mathbf{N}-\omega^2\mathbf{L})-\omega\mathbf{M}| = 0.$$

When $$\mathbf{M} = \begin{bmatrix} b_1 & 0 & \dots & 0 \\ 0 & b_2 & \dots & 0 \\ \dots & \dots & \dots & \dots \\ 0 & 0 & \dots & b_n \end{bmatrix}$$

this reduces to $$\tan\theta_r = \frac{\omega b_r}{c_r-\omega^2 a_r} \quad (r = 1, 2, \dots, n).$$

The modal equation is $$[\tan\theta_r(\mathbf{N}-\omega^2\mathbf{L})-\omega\mathbf{M}]\,\boldsymbol{\varkappa}^{(r)} = \mathbf{0},$$

and therefore, referred to the principal coordinates, only the rth element in the forced mode is non-zero; in other words, the forced modes are the same as the principal modes. This holds whatever the driving frequency.

5.3.4 $$\theta_1 = 143^\circ\,48'; \quad \boldsymbol{\varkappa}^{(1)} = \begin{bmatrix} 1 \\ 1{\cdot}366 \end{bmatrix}; \quad \boldsymbol{\Sigma}^{(1)} = k\begin{bmatrix} 0{\cdot}454 \\ 1{\cdot}240 \end{bmatrix}.$$

$$\theta_2 = 69^\circ\,54'; \quad \boldsymbol{\varkappa}^{(2)} = \begin{bmatrix} 1 \\ -0{\cdot}366 \end{bmatrix}; \quad \boldsymbol{\Sigma}^{(2)} = k\begin{bmatrix} 3{\cdot}974 \\ -2{\cdot}909 \end{bmatrix}.$$

5.4.2 $$\mathbf{q} = \{-0{\cdot}170\boldsymbol{\varkappa}^{(1)}\sin(\omega t-143^\circ\,48')+0{\cdot}271\boldsymbol{\varkappa}^{(2)}\sin(\omega t-69^\circ\,54')\}\,(F/k).$$

5.4.3 $$q_i = \sum_{r=1}^{n}\left\{\frac{\kappa_i^{(r)}\kappa_j^{(r)}\sin(\omega t-\theta_r)}{\sqrt{\{[\boldsymbol{\varkappa}'^{(r)}(\mathbf{C}-\omega^2\mathbf{A})\,\boldsymbol{\varkappa}^{(r)}]^2+[\omega\boldsymbol{\varkappa}'^{(r)}\mathbf{B}\boldsymbol{\varkappa}^{(r)}]^2\}}}\right\}\Phi_j.$$

Yes.

5.5.1 $$\mathbf{Z}^{-1} = \frac{-1}{3+2i}\begin{bmatrix} 2i & 1+2i \\ 1+2i & 1+3i \end{bmatrix};$$

this leads to

$$\mathbf{u} = \frac{1}{3+2i}\begin{bmatrix} 1 \\ i \end{bmatrix}\frac{F}{k}e^{i\omega t} \quad \text{and}$$

$$\mathbf{q} = \begin{bmatrix} (0{\cdot}231\sin\omega t-0{\cdot}154\cos\omega t)\,(F/k) \\ (0{\cdot}154\sin\omega t+0{\cdot}231\cos\omega t)\,(F/k) \end{bmatrix}.$$

5.5.2 (*a*) $$\overline{\mathbf{A}}\ddot{\mathbf{w}}+\overline{\mathbf{B}}\dot{\mathbf{w}}+\overline{\mathbf{C}}\mathbf{w} = \mathbf{K}'\boldsymbol{\Phi}\,e^{i\omega t} = \mathbf{W}\,e^{i\omega t},$$

where $$\overline{\mathbf{A}} = \mathbf{K}'\mathbf{AK}, \quad \overline{\mathbf{B}} = \mathbf{K}'\mathbf{BK}, \quad \overline{\mathbf{C}} = \mathbf{K}'\mathbf{CK}.$$

Equations (5.4.4) show that $\overline{\mathbf{C}}-\omega^2\overline{\mathbf{A}}$ and $\overline{\mathbf{B}}$ are both diagonal matrices (whereas, in general, neither $\overline{\mathbf{C}}$ or $\overline{\mathbf{A}}$ is).

(*b*) If the rth element on the leading diagonal of $\overline{\mathbf{C}}-\omega^2\overline{\mathbf{A}}$ is $\bar{c}_r-\omega^2\bar{a}_r$ and the rth element on the leading diagonal of $\overline{\mathbf{B}}$ is $\bar{b}_r$, then

$$w_r = \frac{W_r e^{i\omega t}}{\bar{c}_r-\omega^2\bar{a}_r+i\omega\bar{b}_r}.$$

6.1.1

$$\omega_r = \sqrt{(C/\mathcal{J}l^2)}\, r\pi \quad (r = 0, 1, 2, \ldots)$$

$$\phi_r(x) = \begin{cases} 1/\sqrt{2} & (r = 0) \\ \cos r\pi x/l & (r = 1, 2, \ldots). \end{cases}$$

$$\alpha_{xy} = \frac{2}{\mathcal{J}l}\sum_{r=0}^{\infty}\frac{\phi_r(x)\phi_r(y)}{\omega_r^2-\omega^2}.$$

6.1.2 The total torque is F and

$$P_s = \frac{2F}{l}\left(\frac{\sin\alpha h}{\alpha h}\right)^2 \sin 2\alpha y,$$

where $\alpha \equiv (2s-1)\pi/4l$. That is

$$f(x) = \frac{2F}{l}\sum_{r=1}^{\infty}\left[\left(\frac{4l}{(2r-1)\pi h}\sin\frac{(2r-1)\pi h}{4l}\right)^2 \sin\frac{(2r-1)\pi y}{2l}\sin\frac{(2r-1)\pi x}{2l}\right].$$

All the sines are less than or equal to 1 so that, for any allowable x, y and h, the sum of the series S satisfies

$$S \leqslant \frac{2F}{l}\left(\frac{4l}{\pi h}\right)^2\sum_{r=1}^{\infty}\frac{1}{(2r-1)^2} = \frac{2F}{l}\left(\frac{4l}{\pi h}\right)^2\frac{\pi^2}{8}.$$

The series for $f(x)$ therefore converges. If $h \to 0$, the bracketed quantity in the above expression for P_s tends to 1.

6.1.3 For both beams the frequency equation is $\cos\lambda l \cosh\lambda l = 1$, and the first non-zero root gives

$$\omega_1 = 22{\cdot}3733\sqrt{(EI/A\rho l^4)}.$$

6.1.4 $\phi_r(x) = \cosh\lambda_r x+\cos\lambda_r x-\sigma_r(\sinh\lambda_r x+\sin\lambda_r x)$ where

$$\sigma_r = \frac{\cosh\lambda_r l-\cos\lambda_r l}{\sinh\lambda_r l-\sin\lambda_r l}.$$

Rigid body modes are 1, $\sqrt{(3)}(2x-l)/l$.

6.1.5

$$\alpha_{xy} = \frac{1}{A\rho l}\left\{-\frac{1}{\omega^2}-\frac{3}{l^2}\frac{(2x-l)(2y-l)}{\omega^2}+\sum_{r=1}^{\infty}\frac{\phi_r(x)\phi_r(y)}{\omega_r^2-\omega^2}\right\}.$$

6.1.6

$$\alpha_{00} = \alpha_{ll} = \frac{-(\cos\lambda l\sinh\lambda l-\sin\lambda l\cosh\lambda l)}{EI\lambda^3(\cos\lambda l\cosh\lambda l-1)},$$

$$\alpha_{0l} = \frac{\sin\lambda l-\sinh\lambda l}{EI\lambda^3(\cos\lambda l\cosh\lambda l-1)}.$$

For both representations of α_{00} we have

$$\lim_{\lambda\to 0}\lambda^4\alpha_{00} = \frac{-4}{EIl}.$$

6.2.1 $\alpha_{13} = (k_2k_3)/\Delta$, $\alpha_{23} = k_3(k_1+k_2-I_1\omega^2)/\Delta$,

$$\alpha_{33} = \{I_1I_2\omega^4-[k_1I_2+k_2(I_1+I_2)+k_3I_1]\omega^2+k_1k_2+k_2k_3+k_3k_1\}/\Delta,$$

where $\Delta = -\{I_1 I_2 I_3 \omega^6 - [k_1 I_2 I_3 + k_2(I_1 I_3 + I_2 I_3) + k_3(I_1 I_2 + I_1 I_3)]\,\omega^4$
$+ [I_1 k_2 k_3 + I_2(k_1 k_3 + k_2 k_3) + I_3(k_1 k_2 + k_2 k_3 + k_3 k_1)]\,\omega^2 - k_1 k_2 k_3\}$.

The frequency equation is $\Delta = 0$.

6.2.2 If $k'/k = y$ then the frequency equation is

$$2x^2 - (2+3y)\,x + y = 0,$$

so that
$$x_1 + x_2 = 1 + \tfrac{3}{2}y, \quad x_1 x_2 = \tfrac{1}{2}y, \quad x_1 + x_2 = 1 + 3x_1 x_2;$$

and $x_2 = (1-x_1)/(1-3x_1)$. If $x_1 = 0{\cdot}3$, $x_2 = 7$ then $y = 4{\cdot}2$.

6.2.3 If the receptances of the beam are represented by α, then the frequency equation is

$$\begin{vmatrix} \alpha_{00} + 1/k & \alpha_{0l} \\ \alpha_{l0} & \alpha_{ll} + 1/k \end{vmatrix} = 0,$$

and since $\alpha_{00} = \alpha_{ll}$ this factorises into $\alpha_{00} \pm \alpha_{0l} + 1/k = 0$.

(i) *Positive sign.* The motion is symmetrical about mid-span, i.e. $v(x) = v(l-x)$. The mid-section $x = \frac{1}{2}l$ will behave as if it were sliding, i.e. $v'(\frac{1}{2}l) = 0 = v'''(\frac{1}{2}l)$. The frequency equation will be that for a beam supported by a spring of stiffness k at $x = 0$ and sliding at $x = \frac{1}{2}l$. In fact

$$\alpha_{00} + \alpha_{0l} = -\frac{2\cos(\frac{1}{2}\lambda l)\cosh(\frac{1}{2}\lambda l)}{\cos(\frac{1}{2}\lambda l)\sinh(\frac{1}{2}\lambda l) + \sin(\frac{1}{2}\lambda l)\cosh(\frac{1}{2}\lambda l)},$$

and this is α_{00} for a free-sliding beam of length $\frac{1}{2}l$.

(ii) *Negative sign.* The motion is anti-symmetrical about mid-span, i.e. $v(x) = -v(l-x)$. The mid-section is pinned, i.e. $v(\frac{1}{2}l) = 0 = v''(\frac{1}{2}l)$. The frequency equation will be that for a beam supported by a spring of stiffness k at $x = 0$ and pinned at $x = \frac{1}{2}l$. In fact

$$\alpha_{00} - \alpha_{0l} = \frac{2\sin(\frac{1}{2}\lambda l)\sinh(\frac{1}{2}\lambda l)}{\cos(\frac{1}{2}\lambda l)\sinh(\frac{1}{2}\lambda l) - \sin(\frac{1}{2}\lambda l)\cosh(\frac{1}{2}\lambda l)},$$

and this is α_{00} for a free-pinned beam of length $\frac{1}{2}l$.

6.2.4 $\omega_1 = 1{\cdot}0769\sqrt{(C/Jl^2)}$.

6.3.1 $\Psi_s = -\alpha_{ss} R$ since the applied force is $-R$. But if $p_1, p_2, \ldots, p_n$ are the principal coordinates,

$$\alpha_{ss} = \sum_{r=1}^{n} \frac{(\partial q_s/\partial p_r)^2}{a_r(\omega_r^2 - \omega^2)}.$$

Near $\omega = \omega_r$ the rth term is dominant and will be positive below ω_r and negative above.

6.3.2 Put $\{R\Psi_n/\Sigma\ \text{(col. 5) (col. 7)}\} = \delta$. Then

$$\frac{\omega_c^2}{\omega_1^2} = \frac{1}{1+\delta}, \quad \frac{\omega_2^2}{\omega_1^2} = 1 - \delta; \quad \frac{\omega_c^2 - \omega_2^2}{\omega_1^2} = \frac{\delta^2}{1+\delta} > 0.$$

6.3.3 Hint:
$$R = -\Phi_{sb} - \Phi_{sc} = \sum_{r=1}^{s} I_r \omega^2 \Psi_r + \sum_{r=s+1}^{n} I_r \omega^2 \Psi_r,$$

$$\sum_{m=1}^{s-1} k_m(\Psi_m - \Psi_{m+1})^2 = \sum_{m=1}^{s} I_m \omega^2 \Psi_m^2 - \Psi_s \sum_{r=1}^{s} I_r \omega^2 \Psi_r,$$

$$\sum_{m=s}^{n-1} k_m(\Psi_m - \Psi_{m+1})^2 = \sum_{m=s+1}^{n} I_m \omega^2 \Psi_m^2 - \Psi_s \sum_{r=s+1}^{n} I_r \omega^2 \Psi_r.$$

The result follows at once from these equations.

6.3.4 $0{\cdot}318k/I$; residual $= 0{\cdot}0121k$. $\quad 0{\cdot}32044k/I$; residual $= 0{\cdot}0000k$.

6.3.5 Subtract $0{\cdot}0622I$ from I_1. This gives $I_1' = 0{\cdot}4378I$, $\omega_1'^2 = 0{\cdot}33024k/I$. Correct $I' = 0{\cdot}43929I$.

6.3.6 (*a*) $\{\mathbf{v}\}_i = \{\mathbf{v}\}_{i'}$.

(*b*)
$$\mathbf{T}_i = \begin{bmatrix} C_0 & \beta_i S_1 & \dfrac{\beta_i^2 C_2}{\alpha_i} & \dfrac{\beta_i^3 S_3}{\alpha_i} \\ \dfrac{(\lambda l_i)^4 S_3}{\beta_i} & C_0 & \dfrac{\beta_i S_1}{\alpha_i} & \dfrac{\beta_i^2 C_2}{\alpha_i} \\ \dfrac{\alpha_i(\lambda l_i)^4 C_2}{\beta_i^2} & \dfrac{\alpha_i(\lambda l_i)^4 S_3}{\beta_i} & C_0 & \beta_i S_1 \\ \dfrac{\alpha_i(\lambda l_i)^4 S_1}{\beta_i^3} & \dfrac{\alpha_i(\lambda l_i)^4 C_2}{\beta_i^2} & \dfrac{(\lambda l_i)^4 S_3}{\beta_i} & C_0 \end{bmatrix},$$

where $\alpha_i = EI_i/EI_s$, $\beta_i = l_i/l_s$ and C_0, S_1, etc., are all evaluated for

$$\lambda = \sqrt[4]{\frac{(A\rho)_i \omega^2}{EI_i}}.$$

(*c*) $\mathbf{T} = \mathbf{T}_n \mathbf{T}_{n-1} \ldots \mathbf{T}_1$.

(*d*) (i) $t_{33}t_{44} - t_{34}t_{43} = 0$.

(ii) $t_{12}t_{24} - t_{22}t_{14} = 0$.

6.4.1 The method of induction is convenient. Replace μ in the determinant Δ_n (of equation (6.4.8)) by $2\cos\theta$ and then show that

$$\Delta_1 = \cos\theta, \quad \Delta_2 = \cos 2\theta \quad \text{and} \quad \Delta_{m+1} = \mu\Delta_m - \Delta_{m-1}.$$

Then assume that $\Delta_n = \cos n\theta$ is valid for $n = m-1$ and $n = m$ and show that, under this hypothesis, $\Delta_{m+1} = \cos(m+1)\theta$. If the result is true for $n = m-1$ and $n = m$ then it is also true for $n = m+1$. But, since it *is* true for $n = 1$ and $n = 2$ it is also true for $n = 3, 4, \ldots, n$. Therefore

$$\Delta_n = \cos n\theta.$$

6.4.2 (*a*) $\omega_{rn} = \sqrt{(C/\mathcal{J}l^2)}\, 2n \sin r\pi/2n \quad (r = 0, 1, 2, \ldots, n)$.

$$\Psi_s^{(r)} = A_r \cos\frac{(s-1)r\pi}{n} \quad (s = 1, 2, \ldots, n+1).$$

(*b*) 7.

6.4.3 (*a*)
$$\omega_r = \frac{m^4(p^2-n^2)\,\omega_{rm} + n^4(m^2-p^2)\,\omega_{rn} + p^4(n^2-m^2)\,\omega_{rp}}{m^4(p^2-n^2) + n^4(m^2-p^2) + p^4(n^2-m^2)}.$$

(*b*) $\omega_2 = 4{\cdot}7121\sqrt{(C/\mathcal{J}l^2)}$.

6.4.4 $1{\cdot}732\sqrt{(C/\mathcal{J}l^2)}$ for Horvay–Ormondroyd representation; numerical factor is $1{\cdot}414$ for Rayleigh representation. (The *exact* factor is $1{\cdot}571$.)

6.4.5 (*a*) For Rayleigh's representation the natural frequencies are given by

$$(I/k)\,\omega_{rn}^2 = 2(1-\cos\theta_{rn}),$$

where
$\theta_{rn} = (2r-1)\pi/2n \quad (r = 1, 2, \ldots, n)$ for a clamped-free shaft,
$\theta_{rn} = r\pi/n \quad (r = 0, 1, \ldots, n)$ for a free-free shaft,
$\theta_{rn} = r\pi/n \quad (r = 1, 2, \ldots, n-1)$ for a clamped-clamped shaft.

Therefore for the Horvay–Ormondroyd representation

$$\frac{(\mathcal{J}l/n)\,\omega_{rn}^2}{(nC/l)+(\mathcal{J}l/6n)\,\omega_{rn}^2} = 2(1-\cos\theta_{rn}),$$

which gives
$$\omega_{rn} = \sqrt{\left(\frac{C}{\mathcal{J}l^2}\right)}\,\frac{2n\sin(\theta_{rn}/2)}{\{1-(1-\cos\theta_{rn})/3\}^{\frac{1}{2}}},$$

and
$$\sqrt{\left(\frac{\mathcal{J}l^2}{C}\right)}\,(\omega_{rn}-\omega_r) \doteqdot \frac{n\theta_{rn}^3}{4!}.$$

(b) $\frac{1}{12}$. Then $\sqrt{\left(\frac{\mathcal{J}l^2}{C}\right)}\,(\omega_{rn}-\omega_r) \doteqdot \frac{-n\theta_{rn}^5}{4.5!}$.

6.5.1 $\delta V-\delta W_{in}-\delta W_e \equiv \int_0^1 \{(EIv'')''+A\rho\ddot{v}-F\}\epsilon\eta\,dx+\epsilon[(EIv'')\,\eta'-(EIv'')'\eta]_0^l = 0.$

To find the differential equation and end conditions take arbitrary η satisfying $\eta = 0 = \eta'$ at $x = 0$ together with the following conditions in turn: (i) $\eta = 0 = \eta'$ at $x = l$; (ii) $\eta = 0$, $\eta' \neq 0$ at $x = l$; (iii) $\eta \neq 0$, $\eta' = 0$ at $x = l$.

6.5.2 (i) $\psi_r(\xi) = \frac{1}{6}\{(r+2)\,(r+3)\,\xi^{r+1}-2r(r+3)\,\xi^{r+2}+r(r+1)\,\xi^{r+3}\}$ where $\xi = x/l$.

No. If a load and/or a moment is applied at the tip then

$$\frac{\partial}{\partial x}\left(EI\frac{\partial^2 v}{\partial x^2}\right) \quad \text{and/or} \quad EI\frac{\partial^2 v}{\partial x^2}$$

is not zero at the tip.

(ii) Applied load only $\psi_r(\xi) = \frac{1}{2}\{(r+2)\,\xi^{r+1}-r\xi^{r+2}\}$.

(iii) Applied moment only $\psi_r(\xi) = \xi^{r+1}$.

(iv) Applied load and moment $\psi_r(\xi) = \xi^{r+1}$.

6.5.3 $\omega_1 = \tau\sqrt{(C/\mathcal{J}l^2)}$ where τ has the values: (i) 1·1107, (ii) 1·0946, (iii) 1·1071, (iv) 1·0954. The exact value is given in Ex. 6.2.4; it is 1·0769. At least in this problem it is no advantage to use the exact principal modes of the shaft.

6.5.4 (i) $\omega_1 = 1{\cdot}72\sqrt{(C/\mathcal{J}l^2)}$; (ii) $\omega_1 = 1{\cdot}57\sqrt{(C/\mathcal{J}l^2)}$.

7.1.1 Equations of motion reduce to

$$\begin{bmatrix} 4 & -6 & -3 \\ -6 & 0 & -6 \\ -3 & -6 & -3 \end{bmatrix}\begin{bmatrix} x_1 \\ x_2 \\ x_3 \end{bmatrix} = \begin{bmatrix} 1 \\ 2 \\ 3 \end{bmatrix},$$

where $\mathbf{x} \equiv 16\{\theta, \phi, \psi\}$. Solution is

$$\begin{bmatrix} x_1 \\ x_2 \\ x_3 \end{bmatrix} = \begin{bmatrix} -0{\cdot}2856 \\ -0{\cdot}3333 \\ -0{\cdot}0476 \end{bmatrix}, \quad \text{whence} \quad \begin{bmatrix} \theta \\ \phi \\ \psi \end{bmatrix} = \begin{bmatrix} -0{\cdot}0179 \\ -0{\cdot}0208 \\ -0{\cdot}0030 \end{bmatrix} \text{ radians.}$$

7.2.1 (*a*)

$$\underset{\mathbf{A}}{\begin{bmatrix} 4 & -6 & -3 \\ -3 & 0 & -3 \\ -1 & -2 & -1 \end{bmatrix}};\quad \begin{bmatrix} 4 & -6 & -3 \\ 0 & -4{\cdot}5 & -5{\cdot}25 \\ 0 & -3{\cdot}5 & -1{\cdot}75 \end{bmatrix};\quad \underset{\mathbf{U}}{\begin{bmatrix} 4 & -6 & -3 \\ 0 & -4{\cdot}5 & -5{\cdot}25 \\ 0 & 0 & 2{\cdot}3333 \end{bmatrix}}$$

$$l_{21} = -0{\cdot}75 \qquad l_{32} = 0{\cdot}7778$$
$$l_{31} = -0{\cdot}25$$

whence

$$\mathbf{L} = \begin{bmatrix} 1 & 0 & 0 \\ -0{\cdot}75 & 1 & 0 \\ -0{\cdot}25 & 0{\cdot}7778 & 1 \end{bmatrix}.$$

Note that **LU** = **A**.

(*b*)

4	−6	−3
$-\frac{3}{4}$ $= -0{\cdot}75$	$0-(-0{\cdot}75)(-6)$ $= -4{\cdot}5$	$-3-(-0{\cdot}75)(-3)$ $= -5{\cdot}25$
$-\frac{1}{4}$ $= -0{\cdot}25$	$\dfrac{-2-(-0{\cdot}25)(-6)}{-4{\cdot}5}$ $= 0{\cdot}7778$	$-1-(-0{\cdot}25)(-3)-(0{\cdot}7778)(-5{\cdot}25)$ $= 2{\cdot}333$

The order of working is $\boldsymbol{\rho}_1$, $\mathbf{col}_1$, $\boldsymbol{\rho}_2$, $\mathbf{col}_2$, etc.

7.2.2 The modified right-hand side, **z**, is

$$1,$$
$$1-(-0{\cdot}75)(1) = 1{\cdot}75,$$
$$1-(-0{\cdot}25)(1)-(0{\cdot}7778)(1{\cdot}75) = -0{\cdot}1111,$$

so that

$$x_3 = (-0{\cdot}1111)/(2{\cdot}3333) = -0{\cdot}0476,$$
$$x_2 = \{1{\cdot}75-(-5{\cdot}25)(-0{\cdot}0476)\}/(-4{\cdot}5) = -0{\cdot}3333,$$
$$x_1 = \{1-(-3)(-0{\cdot}0476)-(-6)(-0{\cdot}3333)\}/(4) = -0{\cdot}2856.$$

7.2.3

$$\mathbf{LDV} = \begin{bmatrix} 2 & 0 & 0 \\ -1{\cdot}5 & 2{\cdot}1213 & 0 \\ -0{\cdot}5 & 1{\cdot}6499 & 1{\cdot}5275 \end{bmatrix}\begin{bmatrix} 1 & 0 & 0 \\ 0 & -1 & 0 \\ 0 & 0 & 1 \end{bmatrix}\begin{bmatrix} 2 & -3 & -1{\cdot}5 \\ 0 & 2{\cdot}1213 & 2{\cdot}4748 \\ 0 & 0 & 1{\cdot}5275 \end{bmatrix} \doteqdot \mathbf{A}.$$

7.3.1

	U			z	$\mathbf{r}^0$	x	$\mathbf{x}^0$
	1	2	3	4	5*	22	23*
	6	9	10	11	12*	20	21*
L	7	13	15	16	17*	18	19*
$\mathbf{c}^0$	8*	14*					

7.3.2 See solution of Ex. 7.2.2.

7.4.1

L′			z	r⁰	x	x⁰
2	-3	$-1{\cdot}5$	$0{\cdot}5$	-2	$-0{\cdot}2856$	$0{\cdot}7143$
	$3i$	$3{\cdot}5i$	$-1{\cdot}1667i$	$5{\cdot}3334i$	$-0{\cdot}3333$	$0{\cdot}6667$
		$2{\cdot}6458$	$-0{\cdot}1260$	$2{\cdot}5198$	$-0{\cdot}0476$	$0{\cdot}9524$

7.4.2 (*a*) Denote the 'principal minors' of $\mathbf{A}$, the determinants shown in equation (2.1.7), by $D_1, D_2, \ldots, D_n$; for a positive definite matrix these are all strictly positive. The diagonal elements of $\mathbf{L}'$ are

$$l'_{11} = \sqrt{D_1}, \quad l'_{22} = \sqrt{(D_2/D_1)}, \quad l'_{33} = \sqrt{(D_3/D_2)}, \quad \text{etc.}$$

and are therefore all real. The remainder may be seen to be real by inspecting equation (7.4.4).

(*b*)
$$\mathbf{L}' = \begin{bmatrix} 2 & 3 & 1{\cdot}5 \\ 0 & 1{\cdot}7321 & 0{\cdot}8660 \\ 0 & 0 & 2{\cdot}4495 \end{bmatrix}.$$

7.4.3 (*a*) Let $\mathbf{DL}' = \mathbf{V}$, then

$$v_{rs} = u_{rr}(\mathbf{L}')_{rs} = u_{rr}l_{sr}.$$

Thus, using equation (7.2.15) and the fact that $a_{rs} = a_{sr}$, we find

$$v_{1s} = u_{11}l_{s1} = a_{s1} = a_{1s} = u_{1s},$$
$$v_{2s} = u_{ss}l_{s2} = a_{s2} - l_{s1}u_{12} = a_{2s} - (a_{s1}/u_{11})\,a_{12}$$
$$= a_{2s} - (a_{21}/u_{11})\,a_{1s} = a_{2s} - l_{21}u_{1s} = u_{2s}, \quad \text{etc.},$$

so that $\mathbf{V} \equiv \mathbf{U}$.

(*b*)
$$\mathbf{L} = \begin{bmatrix} 1 & 0 & 0 \\ -0{\cdot}5 & 1 & 0 \\ -1{\cdot}5 & 1 & 1 \end{bmatrix}; \quad \mathbf{D} = \begin{bmatrix} 2 & 0 & 0 \\ 0 & 0{\cdot}5 & 0 \\ 0 & 0 & 0 \end{bmatrix}.$$

7.4.4 (*a*) If $D_1, D_2, \ldots, D_n$ denote the principal minors of $\mathbf{A}$ shown in equation (2.1.7) then equation (7.2.15) shows that, provided the quotients are meaningful, $u_{11} = D_1$, $u_{22} = D_2/D_1$, $u_{33} = D_3/D_2, \ldots, u_{nn} = D_n/D_{n-1}$. For a positive semi-definite matrix, D_n will be zero and $D_1, \ldots, D_{n-1}$ will be positive or zero: hence the result.

(*b*)
$$\mathbf{L}^0 = \begin{bmatrix} 1{\cdot}4142 & 0 & 0 \\ -0{\cdot}7071 & 0{\cdot}7071 & 0 \\ -2{\cdot}1213 & 0{\cdot}7071 & 1 \end{bmatrix}, \quad \mathbf{D}^0 = \begin{bmatrix} 1 & 0 & 0 \\ 0 & 1 & 0 \\ 0 & 0 & 0 \end{bmatrix}.$$

7.4.5 From Ex. 7.4.1, $\mathbf{A} = \mathbf{EE}'$ (say) where

$$\mathbf{E}' = \begin{bmatrix} 2 & -3 & -1{\cdot}5 \\ 0 & 3i & 3{\cdot}5i \\ 0 & 0 & 2{\cdot}6458 \end{bmatrix} = \begin{bmatrix} 2 & 0 & 0 \\ 0 & 3i & 0 \\ 0 & 0 & 2{\cdot}6458 \end{bmatrix}\begin{bmatrix} 1 & -1{\cdot}5 & -0{\cdot}75 \\ 0 & 1 & 1{\cdot}1667 \\ 0 & 0 & 1 \end{bmatrix} = \mathbf{F}'\mathbf{L}' \text{ (say)}$$

therefore
$$\mathbf{A} = \mathbf{L}(\mathbf{FF}')\mathbf{L}' = \mathbf{LDL}',$$

i.e.

$$\mathbf{D} = \begin{bmatrix} 4 & 0 & 0 \\ 0 & -9 & 0 \\ 0 & 0 & 7{\cdot}0003 \end{bmatrix}; \quad \mathbf{L} = \begin{bmatrix} 1 & 0 & 0 \\ -1{\cdot}5 & 1 & 0 \\ -0{\cdot}75 & 1{\cdot}1667 & 1 \end{bmatrix}.$$

7.6.1

$$0{\cdot}000\,014\,90x_1 + 1{\cdot}0000x_2 = 1{\cdot}4550,$$

$$1{\cdot}0000x_1 + 0{\cdot}301\,27x_2 = 4{\cdot}8828,$$

and

$$x_1 = 4{\cdot}445, \quad x_2 = 1{\cdot}454.$$

7.6.2 (*b*) The standardised set of equations is

$$1{\cdot}0000x_1 + 0{\cdot}8333x_2^* + 0{\cdot}7500x_3^* = 130,$$

$$1{\cdot}0000x_1 + 0{\cdot}9375x_2^* + 0{\cdot}9000x_3^* = 141,$$

$$1{\cdot}0000x_1 + 1{\cdot}0000x_2^* + 1{\cdot}0000x_3^* = 148,$$

where

$$x_2^* = \tfrac{4}{5}x_2, \quad x_3^* = \tfrac{2}{3}x_3$$

the calculation is

	U			z		
	1·0000	0·8333	0·7500	130	$x_3^* = 40{\cdot}22$	$x_3 = 60{\cdot}33$
	1·0000	0·1042	0·1500	11	$x_2^* = 47{\cdot}68$	$x_2 = 59{\cdot}60$
L	1·0000	1·5998	0·0100	0·4022	$x_1 = 60{\cdot}10$	

7.7.1

$$\mathbf{P} = \begin{bmatrix} 0 & 0 & 1 \\ 1 & 0 & 0 \\ 0 & 1 & 0 \end{bmatrix}, \quad \mathbf{Q} = \begin{bmatrix} 1 & 0 & 0 \\ 0 & 0 & 1 \\ 0 & 1 & 0 \end{bmatrix}.$$

7.8.1

		L	
$t = 1$	2 [1]	3 − (2) (−1·5) [9] = 6 −1 − (3)) (−1·5) = 3·5 [10]	0 [11] −2 − (3) (−0·25) − (3·5) (0·25) = −2·125 [14]
	3 [2]		
$s = 3$	−4 [3]	0 [4]	0 [5]

1	6/(−4) [6] = −1·5	1/(−4) [7] = −0·25	1/(−4) [8] = −0·25
0	1	$\dfrac{1-(2)(-0{\cdot}25)}{6}$ [12] = 0·25	$\dfrac{13-(2)(-0{\cdot}25)}{6}$ [13] = 2·25
0	0	1	$\dfrac{5-(3)(-0{\cdot}25)-(3{\cdot}5)(2{\cdot}25)}{-2{\cdot}125}$ [15] = 1

The numbers in brackets indicate the order in which the entries are calculated.

7.8.2 (i) $\mathbf{L} = \begin{bmatrix} 1 & 4\cdot2 & -1\cdot75 \\ -2 & -4\cdot4 & 0 \\ 2\cdot5 & 0 & 0 \end{bmatrix}$, $\mathbf{U} = \begin{bmatrix} 1 & -6\cdot2 & 11\cdot6 \\ 0 & 1 & -1\cdot75 \\ 0 & 0 & 1 \end{bmatrix}$.

(ii) No.

N.B. Further examples on matrix factorisation with row interchanges with and without square root factorisation are to be found in Examples 9.6.

7.11.1 Using the values $\|\boldsymbol{\delta}\mathbf{A}\| = (2\cdot13)10^{-6}$, $\|\boldsymbol{\delta}\mathbf{y}\| = (3\cdot91)10^{-6}$, $\|\mathbf{x}+\boldsymbol{\delta}\mathbf{x}\| < 6$, $\|(\mathbf{A}+\boldsymbol{\delta}\mathbf{A})^{-1}\| = 13\cdot5$ in equation (7.11.17) we obtain $\|\boldsymbol{\delta}\mathbf{x}\| \leqslant (226)10^{-6}$. Actually $\|\boldsymbol{\delta}\mathbf{x}\| = (21)10^{-6}$, as may be seen by inspecting the solution given on p. 268.

8.2.1

$\mathbf{A} = \mathbf{LL}'$, $\mathbf{L} = \begin{bmatrix} \sqrt{2} & 0 & 0 & 0 \\ 0 & \sqrt{3} & 0 & 0 \\ 0 & 0 & 1 & 0 \\ 0 & 0 & 0 & \sqrt{5} \end{bmatrix}$, $\mathbf{L}^{-1} = \begin{bmatrix} 0\cdot707\,107 & 0 & 0 & 0 \\ 0 & 0\cdot577\,350 & 0 & 0 \\ 0 & 0 & 1 & 0 \\ 0 & 0 & 0 & 0\cdot447\,214 \end{bmatrix}$,

$$\mathbf{L}^{-1}\mathbf{C}(\mathbf{L}')^{-1} = \begin{bmatrix} 3\cdot500\,000 & -1\cdot224\,745 & -1\cdot414\,214 & -0\cdot316\,228 \\ -1\cdot224\,745 & 2\cdot000\,000 & -1\cdot154\,700 & -0\cdot258\,199 \\ -1\cdot414\,214 & -1\cdot154\,700 & 4\cdot000\,000 & 0 \\ -0\cdot316\,228 & -0\cdot258\,199 & 0 & 0\cdot400\,000 \end{bmatrix}.$$

8.2.2 The sequence of calculations is

$$\mathbf{A} = \mathbf{LL}', \mathbf{L};\quad \mathbf{LD} = \mathbf{C};\quad \mathbf{D} = \mathbf{L}^{-1}\mathbf{C};\quad \mathbf{BL}' = \mathbf{D},\quad \mathbf{B} = \mathbf{L}^{-1}\mathbf{C}(\mathbf{L}')^{-1}.$$

$\mathbf{L} = \begin{bmatrix} 2 & 0 & 0 \\ 1 & 1 & 0 \\ 1 & 2 & 1 \end{bmatrix}$, $\mathbf{D} = \begin{bmatrix} 2 & -1 & -0\cdot5 \\ -4 & 6 & -1\cdot5 \\ 5 & -13 & 0\cdot5 \end{bmatrix}$, $\mathbf{B} = \begin{bmatrix} 1 & -2 & 2\cdot5 \\ -2 & 8 & -15\cdot5 \\ 2\cdot5 & -15\cdot5 & 29 \end{bmatrix}$.

8.2.3 $\mathbf{C} = \mathbf{LL}'$, $\lambda = 1/\omega^2$, $\mathbf{B} = \mathbf{L}^{-1}\mathbf{A}(\mathbf{L}')^{-1}$,

$\mathbf{L} = \begin{bmatrix} 1 & 0 & 0 & 0 \\ -2 & 1 & 0 & 0 \\ 1 & -2 & 1 & 0 \\ 0 & 1 & -2 & 1 \end{bmatrix}$, $\mathbf{L}^{-1} = \begin{bmatrix} 1 & 0 & 0 & 0 \\ 2 & 1 & 0 & 0 \\ 3 & 2 & 1 & 0 \\ 4 & 3 & 2 & 1 \end{bmatrix}$, $\mathbf{B} = \begin{bmatrix} 1 & \frac{1}{2} & \frac{1}{3} & \frac{1}{4} \\ \frac{1}{2} & \frac{1}{3} & \frac{1}{4} & \frac{1}{5} \\ \frac{1}{3} & \frac{1}{4} & \frac{1}{5} & \frac{1}{6} \\ \frac{1}{4} & \frac{1}{5} & \frac{1}{6} & \frac{1}{7} \end{bmatrix}$.

8.2.4 $\mathbf{C}$ is singular. Write $\mathbf{C} = \mathbf{LDL}'$ where

$\mathbf{L} = \begin{bmatrix} 2 & 0 & 0 & 0 \\ -0\cdot5 & 1\cdot936\,492 & 0 & 0 \\ 0 & -1\cdot032\,795 & 1\cdot712\,698 & 0 \\ -1\cdot5 & -0\cdot903\,696 & -1\cdot712\,697 & 1 \end{bmatrix}$, $\mathbf{D} = \begin{bmatrix} 1 & 0 & 0 & 0 \\ 0 & 1 & 0 & 0 \\ 0 & 0 & 1 & 0 \\ 0 & 0 & 0 & 0 \end{bmatrix}$.

The sequence of calculations is

$$\mathbf{LZ} = \mathbf{A}, \quad \mathbf{Z} = \mathbf{L}^{-1}\mathbf{A}; \qquad \mathbf{BL}' = \mathbf{Z}, \quad \mathbf{B} = \mathbf{L}^{-1}\mathbf{A}(\mathbf{L}')^{-1}.$$

$$\mathbf{Z} = \begin{bmatrix} 2 & 1 & 0{\cdot}5 & 0{\cdot}5 \\ 1{\cdot}549\,193 & 2{\cdot}323\,790 & 1{\cdot}161\,895 & 0{\cdot}645\,497 \\ 1{\cdot}518\,072 & 2{\cdot}569\,045 & 2{\cdot}452\,271 & 1{\cdot}556\,997 \\ 7{\cdot}999\,997 & 8{\cdot}999\,995 & 7{\cdot}999\,997 & 6{\cdot}999\,997 \end{bmatrix},$$

$$\mathbf{B} = \begin{bmatrix} 1 & 0{\cdot}774\,597 & 0{\cdot}759\,036 & 3{\cdot}999\,999 \\ 0{\cdot}774\,596 & 1{\cdot}400\,000 & 1{\cdot}522\,632 & 5{\cdot}680\,372 \\ 0{\cdot}759\,036 & 1{\cdot}522\,631 & 2{\cdot}349\,998 & 8{\cdot}096\,381 \\ 3{\cdot}999\,998 & 5{\cdot}680\,372 & 8{\cdot}096\,382 & 31{\cdot}999\,973 \end{bmatrix}.$$

We partition $\mathbf{B}$ in the form

$$\mathbf{B} = \begin{bmatrix} \mathbf{E} & \mathbf{F} \\ \mathbf{F}' & \mathbf{G} \end{bmatrix},$$

then the new characteristic equation is

$$\mathbf{B}^0\mathbf{x}^0 = (\mathbf{E} - \mathbf{F}\mathbf{G}^{-1}\mathbf{F}')\,\mathbf{x}^0 = \lambda\mathbf{x}^0,$$

where

$$\mathbf{G}^{-1} = 0{\cdot}031\,250\,026,$$

$$\mathbf{B}^0 = \begin{bmatrix} 0{\cdot}5 & 0{\cdot}064\,550 & -0{\cdot}253\,012 \\ 0{\cdot}064\,550 & 0{\cdot}391\,667 & 0{\cdot}085\,429 \\ -0{\cdot}253\,012 & 0{\cdot}085\,429 & 0{\cdot}301\,516 \end{bmatrix}.$$

8.3.1 All the characteristic values lie in that part of the complex plane covered by the circles

$$\begin{array}{ll} |\lambda-1| \leqslant 9 & |\lambda-1| \leqslant 11, \\ |\lambda+3| \leqslant 11 & |\lambda+3| \leqslant 8, \\ |\lambda-3| \leqslant 11 & |\lambda-3| \leqslant 8, \\ |\lambda-3| \leqslant 7 & |\lambda-3| \leqslant 11. \end{array}$$

In particular there must be at least one characteristic value in the regions

$$|\lambda-3| \leqslant 7, \quad |\lambda+3| \leqslant 8.$$

8.3.2

$$\mathbf{Bx}_0 = \{875 \quad 539 \quad 399 \quad 319\}/420,$$

$$\mathbf{x}_0'\mathbf{Bx}_0 = 2132/420; \quad \lambda_0 = 2132/1680 = 1{\cdot}269\,048,$$

$$\mathbf{y}_0 = \mathbf{Bx}_0 - \lambda_0\mathbf{x}_0 = \{1368 \quad 24 \quad -536 \quad -856\}/1680,$$

$$\mathbf{y}_0'\mathbf{y}_0 = (2\,892\,032)/2\,822\,400) = 1{\cdot}024\,671,$$

$$\mathbf{y}_0'\mathbf{y}_0/\mathbf{x}_0'\mathbf{x}_0 = 0{\cdot}256\,168; \quad \sqrt{(\mathbf{y}_0'\mathbf{y}_0/\mathbf{x}_0'\mathbf{x}_0)} = 0{\cdot}506\,130.$$

Therefore there is a characteristic value in the range

$$0{\cdot}762\,918 \quad \text{to} \quad 1{\cdot}775\,178$$

The enclosure theorem states that there is a characteristic value in the range

$$0{\cdot}7595 \quad \text{to} \quad 2{\cdot}0834.$$

8.3.3 The rounding error made is symmetric and bounded by

$$10^{-6}\begin{bmatrix} 0 & 0 & 0{\cdot}333 & 0 \\ 0 & 0{\cdot}333 & 0 & 0 \\ 0{\cdot}333 & 0 & 0 & 0{\cdot}333 \\ 0 & 0 & 0{\cdot}333 & 0{\cdot}143 \end{bmatrix}.$$

The row sums of this are bounded by $10^{-6}\times 0{\cdot}7$ so that the perturbations of the characteristic values of the original matrix are bounded by this.

8.4.1 (i) $|\lambda_3 - 2{\cdot}919\,380\,689| \leqslant 10^{-9}\times 416\,78,$

$|\lambda_4 - 73{\cdot}335\,848\,66| \leqslant 10^{-8}\times 3532.$

(ii) $\{\sum_{i\neq 3}\xi_i^2\}^{\frac{1}{2}} \leqslant 10^{-6}\times 26, \quad \{\sum_{i\neq 4}\xi_i^2\}^{\frac{1}{2}} \leqslant 10^{-6}\times 0{\cdot}7.$

8.4.2 $2{\cdot}919\,380\,689 \leqslant \lambda_3 \leqslant 2{\cdot}919\,380\,690.$

8.5.1 $\mathbf{v}^{(5)} = \{1 \quad 0{\cdot}570\,179 \quad 0{\cdot}406\,787 \quad 0{\cdot}318\,148\}.$

Rayleigh quotient is $\lambda_0 = 1{\cdot}500\,214$ and $|\lambda-\lambda_0| \leqslant 10^{-6}\times 11$. This, together with the error $10^{-6}\times 0{\cdot}7$, found in Ex. 8.3.3, gives $|\lambda-\lambda_0| \leqslant 10^{-6}\times 12$ for the characteristic value of the original matrix.

8.5.2 $\lambda = 37{\cdot}4929, \quad \mathbf{x} = \{0{\cdot}097\,673 \quad -0{\cdot}532\,174 \quad 1\}.$

8.5.3 $\lambda = -0{\cdot}506, \quad \mathbf{x} = \{0{\cdot}530 \quad 1 \quad 0{\cdot}480\}.$

8.6.1 $\mathbf{v} = \{1 \quad 0{\cdot}570\,172 \quad 0{\cdot}406\,779 \quad 0{\cdot}318\,141\}.$

8.6.2 $\mathbf{v} = \{0{\cdot}097\,672\,420 \quad -0{\cdot}532\,174\,055 \quad 1\}.$

$$\lambda_R = \frac{48{\cdot}468\,886\,416}{1{\cdot}292\,749\,126} = 37{\cdot}492\,878\,88.$$

$$\mathbf{Bv}-\lambda_R\mathbf{v} = 10^{-9}\{317 \quad 107 \quad 22\}; \quad |\lambda-\lambda_R| \leqslant 10^{-14}\times 8{\cdot}765\,583$$

$$|\lambda - 37{\cdot}492\,878\,88| \leqslant 10^{-7}\times 3; \quad \lambda = 37{\cdot}492\,879.$$

8.6.3 $\lambda = -0{\cdot}506\,521, \quad \mathbf{x} = \{0{\cdot}530\,416 \quad 1 \quad 0{\cdot}480\,367\}.$

8.7.1 $\lambda = 1{\cdot}013\,642, \quad \mathbf{x} = \{1 \quad -0{\cdot}385\,061 \quad -0{\cdot}302\,591\}.$

8.7.2

$$10\mathbf{C}_3 = \begin{bmatrix} 0{\cdot}482\,473 & 0{\cdot}599\,427 & 0{\cdot}574\,570 \\ 0{\cdot}466\,105 & 0{\cdot}644\,070 & 0{\cdot}649\,719 \\ 0{\cdot}409\,295 & 0{\cdot}606\,197 & 0{\cdot}633\,219 \end{bmatrix}.$$

$$\mathbf{y} = \{0{\cdot}941\,088\,634 \quad 1 \quad 0{\cdot}936\,860\,785\}; \quad \mu = 0{\cdot}169\,141\,2.$$

$$\mathbf{x}^{(3)} = \{1 \quad -0{\cdot}636\,519 \quad -0{\cdot}875\,450 \quad -0{\cdot}883\,129\}.$$

8.7.3 Taking **B** and **x** to be the $10\mathbf{C}_3$ and **y** of Ex. 8.7.2, and taking $\lambda = 1{\cdot}691\,412\,37$ we find the new deflated matrix $\mathbf{C}_2$ to be given by

$$10^3\mathbf{C}_2 = \begin{bmatrix} 43{\cdot}826\,882 & -36{\cdot}873\,166 \\ -27{\cdot}380\,496 & 24{\cdot}522\,748 \end{bmatrix}.$$

The **y** and μ for $\mathbf{C}_2$ are

$$\mathbf{y} = \{1 \quad -0{\cdot}638\,835\,391\}, \quad \mu = 0{\cdot}067\,382\,765,$$

and the corresponding vector and value for $\mathbf{C}_3$ are

$$\mathbf{u} = \{0{\cdot}970422594 \quad -0{\cdot}031428927 \quad -0{\cdot}668279920\} \quad \text{and} \quad \mu = 0{\cdot}0067382765.$$

Finally, taking this last $\mathbf{u}$ as the new $\mathbf{y}$ we find

$$\mathbf{u} = \{-0{\cdot}206005994 \quad 0{\cdot}852963744 \quad -0{\cdot}115227839 \quad -0{\cdot}733818878\},$$

and therefore $\quad \mathbf{x}^{(2)} = \{-0{\cdot}241518 \quad 1 \quad -0{\cdot}135091 \quad -0{\cdot}860317\}.$

9.2.1 (*a*) If $\mathbf{Q}_r'\mathbf{Q}_r = \mathbf{I}$ $(r = 1, 2, \ldots, n)$, then

$$(\mathbf{Q}_1\mathbf{Q}_2 \ldots \mathbf{Q}_n)'\mathbf{Q}_1\mathbf{Q}_2 \ldots \mathbf{Q}_n = (\mathbf{Q}_n'\mathbf{Q}_{n-1}' \ldots \mathbf{Q}_1')(\mathbf{Q}_1\mathbf{Q}_2 \ldots \mathbf{Q}_n)$$
$$= \mathbf{Q}_n'\mathbf{Q}_{n-1}' \ldots (\mathbf{Q}_1'\mathbf{Q}_1)\mathbf{Q}_2 \ldots \mathbf{Q}_n = \mathbf{Q}_n'\mathbf{Q}_{n-1}' \ldots (\mathbf{Q}_2'\mathbf{Q}_2) \ldots \mathbf{Q}_n = \ldots = \mathbf{Q}_n'\mathbf{Q}_n = \mathbf{I}.$$

(*b*) If $\mathbf{Q}'\mathbf{Q} = \mathbf{I}$ then $|\mathbf{Q}'\mathbf{Q}| = |\mathbf{Q}'|.|\mathbf{Q}| = |\mathbf{Q}|^2 = 1$

therefore $$|\mathbf{Q}| = \pm 1.$$

9.2.2 Calculation shows directly that $\mathbf{Q}'\mathbf{Q} = \mathbf{I}$.

9.2.3 $\tan 2\theta = \dfrac{(2)\,(15{\cdot}5)}{29-8} = \dfrac{31}{21}$; $\quad 2\theta = 55°\ 53{\cdot}14'$, $\quad \theta = 27°\ 56{\cdot}57'$.

Sum of squares of off-diagonal elements is

$$2(-2\cos\theta + 2{\cdot}5\sin\theta)^2 + 2(2\sin\theta + 2{\cdot}5\cos\theta)^2 = 2(2^2 + (2{\cdot}5)^2).$$

9.3.1
$$\mathbf{B} = \begin{bmatrix} 1 & -2 & 2{\cdot}5 \\ -2 & 8 & -15{\cdot}5 \\ 2{\cdot}5 & -15{\cdot}5 & 29 \end{bmatrix}.$$

$$\boldsymbol{l}'\mathbf{B} = [1{\cdot}5 \quad -9{\cdot}5 \quad 16]; \qquad \boldsymbol{l}'\mathbf{B}\boldsymbol{l} = 8.$$

$$S^2 = 10{\cdot}25.$$

$$S = -3{\cdot}201562 \text{ (same sign as } b_{12}).$$

$$\mathbf{z} = \{0 \quad -5{\cdot}201562 \quad 2{\cdot}5\}; \quad \boldsymbol{l}'\mathbf{z} = -2{\cdot}701562.$$

$$2\mu = 0{\cdot}060048793;$$

$$\mathbf{y} = \{16{\cdot}653124 \quad -80{\cdot}362496 \quad 153{\cdot}124211\}; \quad \boldsymbol{l}'\mathbf{y} = 89{\cdot}414839.$$

$$(\boldsymbol{l}'\mathbf{B})\,\mathbf{z} = 89{\cdot}414839.$$

$$q = 800{\cdot}821033. \qquad \mu q = 24{\cdot}044168.$$

$$\mathbf{w} = \{1{\cdot}000000 \quad 2{\cdot}684465 \quad 5{\cdot}585366\}; \qquad \boldsymbol{l}'\mathbf{w} = 9{\cdot}269831.$$

$$2\mu(\boldsymbol{l}'\mathbf{y} - \mu q\boldsymbol{l}'\mathbf{z}) = 9{\cdot}269831.$$

$$\boldsymbol{l}'\mathbf{B} - (\boldsymbol{l}'\mathbf{w})\,\mathbf{z}' - (\boldsymbol{l}'\mathbf{z})\,\mathbf{w}' = [4{\cdot}201562 \quad 45{\cdot}969849 \quad 7{\cdot}914635].$$

$$\boldsymbol{l}'\mathbf{B}\boldsymbol{l} - 2(\boldsymbol{l}'\mathbf{z})\,(\boldsymbol{l}'\mathbf{w}) = 58{\cdot}086046; \quad \{\boldsymbol{l}'\mathbf{B} - (\boldsymbol{l}'\mathbf{w})\,\mathbf{z}' - (\boldsymbol{l}'\mathbf{z})\,\mathbf{w}\}\,\boldsymbol{l} = 58{\cdot}086046.$$

$$\overline{\mathbf{B}} = \mathbf{B} - \mathbf{w}\mathbf{z}' - \mathbf{z}\mathbf{w}' = \begin{bmatrix} 1 & 3{\cdot}201562 & 0 \\ 3{\cdot}201562 & 35{\cdot}926822 & 6{\cdot}841465 \\ 0 & 6{\cdot}841465 & 1{\cdot}073170 \end{bmatrix}.$$

$$\boldsymbol{l}'\overline{\mathbf{B}} = [4{\cdot}201562 \quad 45{\cdot}969849 \quad 7{\cdot}914635].$$

9.3.2

Stage 1:
$$\begin{bmatrix} 1 & -0{\cdot}650\,853\,969 & 0 & 0 \\ -0{\cdot}650\,853\,969 & 0{\cdot}650\,585\,367 & -0{\cdot}036\,563\,679 & -0{\cdot}052\,420\,213 \\ 0 & -0{\cdot}036\,563\,679 & 0{\cdot}007\,384\,895 & 0{\cdot}011\,344\,838 \\ 0 & -0{\cdot}052\,420\,213 & 0{\cdot}011\,344\,838 & 0{\cdot}018\,219\,737 \end{bmatrix}.$$

Stage 2:
$$\begin{bmatrix} 1 & -0{\cdot}650\,853\,969 & 0 & 0 \\ -0{\cdot}650\,853\,969 & 0{\cdot}650\,585\,367 & 0{\cdot}063\,912\,294 & 0 \\ 0 & 0{\cdot}063\,912\,294 & 0{\cdot}025\,382\,723 & -0{\cdot}001\,148\,878 \\ 0 & 0 & -0{\cdot}001\,148\,878 & 0{\cdot}000\,284\,482 \end{bmatrix}.$$

9.4.1 (*a*)
$$\begin{aligned} P_0 &= 1, \\ P_1 &= 1-\lambda, \\ P_2 &= (2-\lambda)P_1-2P_0 = -3\lambda+\lambda^2, \\ P_3 &= (3-\lambda)P_2-6P_1 = -6-3\lambda+6\lambda^2-\lambda^3. \end{aligned}$$

(*b*)
$$\begin{aligned} P_0 &= 1, \\ P_1 &= 1-\lambda, \\ P_2 &= (2-\lambda)P_1+2P_0 = 4-3\lambda+\lambda^2, \\ P_3 &= (3-\lambda)P_2+6P_1 = 18-19\lambda+6\lambda^2-\lambda^3. \end{aligned}$$

(*c*)
$$\begin{aligned} P_0 &= 1, \\ P_1 &= -2-\lambda, \\ P_2 &= (-2-\lambda)P_1-2P_0 = 6+4\lambda+\lambda^2, \\ P_3 &= (-2-\lambda)P_2-P_1 = -10-13\lambda-6\lambda^2-\lambda^3, \\ P_4 &= (-2-\lambda)P_3-P_2 = 14+32\lambda+24\lambda^2+8\lambda^3+\lambda^4. \end{aligned}$$

(*d*)
$$\begin{aligned} P_0 &= 1, \\ P_1 &= 4-\lambda, \\ P_2 &= (4-\lambda)P_1-2P_0 = 14-8\lambda+\lambda^2, \\ P_3 &= (4-\lambda)P_2-2P_1 = 48-44\lambda+12\lambda^2-\lambda^3, \\ P_4 &= (4-\lambda)P_3-2P_2 = 164-208\lambda+90\lambda^2-16\lambda^3+\lambda^4. \end{aligned}$$

9.4.2 No. $s(\lambda)$ alters by 2 as λ passes through $\lambda = 1$, the zero of P_1.

9.4.3 (*a*) $s(0) = 0$, $s(8) = 4$; there are 4 positive roots and therefore no negative roots.

(*b*) $s(1{\cdot}5) = 0, s(2) = 1 = s(3), s(3{\cdot}5) = 2 = s(4{\cdot}5), s(5) = 3 = s(6)$, $s(6{\cdot}5) = 4$, so that the roots are approximately 2, 3, 5, 6.

9.5.1 Using the sequence

$$\begin{aligned} &P_0 = 1,\ P_1 = 1-\lambda,\ P_2 = (35{\cdot}926\,822-\lambda)P_1-10{\cdot}249\,999, \\ &P_3 = (1{\cdot}073\,170-\lambda)P_2-46{\cdot}805\,643P_1, \text{ we obtain} \\ &\lambda_1 = -0{\cdot}506\,522, \quad \lambda_2 = 1{\cdot}013\,642, \quad \lambda_3 = 37{\cdot}492\,88. \end{aligned}$$

9.5.2 $-0{\cdot}381\,966$.

9.5.3 $\lambda_3 = 0{\cdot}169\,148$.

9.5.4 The iteration proceeds as follows

r	0	1	2	3	4	5	6	7	8
a_r	1	1·5	1·5	1·625	1·688	1·688	1·704	1·704	1·708
b_r	2	2	1·75	1·75	1·75	1·719	1·719	1·712	1·712

so that $\lambda \doteqdot 1{\cdot}71$.

9.5.5 The number, n, of steps, is given by

$$(1/2)^n < 0{\cdot}0001, \quad \text{i.e. } n\log_{10}2 > 4, \quad n > 13, \quad \text{or} \quad n = 14.$$

9.6.1
$$\mathbf{L} = \begin{bmatrix} -1{\cdot}618 & 0 & 0 & 0 \\ 1 & -0{\cdot}999953 & -0{\cdot}617924 & -0{\cdot}0123275 \\ 0 & 1 & 0 & 0 \\ 0 & 0 & 1 & 0 \end{bmatrix},$$

$$\mathbf{U} = \begin{bmatrix} 1 & -0{\cdot}618047 & 0 & 0 \\ 0 & 1 & -1{\cdot}618 & 1 \\ 0 & 0 & 1 & -1{\cdot}618 \\ 0 & 0 & 0 & 0{\cdot}0123275 \end{bmatrix}.$$

Then
$$\mathbf{y} = \{1, 1, 1, 1\}, \quad \mathbf{z} = \{-0{\cdot}618047, 1, 1, 262{\cdot}4957\},$$
$$\mathbf{x} = \{21293{\cdot}45, 34453{\cdot}80, 34453{\cdot}81, 21293{\cdot}45\},$$
$$\mathbf{y} = \{0{\cdot}618047, 1, 1, 0{\cdot}618047\},$$
$$\mathbf{z} = \{-0{\cdot}381982, 1, 0{\cdot}618047, 224{\cdot}2525\},$$
$$\mathbf{x} = \{18191{\cdot}25, 29434{\cdot}1, 29434{\cdot}1, 18191{\cdot}25\}.$$

Thus
$$\mathbf{x} = \{0{\cdot}618033, 1, 1, 0{\cdot}618033\},$$
$$\lambda = -0{\cdot}382 + (29434{\cdot}1)^{-1} = -0{\cdot}381966.$$

9.6.2 $\overline{\mathbf{B}} - \lambda\mathbf{I} = \mathbf{LU}$, where

$$\mathbf{L} = \begin{bmatrix} -0{\cdot}013600 & 3{\cdot}3498708 & -10^{-2}(1{\cdot}059527220) \\ 3{\cdot}201562 & 0 & 0 \\ 0 & 6{\cdot}841465 & 0 \end{bmatrix},$$

$$\mathbf{U} = \begin{bmatrix} 1 & 10{\cdot}90505884 & 2{\cdot}136914730 \\ 0 & 1 & 10^{-3}(8{\cdot}707199409) \\ 0 & 0 & 10^{-2} \end{bmatrix}.$$

After two iterations one obtains

$$\mathbf{x} = \{10^4(2{\cdot}389474), 10^2(1{\cdot}018154), -10^4(1{\cdot}170147)\},$$

whence
$$\mathbf{x}^{(2)} = \{1 \quad 0{\cdot}004261 \quad -0{\cdot}489709\}$$

and
$$\lambda_2 - 1{\cdot}0136 \simeq 10^{-4}(2{\cdot}389474)^{-1} = 0{\cdot}000042; \quad \lambda_2 = 1{\cdot}013642.$$

9.6.3 $\mathbf{B}-\lambda\mathbf{I} = \mathbf{LU}$, where

$$\mathbf{L} = \begin{bmatrix} 0.791 & -0.7685 & 0.245746484 & -10^{-2}(8.889095664) \\ 2 & -1.209 & 0 & 0 \\ 3 & -0.5 & -0.641175352 & 0 \\ 4 & 0 & 0 & 0 \end{bmatrix},$$

$$\mathbf{U} = \begin{bmatrix} 1 & 3.5 & 7.5 & 12.44775000 \\ 0 & 1 & 4.135649296 & 9.011993382 \\ 0 & 0 & 1 & 4.425081688 \\ 0 & 0 & 0 & 10^{-2}(8.889095664) \end{bmatrix}.$$

The iteration proceeds as follows

y	**z**	**x**
1	0.25	747.468783
1	−0.413565	−589.064511
1	−0.067404	280.498466
1	−5.636002	−63.403546
y	**z**	**x**
1	−0.021206000	2636.839156
−0.788079	0.616763440	−2083.678072
0.375264	−1.165459211	994.087956
−0.084824	−19.99263165	−224.911874

so that
$$\mathbf{x}^{(2)} = \{1 \quad -0.790218 \quad 0.377000 \quad -0.085296\}$$

and
$$\lambda_2 - 0.209 \simeq (2636.839156)^{-1} = 0.000379,$$

$$\lambda_2 = 0.209379.$$

9.7.1 $\mathbf{x} = \{1 \quad -0.385060 \quad -0.302592\}$.

9.8.1
$$\mathbf{T} = \begin{bmatrix} 1 & 0.423611 & 0 & 0 \\ 1 & 0.748945 & -0.057058 & 0 \\ 0 & 0.100000 & -0.117991 & -0.042018 \\ 0 & 0 & -0.100000 & -0.045253 \end{bmatrix},$$

$$\mathbf{S} = \begin{bmatrix} 1 & 0.500000 & 0 & 0 \\ 0.847221 & 0.748945 & 0.100000 & 0 \\ 0 & -0.057057 & -0.117991 & 0.100000 \\ 0 & 0 & -0.042013 & 0.045235 \end{bmatrix},$$

$$\mathbf{Q} = \begin{bmatrix} 1 & 0 & 0 & 0 \\ 0 & 0{\cdot}500\,000 & 0{\cdot}088\,608 & -0{\cdot}049\,957 \\ 0 & 0{\cdot}666\,667 & -0{\cdot}326\,317 & -0{\cdot}079\,828 \\ 0 & 0{\cdot}750\,000 & 0{\cdot}120\,991 & 0{\cdot}459\,395 \end{bmatrix},$$

$$\mathbf{P}' = \begin{bmatrix} 1 & 0 & 0 & 0 \\ 0 & 1 & 0{\cdot}333\,334 & 0{\cdot}166\,666 \\ 0 & -1{\cdot}489\,453 & 1{\cdot}086\,832 & 0{\cdot}026\,886 \\ 0 & -0{\cdot}556\,183 & -0{\cdot}010\,188 & 0{\cdot}379\,845 \end{bmatrix}.$$

INDEX